Advances in 21st Century Human Settlements

Indexed by SCOPUS

This Series focuses on the entire spectrum of human settlements – from rural to urban, in different regions of the world, with questions such as: What factors cause and guide the process of change in human settlements from rural to urban in character, from hamlets and villages to towns, cities and megacities? Is this process different across time and space, how and why? Is there a future for rural life? Is it possible or not to have industrial development in rural settlements, and how? Why does 'urban shrinkage' occur? Are the rural areas urbanizing or is that urban areas are undergoing 'ruralisation' (in form of underserviced slums)? What are the challenges faced by 'mega urban regions', and how they can be/are being addressed? What drives economic dynamism in human settlements? Is the urban-based economic growth paradigm the only answer to the quest for sustainable development, or is there an urgent need to balance between economic growth on one hand and ecosystem restoration and conservation on the other – for the future sustainability of human habitats? How and what new technology is helping to achieve sustainable development in human settlements? What sort of changes in the current planning, management and governance of human settlements are needed to face the changing environment including the climate and increasing disaster risks? What is the uniqueness of the new 'socio-cultural spaces' that emerge in human settlements, and how they change over time? As rural settlements become urban, are the new 'urban spaces' resulting in the loss of rural life and 'socio-cultural spaces'? What is leading the preservation of rural 'socio-cultural spaces' within the urbanizing world, and how? What is the emerging nature of the rural-urban interface, and what factors influence it? What are the emerging perspectives that help understand the human-environment-culture complex through the study of human settlements and the related ecosystems, and how do they transform our understanding of cultural landscapes and 'waterscapes' in the 21st Century? What else is and/or likely to be new vis-à-vis human settlements – now and in the future? The Series, therefore, welcomes contributions with fresh cognitive perspectives to understand the new and emerging realities of the 21st Century human settlements. Such perspectives will include a multidisciplinary analysis, constituting of the demographic, spatio-economic, environmental, technological, and planning, management and governance lenses.

If you are interested in submitting a proposal for this series, please contact the Series Editor, or the Publishing Editor:

Bharat Dahiya (bharatdahiya@gmail.com) or

Loyola D'Silva (loyola.dsilva@springer.com)

Mohammed Firoz C · Bharat Dahiya ·
Lakshmi Priya Rajendran · Lalit Kumar Dashora ·
Uday Chatterjee
Editors

Urban Planning and Design for Megacities in the Global South

Smart and Sustainable Development

Springer

Editors
Mohammed Firoz C
Department of Architecture and Planning
National Institute of Technology Calicut
Calicut, Kerala, India

Lakshmi Priya Rajendran
The Bartlett School of Architecture
University College London
London, UK

Uday Chatterjee
Bhatter College
Dantan, West Bengal, India

Bharat Dahiya
Research Center for Sustainable
Development and Innovation, School
of Global Studies
Thammasat University
Bangkok, Thailand

School of Public Leadership
Stellenbosch University
Stellenbosch, Western Cape, South Africa

Lalit Kumar Dashora
School of Global Studies
Thammasat University
Bangkok, Thailand

ISSN 2198-2546 ISSN 2198-2554 (electronic)
Advances in 21st Century Human Settlements
ISBN 978-981-97-8369-4 ISBN 978-981-97-8370-0 (eBook)
https://doi.org/10.1007/978-981-97-8370-0

This Springer imprint is published by the registered company Springer Nature Singapore Pte Ltd.
The registered company address is: 152 Beach Road, #21-01/04 Gateway East, Singapore 189721, Singapore

This book is dedicated to:

- *To our teachers and mentors who have influenced us greatly in making us what we are today*
- *To our parents, family and friends who have given us the necessary time and space so that we could spend time on preparing this anthology for a greater cause serving humanity*
- *To all researchers who are currently conducting research on megacities*
- *To the universities, institutions and organizations, where we work, for giving us the freedom and time to work on this book*

Foreword by Mr. Rajeev Issar, Policy Specialist, UNDP

As we step into the twenty-first century, our world is becoming more urbanized than ever before. Bustling megacities in Global South are becoming home to millions of people from different backgrounds and cultures and transforming into centres of socio-economic development. The unprecedented expansion of cities in recent decades has intensified some of the pressing challenges faced by humanity, like mobility, housing, energy, water as well as systemic risks. Yet, the process of urbanization is also presenting great opportunities to place resilient and sustainable development at the centre of this transition.

In the forthcoming book, 'Urban Planning and Design for Megacities in the Global South - Smart and Sustainable Development', esteemed editors and contributors delve into the core of this discourse with depth, insight, and innovation. Edited within the esteemed Springer Book Series titled 'Advances in 21st Century Human Settlements', the Book offers deep insights reflecting cross-practice expertise and knowledge, presenting both the theoretical frameworks and practical strategies to navigate the multifaceted challenges facing megacities.

The present book is one of its kind attempt to document challenges faced by megacities from Global South and strategies for smart and sustainable development. The chapters in **Part I: Sustainable Megacities and Planning for Environmental and Disaster Risk Resilience** delve into the critical importance of sustainable practices and environmental resilience in the planning and design of megacities. The chapters also explore various case studies and strategies aimed at addressing environmental challenges in rapidly growing megacities. The chapters in **Part II: Sustainable Megacities and Spatial Planning for Urban Centres** explore the complex dynamics of urban development and spatial planning in megacities, focusing on the imperative of sustainability and inclusivity. The chapters offer a comprehensive exploration of diverse strategies and case studies aimed at fostering smart and sustainable urban development. The chapters in **Part III: Sustainable MegaCities and Planning for Urban Expansion and Regeneration** explore complexities of urban expansion and regeneration in megacities, emphasizing the imperative of sustainability and resilience. The chapters emphasize the critical importance of sustainable urban expansion and regeneration in megacities of the Global South. The chapters also explore diverse strategies and case studies aimed at fostering smart and sustainable urban growth. The chapters in **Part IV: Sustainable Megacities and Planning for Social Resilience** present diverse perspectives and case studies aimed at fostering social resilience and inclusivity in rapidly growing urban environments. The chapters also emphasize the importance of social resilience in megacities of the Global South. Through community engagement, cultural preservation, and social empowerment, megacities can become vibrant, resilient, and inclusive spaces that promote the well-being and prosperity of all residents. The chapters in **Part V: Sustainable Megacities with Energy Planning and Services** explore innovative strategies and case studies aimed at enhancing energy resilience and promoting sustainable urban development. The chapters also emphasize the critical importance of energy planning and service provision in the sustainable development of megacities in the Global South. The chapters in **Part VI: Branding Sustainable Megacities with Innovative Governance and Management Strategies** emphasize the critical role of governance, management, and branding in shaping the sustainable development of megacities in the Global South. The chapters also highlight that through collaborative governance, creative place-making, and strategic branding initiatives, cities can strengthen their resilience, enhance their global visibility, and improve the quality of life for residents, positioning themselves as models of sustainable urbanization in the twenty-first century.

May this book inspire the transformative actions necessary to shape megacities in the Global South into beacons of hope and prosperity, resilience and sustainability for generations to come. I am sure the book will offer a valuable resource characterized by deep research, erudition, and futuristic vision as it embellishes the analysis with contextual case studies, pertinent strategies, and relevant frameworks and will hence become a reference material to guide and share the cities and megacities of the future.

Mr. Rajeev Issar
Policy Specialist, Thematic
Lead—Urban Risk Management
and Resilience, UNDP, Geneva,
Switzerland

Foreword by Prof. Chetan Vaidya

Welcome to the world of urban planning and design, where megacities come alive with the pulse of millions of people. In today's rapidly changing world, megacities stand as beacons of progress, bustling with energy and diversity. They are the epicentres of opportunity, drawing people from all corners of the globe in search of a better life. However, with great opportunities come great challenges, especially when it comes to planning and designing these vast urban landscapes.

"Urban Planning and Design for Megacities in the Global South - Smart and Sustainable Development" is a comprehensive book crafted to address the unique needs and complexities of urban environments in the Global South. It is a testament to the collective effort of experts, researchers, and practitioners dedicated to shaping a more sustainable future for our cities.

The challenges facing megacities in the Global South are as diverse as the communities that inhabit them. Rapid urbanization, inadequate infrastructure, environmental degradation, and social inequalities are just a few of the hurdles that cities must overcome. Yet, amidst these challenges lies an opportunity for transformation and innovation. Through the lens of smart and sustainable development, this book illuminates pathways towards inclusive growth and prosperity. It underscores the importance of collaboration and co-creation in reimagining urban spaces that are equitable, accessible, and environmentally sound.

Megacities are most vulnerable countries to climate change impacts. As a combined effect of climate challenges and inadequate urban management, these cities are frequently facing climatic hazards such as urban flooding, heat waves, cyclonic storms, severe and prolonged dry spells, water scarcity.

In the chapters of this book, you will embark on a journey through the vibrant landscapes of current and future to be megacities of the Global South. As we navigate through the intricate tapestry of urban life, we explore the essence of smart and sustainable development. In this book, you will discover the principles and practices that underpin smart and sustainable urban development. From the intricate network of transportation systems to the harmonious integration of green spaces, each aspect of urban planning and design is meticulously examined to foster resilience and livability. As you delve deeper into the chapters, you will encounter a wealth of case studies, best practices, and visionary insights from Global South. These examples shall serve as beacons of inspiration, illustrating the transformative power of urban planning and design in shaping the destiny of our cities.

In this book, authors from different regions have compressively covered wide range of topics related with megacities such as disaster management, spatial planning, regeneration, social resilience, energy planning, innovative governance and management.

Moreover, this book serves as a call to action for policy and decision-makers, urban planners and practitioners, architects and engineers, and citizens alike. It reminds us of our collective responsibility to nurture our cities as vibrant hubs of human ingenuity and cultural exchange. By embracing innovation and sustainability, we can forge a path towards a brighter future for generations to come.

"Urban Planning and Design for Megacities in the Global South - Smart and Sustainable Development" is more than just a book; it is a manifesto for change. It challenges us to think boldly, act decisively, and embrace the transformative potential of our urban landscapes. As you embark on this enlightening journey, may you be inspired to become champions of sustainable urban development in your own communities. Together, let us build cities that thrive, resonate, and endure as testaments to the resilience and creativity of the human spirit.

In closing, the book makes very important contributions in improving urban planning and its implementation in Global South. Urban and regional planners, disaster planners, climate experts, practising professionals, policymakers, and academicians must read this book. I congratulate the editors and contributors for bringing out this valuable book.

Welcome to the future of urban living.

Sincerely,

Prof. Chetan Vaidya
Independent Director (Non-Executive)
GIFT City Gujarat, Former Director
School of Planning and Architecture
(SPA) Delhi and Former Director
National Institute of Urban Affairs
(NIUA), India

Preface

Globalization, improved mobility and economic development, has increased the urbanization process around the world. Until the mid-twentieth century, urbanization was a major spatial phenomenon only in the Global North, especially the developed regions of Europe and North America. Since 1950s, urbanization started to gain rapid pace in the Global South as well. Thus, by the year 2007, the global urban population exceeded the rural. In the next few decades, the cities of the Global South, particularly in Africa and Asia, are expected to witness higher rates of urbanization. It is now estimated that by the year 2050, at least 66 per cent of the world's population might live in cities. The number of megacities, which was just two in the year 1950, increased to 32 by 2017. By 2030, there may be 41 megacities in the world. Most of these new megacities are likely to be in the larger countries of Africa and Asia, namely China, India, and Nigeria.

The burgeoning expansion of megacities in the Global South stands as a phenomenon of immense social, economic, spatial and environmental significance, reflecting unprecedented levels of urbanization and demographic transformation. Over recent decades, urban areas across Africa, Asia, and Latin America have witnessed exponential population surges, resulting in the emergence of megacities—urban conglomerates with populations exceeding ten million inhabitants. Understanding this phenomenon holds paramount importance for several reasons. Firstly, the study of megacity growth sheds light on the intricate dynamics of urbanization in the Global South. By delving into the underlying factors propelling this rapid urban expansion, such as economic opportunities, rural-to-urban migration, and natural population growth, researchers can discern the complex interplay of socio-economic, environmental, and political forces shaping city-regional landscapes.

Moreover, examining the growth of megacities offers invaluable insights into the challenges and opportunities associated with urbanization in the twenty-first century. While megacities serve as engines of economic growth, innovation, and cultural exchange, they also grapple with pressing issues such as inadequate infrastructure, housing shortages, environmental degradation, and social inequality (Arfanuzzaman and Dahiya, 2019; Dahiya, 2012a, 2012b, 2014). By dissecting these challenges and exploring potential solutions, policymakers, professionals, and experts can develop strategies to promote sustainable urban development and improve the quality of life for urban dwellers in the Global South. Furthermore, studying megacity growth facilitates a deeper understanding of the role of Global South in shaping the future of urbanization and sustainable development in the twenty-first century. As megacities in the Global South continue to expand, they exert increasing demographic and spatial influence on regional and global socio-economic dynamics, environmental challenges, and climate emergencies. By analysing the unique challenges and opportunities faced by such large and growing cities in the Global South, researchers can contribute to d designing policies and development interventions aimed at making cities and human settlements inclusive, safe, resilient, and sustainable, as underlined by the *New Urban Agenda* (United Nations, 2017), the *2030 Agenda for Sustainable Development* (United Nations, 2015), and the *Sustainable Development Goal (SDG) 11* (United Nations, 2024). While presenting opportunities for economic growth and human development, megacities also pose significant challenges that require comprehensive and sustainable solutions to ensure the well-being and prosperity of their inhabitants. The various megacities may require different growth models that are innovative and suited to l ocal planning and development challenges.

This anthology, *Urban Planning and Design for Megacities in the Global South: Strategies for Smart and Sustainable Urban Development*, has made an attempt to compile a diverse array of evidence-based case studies, fostering discussions on emerging issues, strategies, and solutions for the enhanced planning and development of megacities in the Global South.

Furthermore, some cities that have not yet achieved megacity status have also been included in this book to provide a comprehensive understanding of urban growth dynamics and the related factors influencing this growth. By examining these rapidly expanding cities, we can identify the challenges they face in urban planning and policy implementation. These challenges often include inadequate infrastructure, insufficient public services, environmental degradation, and socio-economic disparities.

Studying these emerging cities is crucial for several reasons. First, it allows for the identification of patterns and trends that can predict which cities are likely to become the next megacities. Understanding these patterns helps in forecasting future urbanization trends and preparing for them proactively. Second, it sheds light on the specific issues that need to be addressed to ensure sustainable and equitable growth. For instance, strategies to improve housing, transportation, and waste management can be tailored to the unique needs of these large cities.

Structured into six thematic parts, this co-edited volume offers a comprehensive exploration of sustainable urban development for megacities.

The part one, 'Planning for Environmental and Disaster Risk Resilience', delves into strategies that address environmental challenges and enhance resilience to disasters, ensuring the sustainability of megacity environments. Four case examples, namely planning for landslide and hazards, planning for a resilient city, urban flood risk zonation, and planning for urban floods using a time series mapping using Geographic Information System (GIS), are discussed in this part.

The second part of the book addresses the theme of 'Spatial Planning for Urban Centres'. This part focuses on spatial dynamics and examines innovative approaches to urban planning that optimize land use, transportation networks, and infrastructure development in megacity centres. Four different cases namely, reclaiming streets through place-making in an inner urban core, making Indian religious cities accessible, achieving compact urban development and analysis of commercial centres at the neighbourhood level, are discussed here.

The third part of the book explores 'Planning for Urban Expansion and Regeneration' by focusing on the complexities of urban growth. This part also investigates strategies for managing expansion while revitalizing existing urban areas to accommodate growing populations in a sustainable manner. Accordingly, four case examples namely achieving resilience through sustainable management of brownfield industrial heritage, addressing the re-opening a shipyard to the local community and the subsequent spatial aspects, measuring utility of vertical expansion in Kolkata region, and its future prospect and urban growth patterns and spatial sustainability in coastal cities are discussed.

'Urban Planning for Social Resilience' as a theme is discussed in the fourth part of the book. This part emphasizes the importance of social cohesion, equity, and community resilience and highlights inclusive urban planning approaches tailored to the diverse needs of megacity populations. Accordingly, three different case examples namely housing preferences of slum dwellers in slum relocation projects, deriving a conceptual framework for designing a socio-cultural, relational, and educational ecosystem of cities, and exploring the community vitality of ethnic migrants are discussed here.

The fifth part is composed of the theme 'Energy Planning and Services'. Accordingly, this part addresses energy sustainability and services and explores strategies to enhance energy efficiency, integrate renewable energy sources, and expand access to clean energy services within megacity contexts. Three case examples, namely institutional approach for sustainable service provision in peri-urban areas, city planning approaches through energy retrofit, and an agent-based modelling for energy planning and policy, are discussed under this part.

The last and final part of this book is grouped under the theme 'Innovative Governance and Management Strategies'. This part focuses on governance and management innovations, including stakeholder engagement, participatory decision-making, and institutional reforms to improve the effectiveness and efficiency of megacity governance. Accordingly, three cases namely, assessing water security through improved planning, policies, and governance, repositioning cities as development model based on the concept of creative city and exploring city branding through place-making in megacities of Indian subcontinent are discussed.

By presenting evidence-based case examples and in-depth analyses, the edited volume aims to provide valuable insights for urban practitioners, policymakers, decision-makers, development managers, administrators, experts, scholars, and researchers. This collection of rich case studies serves not only to shed light on the challenges and prospects faced by megacities but also to offer practical solutions and innovations that can inform and guide sustainable urban development efforts.

Ultimately, the book seeks to contribute to the academic discourse and field-based knowledge, offering a valuable resource for addressing the complex and evolving needs of megacities and other emerging large cities in the Global South.

The last chapter of the book provides some concluding thoughts 'Towards Sustainable and Resilient Megacities in the Global South'.

For contributing their innovative and evidence-based research, rich insights, and well-written chapters to this co-edited volume, we are grateful to the authors all chapters.

For writing the Forewords to this anthology on megacities in the Global South, we express our sincere gratitude to Mr. Rajeev Issar, Policy Specialist, Thematic Lead—Urban Risk Management and Resilience, United Nations Development Programme, (UNDP) Geneva, Switzerland, and Prof. Chetan Vaidya, Independent Director (Non-Executive) GIFT City Gujarat, Former Director School of Planning and Architecture (SPA)—New Delhi, and Former Director National Institute of Urban Affairs (NIUA), New Delhi, India.

We would like to thank Dr. Loyola D'Silva, Executive Editor, together with Springer Nature's entire team, for their support and cooperation throughout the preparation and publication of this co-edited volume.

Calicut, India — Mohammed Firoz C
Bangkok, Thailand and Stellenbosch, South Africa — Bharat Dahiya
London, UK — Lakshmi Priya Rajendran
Bangkok, Thailand — Lalit Kumar Dashora
Dantan, India — Uday Chatterjee

References

Arfanuzzaman, M. and Dahiya, B. (2019) Sustainable urbanization in Southeast Asia and beyond: Challenges of population growth, land use change, and environmental health, *Growth and Change: A Journal of Urban and Regional Policy*, Vol. 50, No.2, pp. 725-744, DOI: https://doi.org/10.1111/grow.12297

Dahiya B (2012a) 21st century Asian cities: unique transformation, unprecedented challenges. *Global Asia* 7(1):96–104 (ISSN: 1976-068X). https://www.globalasia.org/v7no1/feature/21st-century-asian-cities-unique-transformation-unprecedented-challenges_bharat-dahiya

Dahiya B (2012b) Cities in Asia, 2012: demographics, economics, poverty, environment and governance. *Cities*, 29(Supplement 2):S44–S61. Retrieved from https://doi.org/10.1016/j.cities.2012.06.013

Dahiya B (2014) Southeast Asia and sustainable urbanization. *Global Asia* 9(3):84–91 (ISSN: 1976-068X). https://www.globalasia.org/v9no3/feature/southeast-asia-and-sustainable-urbanization_bharat-dahiya

United Nations (2015) Transforming Our World: The 2030 Agenda for Sustainable Development, A/RES/70/1. https://sustainabledevelopment.un.org/post2015/transformingourworld/publication

United Nations (2017) The new urban agenda, A/RES/71/256, Habitat III and United Nations

United Nations (2024) Sustainable Development Goal 11: Make cities and human settlements inclusive, safe, resilient and sustainable. https://sdgs.un.org/goals/goal11

Contents

Editors and Contributors

About the Editors

Mohammed Firoz C currently serves as the Head of the Department of Architecture and Planning, at the National Institute of Technology (NIT) Calicut, India. Besides NIT Calicut, he has teaching experience as a visiting faculty at the AIT Bangkok and at the Architectural Association London. Dr. Firoz graduated from NIT Calicut as a university rank holder, completed his post-graduation in Urban and Regional Panning from CEPT university, a postgraduate Diploma in Urban Environment Management and Law from the National Law University, Delhi. India and Ph.D. degree from Indian Institute of Technology (IIT) Kharagpur. He is actively involved in funded research projects, publishing as well as consultancy and university services. He has written 37 papers in Indexed journals and book chapters. He had also guided 3 PhDs and is currently guiding 8 more. He was also in the advisory panel of more than 21 PhD thesis as its chairperson/member, principal guide for 30 Master's thesis and 70 Undergraduate Thesis. He is associated as the Editorial board of 3 reputed journals besides being a reviewer for several journal papers.

Bharat Dahiya is Director of the Center for Sustainability Research at the School of Global Studies, Thammasat University, Thailand. He is an Extraordinary Professor at the School of Public Leadership, Stellenbosch University, Western Cape, South Africa.

An award-winning urbanist, Bharat combines cutting-edge research, policy analysis and development practice aimed at examining and tackling socio-economic, environmental and governance issues in the global urban context of sustainable development. Since the early-1990s, his research and professional work has focused on sustainable cities and urbanization, strategic urban planning and development, urban infrastructure, urban environment, climate change and urban resilience, post-disaster recovery and reconstruction, and cultural heritage and landscapes.

Working with the World Bank, UN-HABITAT, Asian Development Bank, UNDP, United Nations University, and Metropolis, he has initiated, led and/or contributed to international development projects and programmes on sustainable urban development around the world. Some of the projects that Bharat led include the preparation of the first-ever '*City-wide Pro-poor Ger-area Upgrading Strategy and Investment Plan for Ulaanbaatar', and 'Community-Led Ger Area Upgrading in Ulaanbaatar City'* (Mongolia); 'Earthquake-resistant and Energy-efficient School Reconstruction in Sichuan Province', and 'Getting Children Back to School: Providing Prefabricated Classroom Units to the Earthquake-Affected Children of Gansu Province' (China); and 'City Development Strategy for Thanh Hoa City in a Regional Development Context' (Viet Nam).

Throughout his career, Bharat has been involved in the preparation and dissemination of cutting-edge knowledge products. At the World Bank headquarters in Washington DC, USA, he conducted the first-ever systematic review of the Bank's investments for improving urban liveability, published as a co-authored book, *Urban Environment and Infrastructure: Toward Livable Cities* (2004). At UN-HABITAT, he led, conceptualized, and coordinated the preparation of United Nations' inaugural report on *The State of Asian Cities 2010/11* (2010). In 2018, he co-authored *Partnering for*

Sustainable Development: Guidelines for Multi–stakeholder Partnerships to Implement the 2030 Agenda in Asia and the Pacific (UNU-IAS, and UN-ESCAP). He was the lead author of Metropolis's first-ever *Asian Metropolitan Report (2021).* He has co-edited four academic anthologies, *New Urban Agenda for Asia-Pacific: Governance for Sustainable and Inclusive Cities* (Springer, 2020), *Practising Cultural Geographies: Essays in Honour of Rana PB Singh* (Springer Nature, 2022), *Disaster Resilience and Human Settlements: Emerging Perspectives in the Anthropocene* (Springer Nature, 2023), and *City Responses to Disruptions in 2020: From Lockdowns to Aftermath* (Springer Nature, 2023).

Bharat advises several professional, academic, private sector, and non-profit organisations around the world. Since 2019, he has been a Member of the International Advisory Board for the UN-HABITAT's *World Cities Report* (Nairobi, Kenya). Since early-2014, he has been the Series Editor for the SCOPUS-indexed Springer book series, *Advances in 21st Century Human Settlements*, which has a growing collection of over 40 monographs and edited volumes. He serves as a member of the editorial board of several international journals as well as the International Scientific Board of the book series, 'Geographies of the Anthropocene', published by Il Sileno Edizioni, Italy. He has held academic positions in Australia, Indonesia, South Africa, and Thailand.

Bharat holds a Ph.D. in *Urban Governance, Planning and Environment* from the University of Cambridge, United Kingdom. He read for his M.A. in Geography from Jawaharlal Nehru University, and Master of Planning from School of Planning and Architecture, both based in New Delhi, India.

Dr. Lakshmi Priya Rajendran is currently a Lecturer in Environmental and Spatial Equity and Co-Director of Equality, Diversity and Inclusion at the Bartlett School of Architecture, University College London, UK. She is also the Co-chair for the UCL's Environment Research domain Academic co-chair for the UCL Advisory Group on Climate Change and Health and UCL's Climate Crisis Executive Committee member. Her teaching and research interests' deal with equitable futures and climate justice, alternative city imaginaries, decoloniality, critical social and digital media, spatial representation and practice, identity negotiations, cultural encounters in contemporary cities. She is particularly interested in an interdisciplinary understanding of social, spatial, temporal and material practices in cities and a comparative study of these practices in the Global South. Her doctoral research (at Sheffield School of Architecture, University of Sheffield, UK) on inclusive public space planning and design won the International Best Ph.D. Award 2015 from Society for Research on Identity Formation, US. Lakshmi has completed several externally funded projects, some of them include collaborative project in India (PI, ESRC-GCRF funded, Inclusive and Resilient Peri-urban Planning and development (ongoing), Turkey (PI, Newton Funded, Heritage -led Urban Resilience Planning for Smart-Cities), Brazil (Co-I, Research England-GCRF funded, Socio-ecological sustainability planning and management), UK(Co-I, NERC funded, Flood Resilience Planning and Environmental Assessment Transnational (Royal Academy of Engineering funded Frontiers Champion Award (Peri-urban and Environment Resilience in the Global South). She is currently leading a two-year research project Transgressing 'Good' Cities: Decolonising City Narratives for Just Urban Futures (2023-2025) funded by the British Academy's Knowledge Frontiers: International Interdisciplinary Research 2022, in collaboration with IIT Chennai, India, and Radboud University, Netherlands which aims to develop alternative city imaginaries grounded in residents' place narratives in relation to local heritage, culture, and values.

Lalit Kumar Dashora is currently associated with the School of Global Studies, Research Center for Sustainable Development and Innovation, Thammasat University, Bangkok, Thailand. He is a specialist in Urban Risk and Resilience . Formerly, he worked as a Senior Technical Specialist at Asian Disaster Preparedness Center (ADPC) Bangkok Prior to joining ADPC, he worked with the Asian Development Bank (ADB) in Dushanbe (Tajikistan), and Government of Gujarat (India). He holds Master's degree in Geo-Information Science for Earth Observation, Environmental Modelling & Management (ITC, The Netherlands), and Urban & Regional Planning (School of Planning, CEPT University, Ahmedabad). Risk-informed, climate-compatible planning and development is the forefront of his experience. Over the last 18+ years, he has been actively engaged in various high impact and strategic programmes and projects, such as for the WMO supported Climate Risk and Early Warning Systems (CREWS) in the Southeast Asia and Pacific; for DFID (with UK Met Office) supported Evidence Review: Options Analysis for a Regional Climate Change Programme to Deliver More Effective Climate Services, Early Warning and Disaster Risk Reduction; for Norad supported Urban Resilience to Climate Extreme in Southeast Asia quote a few notable ones.

Dr. Uday Chatterjee is an Assistant Professor at the Department of Geography, Bhatter College, Dantan, Paschim Medinipur, West Bengal, India and an Applied Geographer with a Doctoral Degree in Applied Geography at Ravenshaw University, Cuttack, Odisha, India. His areas of research interest cover Urban Planning, Social and Human geography, Applied Geomorphology, Hazards & Disasters, Environmental Issues, disaster governance, community-based disaster risk management, climate change adaptation, urban risk management, and disaster. He has delivered 10 invited lectures in University Grants Commissions (UGC) sponsored national seminars and various academic departments of different colleges in India. In addition, he presented 20 papers in national and international seminars/conferences held in India as well as chaired and co-chaired more than 5 technical sessions. He has successfully guided project dissertations to undergraduate students. He has also conducted (Convener) one Faculty Development Programme on 'Modern methods of teaching and advanced research methods' sponsored by Indian Council of Social Science Research (ICSSR), Govt. of India. Currently, Dr. Uday Chatterjee has completed the Special Issue (S.I) of Urbanism, Smart Cities and Modelling, Geojournal, Springer as a Lead Editor, book series editor Development in Environmental Science, Elsevier (https://www.elsevier.com/books-and-journals/book-series/developments-in-environmental-science). His research work has been funded by the West Bengal Pollution Control Board (WBPCB) Govt. of West Bengal, India. He has served as a reviewer for many International journals. Currently, Dr. Chatterjee doing an international project, in collaboration with Indonesia, Malaysia, and Japan funded by APN (Asia Pacific- Global Change Research) and he is the guest editor of the special issue Social Ecology, Human-Well-being and Sustainability, Global Social Welfare journal, Springer. He has published 35 research papers, 12 edited books, and 2 author books, 15 book chapters, and 2 conference proceedings.

Contributors

Abdillah Abdillah Graduate Program in Administrative Science, Faculty of Social and Political Sciences, Universitas Padjadjaran, Bandung, Indonesia;
Faculty of Social and Political Sciences, Center for Decentralization & Participatory Development Research, Universitas Padjadjaran, Bandung, Indonesia

Hudha Abdul Salam Department of Architecture and Planning, Indian Institute of Technology Roorkee, Roorkee, India;
Department of Architecture, TKM College of Engineering, Kollam, India

Kumar Abhinay Department of Architecture and Planning, Indian Institute of Technology Roorkee, Roorkee, India

Ranjani Balasubramanian Glocal Colab, Bangalore, India

Sumona Bandyopadhyay Institute of Geographical Sciences (Physical Geography), Freie Universität Berlin, Berlin, Germany

Ankhi Banerjee Ranbir and Chitra Gupta School of Infrastructure Design and Management, IIT Kharagpur, Kharagpur, India

Haimanti Banerji Department of Architecture and Regional Planning, Indian Institute of Technology, Kharagpur, India

Vidhu Bansal School of Architecture, Anant National University, Ahmedabad, India

Alara Bilgic Istanbul, Turkey

Archiman Biswas Indian Institute of Technology Kharagpur, Kharagpur, West Bengal, 721302 India

Rabidyuti Biswas Department of Physical Planning, School of Planning and Architecture, New Delhi, 110002 India

Jinal Boricha Department of Architecture & Planning, BIT Mesra, Ranchi, India

Rd Ahmad Buchari Faculty of Social and Political Sciences, Department of Public Administration, Universitas Padjadjaran, Bandung, Indonesia;
Faculty of Social and Political Sciences, Center for Decentralization & Participatory Development Research, Universitas Padjadjaran, Bandung, Indonesia

Uday Chatterjee Department of Geography, Bhatter College Dantan, Paschim Medinipur, West Bengal, India

Subrata Chattopadhyay Indian Institute of Technology Kharagpur, Kharagpur, West Bengal, 721302 India

Susan Cyriac Department of Architecture and Planning, National Institute of Technology Calicut, Kozhikode, India

Bharat Dahiya Research Center for Sustainable Development and Innovation, School of Global Studies, Thammasat University, Bangkok, 10200 Thailand;
School of Public Leadership, University of Stellenbosch, Stellenbosch, 7602 Western Cape, South Africa

Lalit Kumar Dashora Research Center for Sustainable Development and Innovation, School of Global Studies, Thammasat University, Bangkok, Thailand

Marina Demchenko Living Cities Earth (LCE), Zurich, Switzerland

Priyanka Dey Department of Architecture and Regional Planning, Indian Institute of Technology Kharagpur, Kharagpur, West Bengal, India

Mohammed Firoz C Department of Architecture and Planning, NIT Calicut, Kozhikode, India

Dipanwita Ghosh Department of Geography, University of Calcutta, Kolkata, India

Maheshwari Gupta Department of Physical Planning, School of Planning and Architecture, New Delhi, 110002 India

Ruchita Gupta Department of Housing, School of Planning and Architecture, New Delhi, India

Marilyn Hamilton Integral City Meshworks Inc, Findhorn, Scotland, UK

Rajrani Kalra Department of Geography and Environmental Studies, California State University, San Bernardino, CA, USA

Vijay Kapse Department of Architecture and Planning, Visveswaraya National Institute of Technology, Nagpur, Maharashtra, India

Surajit Kar Department of Geography, University of Calcutta, Kolkata, India

Shalini Keshri Department of Architecture and Regional Planning, Indian Institute of Technology Kharagpur, Kharagpur, West Bengal, India

Shabnam Khan Department of Architecture and Planning, National Institute of Technology Calicut, Kozhikode, India

Rachna Khare Department of Architecture, School of Planning and Architecture, Bhopal, India

Ashok Kumar Department of Physical Planning, School of Planning and Architecture, New Delhi, 110002 India

Atul Kumar Department of Architecture and Planning, Indian Institute of Technology Roorkee, Roorkee, India

Paulose N. Kuriakose Department of Urban & Regional Planning, School of Planning and Architecture Bhopal, Bhopal, Madhya Pradesh, India

Alexander Laszlo The Bertalanffy Center for the Study of Systems Science, Buenos Aires, Argentina

Abantika Mukherjee Indian Institute of Technology Kharagpur, Kharagpur, West Bengal, 721302 India

Mahua Mukherjee Department of Architecture and Planning, Indian Institute of Technology Roorkee, Roorkee, India

Subham Mukherjee Institute of Geographical Sciences (Physical Geography), Freie Universität Berlin, Berlin, Germany

Heru Nurasa Faculty of Social and Political Sciences, Department of Public Administration, Universitas Padjadjaran, Bandung, Indonesia

Mario Paris DISA – Università degli Studi di Bergamo, Bergamo, Italy

Shivangi Singh Parmar School of Architecture, Anant National University, Ahmedabad, India

Apurv Bhogibhai Patel Department of Urban & Regional Planning, School of Planning and Architecture Bhopal, Bhopal, Madhya Pradesh, India

Akshay Patil Department of Architecture and Planning, Visveswaraya National Institute of Technology, Nagpur, Maharashtra, India

Prashant Prasad Department of Architecture & Planning, BIT Mesra, Ranchi, India

Varun Prasad Department of Architecture and Planning, National Institute of Technology Calicut, Kozhikode, India

Ajay Raghava Ministry of Environment, Forest, and Climate Change, New Delhi, India

Lakshmi Priya Rajendran Bartlett School of Architecture, University College London, London, UK

Archita Saha Department of Architecture and Planning, Indian Institute of Technology Roorkee, Roorkee, India

Debanjali Saha Department of Architecture and Regional Planning, Indian Institute of Technology, Kharagpur, India

Deepanjan Saha Department of Planning and Architecture, National Institute of Technology Rourkela, Rourkela, India

Kakoli Saha Department of Architecture, School of Planning and Architecture, Bhopal, India

Purbita Samanta Indian Institute of Technology Kharagpur, Kharagpur, West Bengal, 721302 India

Nunna Tagore Sai Priya Ranbir and Chitra Gupta School of Infrastructure Design and Management, IIT Kharagpur, Kharagpur, India

Somi Sareen Department of Architecture and Planning, National Institute of Technology Patna, Patna, India

Satyaki Sarkar Department of Architecture & Planning, BIT Mesra, Ranchi, India

Anjana Sasi School of Planning and Architecture, New Delhi, India

Jayeeta Sen Department of Environmental Planning, School of Planning and Architecture, New Delhi, India

Chetna Singh School of Planning and Architecture, New Delhi, 110002 India

Shweta Srivastav The Hague Center for Global Governance, Innovation and Emergence, Vreeland, Netherlands

Manas Vijayan Department of Architecture and Planning, Visveswaraya National Institute of Technology, Nagpur, Maharashtra, India

Anne-Marie Voorhoeve The Hague Center for Global Governance, Innovation and Emergence, Vreeland, Netherlands

Pieter Wackers The Hague Center for Global Governance, Innovation and Emergence, Vreeland, Netherlands

Ida Widianingsih Faculty of Social and Political Sciences, Department of Public Administration, Universitas Padjadjaran, Bandung, Indonesia;
Faculty of Social and Political Sciences, Center for Decentralization & Participatory Development Research, Universitas Padjadjaran, Bandung, Indonesia

Brijesh Kumar Yadav Department of Hydrology, Indian Institute of Technology Roorkee, Roorkee, India

Sustainable Megacities and Planning for Environmental and Disaster Risk Resilience

Landslide Hazard in Bandung City, Indonesia: Urban Planning Toward Urban Resilience

Abdillah Abdillah, Ida Widianingsih, Rd Ahmad Buchari, and Heru Nurasa

1 Introduction

In geological terms, "slide" describes the various processes that cause slope-forming material to fall, roll over, slide, spread, or flow down the side of a mountain (Highland and Bobrowsky 2008; Santamarta Cerezal et al. 2014). While major and acute erosion tends to attract public attention because of its immediate damaging nature, smaller seasonal erosion also has long-lasting adverse implications for people in areas at risk of landslides (Santamarta Cerezal et al. 2014; Glade and Crozier 2005; Kusuma et al. 2021). In addition to the direct risks to people's lives and health, seasonal landslides also result in the gradual loss of land and property that can seriously threaten urban areas as well as the food security of mountain communities (Cieslik et al. 2019; Lacasse et al. 2009; Malik et al. 2021b).

A. Abdillah (✉)
Graduate Program in Administrative Science, Faculty of Social and Political Sciences, Universitas Padjadjaran, Bandung, Indonesia
e-mail: abdillah18001@mail.unpad.ac.id

I. Widianingsih · R. A. Buchari · H. Nurasa
Faculty of Social and Political Sciences, Department of Public Administration, Universitas Padjadjaran, Bandung, Indonesia
e-mail: ida.widianingsih@unpad.ac.id

R. A. Buchari
e-mail: ahmad.buchari@unpad.ac.id

H. Nurasa
e-mail: heru.nurasa@unpad.ac.id

A. Abdillah · I. Widianingsih · R. A. Buchari
Faculty of Social and Political Sciences, Center for Decentralization & Participatory Development Research, Universitas Padjadjaran, Bandung, Indonesia

M. Firoz C et al. (eds.), *Urban Planning and Design for Megacities in the Global South*, Advances in 21st Century Human Settlements,
https://doi.org/10.1007/978-981-97-8370-0_1

At the same time, the chronically recurring hazard of landslides presents opportunities for learning at government and community levels (Cieslik et al. 2019; Pelling 2003; Baker 2012; McIntyre-Mills et al. 2023; Widianingsih et al. 2023). Governments whose areas are exposed to chronic landslide hazards or have landslide risk zones often apply various mitigation and adaptation strategies (Pelling 2003; Baker 2012; National Research Council 2004). For this reason, this study analyzes a participatory and investigative basis to identify the strengths and weaknesses of local policies in the City of Bandung, Indonesia that govern the integration of landslide hazards into urban planning. It's intended to capitalize on these local capacities, increasing the resilience of communities and cities.

Disaster hazard conditions in Bandung that have frequently occurred in the last 10 years include earthquakes, floods, droughts, hurricanes, and landslides (Pemerintah Kota Bandung 2018). The threat of this geological disaster in Bandung City is huge because it is surrounded by faults from 3 (three) directions, each fault has the potential for seismicity and landslides (Pemerintah Kota Bandung 2018). In the north of the fault (Lembang), in the west of the fault (Cimandiri), and the south of the fault with the (Baleendah) and (Ciparay) to (Tanjung Sari). In addition, the base of the Bandung Basin has a high sedimentation rate so it will have a greater effect if it is hit by an earthquake wave propagation. The highest sedimentation is in the Cibiru, Gedebage, Soekarno-Hatta, and Purbaleunyi Toll Roads. Generally, landslides occur in the city of Bandung when rainfall is high. And areas prone to landslides based on the 2018–2023 Bandung City Regional Medium-Term Development Plan (*Rencana Pembangunan Jangka Menengah Daerah,* RPJMD) include Cibiru Sub-District, Mandalajati Sub-District, Ujungberung Sub-District, Cibeunying-Kaler Sub-District, Cidadap Sub-District, and Coblong Sub-District (Pemerintah Kota Bandung 2018).

Based on data from the Regional Disaster Management Agency (*Badan Penanggulangan Bencana Daerah,* BPBD), Bandung City experienced 2 landslides throughout 2014, while 2 landslides occurred due to flooding. The following is data on landslide events that occurred in the city of Bandung based on Indonesian Disaster Data and Information from the National Disaster Management Agency (*Badan Nasional Penanggulangan Bencana,* BNPB) and data based on the Bandung City RPJMD 2018–2023 as follows: (1) Landslide on 11 June 2014, no casualties; (2) April 10 2014 landslide, no casualties; (3) Landslide in 2012, causing loss of 1 house and 1 mini market; (4) Landslide in 2012, resulted in the collapse of the earthen rampart and bedroom walls, the collapse of the fort of 1 house, and the collapse of the fort and house in Hagarmanah Village, Ciumbuleuit Village, Cidadap Sub-District; (5) Landslide in 2012, resulted in the collapse of the Citepus River kirmir, the collapse of the Sirna Raga TPU kirmir, the cracked kirmir of the Citepus River in RW 8, flooding occurred due to narrowing of the waterways in Pajajaran Village, Arjuna Village, Sukaraja Village, Cicendo Sub-District; (6) Landslides in 2012, causing losses to 6 houses, landslides with Kirmir collapsing with a volume of 191 m; (7) Landslide in 2012, resulted in the collapse of the walls of 2 houses in Kebon-Jeruk Village, Andir Sub-District; (8) Landslides in 2012, causing two houses to be swept away by the river currents in Jatihandap Village, Mandalajati

Sub-District; (9) Landslide in 2012, resulted in the death of victims who were struck by landslides in Kp. Forward Cisurupan Village, Cibiru Sub-District; and (10) Landslide in 2011, causing 1 fatality in Cidadap Sub-District, Batununggal Sub-District and Bojongloa-Kidul Sub-District (Pemerintah Kota Bandung 2018).

Landslide events in these areas slow soil movement, and continuous land subsidence has damaged agricultural land and houses in built-up areas (Pemerintah Kota Bandung 2018). The increase in frequency and severity has hurt the livelihoods of urban communities, in the form of silent disasters, health and environmental damage (Pemerintah Kota Bandung 2018). In such a fragile environment, rapid population growth and human settlements are reshaping landslide risks that raise questions about sustainable development (Pelling 2003; Baker 2012; National Research Council 2004; Abdillah et al. 2022). Although the problem of landslides in Bandung City is common and is expected to cause future disasters, disaster risk reduction planning is inadequate in this area (Pemerintah Kota Bandung 2018). This is less highlighted in disaster management and not disseminated adequately among disaster risk-affected communities and areas when landslides occur as a single event despite having a much wider impact.

1.1 Objective

This study aims to explore disaster mitigation programs and policy support regarding the handling of landslide natural disasters in realizing urban resilience in Bandung City, Indonesia. We will also outline strategies used by local government and community participation support as a way of adapting to living with landslide hazards. This includes demographic shifts and patterns, land-use changes, and alternative employment options as well as experience, and knowledge generation from various mitigation measures. We argue that although these practices allow people to continue living in hazard-prone areas with disaster-adaptive capabilities. If you look at it now, the community has little potential to improve long-term livelihoods in the disaster-hazard zone.

This study will demonstrate that disaster policy and local knowledge can provide good-quality data for scientific modeling and forecasting and promote community sensitivity and involvement (Abdillah et al. 2020; Weichselgartner and Pigeon 2015; Abramson et al. 2015; Malik et al. 2021a). For this reason, in our theoretical model, we view regional policies and planning as urban planning concepts in realizing urban resilience from the threat of landslides (Levy 2017; Naess 2001; Kurniasih et al. 2023; Kusuma et al. 2021). According to J. M. Levy (2017) Urban Planning is planning political activities that are closely related to policies, big money, and decisions in public affairs. Not much different from what was said by P. Naess (2001) urban planning is that planning for sustainable urban development must be oriented toward long-term goals and utilize knowledge of the environmental consequences of different solutions, but must not be based solely on rationality of the final goal. What we mean by urban resilience is the ability of the urban system and its people to face all kinds of disturbances and threats in the future (Prianto and Abdillah 2023; Kurniasih et al. 2023; Abdillah et al. 2023a). Here we draw a dialogue between landslide hazard, urban resilience, local policies, and urban planning to compare and contrast them for the context of this study.

The contribution of this study we divided it into two points. First, we identify, describe, and categorize locally developed strategies that enable local people to live safely in landslide-prone areas. Second, we discuss the analysis of regional planning through which city policies and strategies help navigate the daily uncertainties that society faces.

2 Materials and Methods

2.1 Research Design

This research method uses a qualitative-explorative case study approach (Creswell and Poth 2016) which explores factual research on incidents or actual events that occurred in the field regarding landslide hazards and urban resilience in the city of Bandung, Indonesia. This research also explores urban governance and policies in dealing with these problems. This researcher adopts a social dynamics policy approach to the concept of urban resilience (Surjan et al. 2011; Meerow et al. 2016). This study focuses on urban resilience to landslide hazards by looking at the position of community vulnerability which is summed up in several places in the city of Bandung, Indonesia.

2.2 Site Selection

The focus of this research is the city of Bandung which is the largest metropolitan city in West Java, Indonesia. The city of Bandung has become the center of government for West Java Province but has also grown to become a local and regional trade center, education center, tourism and cultural city, and industrial center. Based on the 2018–2023 Bandung City RPJMD report, Bandung City is a landslide-prone area in various areas including the sub-districts of Cibiru, Mandalajati, Ujungberung, Cibeunying Kaler, Cidadap, and Coblong (Pemerintah Kota Bandung 2018). As shown on the map of Bandung City, Indonesia Fig. 1.

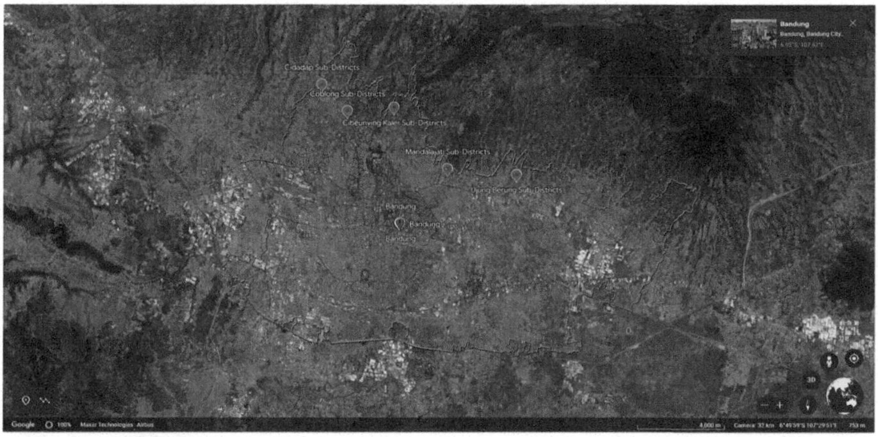

Fig. 1 Geographical location of landslide-prone areas in Bandung City, Indonesia (*Source* Processed via google earth, 2023)

2.3 Data Collection and Data Analysis Techniques

Data collection techniques in this study were from Field Observations, Regulatory Surveys, Literature Studies (digging electronic and physical data on the topics studied), Data from online media needed, and Documentation through documentary data collection obtained from books, journal articles, websites government, government social media, government letters, and related policies.

This technique is used to make it easier for researchers to get data sources, without requiring a lot of time, money, and effort while maintaining the validity of the data. To maintain the quality of this qualitative research, data triangulation was carried out (Flick 2004). Although this technique has its drawbacks, however, these deficiencies are minimized in in-depth interpretation and analysis through Nvivo 12 Plus because Machine Learning (AI)-based identification and analysis is sufficient to generate the best conclusions and show new insights on the topic being studied. The research data analysis technique refers to what was proposed by M. B. Miles et al. (2019) regarding interactive analysis through research data collection, data identification, research data reduction, research data presentation, data verification, and drawing conclusions. The analysis technique in this study was also assisted by the Nvivo 12 Plus qualitative data analysis tool (Woolf and Silver 2017) when managing research data, such as research observations, required social media data, government websites, and government documents. The steps of methods in this study are in Fig. 2.

The sequence of steps in this study is as shown in Fig. 2, namely by deciding to categorize them into three fields (collection of data through online media (a), data observation (b), and related government documents (c)). At the beginning of identifying data containing facts and information from online media. Then processed and processed through Nvivo 12 Plus to categorize the data/information. It is then assessed in relation to observational data (e.g., documents analyzed), and how

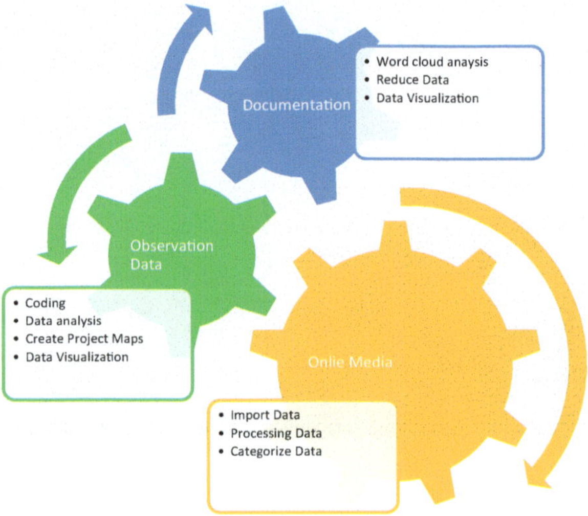

Fig. 2 Research method via Nvivo 12 Plus (*Source* Processed by researchers, 2023)

often this concept occurs is also assessed (in various Bandung City government documents). Combining these factors allows for identifying problems, cases, and phenomena studied to get the best findings and conclusions.

3 Results and Discussions

3.1 Empirical Conditions of Disaster and Landslide Hazards in the City of Bandung: An Empirical Review

Bandung City is one of the areas prone to landslides. According to the Disaster Management Agency, landslides are generally caused by two driving factors and triggering factors. Push factors are factors that affect the condition of the material itself, while trigger factors are factors that cause the material to move. Apart from that, there are also other factors such as erosion caused by runoff of surface water or rain. Heavy rainfall causes soil and rock slopes to become saturated, increasing the likelihood of landslides. Contributing factors include prolonged rainfall and dust-laden streams, as well as vibrations from machinery, traffic, explosives, and lightning. Additionally, excessive load from accumulated rainwater or snow can further destabilize slopes. In recent years, many landslides have occurred in the city of Bandung. Areas prone to landslides based on the Bandung City RPJMD 2018–2023 include Cibiru Sub-District, Mandalajati Sub-District, Ujungberung Sub-District, Cibeunying Kaler Sub-District, Cidadap Sub-District, and Coblong Sub-District (Pemerintah Kota Bandung 2018). The condition of the vulnerability of the landslide area in Bandung City can be seen in Fig. 3.

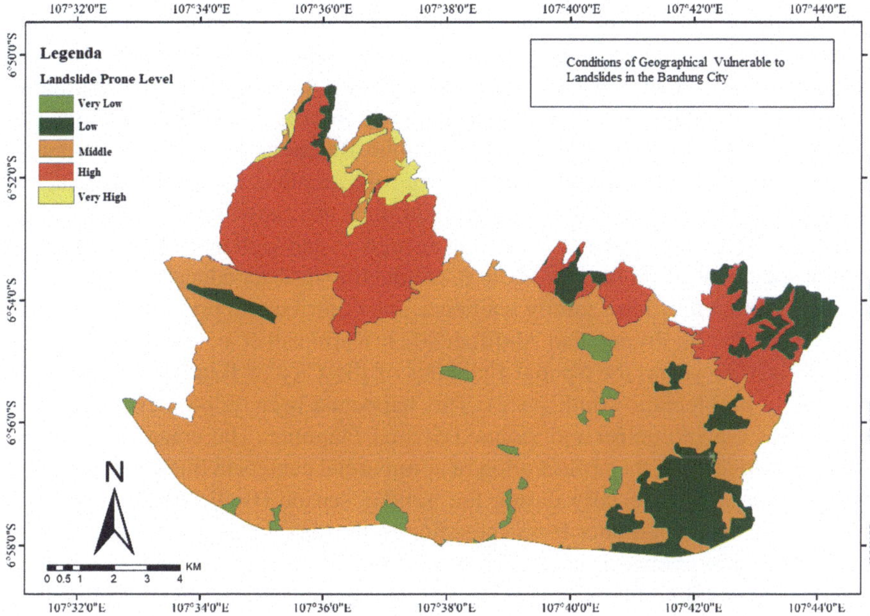

Fig. 3 Landslide prone level conditions in the city of Bandung (*Source* Processed from Kurniasih et al. (2023), 2023)

In 2021 there were recorded 2 landslides in Bandung City, although they did not cause any fatalities, they did cause environmental, social, and economic damage to the community. In 2022, 22 incidents affected 621 people. Coordination from various parties is built to anticipate future disasters, one of which is through disaster preparedness villages. Two landslides were recorded which affected 13 people and 11 tornadoes in 105 people. There were no fatalities as a result of all these incidents. Compared to last year, there has been an increase in the number of natural disasters. In 2021, 6 floods, 3 landslides, and 4 strong winds were reported (Ritonga 2022; Prilatama 2022). One of the areas with a high level of vulnerability to landslides is the Dago area, throughout 2021 there have been several landslides in the area. On November 2, 2021, there was an avalanche on a 10-m cliff in the Dago area which caused severe damage to several houses. By this study mapping and identifying the level of vulnerability to landslides in several areas in the city of Bandung, see Fig. 3 (Warsudi 2021; Fitri 2021; Kurniasih et al. 2023; Abdillah et al. 2022).

Figure 3 shows areas with a landslide vulnerability level in Bandung City. The distribution of landslide vulnerability levels is spread into several categories at certain points in the Bandung City area. Areas that are most prone to landslides are shown in areas with color labels (Yellow) and the lowest level of vulnerability to landslides is shown with color labels (Light green). Residents of Bandung City are encouraged to adopt disaster preparedness measures due to the city's significant geographic vulnerability to natural disasters (Budi 2022; Suharyadi 2016). The condition of Bandung

City is vulnerable to disasters. Making urban planning, supporting policies, and collaboration between stakeholders are very much needed in managing sustainable city governance (Prianto and Abdillah 2023; Kurniasih et al. 2023; Abdillah et al. 2023a; Abdillah et al. 2023b). This is in line with the Bandung City Government's planning that according to Bandung City Public Relations that disaster preparedness needs to be built early and starting from the smallest environment (Citarum Harum Juara 2022a). Community readiness to face threats will determine the risks and impacts of disasters, so that increasing awareness and vigilance of disasters at all levels of society needs to be carried out to conducting disaster education to all levels of society so that they have family resilience and contribute to prevention (Citarum Harum Juara 2022a; Prianto and Abdillah 2023; Kurniasih et al. 2023).

In the last 20 years, the Spatial Planning of the City of Bandung has become increasingly problematic (Budi 2022). This happened because the development of the City was out of control and violated Spatial Planning. The emergence of new settlements along with the proliferation of commercial centers in this city has caused severe traffic jams, especially during the holiday season (Budi 2022). In order to prevent environmental clutter from spreading to the surrounding areas of Bandung City, collaboration between the regional government and the national government is needed in providing the best solution so that environmental damage to Bandung City in the future can be prevented in accordance with spatial conditions (Budi 2022; Suharyadi 2016).

When there was heavy rain that flushed Bandung City, it caused landslides in various areas of the Kebon Waru sub-district, Batununggal District, Bandung City (Citarum Harum Juara 2021). The landslide material at that location almost blocked the flow of the Cibeunying River so that the flow of the river flow was stagnant, considering that the location of the landslide was on the riverbanks of Bandung City (Citarum Harum Juara 2021). In addition, the landslide resulted in the collapse of four houses belonging to residents of the area (Citarum Harum Juara 2021). The landslide occurred allegedly because the Kirmir was unable to withstand the load, the Kirmir that collapsed occurred due to being unable to withstand the load because the house was on the riverbank in the Batununggal District area, Bandung City (Citarum Harum Juara 2021). In the Hegarmanah Village, Cidadap Sub-District, Bandung City, was cut off due to a landslide. Landslide handlers in the area are constrained by access to heavy equipment so the process of evacuating landslide materials takes quite a long time (Putra 2023). Even though there were no casualties, it is necessary to anticipate support and cooperation between various parties such as the Bandung City Government, Satpol PP, Gober (the idea of the Clean Culvert movement), and residents to clean up the landslide material. Due to this incident, the Bandung City government appealed to the community to remain vigilant about the arrival of regional disasters, both strong winds, floods, landslides, and fallen trees (Citarum Harum Juara 2021; Putra 2023; Citarum Harum Juara 2022a; Budi 2022; Suharyadi 2016). The report shows that the level of vulnerability in Bandung City is still high due to the impacts of climate change and environmental damage. So environmental issues in the city of Bandung are very important to be considered to encourage development that is sustainable, inclusive, and resilient (Muhtar et al. 2023; Abdillah et al. 2023a, b).

Several studies admit that it is not easy to improve environmental conditions that are already in shambles (Noguera 2003; Hulme 2009). Because the core of the problem lies in the high will and awareness of all parties involved in natural disaster cases such as landslides (Hart 2013; Hulme 2009). Another factor that also has a big influence, namely funding and priority scale is very decisive in dealing with environmental damage in Bandung City due to landslides and floods (Citarum Harum Juara 2022a; Budi 2022; Suharyadi 2016). If only all components including stakeholders understand the importance of enforcing spatial planning regulations, it is not impossible that all environmental damage caused by landslides and flooding in Bandung City can be overcome (Citarum Harum Juara 2022a; Budi 2022; Suharyadi 2016). Bandung City, which is located in West Java Province, is a disaster-prone location. This means that intensive use of space as a built-up area and cultivation area will make West Java's physical condition even more vulnerable. And if it is not anticipated from now on, it is possible that in 20 years it will disrupt the comfort of living (Citarum Harum Juara 2022a; Budi 2022; Citarum Harum Juara 2021).

3.2 Urban Planning Toward Urban Resilience: Policy Analysis and Regional Planning for Bandung City, Indonesia

Geographically, Bandung City has a large enough disaster potential (Citarum Harum Juara 2022a, b) so Bandung City residents must be prepared for disasters. This is what the Bandung City government is concerned about, which could increase the vulnerability of the community and urban resilience in Bandung City (Citarum Harum Juara, 2021, 2022a, b). The threat of disaster in Bandung City makes disaster preparedness need to be built early and starting from the smallest environment. Community readiness to face threats will determine the risks and impacts of disasters. This preparedness is necessary in increasing disaster awareness and alertness to all levels of society. Of course by educating all levels of society so that they have family resilience and contribute to prevention. During the commemoration, a simulation was also held on using Kentongan (Bamboo Pieces) in an emergency. It is necessary to know the meaning of the sound of "Kentongan" (a type of gong), such as murder (1 beat), robbery (2 beats), fire (3 beats), natural disaster (4 beats), animal theft (5 beats), and security conditions (5 beats). The commemoration event was also attended by a number of communities, community organizations, and volunteers. These efforts have become a positive trend regarding disaster preparedness being pursued in Bandung City (Citarum Harum Juara 2021, 2022a, b).

The implementation of disaster management is carried out by taking into account social, economic, cultural, and environmental sustainability, effectiveness, and area size in Bandung City, regulated in Law Number 24 of 2007 concerning Disaster

Management (Article 31). Apart from that, the Implementation of Disaster Management in Government Regulation Number 21 of 2008 concerning the Implementation of Disaster Management regulates situations when a disaster does not occur or when a disaster occurs, including (1) Pre-disaster; (2) during an emergency response, and (3) Post-disaster. In addition, disaster management in Bandung City is also based on Minister of Home Affairs Regulation Number 33 of 2006 concerning General Guidelines for Disaster Mitigation in the management, prevention, and adaptation of disasters such as landslides. There are also regional policies that regulate disaster management in Bandung City, such as Bandung Mayor Regulation Number 1400 of 2016 concerning Position, Organizational Structure, Duties and Functions, and Work Procedures of the Bandung City Fire and Disaster Management Service and Bandung Mayor Regulation Number 626 of 2017 concerning Formation of the Organization and Work Procedure of the Bandung City Disaster Management Unit (see Fig. 4). These policies serve as legality in disaster management as urban planning creates urban resilience in Bandung City from disasters.

With the support of policies on disaster prevention and management, it needs to be accompanied by preparedness for disaster adaptation by the citizens of the city of Bandung. If you look at the area, now there are more floods in South Bandung due to changes in the ecosystem and because six rivers are fed by it. Meanwhile, landslides tend to be in East Bandung due to the lack of absorption areas (Citarum Harum Juara 2021, 2022a, b). To prevent various disasters from happening, the Bandung City government continues to monitor areas that frequently occur with regional officials. The Bandung City Government mitigates this by educating the public to learn to see the potential and symptoms of social change as well. For example, people have to take care of the gutters, don't throw garbage anywhere (Citarum Harum Juara 2021, 2022a, b). The Bandung City government helped by building a retention pond to prevent flooding. During January–September 2022, there were three hydrometeorological

Fig. 4 Policies in handling landslide disasters in Indonesia and also carried out by the Bandung City government as urban planning in creating urban resilience environmentally, socially, and economically in Bandung City, Indonesia (*Source* Processed from various sources, 2023)

disasters in Bandung City, Indonesia. One case of flooding in Rancasari, and two cases of landslides (Citarum Harum Juara 2021, 2022a, b).

The level of landslide cases in Bandung City from 2021 to 2022, although there were no fatalities, there were material losses, such as a house whose walls had collapsed (Citarum Harum Juara 2021, 2022a, b). As many as 21 houses were affected by flooding, in April 2021. Meanwhile, in 2021, the number of flood cases was 6 incidents, landslides were 3 incidents, and strong winds were 4 (four) incidents (Citarum Harum Juara 2021, 2022a, b). Disasters are a common concern of all Penta Helix sectors. Thus, during this year the Fire Department-PB has prepared disaster management together with other tools (Citarum Harum Juara 2021, 2022a, b). Community vigilance is still lacking in Bandung City. It must massively provide understanding to the community if the disaster continues to lurk so that it can encourage resilience which has a positive impact on development in Bandung City. What's more, the 2022 disaster in Bandung City will tend to be landslides when viewed from the movement of the ground. This has made the regional apparatus of the Bandung City Government push to be aware of the potential or natural phenomena that occur (Citarum Harum Juara 2021, 2022a, b).

This is in line with research from M. R. Masnavi et al. (2019) and K. Ferreira and A. Abiko (2019) who argued that the preparedness and adaptive attitude of urban communities toward urban systems as urban planning is something that must be done in supporting urban resilience from landslides. Because the concept of urban resilience is a concept of how cities can prepare, adapt, and overcome natural disaster problems (Ferreira and Abiko 2019; Shaw and IEDM Team 2009; Masnavi et al. 2019). Because urban resilience is understood as a process that involves increasing the ability to learn about changing conditions and to adapt to those conditions to reduce disaster risk, restore essential urban services, and improve quality of life (Mehmood 2016; Bush and Doyon 2019; McIntyre-Mills et al. 2023; Widianingsih et al. 2023). The large number of cases of natural disasters such as landslides in Bandung City, Indonesia, and the lack of public awareness of the threat of disasters is a challenge that must be resolved by each stakeholder in overcoming vulnerability due to disasters and encouraging resilience in Bandung City, Indonesia.

3.3 Implications of Landslide Risk Management as Urban Planning Toward Urban Resilience in Bandung City, Indonesia

The spatial structure plan for Bandung City from the report on the results of a study on the resilience of Bandung City (Endarwati et al. 2016) from the results of integration and special spatial allocation provisions is intended to increase capacity in dealing with disasters and the impacts of climate change in Bandung City. The spatial pattern plan resulting from this integration is intended to reduce the level of vulnerability. Integration into the urban spatial plan is aimed at improving quality in dealing with disasters by reducing risks both by increasing capacity and reducing vulnerability (Endarwati et al. 2016). Besides that, the integration of the program carried out by Bandung City in landslide-prone areas is by integrating the program to increase catchment areas in landslide-prone disaster areas (*Kawasan Rawan Bencana*, KRB) in the spatial pattern plan (Endarwati et al. 2016). This is expected to reduce the level of vulnerability to landslides, especially in areas previously designated for cultivation activities such as trade and services. Program integration carried out by Bandung City in flood-prone areas is by integrating the relocation of industrial areas in High Flood Prone Areas into the Green Open Space (*Ruang Terbuka Hijau*, RTH) program in the High Flood KRB in the spatial pattern plan and alternative relocations. Integration of the spatial pattern plan, and the integration of the spatial structure plan is carried out by integrating plans for evacuation routes, evacuation sites, and alternative relocation locations. Capacity building can also be increased by establishing special provisions in the allotment of space in areas prone to flooding and landslides in Bandung City (see Fig. 5). Based on this, it is known that after integration with the program in these special provisions, capacity can be increased in the aspect of evaluating the supply of equipment in the disaster risk reduction program (Endarwati et al. 2016; Pemerintah Kota Bandung 2018; Kurniasih et al. 2023).

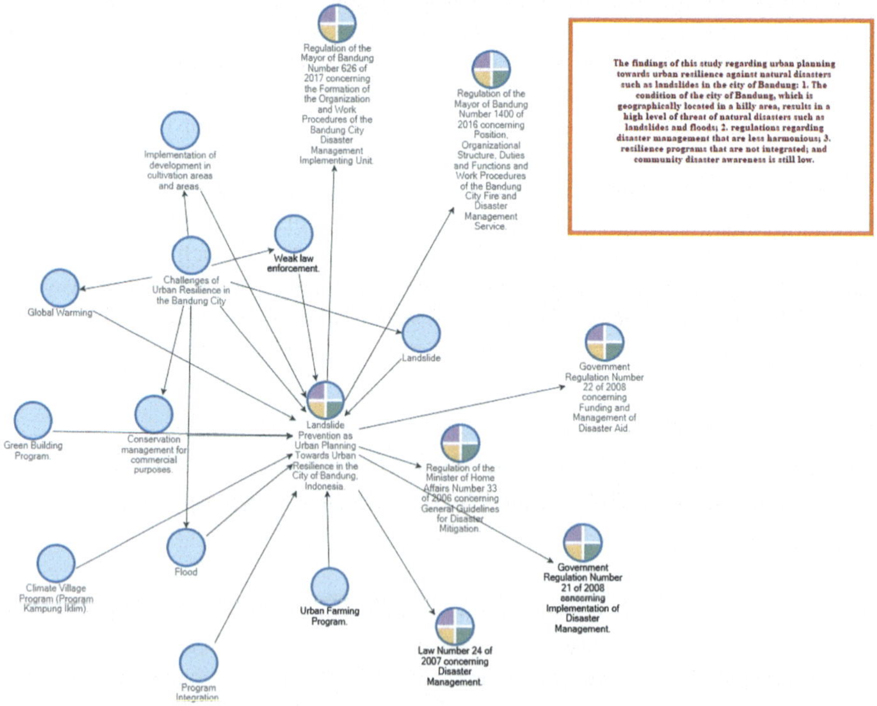

Fig. 5 Landslide prevention as urban planning toward urban resilience in the city of Bandung, Indonesia (*Source* Processed via Nvivo 12 Plus, 2023)

Efforts to increase resilience to climate change (temperature and rainfall projections) are carried out by strengthening food through the Urban Farming program (Abdillah et al. 2023a). In addition, the Green Building program has also been increased, especially in trade and service areas, as well as industry to reduce the effects of greenhouse gases and global warming and in residential areas through the climate village program (Endarwati et al. 2016; Pemerintah Kota Bandung 2018; Kurniasih et al. 2023). Based on the efforts that have been integrated between programs into the city spatial plan, it is expected to increase the resilience of Bandung City in facing disasters and adapting to climate change (Endarwati et al. 2016; Pemerintah Kota Bandung 2018; Kurniasih et al. 2023).

Many cases of landslides in Bandung City are caused by high rainfall, topographically the areas that are often inundated include Baleendah and Bojongsoang, and Dayeuhkolot are lowlands (Pemerintah Kota Bandung 2018; Kurniasih et al. 2023). Another factor is that it is located in the Citarum, Cisangkuy, and Cikapundung river basins. From the community's point of view, there are those who throw garbage into rivers and waterways, which causes disturbances to the flow of water. This is in line with the Bandung City RPJMD 2018–2023 document report that for the Bandung area itself, floods often occur in the South Bandung area which is topographically in the form of lowlands, and an analysis which suggests physical development in the North Bandung Area (*Kawasan Bandung Utara*, KBU) which incidentally is plains High altitude causes water runoff from the highlands to flow swiftly and cannot be accommodated in the lowlands (Pemerintah Kota Bandung 2018; Kurniasih et al. 2023; Endarwati et al. 2016).

A study by community organizations concerned with the environment, such as WALHI (Wahana Lingkungan Hidup Indonesia (WALHI)) in West Java, Indonesia said that the cause of the disaster in the city of Bandung was due to consistency and weak law enforcement and the implementation of development in cultivation areas and conservation management areas for commercial purposes (Sari 2017; Pemerintah Kota Bandung 2018; Kurniasih et al. 2023). In addition, the report on the Strategic Plan (*Renstra*) document for the Spatial Planning Office of the Bandung City Government (Sari 2017) shows that Bandung City spatial planning policies can no longer stand alone, it is very important to start planning in collaboration with surrounding cities and districts. In addition, it is hoped that there will not be conflicts over the use of space due to disharmony of regulations (Sari 2017; Endarwati et al. 2016; Pemerintah Kota Bandung 2018). Urban community awareness of disasters needs to be supported by accurate information and easy-to-obtain disaster education.

4 Conclusion

The condition of Bandung City with high cases of natural disasters in various regions, the condition of public awareness that is still low and various policies that are still not harmonious are serious challenges in realizing urban resilience in Bandung, Indonesia. In addition, Bandung City is an area with a moderate vulnerability to landslides. Bandung City Government has attempted various resilience programs such as urban farming programs, green building, and integration of various programs which are ways of thinking about resilience that can help plan and establish city resilience. Seeing urban planning as a complex adaptive system in resilience thinking also helps in seeing communities and places as subsystems that can be intertwined to formulate, adapt, or even mitigate and mediate change mechanisms from sharing legal support, to governance that refers to social, political, and economic conditions urban (Mehmood 2016; Bush and Doyon 2019).

As with the research conducted by M. R. Masnavi et al. (2019) and K. Ferreira and A. Abiko (2019), increasing urban resilience from landslides requires the development of preparedness and adaptive attitudes of urban communities toward urban systems as urban planning. This is because urban resilience is a process that involves the ability to learn about changing conditions and adapt to those changes. The challenge of community awareness of the threat of disaster is still low. Making various efforts to increase community preparedness and awareness of disasters. It becomes a stakeholder responsible for addressing vulnerability due to disasters and increasing resilience from natural disasters such as landslides in Bandung City, Indonesia. The recommendation in this study for the government and the surrounding community is to pay attention again to areas that have a high level of vulnerability to landslides. In addition, it is necessary to provide education on the prevention or mitigation of landslides, such as things that were done before the landslide disaster according to the Disaster Mitigation Agency, namely: (1) Be aware of high rainfall; (2) Prepare logistical support—Prepared food and drinks—Flashlights and spare batteries—Sufficient cash—Special medicines according to the user; (3) Listen to information from the radio or other media regarding information on rain and the possibility of landslides; (4) If the authorities give instructions to evacuate, do so immediately. In addition, technology must become part of a culture based on disaster awareness. Social media platforms are important for the community because they are a source of reference in accessing information. The delivery of educative content may be a cultural approach to increasing disaster awareness through the delivery of educative messages from community leaders.

References

Abdillah A, Deliarnoor NA, Yuningsih NY, Fatmawati F (2020) The position of auxiliary organ in the government system of West Java Provincial Government. J Contemp GovAnce Public Policy 1(2):67–81. https://doi.org/10.46507/jcgpp.v1i2.11

Abdillah A, Widianingsih I, Buchari RA, Nurasa H (2023a) Implications of urban farming on urban resilience in Indonesia: systematic literature review and research identification. Cogent Food & Agriculture 9(1):2216484. https://doi.org/10.1080/23311932.2023.2216484

Abdillah A, Widianingsih I, Buchari RA, Rusliadi R (2023b) Collaborative communication models in non-cash food assistance (Bantuan Pangan Non-Tunai, BPNT) program: toward community resilience. In: Intelligent communication technologies and virtual mobile networks. Springer Nature Singapore, Singapore, pp 75–91. https://doi.org/10.1007/978-981-99-1767-9_6

Abdillah A, Widianingsih I, Buchari RA, Mustari N, Saleh S (2022) Governance and Quintuple Helix innovation model: Insights from the local government of East Luwu Regency, Indonesia. Front Clim 4:1012108. https://doi.org/10.3389/fclim.2022.1012108

Abramson DM, Grattan LM, Mayer B, Colten CE, Arosemena FA, Bedimo-Rung A, Lichtveld M (2015) The resilience activation framework: a conceptual model of how access to social resources promotes adaptation and rapid recovery in post-disaster settings. J Behav Health Serv Res 42:42–57. https://doi.org/10.1007/s11414-014-9410-2

Baker JL (ed) (2012) Climate change, disaster risk, and the urban poor: cities building resilience for a changing world. World Bank Publications

Balbi M, Petit EJ, Croci S, Nabucet J, Georges R, Madec L, Ernoult A (2019) Ecological relevance of least cost path analysis: an easy implementation method for landscape urban planning. J Environ Manage 244:61–68. https://doi.org/10.1016/j.jenvman.2019.04.124

Budi A (2022) Pemerhati Tata Ruang Sebut Bandung Semrawut Akibat Tata Ruang Tidak Matang. http://koranprogresif.co.id/pemerhati-tata-ruang-sebut-bandung-semrawut-akibat-tata-ruang-tidak-matang/. Accessed June 2023

Bush J, Doyon A (2019) Building urban resilience with nature-based solutions: How can urban planning contribute? Cities 95:102483. https://doi.org/10.1016/j.cities.2019.102483

Cieslik K, Shakya P, Uprety M, Dewulf A, Russell C, Clark J, Dhakal A et al (2019) Building resilience to chronic landslide hazard through citizen science. Front Earth Sci 7:278. https://doi.org/10.3389/feart.2019.00278

Citarum Harum Juara (2021) Longsor Di Kota Bandung, Sektor 22 Harap Warga Waspada. https://citarumharum.jabarprov.go.id/longsor-di-kota-bandung-sektor-22-harap-warga-waspada/. Accessed May 2023

Citarum Harum Juara. (2022a). Rawan Bencana, Warga Kota Bandung Harus Tingkatkan Kesiapsiagaan. https://citarumharum.jabarprov.go.id/rawan-bencana-warga-kota-bandung-harus-tingkatkan-kesiapsiagaan/. Accessed May 2023

Citarum Harum Juara (2022b) Musim Hujan, Warga Kota Bandung Diimbau Waspada. https://citarumharum.jabarprov.go.id/musim-hujan-warga-kota-bandung-diimbau-waspada/. Accessed May 2023

Creswell JW, Poth CN (2016) Qualitative inquiry and research design: choosing among five approaches. Sage Publications

Endarwati MC, Imaduddina AH, Widodo WHS, Fitria LM, Giffari RA (2016) Kota Bandung Menuju Kota Tanggu Bencana Dan Berketahanan Perubahan Iklim. Direktorat Jenderal Tata Ruang Kementerian Agraria dan Tata Ruang/ Badan Pertanahan Nasional. http://eprints.itn.ac.id/5046/1/Buku-3%20Kota%20Bandung%20Menuju%20Kota%20Tangguh%20Bencana%20dan%20Berketahanan%20Iklim_reduce%20OK.pdf. Accessed July 2023

Ferreira K, Abiko A (2019) Urban resilience and landslide risk management: the case of Santos (Brazil). Urban Resilience for Risk and Adaptation Governance: Theory and Practice, 207–228. https://doi.org/10.1007/978-3-319-76944-8_12

Fitri NR (2021) Peta Rawan Longsor Kota Bandung. https://geo.mapid.io/blog_read/61b2bf01a29c7305d3750afc. Accessed May 2023

Flick U (2004). Triangulation in qualitative research. A Companion to Qualitative Research 3:178–183.

Glade T, Crozier MJ (2005) The nature of landslide hazard impact. Landslide Hazard and Risk, 41–74. https://doi.org/10.1002/9780470012659

Hart RA (2013) Children's participation: the theory and practice of involving young citizens in community development and environmental care. Routledge.

Highland L, Bobrowsky PT (2008) The landslide handbook: a guide to understanding landslides. US Geological Survey, Reston, VA, USA, p 129

Hulme PE (2009) Trade, transport and trouble: managing invasive species pathways in an era of globalization. J Appl Ecol 46(1):10–18. https://doi.org/10.1111/j.1365-2664.2008.01600.x

Hussein A (2009) The use of triangulation in social sciences research: can qualitative and quantitative methods be combined? J Comp Soc Work 4(1):106–117. https://doi.org/10.31265/jcsw.v4i1.48

Kurniasih D, Prianto AL, Abdillah A, Congge U, Akib E (2023) Strengthening climate change governance & smart city through smart education in Bandung Indonesia. In: Urban environment and smart cities in asian countries: insights for social, ecological, and technological sustainability. Springer International Publishing, Cham, pp 369–386. https://doi.org/10.1007/978-3-031-25914-2_16

Kusuma RR, Widianingsih I, Ningrum S, Myrna R (2021) Five clusters of flood management articles in Scopus from 2000 to 2019 using social network analysis. Sci Ed 8(1):85–92. https://doi.org/10.6087/kcse.234

Lacasse S, Nadim F, Lacasse S, Nadim F (2009) Landslide risk assessment and mitigation strategy. Landslides–Disaster Risk Reduction, 31–61. https://doi.org/10.1007/978-3-540-69970-5_3

Levy JM (2017) Contemporary urban planning. Routledge-Taylor & Francis, New York

Malik I, Abdillah A, Rusnaedy Z, Khaerah N (2021a) Coastal women's resilience strategy against climate change vulnerability in Makassar, Indonesia. In: E3S Web of Conferences. EDP Sciences, Vol 277, p 01003. https://doi.org/10.1051/e3sconf/202127701003

Malik, I., Prianto, A. L., Abdillah, A., Rusnaedy, Z., & Amalia, A. A. (2021b). Urban resilience strategy in the climate change governance in Makassar City, Indonesia. J GovMent Civ Soc 5(1):31–50. https://doi.org/10.31000/jgcs.v5i1.3884

Masnavi MR, Gharai F, Hajibandeh M (2019) Exploring urban resilience thinking for its application in urban planning: a review of literature. Int J Environ Sci Technol 16:567–582. https://doi.org/10.1007/s13762-018-1860-2

McIntyre-Mills JJ, Makaulule M, Lethole P, Pitsoane E, Arko-Achemfuor A, Wirawan R, Widianingsih I (2023) Ecocentric living: a way forward towards zero carbon: a conversation about indigenous law and leadership based on custodianship and praxis. Syst Pract Action Res 36(2):275–319. https://doi.org/10.1007/s11213-022-09604-0

Meerow S, Newell JP, Stults M (2016) Defining urban resilience: a review. Landsc Urban Plan 147:38–49. https://doi.org/10.1016/j.landurbplan.2015.11.011

Mehmood A (2016) Of resilient places: planning for urban resilience. Eur Plan Stud 24(2):407–419. https://doi.org/10.1080/09654313.2015.1082980

Miles MB, Huberman AM, Saldana J (2019) Qualitative data analysis: A methods sourcebook (fourth). Arizona State University

Muhtar EA, Abdillah A, Widianingsih I, Adikancana QM (2023) Smart villages, rural development and community vulnerability in Indonesia: a bibliometric analysis. Cogent Social Sciences 9(1):2219118. https://doi.org/10.1080/23311886.2023.2219118

Naess P (2001) Urban planning and sustainable development. Eur Plan Stud 9(4):503–524. https://doi.org/10.1080/713666490

National Research Council (2004) Partnerships for reducing landslide risk: assessment of the national landslide hazards mitigation strategy. National Academies Press

Noguera PA (2003) The trouble with Black boys: The role and influence of environmental and cultural factors on the academic performance of African American males. Urban Education 38(4):431–459. https://doi.org/10.1177/0042085903038004005

Pelling M (2003) The vulnerability of cities: natural disasters and social resilience. Earthscan.

Pemerintah Kota Bandung (2018) Rencana Pembangunan Jangka Menengah Daerah (RPJMD) Kota Bandung Tahun 2018–2023. Pemrintah Kota Bandung. https://peraturan.bpk.go.id/Download/127566/AUTENTIFIKASI%20PERDA%203%20TAHUN%202019%20RPJMD%202018%20-%202023%20%5BEDIT%206%20MEI%202019%5D%20CETAK%20(4).pdf. Accessed on July 2023.

Prianto AL, Abdillah A (2023) Resilient cities, vulnerable communities: disaster governance in the coastal cities in Indonesia. In: Singh A (eds) International handbook of disaster research. Springer, Singapore. https://doi.org/10.1007/978-981-16-8800-3_193-1

Prilatama MN (2022) Waspada, 13 Kecamatan di Kota Bandung ini Berpotensi Longsor dan Banjir di Peralihan Musim. https://jabar.tribunnews.com/2022/09/21/waspada-13-kecamatan-di-kota-bandung-ini-berpotensi-longsor-dan-banjir-di-peralihan-musim. Accessed May 2023

Putra W (2023) Penanganan Lonsor di Cidadap Bandung Terkendala Akses Alat Berat. https://www.detik.com/jabar/berita/d-6705032/penanganan-longsor-di-cidadap-bandung-terkendala-akses-alat-berat. Accessed June 2023

Ritonga MW (2022) Tahun ini, Kejadian Bencana Alam di Bandung Berpotensi Lebih Tinggi. https://www.kompas.id/baca/nusantara/2022/10/10/tahun-ini-kejadian-bencana-alam-di-bandung-berpotensi-lebih-tinggi. Accessed May 2023

Santamarta Cerezal JC, Hernández-Gutiérrez LE, Arraiza Bermúdez-Cañete M (2014) Natural hazards & climate change Riesgos Naturales y Cambio Climático

Sari D (2017) Menyoroti Penyebab dan Upaya Penanganan Banjir di Bandung dan Sekitarnya. https://balitbangsdm.kominfo.go.id/upt/bandung/berita-menyoroti-penyebab-dan-upaya-penanganan-banjir-di-bandung-dan-sekitarnya-5-38#:~:text=Pandangan%20Walhi%20Jawa%20Barat%2C%20hal,pengelolaan%20konservasi%20untuk%20tujuan%20komersial. Accessed July 2023

Shaw R, IEDM Team (2009). Climate disaster resilience: focus on coastal urban cities in Asia. Asian J Environ Disaster Manag 1(1):101–116.

Suharyadi I (2016) Peran Penting Asia Africa Smart City Summit (AASCS) 2015 terhadap Perkembangan Paradiplomasi Kota Bandung. Global: Jurnal Politik Internasional 18(1):95–107. https://doi.org/10.7454/global.v18i1.37

Surjan A, Takeuchi Y, Shaw R (2011) From disaster and climate risk to urban resilience: Approaching through community based environmental improvement. Research Publishing Service. Accessed on https://cir.nii.ac.jp/crid/1130282269894960256

Warsudi A (2021) Tanah longsor Terjang Dago Bandung, 2 Rumah Warga Rusak Parah. https://jabar.inews.id/berita/tanah-longsor-terjang-dago-bandung-2-rumah-warga-rusak-parah. Accessed May 2023

Weichselgartner J, Pigeon P (2015) The role of knowledge in disaster risk reduction. International Journal of Disaster Risk Science 6:107–116. https://doi.org/10.1007/S13753-015-0052-7

Widianingsih I, McIntyre JJ, Rakasiwi US, Iskandar GH, Wirawan R (2023) Indigenous sundanese leadership: eco-systemic lessons on zero emissions: a conversation with indigenous leaders in Ciptagelar, West Java. Syst Pract Action Res 36(2):321–353. https://doi.org/10.1007/s11213-022-09606-y

Woolf NH, Silver C (2017) Qualitative analysis using NVivo: The five-level QDA® method. Routledge

Environmental Planning Strategies for Resilient City: Case Study of Thiruvananthapuram, Kerala

Anjana Sasi, Ajay Raghava, and Jayeeta Sen

1 Introduction

In "The Global Climate Risk Index by Germanwatch", India ranked 7th out of 181 countries, indicating its high sensitivity to climate change with considerable economic losses and fatalities, with population pressure and strain on natural resources being the key contributing factors (Germanwatch 2021). Kerala State Action Plan on Climate Change has concluded Trivandrum to be under highly vulnerable category (Department of Environment and Climate Change 2014). Thiruvananthapuram city is home to nearly 9.58 Lakhs inhabiting 215 km^2 of land, with the average population density of 4454 persons/km^2, with higher concentration towards the coastal areas. Thiruvananthapuram district has one among the highest pace of urbanization at 10.5% between 1961 and 2001 (Department of Town and Country Planning 2012). City has a historical susceptibility to natural disasters and faces added risks from climate change, including sea level rise and heightened community vulnerability along the 35 km long coastal areas (Government of Kerala et al. 2015). Liquid waste management in metropolis is only 30%, which coupled with the poor solid-waste management is a fundamental cause to urban flooding. Thiruvananthapuram has reported its highest number of waterlogging/flooding situations during 2012–2015 (Government of Kerala et al. 2015).

A. Sasi (✉)
School of Planning and Architecture, New Delhi, India
e-mail: ar.anjana.1295@gmail.com

A. Raghava
Ministry of Environment, Forest, and Climate Change, New Delhi, India
e-mail: ajay.raghava@nic.in

J. Sen
Department of Environmental Planning, School of Planning and Architecture, New Delhi, India

M. Firoz C et al. (eds.), *Urban Planning and Design for Megacities in the Global South*,
Advances in 21st Century Human Settlements,
https://doi.org/10.1007/978-981-97-8370-0_2

The smart city proposals and master plan reiterate the need to improve the basic services (Ministry of Urban Development & GoI 2020). The city lacks a comprehensive study that ties together multiple challenges of climate change. The research tries to address aspects of water resilience which has emerged as a major challenge by establishing a comprehensive baseline for identifying and analysing the critical watersheds. Thiruvananthapuram metropolis which lies in the critical watershed was evaluated based on the identified parameters, facilitating a nuanced understanding of acute shocks and stresses within these areas, to delineate critical zones. Lastly, the project proposes specific strategies geared towards integration into the broader development plan for enhancing water resilience. Water resilience initiatives will help align with national initiatives like the *National Mission on Sustainable Habitat, National Water Mission, and Nationally Determined Contributions* that emphasize resilient cities. By fostering resilience, Thiruvananthapuram can work towards achieving Sustainable Development Goals, particularly *SDG 11*, focused on safe, inclusive, resilient, and sustainable cities, and *SDG 6*, aimed at ensuring clean water and sanitation for overall human development (Government of India 2015; United Nations 2021).

2 Conceptual Framework

This paper aims to delve into understanding resilience and explore the associated toolkits within the context of spatial planning. In this section, a set of national and international case studies were examined, to gain insights into the methodologies and frameworks utilized for evaluating the resilience of cities. These critical parameters have enabled the research to identify the spatial distribution of risks and vulnerabilities within the planning context.

2.1 Evolution of Concept of Resilience

Resilience serves as a vital intersection among diverse yet fundamentally similar paradigms in urban development. According to the UN Habitat's World Cities 2017 report, it is projected that by 2050, a staggering 70% of the global population will reside in urban areas. This urbanization trend is expected to trigger a substantial surge in the demand for energy and water, reiterating the need for cities to be resilient. The concept of resilience started in the early 1960s in the ecological systems. It was referred to as the ability of a system to bounce back to its original state (UN-Habitat 2017). Later, it was adapted to different fields. In the early 2000s the concept of resilience was discussed in UN conference on Disaster Risk Reduction. In 2014, the International Organization for Standardization (ISO) released the ISO 37120 standard titled "Sustainable Development of Communities – Indicators for City Services and Quality of Life". In the years 2015 and 2016, resilience emerged as a prominent theme within six significant global agendas. These included the Sendai Framework, the Addis Ababa Action Agenda, the Sustainable Development Goals, the Paris Agreement, the World Humanitarian Summit Commitments to Action, and the New Urban Agenda (Nations Office for Disaster Risk Reduction 2015; UN 2015, 2016; United Nations Sustainable Development 2015). In the national context Urban and Regional Plans Formulation and Implementation (URDPFI) guidelines, which are the governing document to formulate the city or regional level plans in India, discuss the need for cities to be resilient by integrating sustainability principles, climate change mitigation, and adaptation efforts in the planning process and development plans (Ministry of Urban Development 2014). This is also in line with the National Action Plan on Climate Change (NAPCC) which was established by the Government of India in 2008, with the primary objective of outlining a sustainable developmental pathway while prioritizing ecological integrity (MoEFCC 2021).

2.2 *City Resilience Profiling Toolkit*

UN Habitat developed the City Resilience Profiling Tool (CRPT) to be a robust and comprehensive approach for cities to build resilience. This approach is implemented using a 5-step process that involves the training of the city and regional level staff and stakeholders to access the data required. The second step is to perform the data collection and analysis of the same across 4 sets. 1st set investigates the city development narrative through its historical background and spatial context into the aspects relevant to climate, ecosystems, urban areas, and physical assets. It introduces the city's administrative structure, characteristics, and strategies, and describes the city's inhabitants through their composition through information on the economy and livelihoods. It also evaluates the hazard profile of the city. Through set 2, the tool attunes to the local governments and stakeholders, looking into the pros and cons of the governance at city level. Set 3 investigates the shocks, stresses, and challenges. Stresses are categorized into 3, firstly internal stress, an example being unemployment. Second type of stress addressed is external, for example, a natural calamity like flooding. The third type of stress is a combination of the first two, for e.g., coastal inundation can be due to sea level rise as well as unplanned development. Set 4 urban elements includes all the elements that frame the urban area, from a close-up in the built environment to the broader scale of the ecology, the supply chain, basic infrastructure, and mobility, all the elements at the service of inhabitants, or enabling life in the urban system (UN Habitat 2018) (Fig. 1).

2.3 *Chennai Resilient City Strategy*

Chennai is one among the most populated urban agglomerations in India. Chennai Resilient City initiative is supported by Rockefeller Foundation Resilience Centre and the Resilient Cities Network. After understanding the city's context and identifying

Fig. 1 City resilience profiling toolkit framework (UN Habitat 2018)

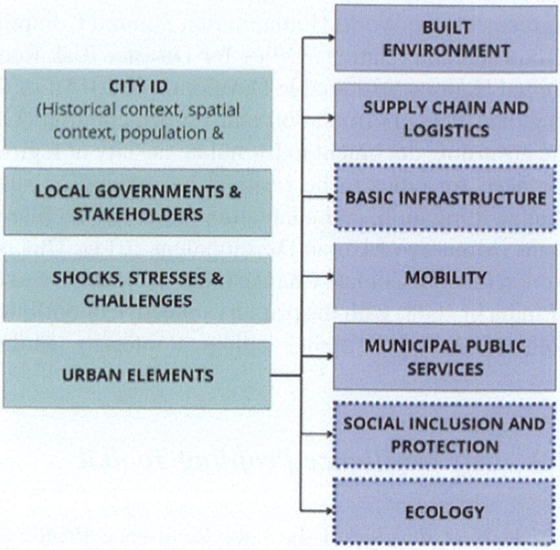

the key resilience challenges, five broad areas including healthy and planned urbanization, water systems restoration, disaster preparedness for vulnerable communities, governance framework improvement, and empowering vulnerable communities have been included in the final resilience strategy. Detailed aspects to the pillars of resilience for the Chennai city have been captured in Fig. 2.

2.4 *Master Plan for Thiruvananthapuram—2012*

Thiruvananthapuram's main transit centres experience triggered floods. This issue is exacerbated due to water body degradation and overexploitation, leading to physio-chemical parameter exceeding the limits. Inadequate water retention areas worsen the problem. 46% of wards lack solid-waste management. Coastal regions face saline water intrusion, and green spaces are declining (Thiruvananthapuram Corporation et al. 2012).

3 Methodology and Rationale

The study started with a Literature analysis to understand the frameworks of resilience, and to identify the shocks and stresses in the study area. Further evaluation was conducted on water resilience, which has emerged as a critical challenge

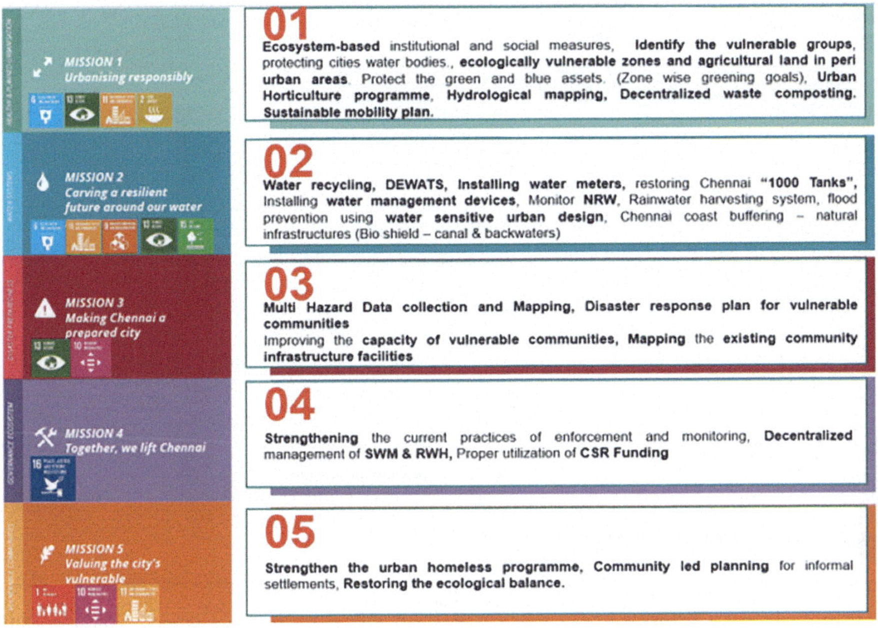

Fig. 2 City resilience framework for Chennai (Referred from Okapi et al. 2019)

in the study area. Analysis of different parameters relevant to water resilience was done, to identify the critical watersheds in the study area. Thiruvananthapuram city limits which fall within the critical watershed were further studied using a vulnerability assessment to understand the critical zones within the planning area. Furthermore, research focused on proposing strategies for integrating water resilience into the development plan process. This will contribute to the enhancement of adaptive capacity and the ability to cope with water-related challenges in the identified region. The study methodology that has been outlined for the paper is shown in Fig. 3.

4 Materials and Methods

Kerala is the southernmost state on India's Western Coast. The state of Kerala shares its border with Karnataka and Tamil Nadu. Kerala is blessed with a unique set of geographical features; the Western Ghats binds it to the east and the Arabian Sea to the west. *The capital city of Kerala is Thiruvananthapuram*, as shown in Fig. 4. The district covers 2,189 sq.km, accounting for 5.63% of the state's total area. It ranks 11th in size among all districts and is bordered by Kollam, Tirunelveli, and Kanyakumari.

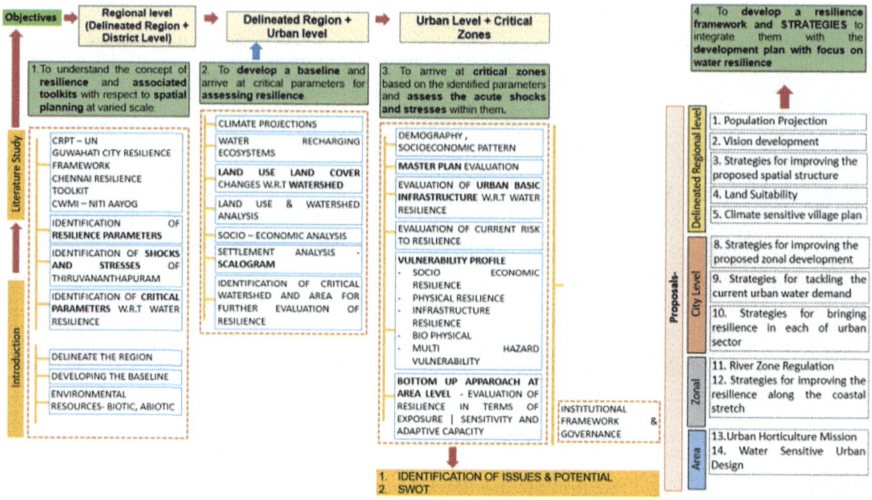

Fig. 3 Methodology for the paper

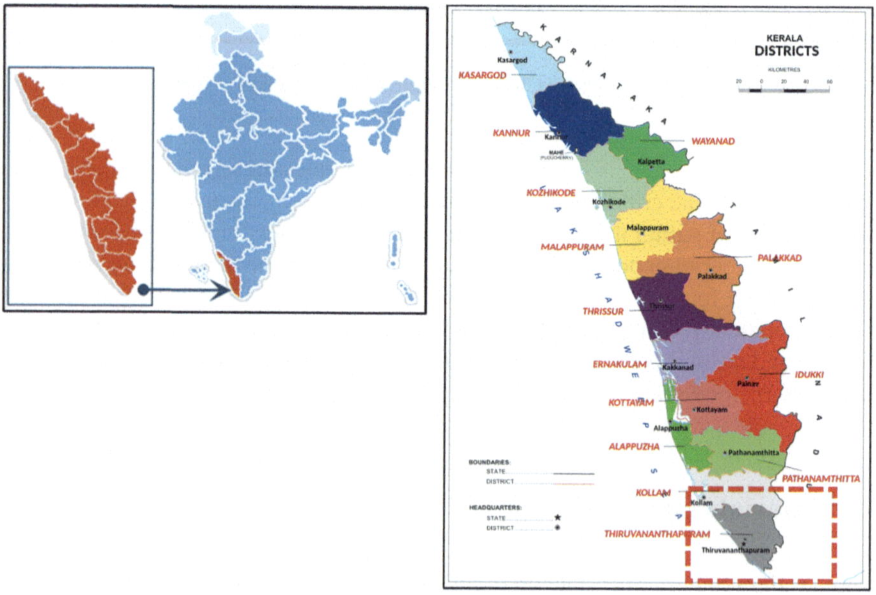

Fig. 4 State of Kerala and capital city of Thiruvananthapuram (highlighted) (Kerala State Land Use Board 2013)

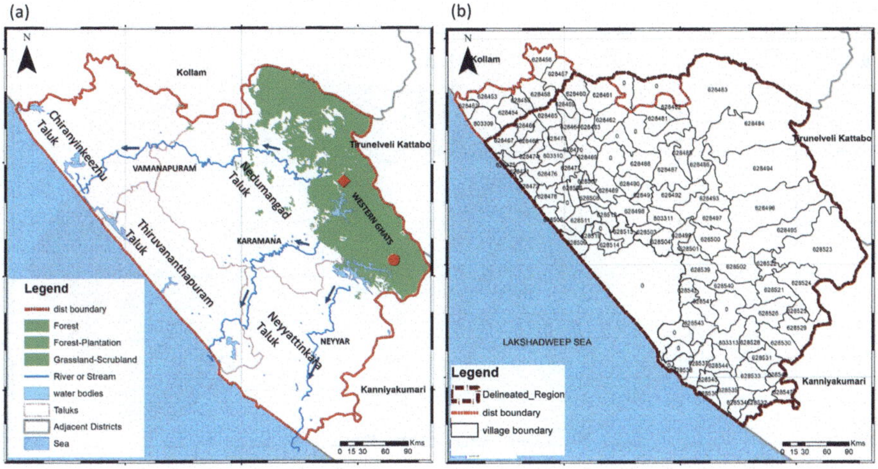

Fig. 5 **a** Delineated boundary with watershed and taluk boundaries. **b** further delineated boundary with the village boundaries (Directorate of Census Operations 2014; Kerala State Land Use Board 2013)

4.1 Regional Delineation

Water resilience being the area of focus, the method followed for regional delineation was to overlay the maps using Geographic Information System (GIS). Watershed map as per the Soil and Land-Use Survey of India was overlayed to the planning boundary at the regional level. These planning units are referred to as "Talukas" and Thiruvananthapuram district comprises of four of them. There are three river bodies, namely Vamanapuram, Karamana, and Neyyar that originate in the Western Ghats, and drain into the Arabian Sea Fig. 5a. Further, village layer was overlayered to the watershed-taluka layer, to obtain the final boundary. The regional boundary was further attuned for the inclusion of village whose 50% was falling within the watershed-taluka boundary as shown in Fig. 5b. The total delineated area in the region is 1977 sqkm, comprising of 63 villages from a total 68 villages in the district.

4.2 Environmental Resources

The district's landforms have been shaped by a combination of marine, fluvial, and denudational processes. These landforms can be classified into three main units: lowlands, midlands, and highlands as shown in Fig. 6b (Department of Town and Country Planning & Kerala Government 2017). Slope of the district falls under 0–5%, which is a gentle slope. Followed by a 5–15% slope which is moderate. Both these are ideal for residential and agriculture development. Then it increases up to 70% towards the Western Ghats as shown in Fig. 6b. The soil characteristics of

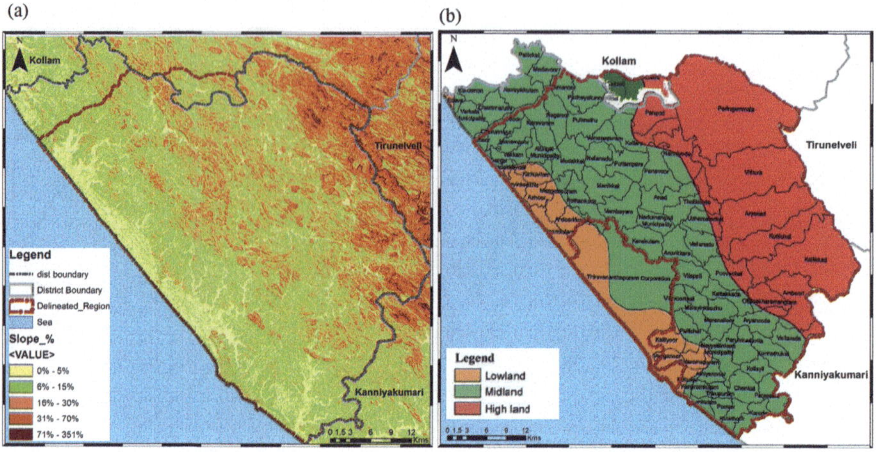

Fig. 6 **a** Thiruvananthapuram district slope map, **b** Physiography map adapted from Kerala state land-use board publication (Department of Town and Country Planning 2017)

the district can be broadly categorized into three, the low land soils are imperfectly drained hydromorphic soils, depth varies from 75 cm to more than 150 cm. The midland soils have a depth of less than 100 cm. Upland soils with depth less than 75 cm developed from gneissic rock (Soil and Land Use Survey of India 2021). As per the natural resource bank, the laterite capping along the midland region is the reason for slow seepage of water in these areas, making the peak discharge higher (Kerala State Land Use Board 2013). There is high dependence of groundwater from the domestic and agriculture sector. From the studies, 7 blocks are falling under Semi critical category as shown in Fig. 7. The increasing density of built-up, the nature of the crop, and the water demand are few reasons for stressing the water resource (Government of India et al. 2017).

The climate of Thiruvananthapuram district is generally *hot tropical*. The mean maximum temperature is 35 °C and the mean minimum temperature is 20 °C (Thiruvananthapuram Corporation et al. 2012). Kerala is situated in the high rainfall humid tropical region. The total annual average rainfall in the district is about 1,500 mm per annum. The southwest monsoon, from June to September, is the principal rainy season. The district receives most of its annual rainfall in this season. The second rainy season is the Northeast monsoon. It is from Oct to Nov. Temperature projections as per IISC report and state climate action plan foresees an increase of 0.5–4 °C rise in temperature. Climate change projected changes for the Western Ghats and Kerala over the next two decades include factors such as reduced rainfall, higher atmospheric temperatures, and increased flooding due to rising sea levels. These changes can impact crop resilience, crop yield, evapotranspiration, and water requirements (Department of Environment and Climate Change 2014; Indian Institute of Science & Ministry of Environment 2021).

Fig. 7 Critical blocks w.r.t groundwater (Reproduced from [Government of India et al. 2017])

4.3 Water Recharging Ecosystems

The district has *two protected areas viz., Peppara and Neyyar* which together accounts for 181 km^2 of forests. The wetland system of the study region can be classified into three: Coastal Wetlands, Inland Wetlands, and Unique Wetlands. Thiruvananthapuram has several freshwater lakes linked via canals. Another important typology found here is the Myristica swamps which is IUCN red-listed type, it is found along Kulathupuzha range. Paddy fields are also excellent for water retention. The analysis for change in paddy over the years 2008 to 2019 shows a decline of 10 sqkm as shown in Table 1 (Economics and Statistics Department 2020).

Table 1 Area under paddy over the years

Area under paddy	
Year	Total
2018–2019	19.68 sqkm
2010–2011	29.19 sqkm
2009–2010	29.40 sqkm
2008–2009	29.95 sqkm

Reproduced from Economics and Statistics Department (2020)

4.4 *Land-Use Land Cover Analysis (LULC) in the Delineated Region*

Land use is a crucial indicator for understanding and characterizing ecosystems, the LULC maps were developed with the use of National Remote Sensing Centre (NRSC) and the U.S. Geological Survey (USGS). Human land-use practices have substantially altered over one-third of the Earth's terrestrial surface, resulting in both positive social and economic benefits as well as negative impacts on the natural environment. The land-use changes were identified across three different years: *2005, 2015, and 2020*, and were categorized into four broad land-use categories: *Agriculture, Built-up, Barren land/wasteland, Forest, and Water bodies* as shown in Fig. 8. When examining the change in the built-up category over the years, it was found that the area under built-up increased from 8.66% in 2005 to 10.30% in 2015, and further to 13.66% in 2020, indicating a close to *5% increase in the built-up area.* Conversely, the area under *agriculture/plantation experienced a 5% decline* during the same period. The area under water bodies also decreased from 2.07% in 2005–2006 to 1.72% in 2020. These changes reflect a decline in the area under impervious land, which can have implications for water resilience in the study area (Thiruvananthapuram Corporation et al. 2012). The findings from the analysis of LULC reveal a concerning trend in the Vamanapuram and Karamana Watersheds, with a more pronounced decline observed in the sinks. *The area under Karamana and Vamanapuram watersheds has seen a major upsurge in the built areas, these are also the watershed basins that encompass the TMC downstream. The District Development Plan also cites as to how there is an increased conversion in the land use from agriculture to wasteland and eventually to build up areas* (Department of Town and Country Planning & Kerala Government 2017).

4.5 *Runoff Evaluation and Morphology Analysis*

As observed in the LULC changes, there is an upsurge in the built-up areas. This increase can lead to increased impervious surfaces, resulting in enhanced surface runoff. These runoff changes alter the hydrologic system and reduce natural water

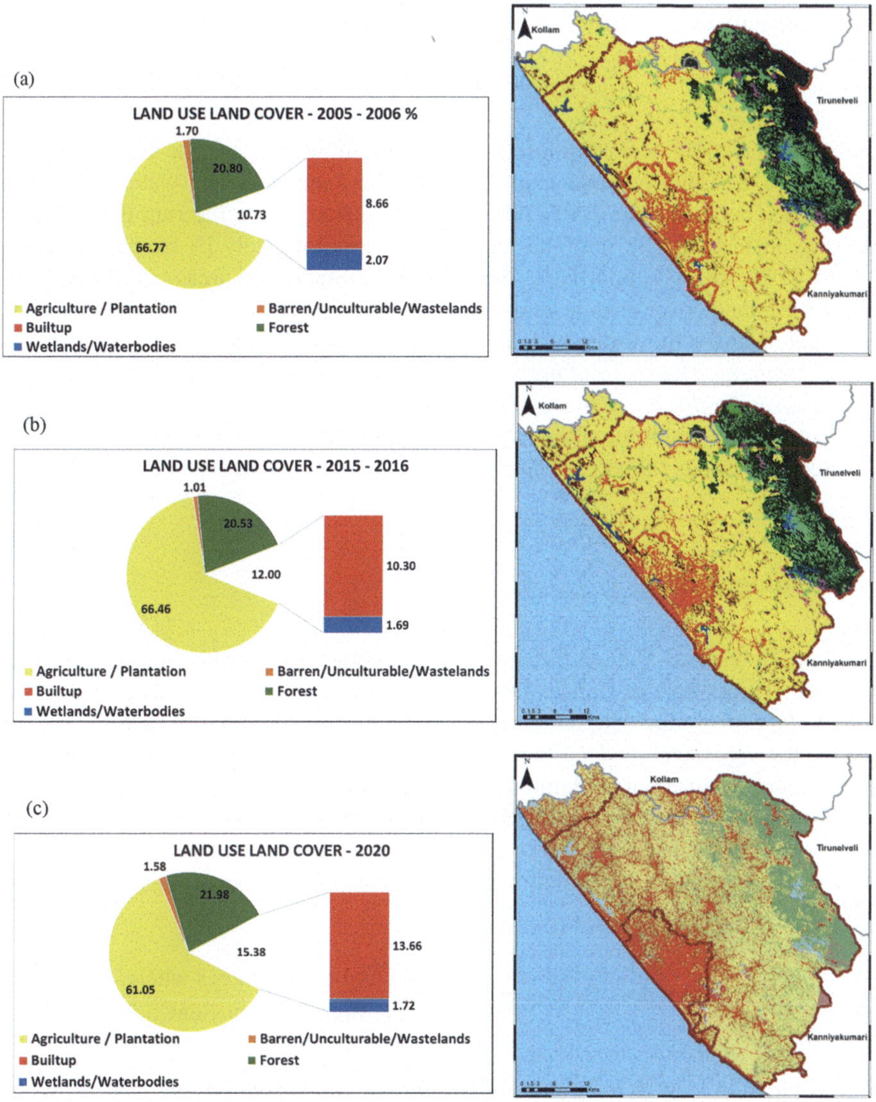

Fig. 8 **a** Thiruvananthapuram district land-use land cover changes 2005–2006. **b** Thiruvananthapuram district land-use land cover changes 2015–2016. **c** Thiruvananthapuram district land-use land cover changes 2019–2020 (National Remote Sensing Centre and ISRO 2021; Thiruvananthapuram Corporation et al. 2012)

storage capacity due to decreased infiltration. To calculate runoff (Q), the formula

$$Q = C \times R \times A,$$

where C is the coefficient of runoff, R is annual rainfall, and A is the area of rainfall (Khanna 2012). This formula helps estimate the annual surface water yield in a watershed. The runoff for 2005–2006 was 3.67 MLD, while the runoff for 2021–2020 was 3.81 MLD. The runoff has increased by approximately 3% from 2005–2006 to 2021–2020. This is supportive to the decrease in impervious areas as inferred from the land-use land cover change analysis. The drainage density and bifurcation ratios were evaluated for the watersheds in the study region through secondary sources. *Higher Drainage Density and higher bifurcation ratio indicate that shorter will be the time taken for discharge to reach the outlet, and higher will be the peak discharge.* Drainage density and bifurcation ratios are highest for Karamana watershed among the three watersheds (Dinagara Pandi et al. 2017; Gopal et al. 2018; Sreeja 2011).

4.6 Scalogram

Based on the Census Village directory, infrastructure-based adaptive capacity was evaluated (Directorate of Census Operations 2014). A scalogram was prepared for all 63 villages from the delineated region. Composite Functionality Index (CFI) for the villages were developed. CFI helps to assess and rank the level of functionality or development of different settlements or areas. 18 parameters were considered, and their availability and non-availability was marked from Census Village directory. Their presence/absence were marked in binary (0 and 1) and then the CFI calculation was carried out. Detailed calculations can be viewed in the sheet (https://shorturl.at/fBKLO)

$$\text{CFI} = (\text{Total no. of villages})/\left(\begin{array}{c}\text{Total number of villages where}\\ \text{the function is present}\end{array}\right).$$

This will help in identifying those villages with better CFI index and availability of functions which in turn have the potential to become future towns. This will also help in identifying those villages faring poorly w.r.t access to these functions and hence need planning solutions. Villages are performing well, due to the rural urban continuum observed in Thiruvananthapuram. Aspects evaluated are shown in Table 3. Based on the CFI, the villages are grouped into 4 orders. 1st and 2nd order villages have higher CFI scores thus indicating better development and 3rd and 4th order settlements have lesser functions and fare poorly. *Karamana and Vamanapuram watershed that encompass the major portion of city has the major share of 3rd and 4th order settlements making the city downstream further vulnerable* as shown in Fig. 9. Among the three rivers draining from Western Ghats, Karamana River flows into the city downstream. The designation of *this river as the Priority 1 polluted stretch* by the Central Pollution Control Board further adds to the challenge of water resilience (Government of India et al. 2017) (Table 2).

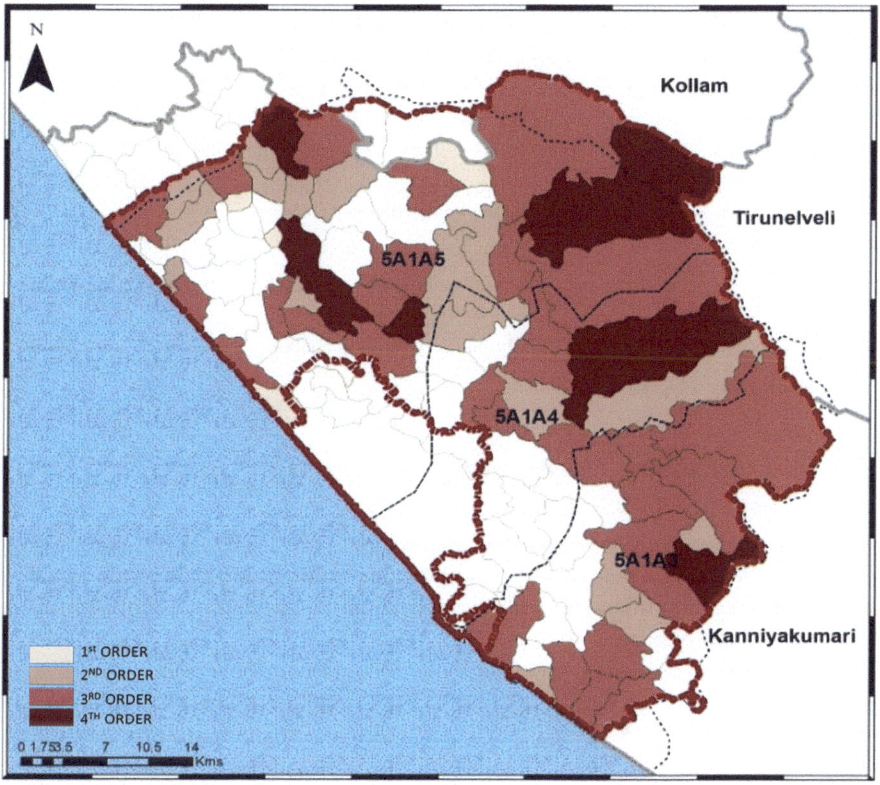

Fig. 9 Map with CFI ranking

Table 2 Number of settlements watershed wise

Watersheds	1st order	2nd order	3rd order	4th order
Vamanapuram (5A1A5)	4	12	13	5
Karamana (5A1A4)		3	7	1
Neyyar (5A1A3)		4	13	1

Table 3 Functions considered for the calculation of CFI

Functions considered	
Education	Secondary schools colleges
Health	Community health centre hospitals
Drinking water source	Improved drinking water source like hand pump, covered well, protected spring
Drainage	Presence of drainage facility
Sewer plant	Presence of sewer plant
Waste disposal	Presence of waste collection mechanism
Transportation and communication	Railway station (roads, mobile network, post offices present in all villages)
Electricity	Presence of power supply
Banking	Agricultural credit society, agricultural marketing society, Cooperative Bank
Others	Public library, weekly haats

Directorate of Census Operations (2014)

4.7 *Baseline of the Thiruvananthapuram City*

Thiruvananthapuram city evolved from a municipality in 1920 to a corporation with 100 wards. With abundant natural resources and skilled workforce, the city has become an important historical and tourist destination, aiming to establish itself in the global system of cities.

Thiruvananthapuram has a population of 9.58 Lakhs and covers an area of 214.86 sqkm, resulting in an average density of 4454 persons per sqkm, much higher than the district's value of 1500 persons per sqkm (Thiruvananthapuram Corporation et al. 2012). The corporation's population growth rate has decreased over time, with the highest growth observed during 2001–2011. The city exhibits a high percentage share of non-workers, although it has declined between 2001 and 2011. Peripheral wards show higher concentrations of marginal workers, while non-workers are concentrated in coastal wards. Other workers constitute most of the workforce (94.7%), attributed to the presence of administrative and research institutions in Thiruvananthapuram (Thiruvananthapuram Corporation et al. 2012).

4.8 *Land-Use Changes at City Level*

The land-use breakup of the city in 1966, 1990, and 2012 are depicted in Fig. 10. The land-use breakup of the city in 1966 shows that major single land use was residential at 71% followed by agriculture at 12%. The land-use breakup of 1990 shows that the residential percentage was 63%, followed by agriculture at 11%, the area of the

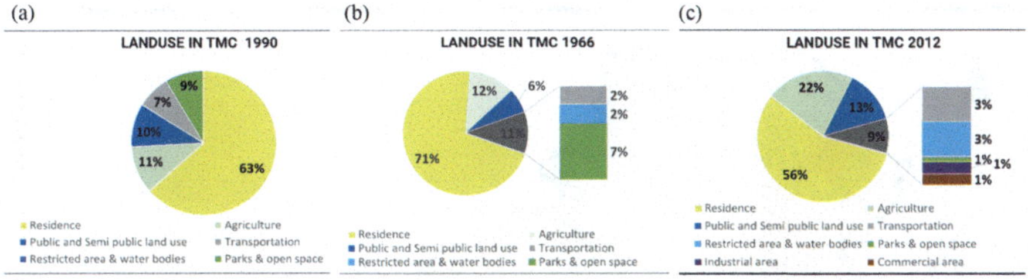

Fig. 10 Thiruvananthapuram Municipal Corporation land use, **a** 1990, **b** 1966, **c** 2012 (Reproduced from Thiruvananthapuram Corporation et al. 2012)

city increased from 74.9 sqkm in 1966 to 141.74 sqkm in 1990. The land-use plan of 2012 shows the residential land use at 56%, followed by agriculture at 22% and the area under the city boundary has increased to 214 sqkm (Fig. 11).

4.9 *Infrastructure Resilience*

The presence of robust infrastructure, particularly in the realm of water, plays a pivotal role in building resilience. It not only improves the quality of life for residents but also attracts investments. Furthermore, it mitigates vulnerability to disasters and lessens the impact of climate change on essential infrastructure, such as water supply, sanitation, drainage, and waste management. Haphazard development has resulted in crowded landmarks and lack of open spaces within Trivandrum (Thiruvananthapuram Corporation et al. 2012). The city of Thiruvananthapuram is densely populated and has two major rivers, Karamana and Killi, along with several canals and natural ponds as shown in Fig. 13. But currently in the Trivandrum city, these canals and thodus are clogged with solid waste. It has a stenching smell due to the extremely poor water quality as seen in Fig. 12. The water quality parameters in terms of Biological Oxygen Demand, Feacal Coliform, Dissolved Oxygen at different locations in Karamana River are exceeding the permissible limits (Kerala State Pollution Control Board et al. 2019). Proper management of these water bodies and drainage system is essential for the city's sustainable development and water resilience (R et al. 2018).

Thiruvananthapuram city relies on Peppara and Aruvikkara reservoirs for water supply. As per the studies published under smart city the status of water infrastructure in comparison with the benchmark are depicted in Tables 4 and 5. For non-revenue water existing service level benchmark is 35%. 107 MLD sewage treatment plant covers 37% area but overflows untreated in some water bodies as shown in Fig. 13. Stormwater network covers 60% and needs improvement (Thiruvananthapuram Municipal Corporation 2014). Out of the six diseases included under National Vector Borne Disease Control Programme viz, Malaria, Dengue fever, Filariasis, Chikungunya, Kala-azar, and Japanese encephalitis, the first 5 are being reported in the city (Department of Environment and Climate Change 2014).

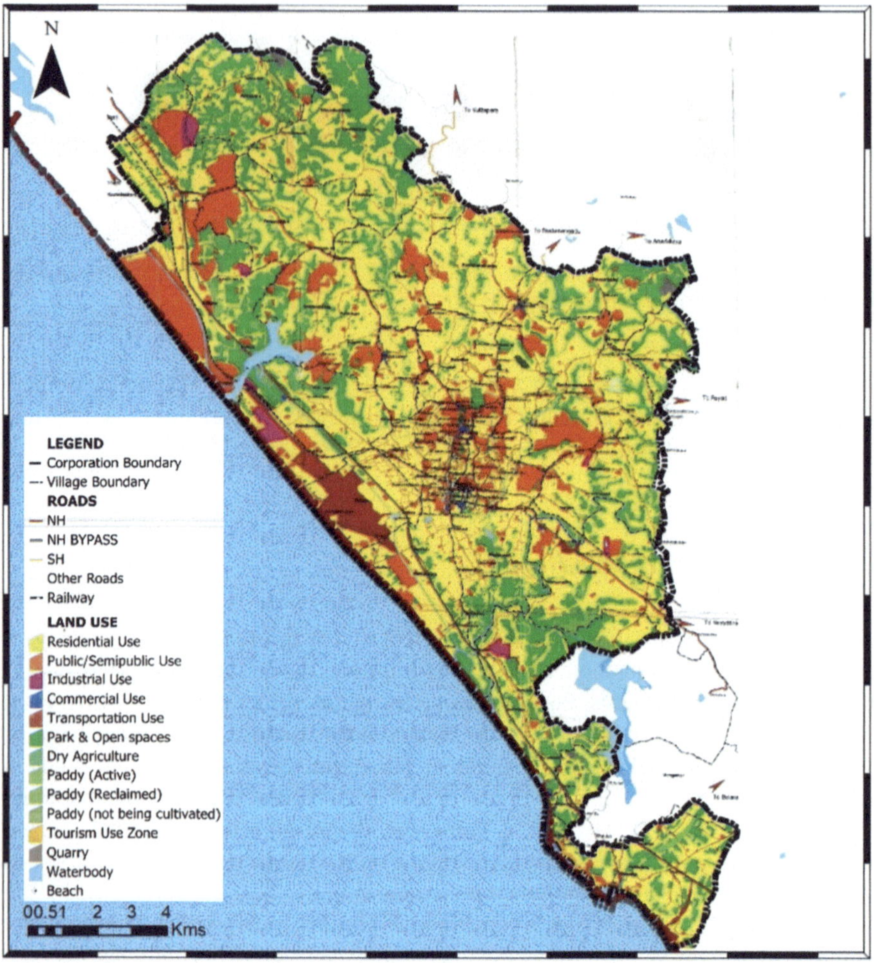

Fig. 11 Land-use map of Thiruvananthapuram Municipal Corporation (Reproduced from Thiruvananthapuram Corporation et al. 2012)

4.10 Water Balance for the City

As shown in Fig. 14, the water balance for the city shows that Thiruvananthapuram city draws 276 MLD water from Peppara and Aruvikkara dams. Annual rainfall is at 1800 mm. Treated water from the dams is stored in 11 overhead tanks and 15 ground-level reservoirs. The water lost before it reaches the end consumer (NRW) is 94 MLD. The remaining water is distributed in different sectors with Residential at 95 MLD, Agriculture at 39 MLD, Commercial at 9 MLD, and other sectors at 31 MLD. 10% of water demand is met by groundwater. Out of 104 MLD of wastewater generated, 44 MLD is treated, and the rest goes to septic tank, boreholes and open

Fig. 12 Condition of Parvathy Puthanar and Aamayizhanjanthodu (Reproduced from Abraham 2019)

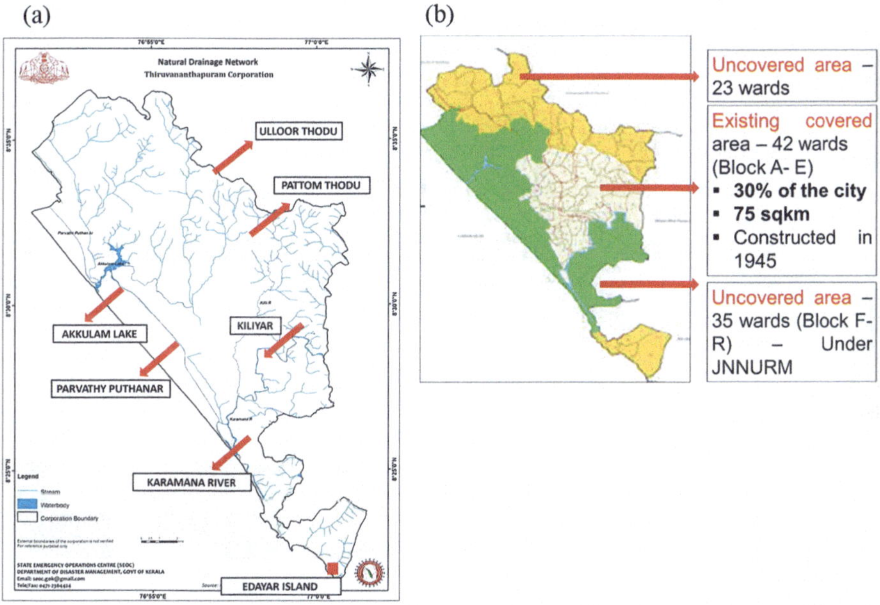

Fig. 13 **a** Drainage network, **b** Sewerage network coverage (Reproduced from Government of Kerala et al. 2015; Thiruvananthapuram Corporation et al. 2012)

drains. The effluents even after sewerage treatment are found to have poor physio-chemical properties as per a study carried out by Mahatma Gandhi University in 2018 (CS 2016). The treated water flows into Karamana River and finally empties into the Arabian Sea. Thus, it can be concluded that there is major water-related infrastructure gap that the city is facing.

Table 4 Status of water supply

Sl. No	Indicators	Existing SLB	MOUD benchmarks
1	Coverage of water supply connections	78%	100%
2	Per capita supply of water	100lpcd	135lpcd
3	Extent of NRW	35%	20%
4	Quality of water supplied	92%	100%
5	Cost recovery in water supply services	100%	100%
6	Efficiency in collection of water supply-related charges	60%	90%

Kerala Water Authority and Thiruvananthapuram Municipal Corporation (2014)

Table 5 Status of sewerage supply

Sl. No	Indicators	Existing SLB (%)	MOUD benchmarks (%)
1	Coverage of Latrines (Ind & Community)	99	100
2	Coverage of sewerage network services	37	100
3	Efficiency of collection of sewerage	41	100
4	Adequacy (TP)	95	100

Kerala water Authority and Thiruvananthapuram Municipal Corporation (2014)

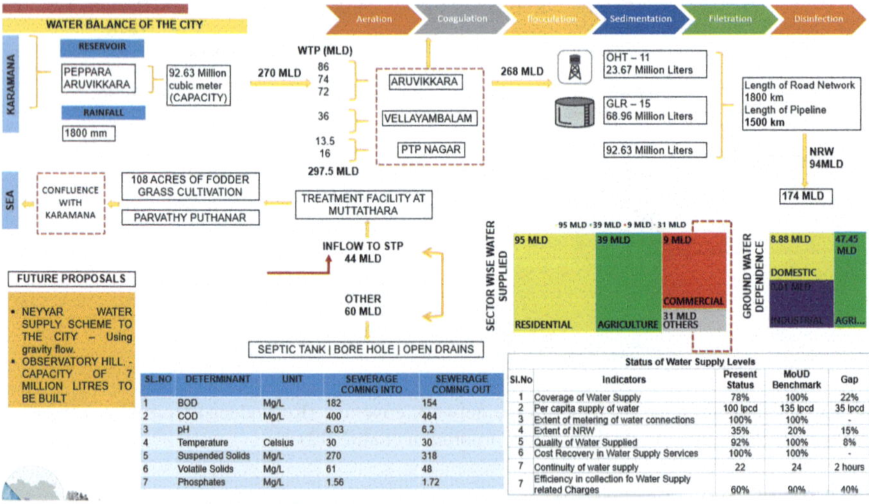

Fig. 14 Water balance for TVPM

4.11 Vulnerability Profile

Going by the definition for resilience, it is very important for the city to tackle the vulnerabilities to effectuate resilience. Five broad themes having an implication on water resilience, i.e. *socio-economic vulnerability* (% of illiterates, % of non-workers), *physical vulnerability* (% of -dilapidated housing, vulnerable roofing material, vulnerable wall materials, with no exclusive dwelling room/one dwelling room), *infrastructure vulnerability* (% household with—unimproved source of drinking water, unimproved sanitation, unclean cooking fuel, lack of accessibility to services), *bio-physical vulnerability (%* under wasteland,% area falling within 500 m buffer of river, % falling with CRZ) and *multi-hazard vulnerability* were considered (Thiruvananthapuram Municipal Corporation 2015). Post which the scores were normalized using Z-score (where Z is standard score, x is observed value, μ is mean of the sample σ is standard deviation of the sample). Based on the z-score computed a composite vulnerability map was derived.

$$Z = (x - \mu)/\sigma$$

5 Results and Discussions

The study reveals several concerning aspects related to water resources and urban development in the area. LULC reveal the stress on water resources due to factors like increasing built-up density, laterite capping, and growing groundwater depletion. A decline in paddy cultivation is observed over the years, while built-up areas have increased by nearly 5%, accompanied by a decrease in areas of agriculture and water bodies. The Karamana and Vamanapuram watersheds, vital for the city downstream, experience significant land-use changes, posing vulnerability to settlements. Runoff has increased by approximately 3.81%, with Karamana watershed showing the highest drainage density and bifurcation ratio. These changes, along with the priority 1 status of Karamana River by CPCB, highlight the vulnerability of the downstream areas. The city reports vector-borne diseases, and the water infrastructure faces challenges with 35% non-revenue water and poor-quality treated sewage draining into rivers, collectively posing a significant resilience challenge.

Vulnerability assessment reveals that, regarding socio-economic and physical resilience, for the parameters considered at ward level, coastal, and old city areas is particularly vulnerable. Fort area, Chala, Bheemapally East, Puthanpally, and Poonthura are few of the wards in this area. These wards together with Kulathoor, Karikkakom, & Akkulam areas are particularly vulnerable to Infrastructure resilience. Along the coastal areas, predominant land use is residential, whereas in the city centre it is dominated by commercial land use. As depicted in Fig. 15a

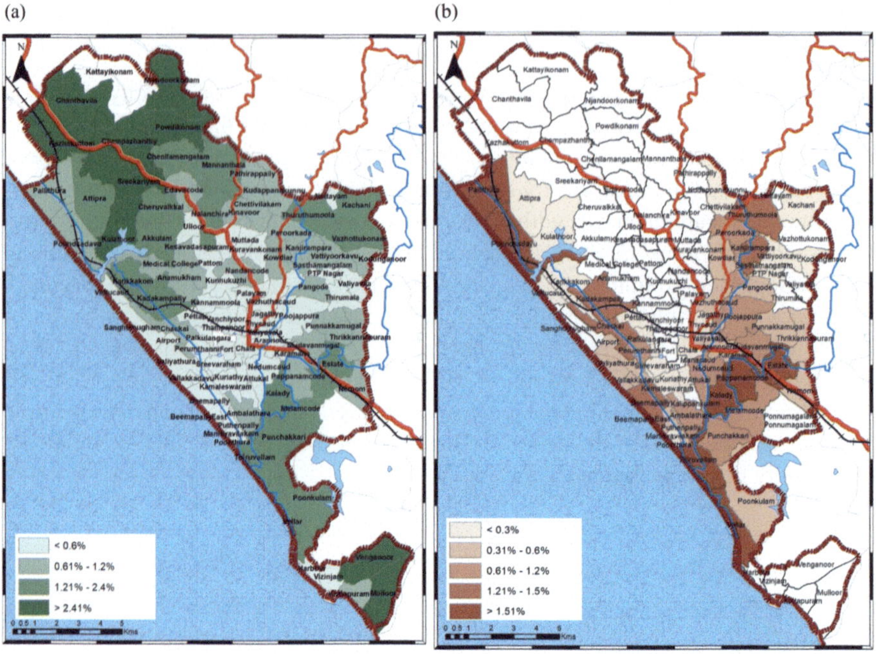

Fig. 15 **a** % of green sinks (Referred from Thiruvananthapuram Municipal Corporation 2015). **b** Area falling within 500 m buffer of the river body (Reproduced from Thiruvananthapuram Corporation et al. 2012)

the per cent of green sinks, wards in the 500 m buffer zone of the river and the wards along the CRZ zones were also mapped.

The normalized Z-score as mentioned above helped arrive at the composite vulnerability map. As observed in Fig. 15b, the very highly vulnerable wards are along the coastal areas, river, canals, and the core city area. The key factors challenging the resilience as per the analysis in these wards are their source of drinking water, lack of drainage system, solid-waste management, % share of non-workers, and unclean cooking fuel. The outcome can further be supported through the finding as per (Government of Kerala et al. 2015) where the core city area of Thiruvananthapuram, encompassing Thampanoor, Pazhavangadi, Thakaraparambu, Vanchiyur, Pattoor, and other major intermodal transit nodes is outlined to have significant flood risks. Wards like Thampanoor and Fort also have an exceptionally high population density, exceeding 15,000 persons/km^2. Close to 50 km^2 of the area is identified as flood-prone, attributed to both natural factors, such as slopes below 3%, and human activities (Fig. 16).

The composite vulnerability map was compared with the zone-wise development as outlined in the master plan to understand the loopholes and drawbacks in the projected development. Table 6 shows the zone-wise recommendation and the comparison. Zones 1 and 4 where restricted development is suggested are in line with

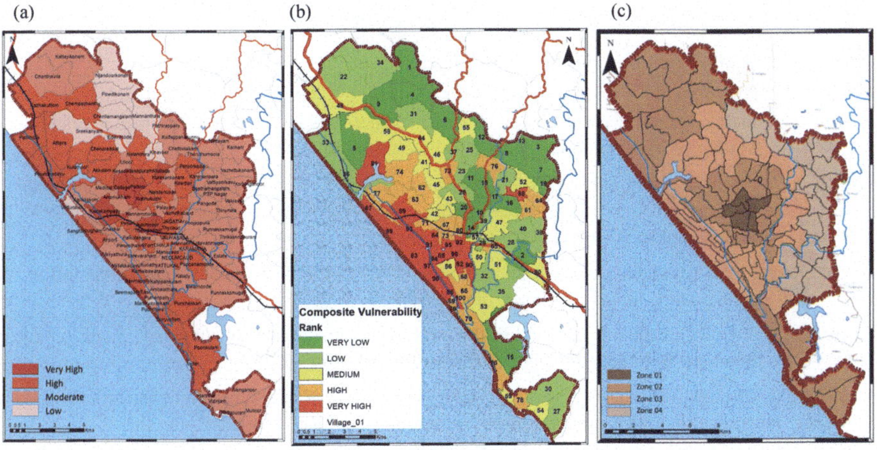

Fig. 16 **a** Multi Hazard vulnerability map (Referred from Thiruvananthapuram Municipal Corporation 2015). **b** Composite vulnerability map. **c** Zonal development plan (Reproduced from Thiruvananthapuram Corporation et al. 2012)

the assessment. However, in Zone 2 and Zone 3, the density outlined has already been exceeded and further intensification of development in these zones should only be after sufficient interventions to support the infrastructure in this area. The results depicted help in concluding that the city suffers from deficiencies in sewerage and solid-waste management, leading to potential health hazards. Additionally, the lack of green spaces and planned open areas impacts the urban environment. The city's economic value is at risk due to flooding and environmental degradation (Fig. 17).

6 Recommendations

6.1 Land Suitability

It is important to investigate the environmentally sensitive parameters and base the development accordingly for long-term resilience, hence land suitability has been done at the regional level. The parameters considered have been divided into 4 categories as, no development zone, least suitable, moderately suitable, and highly suitable as indicated in Table 7. Accordingly, the outcome of land suitability is shown in Fig. 18, 31% area falls under no development zone, 8% under least suitable, 45% under moderately suitable, and 16% under highly suitable land area.

Table 6 Comparison of zonal development with composite vulnerability

Zone	Characteristics in master plan 2031
Zone 1	The city core with the highest built-up coverage, suggested to have restricted development. Recommended to avoid further densification, the existing average density of 7500 persons/sqkm is to be continued
Zone 2	The area surrounding the city core (central city). Area having comparatively lesser built-up coverage. Suggested for regulated development. Density of 4000 to 7500 persons/sqkm is suggested
The intervention of regulated development is apt as some of the highly vulnerable areas fall in this zone, this can ensure future resilience in these areas	
Zone 3	The city extension adjacent to the central city. Proposed for intensive development, proposed to have a density of 2000–4000 persons/sqkm
As per the evaluation, especially regarding water-related disasters and different aspects to resilience, many wards in this area are highly vulnerable and hence intensifying the development will make these areas highly susceptible to hazards	
Zone 4	The peripheral area of the city (Most area under dry agriculture use), suggested to be retained as such. Conservation of these area is recommended. Proposed to have a density of 2500 persons/sqkm

Referred from Thiruvananthapuram Corporation et al. (2012)

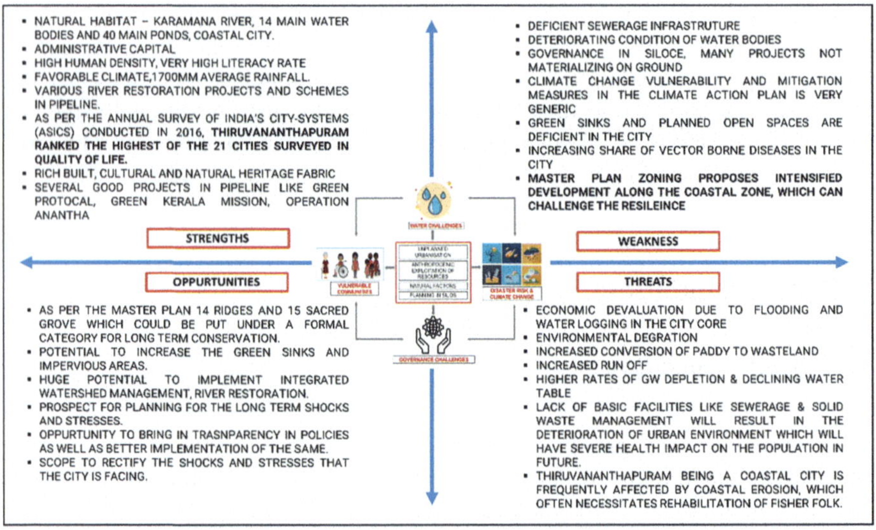

Fig. 17 Strength, opportunities, weakness, and threats

6.2 Proposals

At the regional level, there is a need to investigate environmentally sensitive parameters which are currently overlooked in the proposed spatial hierarchy in district

Table 7 Parameters for land suitability

Parameters	**No development**	**Least suitable**	**Moderately suitable**	**Highly suitable**
Geomorphology	Dams and reservoirs	Valley fill, Coastal plain, Denudational hill, Flood plain	Structural hill, Quarry and Mine	Lower Plateau
Slope	–	Greater than 15%	Between 5–15%	Less than 5%
Water bodies (WB) (Considered WB>1 ha)	Water bodies	Up to 50 m from water bodies	Beyond 50 m from waterbodies	
Major rivers (Karamana, Neyyar, Vamanapuram)	Rivers	Up to 500 m from rivers	Beyond 500 m from Rivers	
Rivers (tributaries)	Rivers (Tributaries)	Up to 50 m from rivers	Beyond 50 m from rivers	
CRZ	Up to 500 m from sealine	Beyond 500 m		
Forest (RF, PF),	Forest	Non-forested areas		
Land use land cover		Built up | Paddy land with double cropping	Agriculture land	Barren
Roads (NH & SH)	NH&SH	Up to 60 m from Roads	Beyond 60 m from roads	
Railway	Railway	Up to 30 m from Railway	Beyond 30 m from Railway	

Department of Town and Country Planning and Kerala Government (2017), Environmental Planning Studio (2019), and Thiruvananthapuram Corporation et al. (2012)

development plan. As per the scalogram analysis done for villages, few of the 3rd order villages like Vellarada, Kilimanoor, Parasala, and Vithura lack basic amenities which need to be addressed. To improve the adaptive capacity of identified villages, there could be focus on establishing *climate-smart villages* focussed on promoting resilient agriculture practices, forest protection, integrated resource, and waste management (*climate-smart villages*, n.d.).

The recommendation at the city level, is the scope of implementing *Integrated Urban Water Management* approach, stressing integration of ecological and social dimensions to water management. Further, other alternate sources like rainwater harvesting, the revival of local water sources, ground water recharging, reusing the wastewater, etc. should be investigated. By tapping even 50% of the runoff close to 20,000 lpcd can be generated (Chattopadhyay & Harilal 2017; Niti Aayog 2021).

At city level the scope of implementing a river regulation zone (RRZ) should be investigated. It can be composed of riparian buffer extending up to 20 m, followed by a no development zone up to 50 m and regulated development up to 500 m. This can significantly enhance the current scenario of the water bodies in the city limits (NIUA

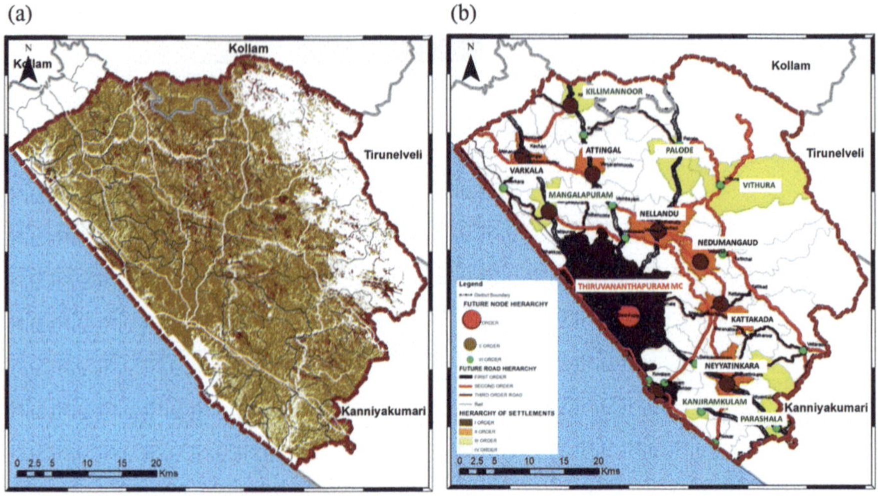

Fig. 18 **a** Land suitability map. **b** Proposed district spatial structure (Reproduced from Department of Town and Country Planning & Kerala Government 2017)

2020). There is a need to enhance the coastal resilience through long-term measures such as nature-based solutions, aimed at enhancing the degraded ecosystem and mangrove (Rohilla, Jainer et al. 2017a, b). Another aspect that can be implemented in Thiruvananthapuram is *Water Sensitive Urban Design* focusing on sustainable urban drainage, decentralized wastewater management, and potential water tapping using building footprint, along with streetscape design, avenue plantation, infiltration trenches. Replicating the successful Alappuzha waste management model recognized by UNEP can help tackle the issues related to solid-waste management (Mazingira 2017; Rohilla, Matto et al. 2017a, b) (Fig. 19).

7 Conclusions

The questions that led to this study, regarding the stresses and shocks that the city of Thiruvananthapuram is succumbing to, have been supported with detailed analysis at regional and city levels. The study boundary delineated using watershed-taluka-village boundary helped in understanding the LULC changes in the region which indicated a 5% increase in built-up area between 2005 and 2020, conversely a decline in the area under impervious categories including agriculture areas, paddy fields and water bodies. Other parameters like laterite capping along the midland region and depletion of groundwater add to the challenge of water resilience. The area under Karamana and Vamanapuram watersheds has seen a major upsurge in the built-up areas, these are also the watershed basins that encompass the TMC downstream. The District Development Plan also cites as to how there is an increased conversion in the

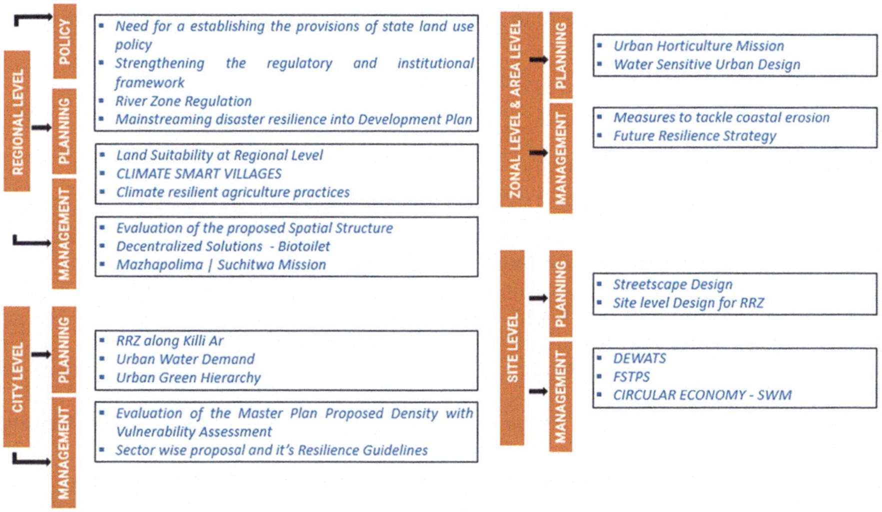

Fig. 19 Proposals at multiple spatial scales to enhance water resilience

land use from agriculture to wasteland and eventually to build up areas. The runoff from the region has increased by approximately 3.81% from 2005 to 2020. Karamana River being priority 1 polluted river by CPCB, coupled with its high drainage density, bifurcation ratio adds to the stress in the city downstream.

Further, a detailed evaluation of different facets of resilience was done at the city level. With its exceptionally high urban population and density, Thiruvananthapuram Municipal Corporation counters multiple challenges on the infrastructure front, regarding the dearth of green, ageing, and inadequate sewerage system, and water supply. The water infrastructure for the city has major loopholes in terms of NRW which is at 35%. The stormwater system has exceeded its designed capacity resulting in the discharge of a large quantity of raw sewage as overflow into the Karamana River. Also, the treated sewerage has very poor physio-chemical parameters, which are then drained into the river channels. Out of the six diseases included under National Vector Borne Disease Control Programme viz, Malaria, Dengue fever, Filariasis, Chikungunya, Kala-azar, and Japanese encephalitis, the first 5 are being reported in the city. The composite vulnerability profile done for the study, assessing the social, economic, physical, and infrastructural challenges that the city is facing shows increased risk for wards along the coastal areas, and the core city area.

The gaps identified should be plugged effectively by ingraining water resilience solutions in the development plan and adopting an effective use of grey and green infrastructure. Land suitability at the regional and city levels integrating the environmental parameters should be considered for the sustainable and long-term resilience of the proposed spatial structure in the District Development Plan. Area and site-level interventions suggested for the critical zones for improving water resilience like the green hierarchy, river regulation zone, water sensitive urban design should be addressed within the master plan and regional development plan for the effective implementation on the ground.

References

Abraham K (2019) Kerala govt begins mammoth task of cleaning up the Aamayizhanjanthodu canal in Thiruvananthapuram. The News Minute. https://www.thenewsminute.com/article/kerala-govt-begins-mammoth-task-cleaning-aamayizhanjanthodu-canal-t-puram-101676
Chattopadhyay S, Harilal KN (2017) Emerging challenges of urban water governance case of Thiruvananthapuram City, India
Climate-Smart Villages (nd) Retrieved July 27, 2023, from https://ccafs.cgiar.org/climate-smart-villages.
Department of Environment and Climate Change (2014) Kerala State Action Plan on Climate Change
Department of Town and Country Planning (2012) State urbanization report, Kerala
Department of Town and Country Planning, Kerala Government (2017) District development plan
Dinagara Pandi P, Thena T, Nirmal B, Aswathy MR, Saravanan K, Mohan K (2017) Morphometric analyses of Neyyar River Basin, southern Kerala, India. Geology, Ecology, and Landscapes 1(4):249–256. https://doi.org/10.1080/24749508.2017.1389494
Directorate of Census Operations, Kerala (2014) India—census of India 2011—Kerala—Series 33—Part XII B—District Census Handbook, Thiruvananthapuram. https://censusindia.gov.in/nada/index.php/catalog/665
Economics and Statistics Department (2020) Agricultural statistics
Environmental Planning Studio (2019) Studio report of Chandrapur
Germanwatch (2021) Global Climate Risk Index. https://www.germanwatch.org/en/19777
Gopal VV, Dhanya G, Joseph S (2018) Drainage character delineation-a remote sensing approach over Vamanapuram River Basin, Kerala. Int J Res Eng Appl & Manag (IJREAM) 3(11):2454-9150. https://doi.org/10.18231/2454-9150.2018.0034
Government of India (2015) India's updated first nationally determined contribution under paris agreement
Government of India, Ministry of Jal Shakti, Department of Water Resources, & RD, GR Central Ground Water Board (2017) National Compilation on dynamic ground water resources of India, 2017. https://cgwb.gov.in/GW-Assessment/GWRA-2017-National-Compilation.pdf
Government of Kerala, Department of Disaster Management, Kerala State Disaster Management Authority (KSDMA), Kerala State Emergency Operations Centre (SEOC) (2015) Hazard and vulnerability assessment report-Thiruvananthapuram City
Indian Institute of Science, Ministry of Environment (2021) Historical climate change trends and climate change projections in Kerala
Kerala Forest Research Institute, Vijayakumaran, PD (2010) Directory of wetlands of Kerala
Kerala State Land Use Board (2013) Natural resource bank
Kerala State Pollution Control Board, Department of Irrigation, Kerala Water Authority, Thiruvananthapuram Corportion, Kerala State Remote Sensing and Environment Centre, Kerala Socio Economic Foundation (2019) Karamana River action plan
Kerala water Authority, Thiruvananthapuram Municipal Corporation (2014) Service level improvement plan—sewerage system
Kerala Water Authority, Thiruvananthapuram Municipal Corporation (2014) Service level improvement plan for water supply
Khanna, PN (2012) Indian practical civil engineers handbook. Engineers Publishers
Mazingira UU (2017) Solid approach to waste: how 5 cities is beating pollution. UNEP. https://www.unep.org/sw/node/19457

Ministry of Urban Development (2014) Urban and regional development plans formulation and implementation (URDPFI) guidelines
Ministry of Urban Development, GOI (2020) India smart city mission, smart city challenges
MoEFCC (2021) National Action Plan on Climate Change (NAPCC)
National Remote Sensing Centre, ISRO (2021) Aerial services & digital mapping. https://www.nrsc.gov.in/Eop_Aerial_Objective
Nations Office for Disaster Risk Reduction (2015) Sendai framework for disaster risk reduction 2015–2030
Niti Aayog (2021) Compendium of best practices in water management 2.0
NIUA (2020) A strategic framework for managing urban river stretches in the Ganga River Basin
Okapi, 100 Resilient Cities, Chennai Municipal Corporation (2019) Chennai city resilience strategy
R RB, Jayan A, Kumar RPB (2018) Analysis of heavy metal pollution on Parvathy Puthanar, An Artificial River Canal in South Kerala. 5. http://ijrar.com/
Rohilla SK, Jainer S, Matto M (2017) Green infrastructure a practitioners guide
Rohilla SK, Matto M, Jainer S, Sharda C (2017) A practitioner's guide water sensitive urban design and planning
Soil and Land Use Survey of India (2021) Soil resource mapping. soil and land use survey of India (SLUSI). http://slusi.dacnet.nic.in/srm/srm_report_wise.html
Sreeja R (2011) Evolution of Karamana river basin and its implications on hydrogeology a geomatics approach using morphometric analysis. University. http://hdl.handle.net/10603/222570
Thiruvananthapuram Corporation, Department of Town, and Country Planning, Government of Kerala (2012) Thiruvananthapuram master plan
Thiruvananthapuram Municipal Corporation (2014) Service level improvement plan storm water
Thiruvananthapuram Municipal Corporation (2015) City disaster management plan
UN (2015) UN world conference on disaster risk reduction. https://www.un.org/sustainabledevelopment/un-world-conference-on-disaster-risk-reduction/
UN (2016) New urban agenda. https://habitat3.org/wp-content/uploads/NUA-English.pdf
UN Habitat (2018) City resilience profiling tool. https://unhabitat.org/sites/default/files/2021/01/crpt-guide.pdf
UN-Habitat (2017) Trends in urban resilience. https://unhabitat.org/trends-in-urban-resilience-2017
United Nations (2021) Sustainable development goal. https://sdgs.un.org/goals
United Nations Sustainable Development (2015) The sustainable development agenda. https://www.un.org/sustainabledevelopment/development-agenda/

Urban Flood Risk Zonation Using GeoSM-NatE Tool: A Case Study of Kochi City, India

Hudha Abdul Salam, Mahua Mukherjee, Brijesh Kumar Yadav, and Atul Kumar

Highlights:

- Application of the Geo-Spatial Mapping for Natural Ecosystem (GeoSM-NatE) Tool.
- Monitoring and assessment of urban flood events.
- Identification of natural and man-made causative parameters during urban floods.
- Development of urban flood Risk Map using Hazard, Vulnerability, and Human Exposure Index

1 Introduction

Urbanization has been a critical indicator of development, and as the cities develop, more rural areas are converted to urban. India's overall percentage of urbanization increased from 27.7% in 2001 to 31.1% in 2011, an increase of 3.3 percentage points from 2001 to 2011 compared to a gain of 2.1 percentage points from 1991 to 2001 (Bhagat 2011). A very high rate of urban population growth is observed in Kerala, among other states, which accounts for 6.5% per annum (Bhagat 2018). Quite a

H. Abdul Salam · M. Mukherjee (✉) · A. Kumar
Department of Architecture and Planning, Indian Institute of Technology Roorkee, Roorkee, India
e-mail: mahuafap@iitr.ac.in

H. Abdul Salam
e-mail: hudhaa_salam@ar.iitr.ac.in

H. Abdul Salam
Department of Architecture, TKM College of Engineering, Kollam, India

B. K. Yadav
Department of Hydrology, Indian Institute of Technology Roorkee, Roorkee, India

M. Firoz C et al. (eds.), *Urban Planning and Design for Megacities in the Global South*, Advances in 21st Century Human Settlements,
https://doi.org/10.1007/978-981-97-8370-0_3

few new towns also evolved through the rural-to-urban classification in the 2011 census (Nair and CJ 2019; State Planning Board 2017). Climate change impact is an emerging issue globally. Change in rainfall pattern is affected adversely within the states of India (Kulkarni et al. 2020). In August, September, and December of 2020, the Ernakulam district had severe weather that resulted in substantial damage. These included the devastation of more than 250 homes, the displacement and relocation of residents to camps for relief, damage to vehicles and crops, traffic jams, and the deficiency of other essential services (India Meteorological Department 2020). In the subsequent year, Kerala was hit by an unusual amount of rainfall, according to the India Meteorological Department (IMD). The precipitation for 2021 was a record-breaking 3610.2 mm (India Meteorological Department 2021), exceeding the average yearly precipitation of 2924.7 mm by 23.4% (Kallungal 2022). We still lack scientific consensus regarding how climate change affects urban drainage systems. While development is inevitable, unplanned, and unorganized urban development is often associated with unfavourable situations. Diverse problems that arise due to urbanization include congestion in the existing infrastructure, changes in land use, depletion of the natural ecosystem, urban heat island, air pollution, and urban flooding (Kjeldsen 2009). The land use and land cover change due to urbanization brings significant issues in the urban areas (Pal et al. 2022). Surface run-off due to rainfall increases thirty percentage (Subramanian et al. 2023) or greater when natural terrain is turned into paved and tarred roadways due to urbanization, and the same leads to urban flooding (Barron et al. 2011). The permeability of the ground decreases due to the change in land cover, which prevents the water from infiltration and leads to depletion in groundwater storage (Devi and Nair 2021). The intensity of the hazards depends on the physical, social, economic, and environmental aspects of any area. The changing climatic pattern also adds to the severity of the hazards (Pal et al. 2022). The changing rainfall pattern and frequency add to the urban flood effect on the urban area. Urban areas are incapable of accommodating the increased volume of rainfall with a less permeable urban landscape. In the context of a disaster event, risk refers to the probability of experiencing property damage, loss of life, or adverse health effects due to a hazardous occurrence (UNDRR 2004). This risk is determined through a comprehensive assessment of the characteristics of the hazard, the vulnerability of the affected area or community, and the degree of exposure to hazard-prone zones. Integrating these three crucial elements—Hazard, Vulnerability, and Exposure—enables a more profound comprehension and evaluation of the potential impact of a disaster (UNDRR 2009). Urban flood is a hazard characterized by the inundation of urban areas, posing risks to urban environment, infrastructure, lives, and livelihoods. National Disaster Management Authority (NDMA), India, has recognized urban flooding as a distinct hazard due to its increasing frequency and escalating impact, demanding focused attention in technical discourse (National Disaster Management Authority 2023).

Developing countries are the most affected by the urban flooding. Modelling approaches are widely used in the evaluation of flood hazards and in framing mitigation strategies. The various flood modelling techniques used worldwide can be categorized into three: (i) Spatial Model (Zlatanova et al. 2014), (ii) Dimensional

Model or hydrodynamic model (Morante-Carballo et al. 2022), and (iii) Mathematical Model (Morante-Carballo et al. 2022; Nkwunonwo et al. 2020). A different approach of source tracking, the tracer-aided urban flood model is employed (Qi et al. 2022) to prioritize catchment flood management interventions spatially to facilitate stakeholder collaboration and resource-efficient strategies. For Indian cities like Bhubaneswar, Delhi, Chennai, and Bangalore, risk assessment studies utilize spatial analysis tools, with some also incorporating scenario-based analyses (Halder and Saha 2021; Pal et al. 2022). Alisa Sahu's research on urban flood proposes an urban flood-resilient land use plan for Bhubaneswar, integrating GIS and remote sensing to assess risks from hazards, social infrastructure vulnerabilities, and physical infrastructure vulnerability (Sahu et al. 2021). An integrated approach that combines remote sensing, geographic information system (GIS), and field survey-based methods to identify and predict urban flood-prone areas is approached for the Delhi, India (Tomar et al. 2021). Spatialized Urban Flood Resilience Index (S-FRESI), by Louise Bertilsson et al., proposes a scenario-based tool for flood resilience (Bertilsson et al. 2019). The indicators of urban flood comprise of the flood event itself, including its magnitude, extent, and duration. Other indicators includes the expansion of built-up areas and physical infrastructure, population density, presence of vegetation, and water bodies (Diez et al. 2011; Singh et al. 2023). The indirect causative factors of Global climatic change and urbanization, and the direct causes like the unawareness of the community and improper management (Kasdan 2022) contribute to escalate the hazard impact (Gupta and Nair 2011). A comprehensive examination of these indicators leads to understanding of urban flood risk assessment and management strategies.

In context of the study area, Kochi Municipal Corporation with a potential risk of urban flooding, very few studies have been done. Sowmya's urban flood vulnerability study focuses on identifying issues related to drainage infrastructure and social vulnerability factors among different sections of the population in the city, using geo-spatial analysis (Sowmya et al. 2015). Urban community resilience framework focuses on the social resilience, using the survey method and statistical analysis (Ali and George 2022). In this era of increased urbanization and climate change, academic research has not yet undertaken a comprehensive risk assessment for Kochi Municipal Corporation that considers the combined aspects of hazard, vulnerability, and exposure. This chapter aims to fill this gap by conducting a risk assessment in Kochi, incorporating these crucial elements to enhance disaster preparedness and implement effective mitigation strategies.

The chapter aims to identify urban flood hotspot wards in Kochi Municipal Corporation. The objective is to utilize geo-spatial data to identify wards at risk to urban flooding by delineating multiple parameters layers that contribute to flood risk within the study area. A better understanding of the Risk for the study area will help in improving the preparedness capacity of the city and its people. The study explores the importance of GIS and other spatial analysis tools in assessing the Hazard and the Risk itself from a diverse perspective to the root cause of urban flooding issues. The study intends to be an aid to enhance urban resilience, minimize flood-related damages, and foster sustainable urban development in Kochi Municipal Corporation, ultimately leading to a safer and prepared urban environment.

2 Study Area

Kerala state lies in the southwestern part of peninsular India. The state is a narrow strip of land comprising low land, midland, and high land with a maximum width of 125 km (75 Kms average width), sandwiched between the western ghats on the east and the Arabian Sea on the West. The state of Kerala receives roughly 3000 mm of precipitation on a yearly average. Six monsoon months, from June to November, account for temperatures between 20 °C and 35 °C.

Kochi municipal corporation (Fig. 1) characterizes an active central commercial sector that spreads in a linear direction (Department of Town & Country Planning 2014). The Kochi port is one of the few significant hubs of activity that contributed to Kochi's development. With all its historical structures and legacy, the Mattancherry—Fort Kochi area has been a significant tourist destination in the city and in Kerala (George et al. 2005). Towards the northeast part of Kochi is the industrial area of the district. The National Waterway III, from Kochi to Kollam, also had an influence on the development of tourism in the state (Department of Town & Country Planning 2014). The more recent developments, especially the residential and commercial, show a high-rise pattern by achieving a maximum floor area ratio. Though the previously developed low-rise structures still function within Kochi municipal corporation, the increasing trend of population and less land availability is forcing to come up with high-rise development in the city (George et al. 2005).

Kerala's urban population, which increased to 47.7% in 2011 (State Planning Board 2017), faced many difficulties, including congestion, ineffective infrastructure, inadequate street designs, substandard public service delivery, and coordination challenges between local bodies in different sectors (T.R. Raghunandan et al. 2017). Development of informal settlements, housing, poverty, access to transit, water, and jobs, solid waste management and hygiene are other associated issues faced by the state in the phase of urban development (Raghunandan et al. 2017). In Kerala, cities are at the peak of their development. This development and urbanization is an unplanned phenomenon in Kochi, similar to most the Indian cities (George et al. 2005). Kochi is the commercial hub of Kerala, and the migration of people for better job opportunities and lifestyle is choking the city's infrastructure. The census report of 2011 shows an increase of 20.5% in the urban population of Ernakulam district (Department of Town & Country Planning 2012). The increase in urban land area is 11.7%, which clearly shows the increase in the population density in urban areas. The impervious area in Kochi municipal corporation has increased by 36.8% in 20 years (Devi and Nair 2021), and the increase in the built-up area shows an increase of 26.98% (Krishnan and Firoz 2021) in the district.

Kerala has been vulnerable to multiple hazards, experiencing flooding in the recent years of 2018 (National Institute of Disaster Management 2018), 2019 (State

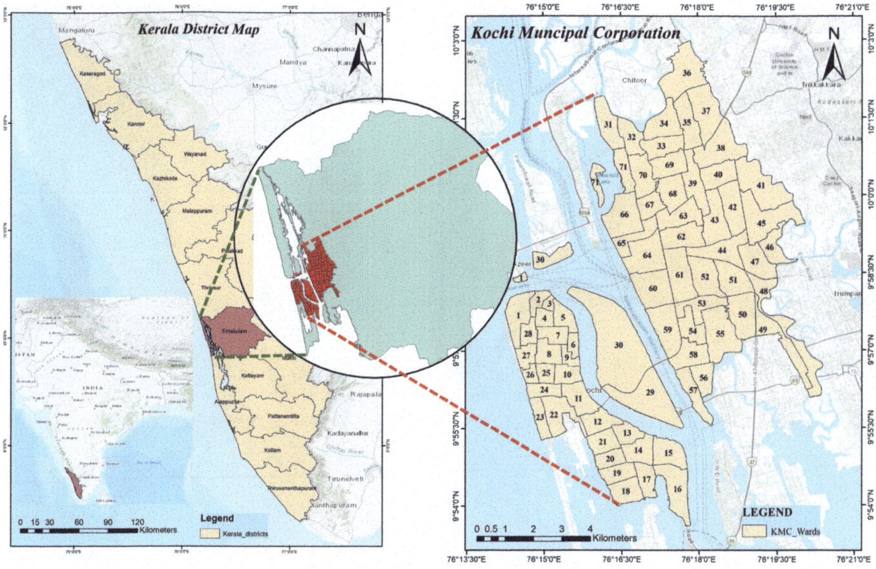

Fig. 1 Location of study area selected here for the urban flood risk assessment

Relief Commissioner 2019), 2020 (India Meteorological Department 2020), and 2021 (India Meteorological Department 2021). In 1924, Kerala state witnessed a great flood due to the unprecedented downpour and windspeed (Sivaraman 2018). More than thousand people lost their life and property in this disaster (Roychowdhury 2018). The next episode of a devastating flood happened in 2018 in Kerala. Rapid urbanization has led to this increased frequency of flooding, particularly in urban areas, within the state in the past few years. The sudden impact and inundation are catching people unprepared and making it difficult for them to cope with the situation (Staff Reporter 2022).

The frequent events of flooding and associated issues are affecting city activities, causing them to stand still for several hours to one or two days. Without prior preparation, people had to abruptly move to relief camps. Recovering from the impacts and aftermath of a flood and to restore normalcy in life is another significant challenge. Returning home after the flood entails facing numerous hardships before resuming daily routines, as illustrated in Fig. 2. For each season, new locations are affected, making preparing for the Hazard challenging (Roychowdhury 2018). The unaffected areas in the previous floods are also experiencing inundation, and the sudden rise of water levels within the buildings escalates the impacts (Staff Reporter 2022). Roads are blocked, which restrains the inhabitants from commuting, for their work and livelihood. Flooding also contaminates sump tanks (which store corporation water at the household level) and open wells, which Kochi municipal corporation inhabitants rely on for domestic water needs. The situation of the informal settlements, those developed mainly near the canal banks, gets affected easily due to the overflow of the canal systems itself. The inadequate waste management system leads to increased

Fig. 2 Post flood (2018) photograph of a household in Ernakulam district—filled with mud and debris

waste dumping in roadside drains and canals, negatively affecting the city's drainage infrastructure.

3 Materials and Methods

Shuttle Radar Topography Mission (SRTM) Digital Elevation Model (DEM), with 30-m resolution elevation data downloaded from the United States Geological Survey (USGS), is used to analyse the topographical features of the study area. Landsat-8 data is acquired for March 11, 2021, the pre-monsoon period, considering a cloud cover of 0.15% is used to delineate LULC and various land cover remote sensing indices maps. Rainfall data is collected from the India Water Resources Information System (India-WRIS) Portal developed by the Central Water Commission (CWC) and the Indian Space Research Organization (ISRO). The daily rainfall data for 2021 is collected for 18 rain gauges available in India-WRIS in and around the Ernakulam district. Information about age, gender, socio-economic class, and population density is obtained from the 2011 census statistics. Critical infrastructure such as hospitals and educational institutions are geotagged on Google Earth and imported into a GIS platform. The road and Railway network data layer is downloaded from the

Open Street Map. Soil hydrological data obtained from World soil data of Global Hydrological Soil Groups (Ross et al. 2018) is used to delineate run-off maps based on SCN curve numbers. The details of the data acquired, and the sources of the data are clearly mentioned in Table 1.

The GeoSM-NatE tool, which stands for Geo-Spatial Mapping for Natural Ecosystem, is employed for the current study. GeoSM-NatE is an urban risk resilience toolbox for the strategic integration of Nature-based Solutions (Blue-Green Infrastructure) for Urban Risk Reduction and Sustainable Development. The tool is developed in the 'Climatology Lab', IIT Roorkee, India. The GeoSM-NatE tool can detect long-term changes in ecosystem coverage and ecological degradation in urban settings through multi-temporal analysis with remote sensing techniques. The tool has been designed with the specific objective of assessing urban issues in different cities and different climates. The parameters for the assessment are identified in consideration of the distinct environments and climates of each city. The tool is particularly useful for studies that focus on urban areas because of its unique capacity

Table 1 Data used and their application in the study, sources, and remarks on the data used

Data	Source	Applications	Maps Prepared for—Hazard (H), Vulnerability (V), and Exposure (E) Index
DEM (Landsat 8) Shuttle Radar Topography Mission (SRTM), Resolution - 30 mts	United States Geological Survey (USGS)	Delineation of Urban Hydrology features—Watershed, Catchment, Slope, Stream Order, Stream Network, and Density	Slope Map (E) Stream Density (V)
Landsat 8 (Landsat Series Data) Date: March 11, 2021 Cloud cover - 0.15% Resolution - 30 mts	United States Geological Survey (USGS)	Delineation of Remote Sensing Indices and LULC	NDVI, NDWI (V) NDBI (E) Runoff Map (H) (SCS – CN Method)
Socio-Economic Data - Ward wise Population	Census 2011, Govt. of India	Population Density Data	Population Density (E)
Critical Infrastructure	Open Street Map	Road Map and Rail map	Street Density (V)
Rainfall Data Year 2021 18 Monitoring Stations	India Water Resources Information System (India-WRIS) Portal	Rainfall distribution map	Run-off Map (H) (SCS – CN Method)
Soil Type	Global Hydrological Soil Groups	Soil Group Antecedent Moisture Content (AMC)	Run-off Map (H) (SCS – CN Method)

to evaluate multi-hazards and its flexibility in adding or removing factors based on site-specific variables. The tool was employed to carry out an urban analysis of Gurugram, Haryana, India, and the results were validated using field data (Kumar et al. 2023; Mukherjee et al. 2022). A similar methodology was adopted in the investigation of Patna City's urban risk hotspots in Bihar, India (Rashiq et al. 2022). Though the GeoSM-NatE tool is for multi-hazard risk assessment, the present study has focused only on urban flood risk assessment. A case study is conducted with the Kochi Municipal Corporation in Southern India as the study area. The urban flood hazard mapping has been done by considering natural as well as man-made causative parameters, urban ecosystem (built-up and vegetation), and urban hydrology. Further, for vulnerability and exposure mapping, human exposure parameters, building and critical infrastructure, natural watershed and drainage were delineated with geo-spatial data. The run-off depth map layer is generated for the understanding of hazard events. Maps designed for vulnerability assessment include the Vegetation Index—NDVI Map, Water Index—NDWI Map, Street Density Map, and Stream Density Map. The exposure index maps consist of the Population Density Map, the Building Index—NDBI Map, and the Slope Map. Finally, a conditional-based assessment is done by assigning equal weightage to each layer to map the urban flood risk zone of the study area. The probability of urban flood events shows a dependency on the urban features (Table 2). The conditional approaches of the tool are as follows:

- The probability of flooding increases with higher surface run-off.
- A decreased presence of vegetation and surface water reduces permeable areas, leading to a higher probability of flooding.
- Increased built-up area leads to greater run-off, resulting in a higher probability of flooding.
- An increase in the density of natural streams and their higher stream order corresponds to a higher probability of flooding.
- Low-lying downstream regions are more prone to flooding.
- As the population density increases, more people get exposed to the flood.

Table 2 Impact of urban features on probability of urban flood event

Urban features	Impact on probability of urban flood event (inundation zone)
NDBI ▲	▲ High
NDVI ▲	▼ Low
NDWI ▲	▼ Low
Stream Density + NDBI ▲	▲ High
Stream Density + NDVI + NDWI ▲	▼ Low
Slope ▲	▼ Low
Population Density ▲	▲ High
Street Density ▲	▲ High

- Areas with higher street density experience increased traffic congestion, leading to greater vulnerability in those locations.
- Low-land areas with moderate to low slopes are prone to more inundation during flood events.

The detailed steps followed for the study are depicted in Fig. 3. The hydrological features delineated with the hydrology tool include the watershed delineation, stream density, and stream order. The Land Use Land Cover (LULC) map, Fig. 4, is prepared using the Iso Cluster Unsupervised classification technique using the Landsat-8 Data. The classification is done for six classes, and two similar classes are reclassified to a final four classes. Built-up, including all the buildings, road infrastructure, and other hard land surfaces, accounts for the maximum land cover of 77.78%. Other classes and the corresponding percentage of land cover for each are mentioned in Table 3.

Several types of land change processes are assessed and monitored using spectral indices in remote sensing (RS), a tool that helps in modelling, predicting, or inferring land surface processes (Montero et al. 2023). The RS indices are determined as the difference between two bands of significance, which is then normalized by

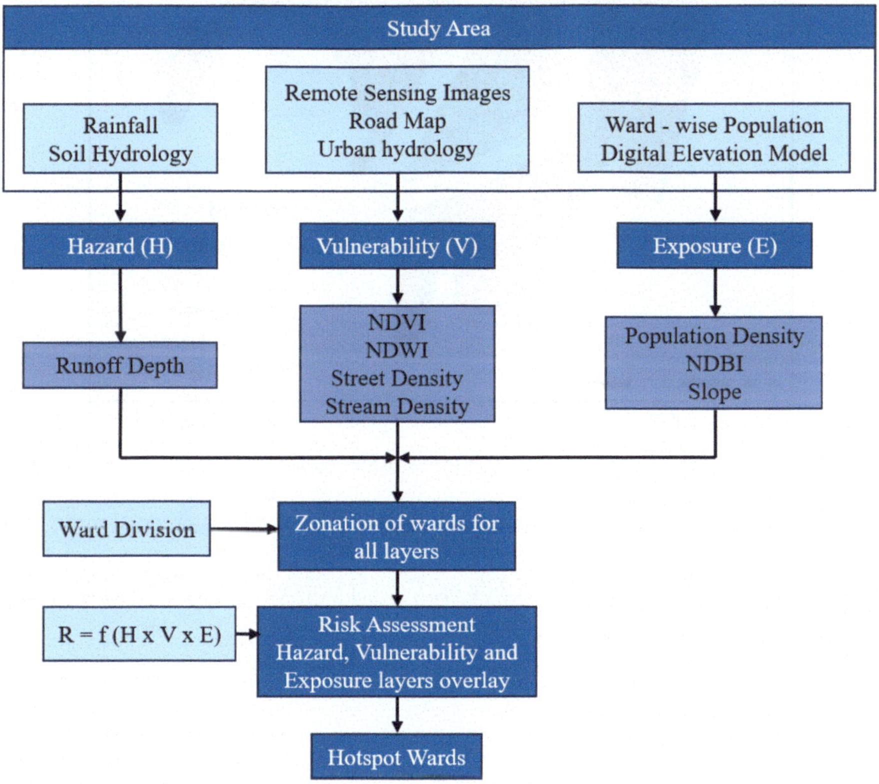

Fig. 3 Methodological framework followed for the risk assessment and hot spot ward identification

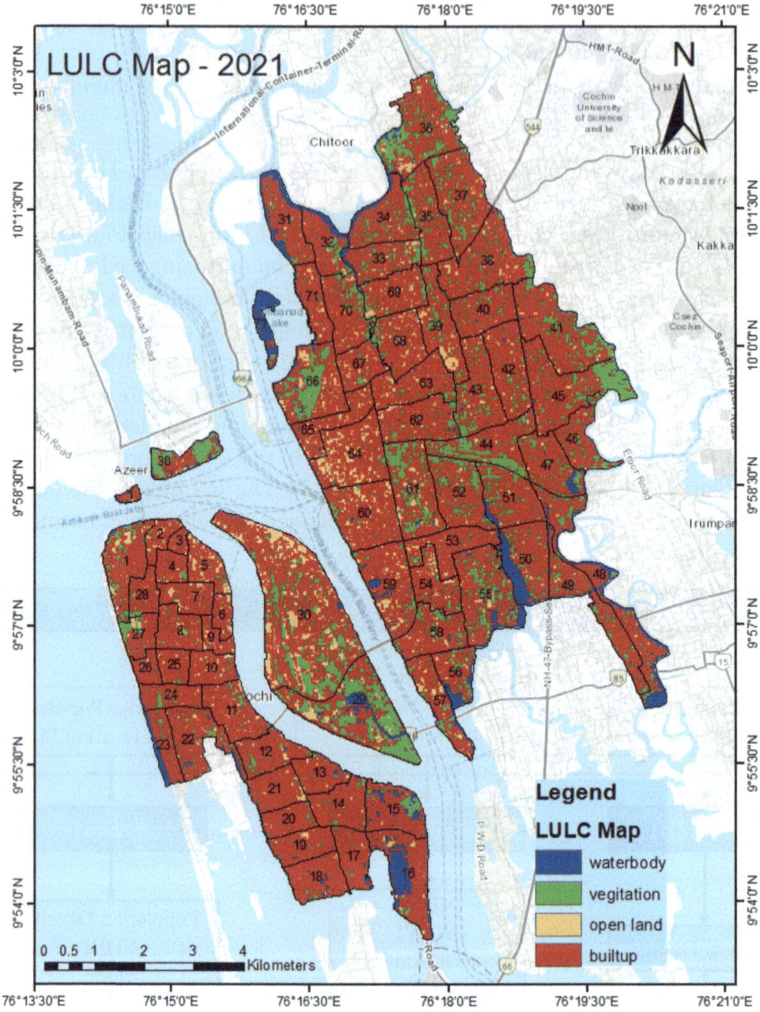

Fig. 4 LULC Map prepared for the year 2021 in ArcGIS using Landsat 8 Data

Table 3 Land use land cover—percentage wise division of each class identified—Year 2021 LULC map

Class	Vegetation	Built-up	Open Land	Waterbody
Area (in %)	12.43	77.78	5.13	4.66

Table 4 Hydrological soil group for the study area corresponding to the pixel values

Pixel values	Description
3	HSG-C: moderately high run-off potential
4	HSG-D: high run-off potential
13	HSG-C/D: high run-off potential unless drained
14	HSG-D/D: high run-off potential unless drained

their total. Normalized Difference Vegetation Index (NDVI), Normalized Difference Water Index (NDWI), and Normalized Difference Built-up Index (NDBI) identified by the normalization of different bands in the multispectral radiation varies for each type. NDVI = (NIR – Red) / (NIR + Red). The range of the NDVI is from -1 to 1. Higher NDVI values indicate high Near Infrared (NIR) reflection, which indicates thick greenery. NDBI = (SWIR – NIR) / (SWIR + NIR). The NDWI also ranges from -1 to + 1. Higher NDBI values indicate built-up regions, while lower values indicate water bodies. NDWI = (Green – SWIR) / (Green + SWIR). The range of the NDWI is from -1 to 1, like the other two indices discussed before. NDWI values for water bodies are often higher than 0.5. Vegetation has substantially lower values, making it simple to identify it from bodies of water.

The Inverse Distance Weightage (IDW) interpolation tool in ArcMap 10.8 is used to extend the data of daily rainfall acquired for the eighteen rain stations to the entire study area. The maximum rainy day of the year is selected to prepare the run-off depth of that rainfall event. The Antecedent Moisture Condition (AMC) of the soil is prepared in Excel. The soil hydrological group falls into four categories in the study area, as in Table 4. The look-up table for the Curve Number (CN) corresponding to the hydrological soil group is adapted from the book "Soil Conservation Service Curve Number (SCS-CN) Methodology". The CN data extracted for the current study is mentioned in Table 5. The SCS-CN formula gives the run-off depth for the selected rainfall events for the entire study area. For the computation of flood depth, it is necessary to have access to both precipitation data and Curve Number (CN) values determined based on the land type and Antecedent Moisture Condition (AMC) class.

The road network map is obtained from the Open Street Map, and the line density tool is employed to analyse the street density, considering the type and quantity of

Table 5 Curve Number (CN) for various soil groups in the study area with respect to three different Antecedent Moisture Conditions (AMC)

Class	AMC I		AMC II		AMC III	
	HSG C, C/D	HSG D, D/D	HSG C, C/D	HSG D, D/D	HSG C, C/D	HSG D, D/D
Vegetation	55	63	74	80	88	91
Built up	85	87	94	95	98	98
Open land	55	63	74	80	88	91
Waterbody	100	100	100	100	100	100

streets and roads within the designated study area. The urban hydrology analysis, based on the Digital Elevation Model, delineated the stream density, similar to the natural drainage patterns. The line density tool is utilized to assess and quantify stream density for the study area.

The hazard parameter under consideration for this study is the run-off depth of rainfall, which is assessed using the SCS-CN Method. A higher magnitude of run-off depth indicates an increased risk associated with disaster. For the vulnerability assessment of the study area, several factors are considered. The presence of vegetation and water bodies tends to reduce the vulnerability to floods, whereas increased street and drainage density are associated with higher vulnerability. Remote sensing images are utilized to derive indices for vegetation and water, while road and drainage density are considered together to explain the vulnerability of the wards in the study area. The study focuses on assessing the exposure level of different wards concerning three critical factors: maximum built-up area, extent of hardscapes, and population density. These factors collectively contribute to increased exposure of both people and property during extreme events.

3.1 Results and Discussion

Delineation of the impact of each factor at the ward level allows us to identify the hotspot wards. The zonation tool identifies the initial fifteen hotspot wards corresponding to each parameter. The fifteen wards with the highest run-off generated, Fig. 5a, during the selected rainfall event, and thus posing the highest risk, are identified as follows: Ward Number 15, 20, 21, 18, 23, 17, 16, 19, 12, 29, 8, 22, 14, 13, and 26. The wards with the lowest values of the Vegetation Index (NDVI) and Water Index (NDWI) indicate higher vulnerability to flooding. The lowest values NDVI and NDWI are highlighted in Fig. 5b and Fig. 6a, respectively. An increased number of streets means a high chance of getting inundated, and this affects people's daily commutes. The wards with higher street density (Fig. 6b) are identified for the vulnerability assessment, as it increases the probability of inundation impacting day-to-day commutation. The Stream Density Map, Fig. 7a, identifies the maximum density of the natural drainage system in the study area. During flood events, water tends to move in its natural path; thus, the wards with more stream density have more vulnerability.

More people will be exposed to a hazard event when the population density is high in a specific location. The wards with the highest population density, as illustrated in Fig. 7b, are significantly more exposed during hazardous events such as floods. The Built-up Index (NBDI) identifies the built spaces within the study area, indicating that higher Built-up Index are more exposed to the hazard event, as highlighted in Fig. 8a. The topographic gradient significantly influences the flow of floodwater. Higher slope values facilitate the movement of water towards lower elevations and adjacent low-lying regions (Dano 2022). The land elevation profile of Kochi city is relatively flat, causing water to accumulate in such regions. Figure 8b, illustrates the

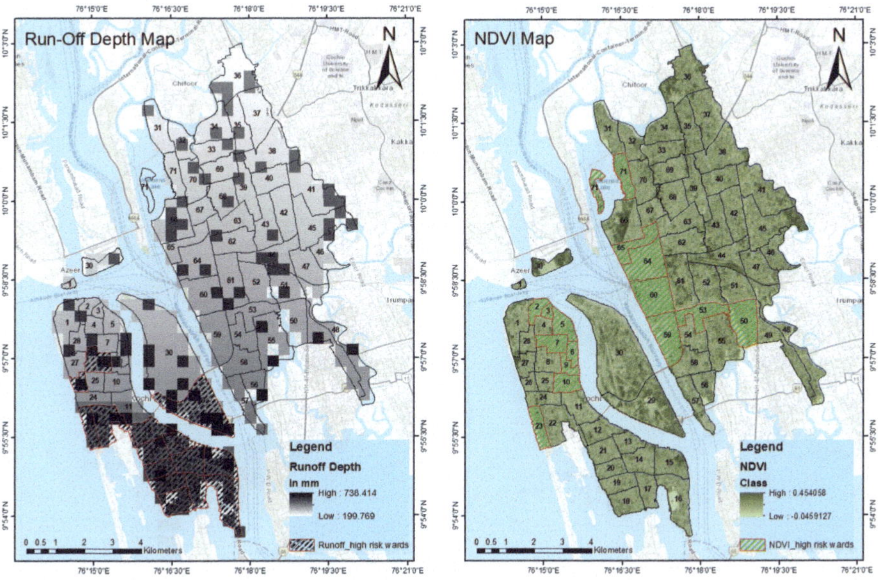

Fig. 5 Identified hotspot wards in the study area. **a** Run-off depth map. **b** Normalized Difference Vegetation Index (NDVI) map

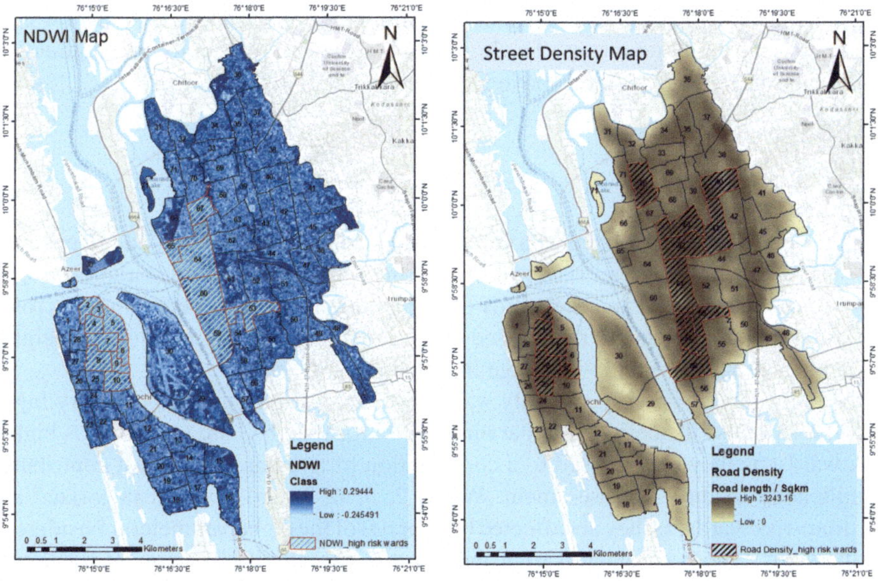

Fig. 6 Identified hotspot wards in the study area. **a** Normalized Difference Water Index (NDWI) map. **b** Street density map

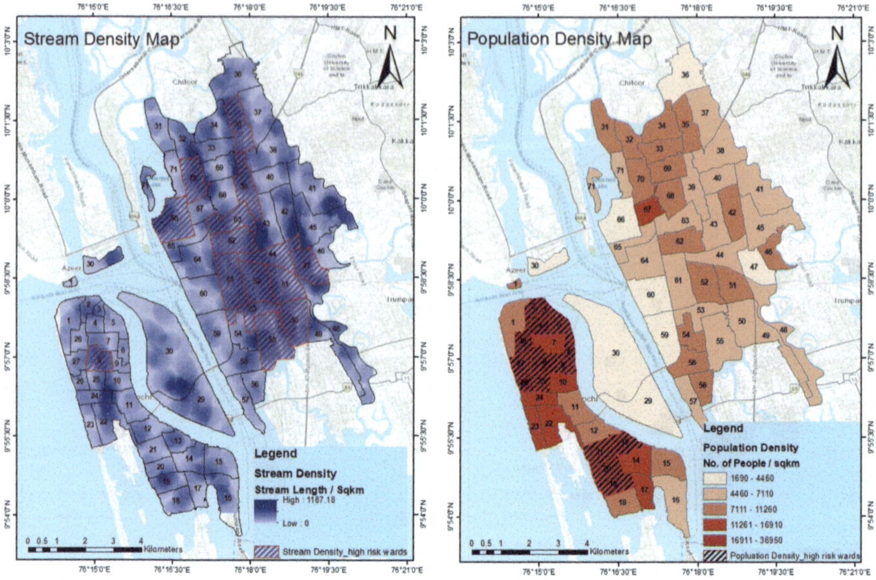

Fig. 7 Identified hotspot wards in the study area. **a** Stream density map. **b** Population density map

slope map, highlighting high-risk wards with minimal slope that face the risk of water inundation during flood events. The high-risk wards with respect to the parameters under consideration explained above are listed in Table 6.

3.1.1 Risk Indexing

Risk is commonly formulated as a function of hazard, vulnerability, and exposure. To calculate the ward-level risk within the study area, all relevant parameters are superimposed (Fig. 9a) and the wards experiencing the highest occurrences across the different parameters are recognized as the hotspot wards. In this study, all eight parameters are given equal weightage in influencing the risk.

Notably, three wards, specifically wards 3, 8, and 9, are found to be at very high risk for six out of the entire eight parameters. Wards 4 and 53 are identified as high-risk wards, as they are exposed to a combination of five parameters that contribute significantly to the risk assessment. Wards 2, 5, 6, 7, and 10 are categorized as medium-risk wards, as they are affected by four out of the eight parameters considered in the study. These parameters contribute to the overall risk assessment for these wards. A total of 10 wards, as shown in Fig. 9b, and listed in Table 7 are identified as the most exposed and vulnerable wards within the total 74 wards. The risk assessment study of Kochi Corporation (The Hindu Bureau 2022) aligns with the findings of the current study. Wards 3, 4, 8, and 53, which were identified as high-risk wards in this research, are also reported to be at high risk in the risk assessment study.

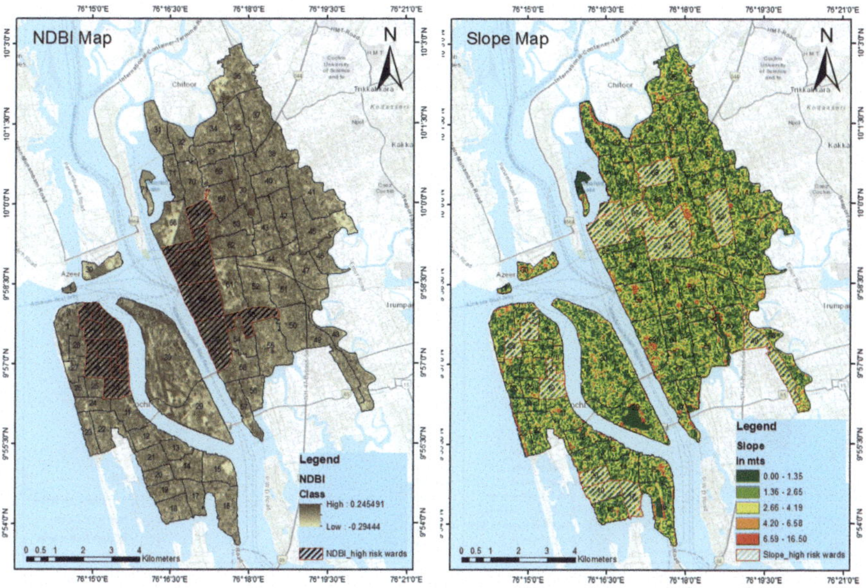

Fig. 8 Identified hotspot wards in the study area. **a** Normalized Difference Built-up Index (NDBI) map. **b** Slope map

Table 6 Wards at risk for the parameters under consideration

Parameter	Wards under Risk	Percentage of area
Run-off Depth	8, 12, 13, 14, 17, 15, 16, 18, 19, 20, 21, 22, 23, 26, 29	17.48
NDVI	2, 3, 5, 6, 7, 9, 10, 23, 50, 53, 59, 60, 64, 65, 71	15.82
NDWI	2, 3, 4, 5, 6, 7, 8, 9, 10, 53, 59, 60, 64, 65, 67	14.68
Street Density	3, 4, 7, 8, 9, 25, 40, 43, 53, 54, 58, 61, 62, 63, 70	16.10
Stream Density	8, 35, 39, 44, 47, 50, 51, 52, 53, 55, 61, 62, 63, 66, 70	23.64
Population Density	2, 3, 4, 5, 6, 8, 9, 13, 19, 20, 21, 25, 26, 27, 28	9.25
NDBI	2, 3, 4, 5, 6, 7, 8, 9, 10, 53, 59, 60, 64, 65, 67	14.6
Slope	3, 4, 9, 10, 17, 19, 20, 28, 42, 49, 62, 63, 66, 67, 69	15.24

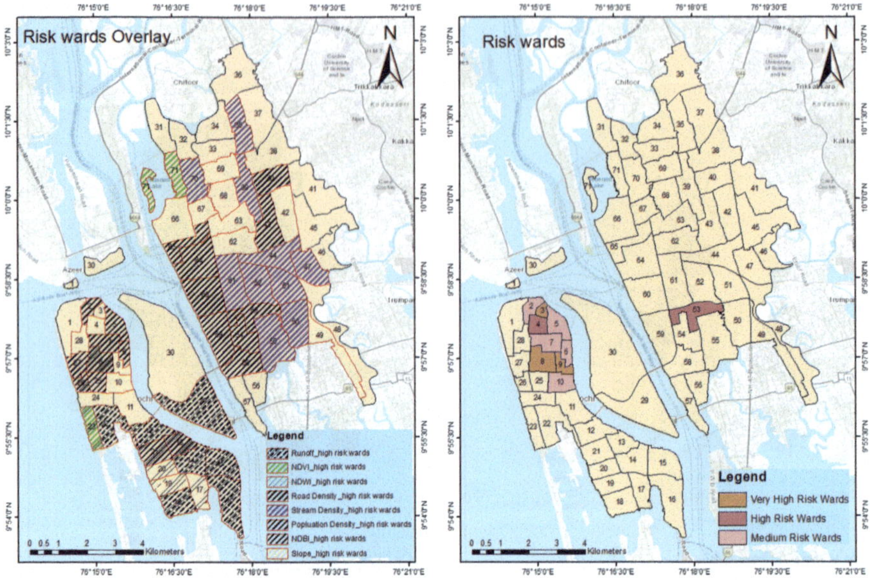

Fig. 9 **a** Composite map of high-risk wards based on eight parameters considered in the study. **b** Risk Map: Delineating Very High Risk, High Risk, and Medium Risk wards based on multiple occurrences among various parameters.

Table 7 Index considered for categorizing the level of risk

Risk type	No of parameters under Risk	Wards affected
Very High Risk	6 or more	3, 8, 9
High Risk	5	4, 53
Medium Risk	4	2, 5, 6, 7, 10

4 Conclusion

The wards situated on the periphery of Kochi City, adjacent to water bodies, exhibit high vulnerability to urban flooding. Among the parameters studied, a consistent pattern emerges where wards with low values of Vegetation Index (NDVI) and Water Index (NDWI), coupled with high values of Built-up Index (NDBI) and population density, stand out as key indicators of risk. Street density and slope, which contribute nearly equally, play significant roles in identifying risk-prone wards. The results highlight the crucial role that the topographical qualities, the amount of urbanization, and land cover variables play in influencing flood risk. Identifying these trends can aid in more focused measures to improve resilience and lessen the effects of urban flooding as urban planning and disaster management efforts advance.

Future studies offer the potential to enhance risk identification comprehensively by integrating additional parameters that influence risk functions. The methodology employed in this study can be utilized for implementation in other locations, where parameters relevant and specific to the characteristics of those areas are thoughtfully chosen. Additionally, this study can explore the utilization of varying weightages for different layers to analyse hotspot regions and underlying reasons. The GeoSM-NatE identifies both urban risk hotspots and potential intervention sites in the study area. The findings of this study will be valuable for various stakeholders. For instance, the Disaster Mitigation Authority can utilize the information to employ early detection systems and interventions in the identified risk hotspots. Similarly, urban planners can use the data to develop Master Plan that incorporates urban risk resilience. In the context of this paper, conducting a pilot study within the identified hotspots in the wards would provide valuable insights into site-specific challenges and opportunities for effective mitigation strategies. Emphasizing and prioritizing these hotspot wards will be crucial during the process of policy formulation and implementation process.

Acknowledgements We acknowledge the climatology laboratory of IIT Roorkee for providing support in using GeoSM-NatE tool. The first author also acknowledges the AICTE—QIP Scheme for providing assitantship during this study.

References

Ali S, George A (2022) Modelling a community resilience index for urban flood-prone areas of Kerala, India (CRIF). Nat Hazards 113(1):261–286. https://doi.org/10.1007/s11069-022-05299-7

Barron OV, Pollock D, Dawes W (2011) Evaluation of catchment contributing areas and storm run-off in flat terrain subject to urbanisation. Hydrol Earth Syst Sci 15(2):547–559. https://doi.org/10.5194/HESS-15-547-2011

Bertilsson L, Wiklund K, de Moura Tebaldi I, Rezende OM, Veról AP, Miguez MG (2019) Urban flood resilience – A multi-criteria index to integrate flood resilience into urban planning. J Hydrol 573:970–982. https://doi.org/10.1016/j.jhydrol.2018.06.052

Bhagat RB (2011) Emerging pattern of urbanisation in India. Econ Pol Wkly 46:10–12

Bhagat RB (2018) Urbanization in India: trend, pattern and policy issues. Working Paper No. 17, International Institute for Population Sciences, Mumbai

Dano UL (2022) An AHP-based assessment of flood triggering factors to enhance resiliency in Dammam, Saudi Arabia. GeoJournal 87(3):1945–1960. https://doi.org/10.1007/s10708-020-10363-5

Department of Town & Country Planning (2012) State urbanisation report, Kerala.

Department of Town & Country Planning (2014) Development Plan for KCR 2031. https://kochimetro.org/wp-content/uploads/2014/11/Vol3_Development_ProposalsControlRegulations.pdf

Devi AB, Nair AM (2021) Effects of urbanisation in a shallow coastal aquifer: An integrated GIS-based case study in Cochin, India. Groundwater for Sustainable Development, 15. https://doi.org/10.1016/j.gsd.2021.100656

Diez JJ, Esteban MD, Paz R, López-Gutiérrez JS, Negro V, Monnot JV (2011) Urban coastal flooding and climate change. J Coast Res, Spec Issue 64:205–209. https://www.researchgate.net/publication/265084339

George A, Sinha A, Pallathucheril VG (2005) The Goshree road link, Kochi, India: A design approach towards creating a legible streetscape. J Urban Des 10(2):275–288. https://doi.org/10.1080/13574800500087202

Gupta AK, Nair SS (2011) Urban floods in Bangalore and Chennai: Risk management challenges and lessons for sustainable urban ecology. Curr Sci 100(11):1638–1645

Halder S, Saha U (2021) Future projection of extreme rainfall for flood management due to climate change in an urban area. J Sustain Water Built Environ 7(3):4021012. https://doi.org/10.1061/jswbay.0000954

India Meteorological Department (2020) Disastrous weather events 2020. Government of India.

India Meteorological Department (2021) Annual report. Government of India.

Kallungal D (2022) Kerala set to witness extreme weather events in 2022. The New Indian Express. https://www.newindianexpress.com/states/kerala/2022/jan/02/kerala-set-to-witness-extreme-weather-events-in-2022-2402137.html

Kasdan DO (2022) Governance quality, administrative values and disaster risk management. Disaster Prev Manag: Int J 31(4):349–360. https://doi.org/10.1108/DPM-09-2021-0252

Kjeldsen TR (2009) Modelling the impact of urbanisation on flood run-off volume. Proceedings of the Institution of Civil Engineers—Water Management 162(5):329–336. https://doi.org/10.1680/wama.2009.162.5.329

Krishnan VS, Firoz CM (2021) Impact of land use and land cover change on the environmental quality of a region: A case of Ernakulam district in Kerala, India. Reg Stat 11(2):102–135. https://doi.org/10.15196/RS110205

Kulkarni A, Sabin TP, Chowdary JS, Rao KK, Priya P, Gandhi N, Bhaskar P, Buri VK, Sabade SS, Pai DS, Ashok K, Mitra AK, Niyogi D, Rajeevan M (2020) Precipitation changes in India. In: Krishnan R, Sanjay J, Gnanaseelan C, Mujumdar M, Kulkarni A, Chakraborty S (eds) Assessment of climate change over the Indian region. Springer Singapore, pp 47–72. https://doi.org/10.1007/978-981-15-4327-2_3

Kumar A, Schreyers L, Satheesh A, Sudmeier-Rieux K, Mukherjee M, Walz Y, Naorem V, Mallawantantri A, Ravan S (2023) Harnessing geo-spatial data to design blue-green infrastructure for building regional and local scale disaster resilience. In Babu SA (ed) Fifth world congress on disaster management. Routledge. https://doi.org/10.4324/9781003342090

Montero D, Aybar C, Mahecha MD, Martinuzzi F, Söchting M, Wieneke S (2023) A standardized catalogue of spectral indices to advance the use of remote sensing in Earth system research. Science Data, 1–20. https://doi.org/10.1038/s41597-023-02096-0

Morante-Carballo F, Montalván-Burbano N, Arias-Hidalgo M, Domínguez-Granda L, Apolo-Masache B, Carrión-Mero P (2022) Flood models: an exploratory analysis and research trends. Water (Switzerland) 14(16):2488. https://doi.org/10.3390/w14162488

Mukherjee M, Wickramasinghe D, Chowdhooree I, Chimi C, Poudel S, Mishra B, Ali ZF, Shaw R (2022) Nature-based resilience: experiences of five cities from South Asia. Int J Environ Res Public Health 19(19):11846. https://doi.org/10.3390/ijerph191911846

Nair SB, CJ PL (2019) Urbanization in Kerala—what does the census data reveal? J Hum Dev II(3):356–386. https://doi.org/10.1177/0973703018763241

National Disaster Management Authority (2023) Urban floods. Govt. of India. https://ndma.gov.in/Natural-Hazards/Urban-Floods

National Institute of Disaster Management (2018) Kerala flood 2018. Ministry of Home Affairs, Govt. of India. https://nidm.gov.in/PDF/pubs/KeralaFlood_18.pdf

Nkwunonwo UC, Whitworth M, Baily B (2020) A review of the current status of flood modelling for urban flood risk management in the developing countries. Scientific African 7:e00269. https://doi.org/10.1016/j.sciaf.2020.e00269

Pal SC, Chowdhuri I, Das B, Chakrabortty R, Roy P, Saha A, Shit M (2022) Threats of climate change and land use patterns enhance the susceptibility of future floods in India. J Environ Manage 305:114317. https://doi.org/10.1016/j.jenvman.2021.114317

Qi W, Ma C, Xu H, Zhao K, Chen Z (2022) A comprehensive analysis method of spatial prioritization for urban flood management based on source tracking. Ecol Ind 135:108565. https://doi.org/10.1016/j.ecolind.2022.108565

Raghunandan, TR, Jose TK, Harilal KN, Kumary NP (2017) Government of Kerala, Kerala State Planning Board, Thirteenth Five-Year Plan. In Working Group On Urban Issues Report.

Rashiq A, Prakash O, Kumar A (2022) Urban flood risk hotspot zonation using GIS—based techniques. 9th Zero Energy Mass Custom Home International Conference
Ross CW, Prihodko L, Anchang J, Kumar S, Ji W, Hanan NP (2018) Global Hydrologic Soil Groups (HYSOGs250m) for Curve Number-Based Run-off Modeling. ORNL Distributed Active Archive Center. https://doi.org/10.3334/ORNLDAAC/1566
Roychowdhury A (2018, August 22). Kerala floods: the deluge of 1924 was smaller, but impact was similar. The Indian Express. https://indianexpress.com/article/research/year-1099-keralas-great-flood-of-1924-too-affected-same-areas-5317677/
Sahu A, Bose T, Samal DR (2021) Urban flood risk assessment and development of urban flood resilient spatial plan for Bhubaneswar. Environ Urban ASIA 12(2):269–291. https://doi.org/10.1177/09754253211042489
Singh H, Nielsen M, Greatrex H (2023) Causes, impacts, and mitigation strategies of urban pluvial floods in India: a systematic review. Int J Disaster Risk Reduct 93(May):103751. https://doi.org/10.1016/j.ijdrr.2023.103751
Sivaraman S (2018, August 17) Do you know about the Kerala Flood of 1924? The Indian Express. https://www.newindianexpress.com/states/kerala/2018/aug/17/do-you-know-about-the-kerala-flood-of-1924-1859072.html
Sowmya K, John CM, Shrivasthava NK (2015) Urban flood vulnerability zoning of Cochin City, southwest coast of India, using remote sensing and GIS. Nat Hazards 75(2):1271–1286. https://doi.org/10.1007/s11069-014-1372-4
Staff Reporter (2022) Flash floods in the wake of unrelenting showers sink Kochi. The Hindu. https://www.thehindu.com/news/cities/Kochi/flash-floods-submerge-kochi-ernakulam-kerala/article65828888.ece
State Planning Board (2017) Economic review 2017—population profile of the state. GOK, India. https://spb.kerala.gov.in/economic-review/ER2017/web_e/ch11.php?id=1&ch=11#:~:text=The share of urban population,The rural population was 17%2C471%2C135.
State Relief Commissioner, DM (2019) Memorandum | Kerala - Floods & Landslides 2019. GOK, India
Subramanian A, Nagarajan AM, Vinod S, Chakraborty S, Sivagami K, Theodore T, Sathyanarayanan SS, Tamizhdurai P, Mangesh VL (2023) Long-term impacts of climate change on coastal and transitional ecosystems in India: an overview of its current status, future projections, solutions, and policies. RSC Adv 13(18):12204–12228. https://doi.org/10.1039/d2ra07448f
The Hindu Bureau (2022) Six divisions of Kochi Corporation identified as facing 'high risk' in a vulnerability and risk assessment study. The Hindu. https://www.thehindu.com/news/cities/Kochi/six-divisions-of-kochi-corporation-identified-as-facing-high-risk-in-a-vulnerability-and-risk-assessment-study/article65935344.ece
Tomar P, Singh SK, Kanga S, Meraj G, Kranjčić N, Đurin B, Pattanaik A (2021) GIS-Based urban flood risk assessment and management—a case study of delhi national capital territory (NCT), India. Sustainability (Switzerland) 13(22):1–20. https://doi.org/10.3390/su132212850
UNDRR (2004) Risk awareness and assessment. Living with Risk: A Global Review of Disaster Reduction Initiatives, 36–78.
UNDRR (2009) UNISDR terminology on disaster risk reduction. In: International Strategy for Disaster Reduction.
Zlatanova S, Tarun G, Kaour A, Neuvel JMM (2014) Integrated flood disaster management and spatial information: case studies of Netherlands and India. The International Archivesof the Photogrmmetry, Remote Sensing and Spatial Information Sciences. https://doi.org/10.5194/isprarchives-XL-8-147-2014

Cause-Effect Time-Series Mapping of Urban Floods Using GIS for Patna, India

Kumar Abhinay, Mahua Mukherjee, Atul Kumar, and Archita Saha

Highlights:

- The average annual rainy days for Patna during 2007–2019 have been around 77 days.
- The average rainfall in the peak monsoon season has shown a rising trend and has increased by 12.69% between the 2007 and 2019 flood events.
- The Built-up in the city has increased by 23.44% in the last two decades.
- The runoff has increased by 3.87% between the 2007 and 2019 flood events.

1 Introduction

According to the UN, the world population has already crossed the mark of 8 billion in 2022 (United Nations 2022), and India has become the most populous country after overtaking China (UN DESA 2023). Most of this population, around 58% globally and 36% in India, live in urban areas (UN DESA 2019). This brings the issue of

K. Abhinay · A. Kumar · A. Saha
Department of Architecture and Planning, Indian Institute of Technology Roorkee, Roorkee, India
e-mail: kumar_a@ar.iitr.ac.in

A. Kumar
e-mail: akumar5@ar.iitr.ac.in

A. Saha
e-mail: archita_s@ar.iitr.ac.in

M. Mukherjee (✉)
Centre of Excellence in Disaster Mitigation and Management, Indian Institute of Technology Roorkee, Roorkee, India
e-mail: mahuafap@iitr.ac.in

M. Firoz C et al. (eds.), *Urban Planning and Design for Megacities in the Global South*, Advances in 21st Century Human Settlements,
https://doi.org/10.1007/978-981-97-8370-0_4

rapid urbanization into the picture. According to geographers, urbanization is the process of changing rural human settlements into urban settlements, mainly driven by social productivity (Li et al. 2013). According to the planners, urbanization is the physical growth of the rural area into urban areas mainly due to the immigration of the rural population to an existing urban area in search of better living standards, employment, better health and education facilities, and other basic infrastructures. Urbanization or urban expansion results in converting existing natural landscapes, such as forests, water bodies, and agricultural lands, into built environments, leading to profound alterations in the natural landscape and ecological balance.

Though the actual shift of population from a rural area to any already existing area brings a lot of improvement in opportunities and adds access to various facilities, it also negatively impacts the living conditions of both the migrants as well as the existing inhabitants of the urban area. Most of the population moving to the urban areas are poor and have to settle in unauthorized places that are not suitable for settlement and are prone to various hazards, thus at higher risk (Kuddus et al. 2020; Williams et al. 2019). One of the primary stresses caused by the migration of rural populations and urbanization is on the land resources for the development of infrastructures like residential units, roads, commercial centers, and other facilities for waste management to cater to higher populations (Assennato et al. 2022). The new population moving to an urban area, primarily people with low incomes, like daily wage laborers and construction workers, are compelled to live in the slums. The raised land demand in the urban area leads to changes in the existing Land Use (LU) of the urban area, which is often termed Land Use Land Cover Changes (LULCC) (Nath et al. 2021). LULC change is a change in an area's existing land use patterns and conversion of that land into a different use category. For example, due to urbanization, there is an increased demand for land for building houses, so the existing sub-urban areas with land being used for agriculture or under forest cover are converted into residential areas (Dewan and Yamaguchi 2009). The change in the existing physical features of the urban regions creates many urban disasters, most commonly urban floods. Other prevalent problems the urban area faces are air pollution, water pollution, Urban Heat Island (UHI), solid waste management, and poor living conditions. Urban areas are the most affected and significant contributors to the changing climate in the form of GHG emissions, solid waste, wastewater, and stress on fresh water. The marginalized section of society is at higher risk due to inadequate living conditions and affordability to maintain a safe and hygienic life. It is expected that the urban poor are the ones who are and will be impacted the most by the adversities of the changing climate (Islam and Winkel 2017).

Urban floods have been a frequent and ongoing problem faced by almost all urban regions leading to substantial economic and life losses (Rubinato et al. 2017; Wang and Fu, n.d.; Willner et al. 2018). According to the IPCC Report 2021, the frequency and intensity of urban flood susceptibility are likely to increase continuously. The loss and damage due to urban floods are expected to be 20 times more than the current amount by the late twenty-first century if proper actions are not taken (Winsemius et al. 2015). The primary reason behind these frequent floods in urban areas is rapid urban growth, changes in precipitation intensity due to climate change, inadequate planning of drainage systems (Hassan et al. 2022; Zhou et al. 2019), and changes in the land use pattern. The urban development in cities under the variation in rainfall intensity affects the urban catchment through flooding problems (Hassan et al. 2022; Mzava et al. 2021; Neupane 2018). Urban development results in the change of existing previous land feature into impervious surfaces and impact the hydrological processes in urban areas, resulting in reduced hydrological response time, which increases flood risk (Wang et al. 2020; Zang et al. 2019). The increase in the impervious surfaces reduces the infiltration rates and increases the surface runoff, i.e., runoff volume in a short period, leading to urban floods (Xu et al. 2020). Apart from urban development, other drivers of an urban flood are rainfall events (no. of rainy days) in the monsoon season, rainfall intensity, and runoff generated after rainfall events (Pradhan-Salike and Raj Pokharel 2017).

The formidable challenge of climate change being faced globally is mainly driven by human activities such as burning fossil fuels and deforestation, leading to the accumulation of greenhouse gases in the atmosphere, resulting in global warming and a range of climate-related adverse events. Climate change's direct impacts can be noticed in rising temperatures and changes in precipitation patterns. The summers are getting hotter and prolonged, whereas the precipitation pattern has changed globally. One of the most critical consequences of changing climate could be traced to the precipitation pattern, with rainfall intensity, frequency, and distribution changes. The increased flooding events are clear evidence of the changing climate apart from the melting of glaciers and rising sea levels that could be perceived and understood by even ordinary people. Flood events have increased by about 23% globally, risking around 2 billion people, 600 million of whom are poor (Rentschler et al. 2022). India, a developing nation with a vast population of about 1.4 billion, is highly vulnerable to floods. According to NDMA, 80% of the precipitation falls in the monsoon period, i.e., June to September, leading to seasonal and riverine floods. Apart from the seasonal floods, events like cloud bursts and floods due to cyclones have been very frequent in India in the last two decades. The Uttarakhand floods (2013), Jammu & Kashmir Flood (2014), Chennai Floods (2015), Kerala floods (2018), and Cyclone Fani (2019) are examples of some recent floods caused by cloud bursts and cyclones.

Several cities that have shown a rising trend in urbanization have frequently witnessed flooding cases. The Surat flood of 2006 is an example, where the population has grown by over ten-fold since 1951 (Bhat et al. 2013), leading to the increase in the Urbanization Index from 14.23% to 24.29% for the year 1968 and 2006, respectively (Waghwala and Agnihotri 2021). In India, the cases of urban floods in the recent past have risen exponentially. Significant flood events were recorded in some prominent cities of India, like floods in Nagpur and Delhi in 2023, Ahmedabad and Bengaluru in 2022, and Hyderabad in 2021 and 2020, leading to substantial economic losses and forcing many residents to abandon and flee their places. One of the main reasons for frequent urban flooding is related to the rising urbanization trend in these cities (Bagyaraj et al. 2023). The main reason behind the urban flooding in Delhi in July 2023 was siltation and encroachment.

Similarly, the main reason behind the Chennai floods in 2023 was also urbanization. The problem of urban flooding is prevalent across the globe, too. The flooding in Bangkok, Thailand, in 2011 resulted in around 815 deaths, and one of the causes behind the flooding in Bangkok was rapid urbanization (Seemuangngam and Lin 2024). Dhaka, Bangladesh, has been facing the issue of frequent urban flooding, and rapid urbanization is the root cause of the problem (Mohuya et al. 2023).

The main objective of this study was to investigate the leading cause of urban flooding in one of the fastest urbanizing cities of India, Patna, the capital city of Bihar. Patna witnessed rapid urbanization at the rate of 44.3% in the last decade, leading to considerable disturbances in the city's original topography. Patna faced severe urban flooding in the previous two decades (2000–2019), with two major flood events in 2007 and 2019. The study began with the null hypothesis (H_0), considering only the natural rainfall events responsible for the urban flooding in Patna. The study investigated the change in rainfall pattern, change in precipitation pattern in the region, and change in the LULC changes to find the change in the runoff generated in Patna city and the actual cause of these two flood events. This study will be helpful for the authorities concerned, especially the planning department, to plan the city strategically to cope with the excess runoff generated during the monsoon season. The study will help understand the need and requirement of changing climate-responsive design considerations for the city of Patna to tackle urban flooding events in the future.

2 Conceptual Framework

It is evident from the above section that urban floods pose significant challenges to cities globally. Patna, with the frequent flood incidents in the city in the last two decades, proves that Patna is no exception. The unpredictable weather events due to climate change, especially precipitation, and rapid urbanization leading to land use change are the major causes of the increased frequency and intensity of urban flooding globally in recent years. Understanding and investigating the causes and effects of urban floods in Patna is essential for effective flood management and mitigation strategies. Studying the interconnections between driving factors, urbanization, land use change, altered rainfall frequency and intensity, and runoff causing floods is crucial.

The urban flooding in Patna in recent times has been a matter of grave concern for the administration and the government for public safety, infrastructure, economic losses, and the overall well-being of the residents of Patna city. This study adopts the time-series approach for investigating the cause-effect relationship between the evolving city's natural landscape, changing climate, and urban floods over time. The study uses the GIS technique for spatial investigation of the changes in the built-up index from 2007 to 2019. The findings of this study will be valuable for policymakers, urban planners, and disaster management authorities to understand the cause-effect of urban floods, formulate strategies to mitigate the flood risk, and improve the city's resilience.

The study aimed to investigate the change in Patna in the last two decades that has led to frequent urban flooding in the city. The primary objective of the study was (i) to analyze the historical rainfall pattern during the southwest monsoon season for the Patna region, (ii) to assess the changes in the built-up area in Patna city, and (iii) to identify the change in rainfall intensity and runoff generated.

The primary change globally due to changing climate leading to altered rainfall patterns and intensity applies to the case of Patna city, too. Still, the changing climate cannot be contained or mitigated at a city level and in a short period. The best humans can do in a short period is to be prepared to cope with urban flood events and build the capacity to mitigate the adversities of these frequent flood events. The study took two significant causes of urban flooding, i.e., changes in rainfall intensity, frequency, and urbanization. The rainfall data were analyzed in the last two decades and compared in 2007 and 2019, two major urban flooding events in Patna. The change in the built-up index was also analyzed to find the relation between the flooding event and the city change. The following steps were adopted for the study:

Step I: A literature review was carried out to understand the phenomenon of urban floods, the causes, and the drivers leading to urban flooding in urban areas. This step thoroughly explored existing scientific articles, research papers, government reports, news articles, and other publications related to the urban flood. The primary aim of the literature review was to gain a deep understanding of the factors and drivers that contribute to flooding in urban areas. The study also includes a thorough analysis of the rainfall pattern of the study area, especially during the monsoon season.

Step II: The numbers of average rainy days and the maximum rainfall in the years 2007 and 2019 during the southwest monsoon season (June–September) were analyzed to check if there has been any significant change in the rainfall events in both years, as a greater number of rainy days has an impact on the flood water volume and the runoff generated.

Step III: The change in the built-up was checked using the NDBI method using ArcGIS for both 2007 and 2019, as the increase in the built-up leads to a reduction of the previous layer in an area, resulting in higher runoff generation.

Step IV: A time-series mapping of the rainfall frequency and intensity was carried out to analyze the precipitation received and runoff generated leading to both flood events. The main aim of this step was to gain insights into the changing patterns of precipitation to understand their impact on the generation of surface runoff after a rainfall event. This step included collecting and analyzing historical data on rainfall events for Patna city for the years 2007 and 2019.

Step V: The primary reason for the flood events was found using the results of steps III and IV. This step included carefully examining the frequency and intensity of rainfall occurrences concerning the built-up increase to identify discernible patterns or trends.

The main focus of the analysis was to assess the relationship between the precipitation, runoff, and the built-up change leading to the flood events in Patna. This relationship is crucial for understanding how precipitation frequency and intensity change leads to flooding under current and changing built-up in the city.

Patna has experienced a rapid urbanization rate of 44.3% in the last decade, resulting in increased demand for land resources for infrastructures like roads, residential buildings, and commercial centers. Patna houses around 5.8 million people, according to the 2011 census, out of which 2.35 million reside in the city area of Patna, making it the 19th largest city in India. Patna has faced several urban floods since 2000, including the major flood events of 2007 and 2011. Despite being situated on the bank of the river Ganga, the city of Patna faces the issue of urban floods regularly. The basic idea behind this study was to investigate the primary cause of these frequent flood events in the city. The two possible causes of the flood formed the hypothesis for this study, i.e., either the rainfall intensity and frequency have increased in the region, leading to a higher volume of rainfall leading to flood, or there could be some other anthropogenic reason behind the flooding. This study will find the actual cause of the flood events and help solve them based on the suggested recommendations in the chapter.

3 Material & Method

3.1 Study Area

Patna is the capital city of Bihar and also the largest city by population. Patna is situated on the banks of the river Ganga and is located between coordinates of 25°35′38.74″ N and 85°8′15.23″ E at an elevation of 53 m (173 feet) above sea level (Sharma et al. 2017). It is the urban center of Bihar with a rapidly expanding population, leading to significant urban development (Kumar and Yadava 2018). The population of Patna has increased in recent decades due to high migration from all over Bihar. Patna's LULC is a critical symbol of its urban development, where the residential areas, commercial zones, industrial sectors, and other infrastructure developments have expanded at the cost of encroachment of natural landscapes and flood plains. The rapid urbanization in Patna has led to significant land use and cover changes, with built-up areas replacing the natural vegetation. Urban expansion has altered the hydrological regime of the city, amplifying the flood risk (Sugianto et al. 2022) (Fig. 1a).

Patna falls under a humid subtropical climate under the Köppen climate classification (Cwa), with hot summers from the end of March to June, the monsoon season from June to the end of September, and cold winters from November to February. Patna witnessed the events of urban floods in 2001, 2002, 2003, 2007, 2008, and 2019 during the monsoon season, leading to water-logging in low-lying areas of the city (Tripathi et al. 2022). The historical data on rainfall patterns in the city is

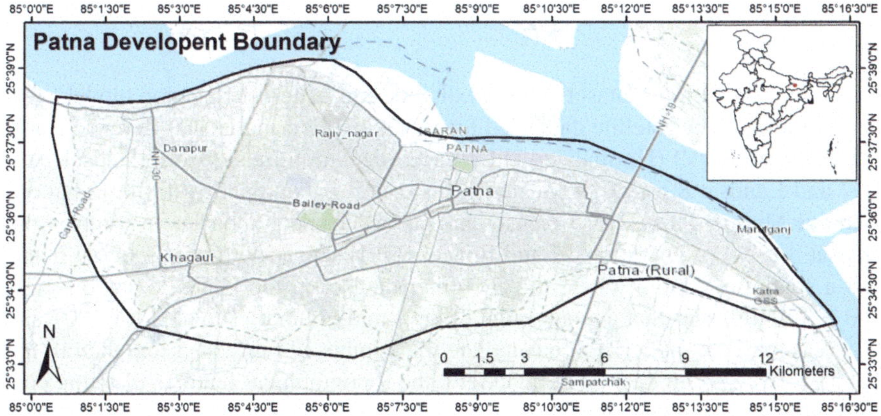

Fig. 1a Map showing the Area of Interest in the urban region of Patna city facing frequent urban floods

critical in understanding the trends and intensity of rainfall events during the southwest monsoon season. Frequent rainfall events, especially toward the end of the southwest monsoon season, have been known to trigger urban floods in Patna (Bhatt et al. 2021). The changing demographics of Patna, including population growth and inward migration, have put additional stress on the city's drainage and sewer system. The demographics, LULC, climate, and recent flood events seem interconnected. Figure 1b shows the contour map of the urban region of Patna city. The natural slope of the area is toward the Ganga River in the north direction of the city. The slope gradually increases while moving toward the south.

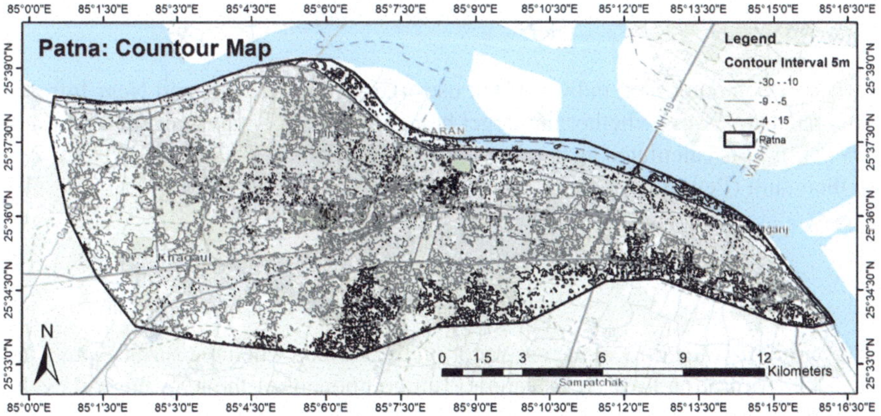

Fig. 1b Contour Map of Area of Interest in the urban region of Patna city facing frequent urban floods

4 Data Used

The study used various datasets on rainfall, land use patterns, elevation models, and soil hydraulics. The satellite data used for the NDBI (Normalized Difference Built-up Index) and NDVI (Normalized Difference Vegetation Index) from Landsat 7 for 2007 and Landsat 8 for 2019 was used. Landsat 7 is equipped with the Enhanced Thematic Mapper Plus (ETM+) sensor and has seven spectral bands. In comparison, Landsat 8 uses an Operational Land Imager (OLI) sensor with nine spectral bands with a spatial resolution of 30 m. The other data from the CHIRPS satellite was used to find the seasonal average rainfall days and volume. The data for 2007 and 2019 was used from CHIRPS to calculate the number of rainy days and the rainfall change between both years. For analyzing the topographical features of Patna city, the Shuttle Radar Topography Mission (SRTM) DEM data from USGS was used. It has a spatial resolution of 30 m. Data for delineation of runoff soil hydrological data was obtained from the World soil data of Global Hydrological Soil Groups. The runoff for the study area was delineated based on SCN curve numbers.

5 Methodology Adopted

The NDBI uses the Near Infrared (NIR) and Short-Wave Infrared (SWIR) bands to emphasize the manufactured built-up features (Zha et al. 2003). NDBI (Eq. 1) is the ratio based to mitigate the effects of terrain illumination differences as well as atmospheric effects.

$$\mathrm{NDBI} = \frac{(\mathrm{SWIR} - \mathrm{NIR})}{(\mathrm{SWIR} + \mathrm{NIR})} \tag{1}$$

The NDVI is a numerical indicator that uses the visible Red (R) and Near Infrared (NIR) bands to assess whether the target being observed contains vegetation or not. NDVI (Eq. 2) is calculated as the difference ratio between the NIR and visible Red and their sum (Pettorelli et al. 2005).

$$\mathrm{NDVI} = \frac{(\mathrm{NIR} - \mathrm{R})}{(\mathrm{NIR} + \mathrm{R})} \tag{2}$$

Surface runoff is the flow of excess water on the surface when the surface strata are at a whole saturation level. It is a significant component of the hydrological cycle. Runoff is usually generated because of higher rainfall than the infiltration rate or any depression storage filled. Heavy runoff can lead to flooding when the watercourse cannot convey the excess water to the downstream or outlet. The runoff can be quantified by measuring total precipitation and subtracting it with the loss that occurs during evapotranspiration (losing to the atmosphere from soil surfaces and plant leaves), storage (as in temporary ponds), and other such abstractions. The National Resources Conservation Services—NRCS (formerly Soil Conservation Services) developed the widely used curve number (CN) method in 1964 (Ramakrishnan et al. 2009). The curve number is a dimensionless parameter that forms the empirical equation to calculate how much runoff (in mm) is generated by any rainfall event (in mm). Maximum potential retention (S) (Eq. 3) is calculated using the CN value, which is obtained by considering the hydrologic soil group (HSG), the land use land cover (LULC) of the area, and the antecedent moisture condition (AMC). So, the CN values account for both soil and land use type while considering soil moisture in terms of AMC condition. Based on precipitation (P) and S values, we can compute the initial abstraction (Eq. 4) (Ia) and runoff (Q) (Eq. 5).

$$S = \frac{25400}{\text{CN}} - 254 \tag{3}$$

$$Ia = 0.2S \tag{4}$$

$$Q = \frac{(P - 0.2S)^2}{P + 0.8S} \tag{5}$$

The algorithm developed by (Jain et al. 2021) to prepare dynamic CN maps in GEE has been used to delineate runoff maps for the study area using OpenLandMap Soil Texture and MODIS land use/land cover (LULC) data through ternary function and CHRIPS grided rainfall collection and creation of antecedent moisture condition. Simple multiplication of depth with the area of the image pixel yields runoff volume, which, normalized by time, will give us discharge.

6 Results & Discussion

According to Indian Meteorological Department, the Coefficient of Variation (CV) values for rainfall for the state of Bihar varies between 15 and 70%. The southwest monsoon season in Bihar falls between June and September, resulting in most of the floods in Bihar in September. The Coefficient of Variation (CV) for the Patna region ranges between 10 and 85% (Department 2023). This shows that the rainfall in Patna varies vastly and is unpredictable primarily, which has been more inconsistent with the changing climate. It was found that the average number of rainy days in the region was around 77 throughout the study period of 2000–2019. The above graphs (Fig. 2) show that the rainfall intensity and number of rainy days in 2007 and 2019 were almost similar, with slightly fewer rainy days. The result (from Fig. 2a and 2b) indicates that the main reason for the severity of the 2019 flood was poor drainage, uncontrolled urban expansion, and reduction of green permeable surfaces in the city.

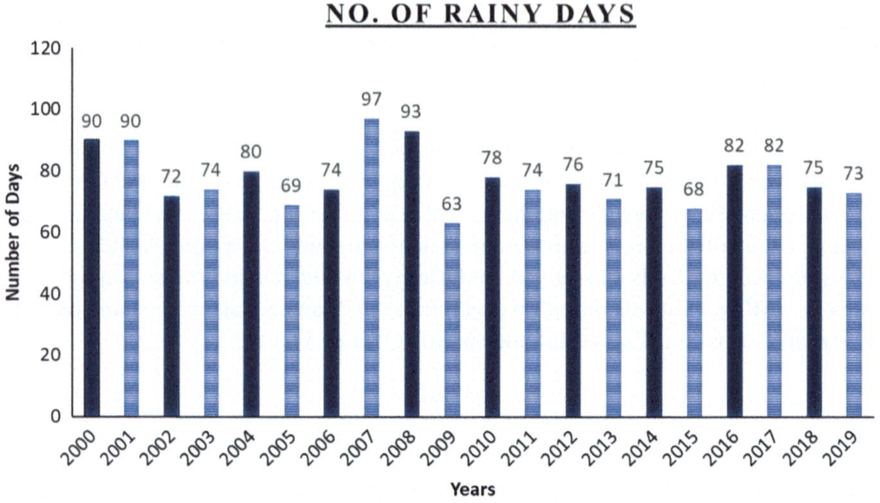

Fig. 2a Rainy days (based on the CHIRPS Data, a rainy day has been defined as a day with rainfall of 2.5 mm or more rainfall, if rainfall is more than 2.5 mm as per IMD)

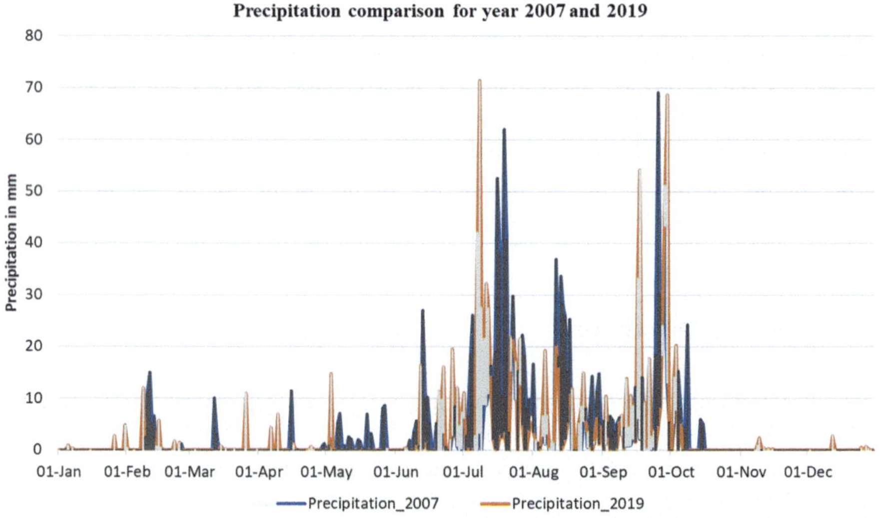

Fig. 2b Precipitation comparison of 2007 and 2019 based on the data from CHIRPS (https://www.chc.ucsb.edu/data/chirps)

The study also found that rainfall runoff showed a rise of 12.69% between the two major floods of 2007 and 2019. This is one of the reasons for the flood event in Patna since the runoff increased by around 12%, which resulted in an increased volume of rainwater received by the city. The study found a sharp increase of 23.44% in the built-up between 2007 and 2019. The built-up coverage in the study area was around 75.87 sq. km in 2007, which increased to around 99.1 sq. km in 2019. Most of the changes in the built-up area were observed in the central part of the city, which is the city's central commercial hub (Fig. 3). The city's outskirts, which used to be open and covered with vegetation previously, was also converted into a built-up area to accommodate the moving population to the city, which led to more surface runoff generation. There was also a sharp decline of 12.24% in the vegetation cover in the study area between 2007 and 2019. The vegetation cover was around 78.27 sq. km in 2007, which reduced to about 69.73 sq. km in 2019. It is a response to the urbanization and infrastructure development in the region. The vegetation cover in the central city, such as the eco-park and zoo, is maintained because of the government and authorities' protection. Otherwise, the private open and vegetated areas have been converted into residential buildings or commercial centers. The table below shows the change in the built-up area and vegetation cover in 2007 and 2019.

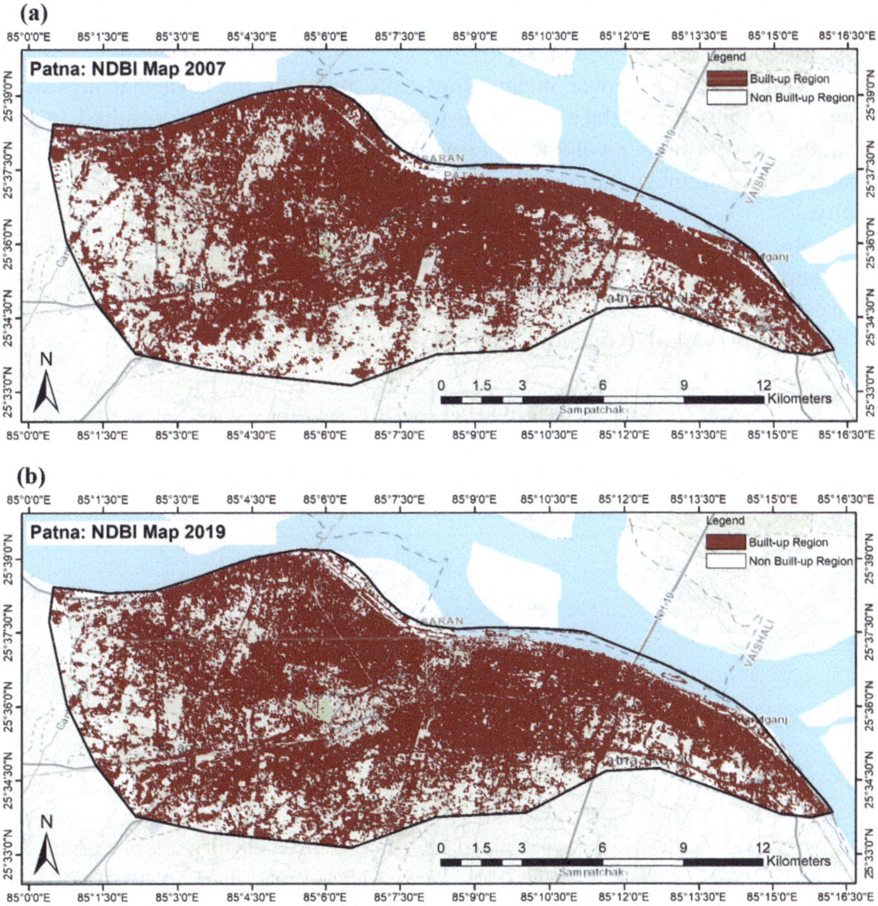

Fig. 3 Map showing the built-up index in the study area in **a** 2007 and **b** 2019

	Built-up area	Vegetation cover
2007	75.87 sq. km	78.27 sq. km
2019	99.1 sq. km	69.73 sq. km
% Change	+23.44%	–12.24%

The result of increased rainfall and change in the built-up and vegetation cover is reflected directly in the runoff generated after the flood. It was found that the runoff generated in the case of the 2019 flood was 3.87% higher than the runoff generated in the 2007 flood case (Fig. 4). Though the increase of around 4% might seem less than the change in rainfall volume and rise in built-up, it is essential to note that the vegetation cover also reduced from 2007 to 2019. The map shows (Fig. 5) the central portion of the city, which is also the major commercial and highly dense residential part of Patna city that has produced higher runoff in 2019 compared to 2007. The higher volume of produced runoff could be directly linked to the built-up index map, and it can be inferred that the areas with higher built-up changes have resulted in higher surface runoffs.

Reducing vegetation cover means drastically reducing the area that acts as a sponge that holds the surface runoff just after a rain event. This resulted in the accumulation of a higher volume of rainwater in a shorter period, surpassing the capacity of city drainage, ultimately resulting in water lodging in the lower areas of the city.

The above results reflect that the frequent urban flooding in Patna city is a cumulative output of the rise in rainfall events, increased built-up area, and reduced vegetation cover, resulting in increased surface runoff. The study began with a hypothesis that a change in rainfall frequency and pattern is responsible for the frequent flooding

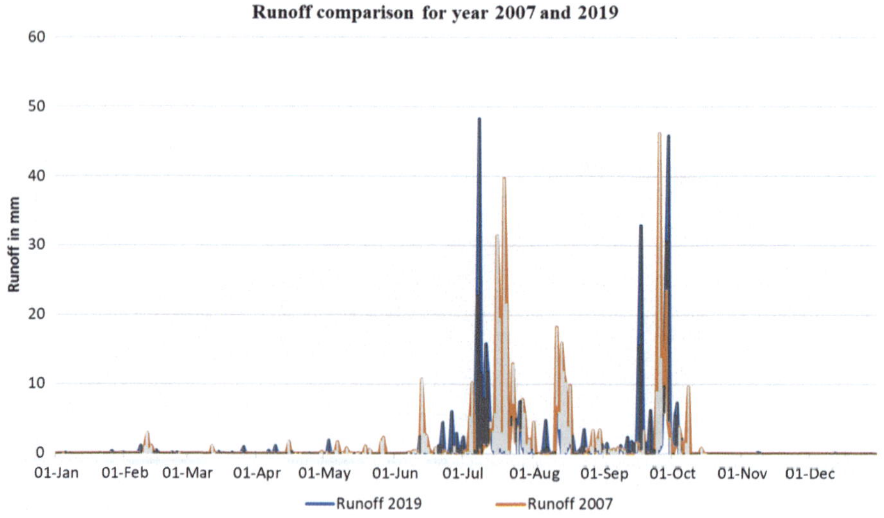

Fig. 4 Runoff comparison between 2007 and 2019 year

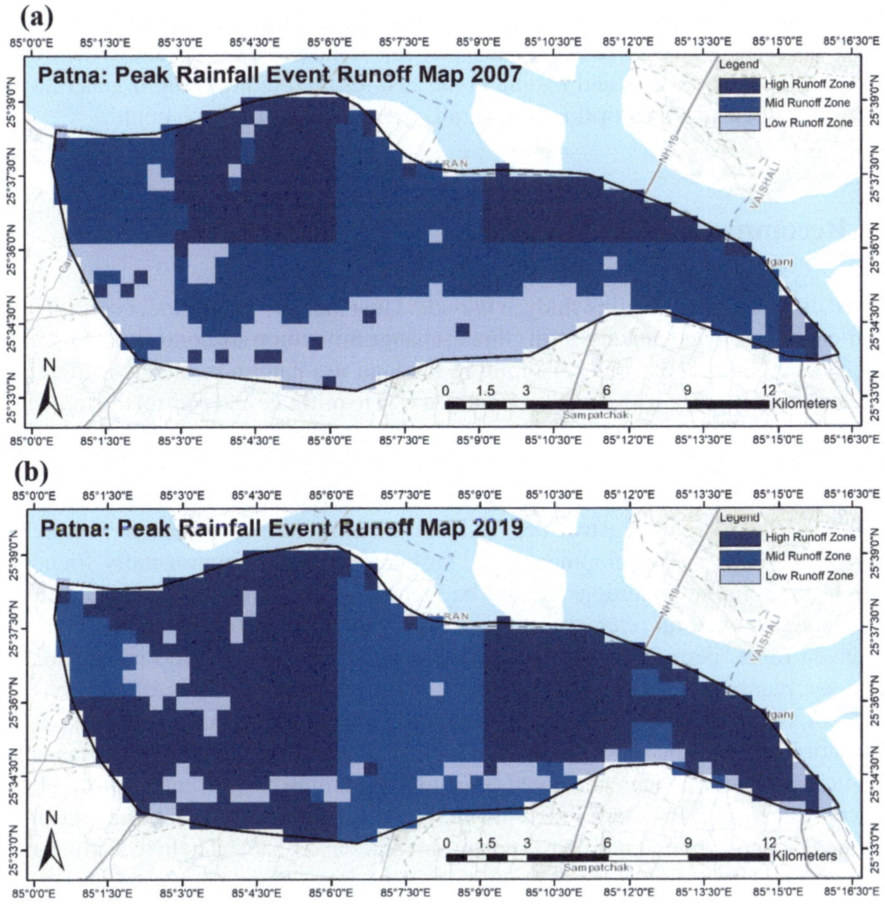

Fig. 5 Map showing peak rainfall event runoff in **a** September 2007 and **b** September 2019

in Patna city. But, in the study, it was found that it is not the case, as the other factors, i.e., increase in built-up area, are a significant contributor to these flood events. Ignorance and violation of the natural drainage pattern in the region are considerable alterations that reduce the city's capacity to cope with the runoff after a rain event. Similarly, a decrease in the vegetation cover in the city reduced the previous and permeable surface cover in the region, reducing the region's capacity to hold the runoff and allow it to be absorbed in the ground. These two factors resulted in increased surface runoff, which is the ultimate reason behind the urban floods in Patna city.

On comparing the peak rainfall event runoff for 2007 and 2019 from Fig. 5, it is evident that the southern part of the city is generating more runoff. This also signifies that the city is expanding and the green porous spaces in the city are reducing. Earlier in 2007, the high runoff zone was only in the northern ends of the city, which were dense settlements of Patna city, Ashok Rajpath, Gandhi Maidan, Golghar, Rajapul, Kurji More, etc. However, in 2019, the expansion of the city toward Bihta in the west, Sampatchak in the south, and Fatuha in the east resulted in high runoff generation, making the new hotspots of floods. Overall, from the runoff map, it could.

7 Recommendation

Based on the results from this study, it is evident that the only things under control in a shorter period other than long-term climate change mitigation are control over anthropogenic activities and proper planning of the land use pattern in the city. Effective urban planning and land use management is a way to mitigate and control the frequent urban flooding in Patna. Effective urban planning measures should be introduced and implemented to regulate the growth of built-up areas and urban sprawl, which is the root cause of the increased surface runoff and flood risk. The Green Infrastructure (GI) and Blue-Green Infrastructure (BGI) in infrastructure development should be devised to make the development more sustainable and environmentally friendly. Green Infrastructure solutions should be incorporated into urban design practices to enhance the natural retention of rainwater. Possible solutions on this path could be green roofs, permeable pavements, and urban green spaces. The use of mixed land use planning should be promoted to reduce the concentration of impervious surfaces. Apart from this, there is a dire need for stormwater management in the city. The stormwater drainage infrastructure must be more robust to effectively manage the increased stormwater generated after excess rainfall to prevent flash flooding in low-lying areas. Practices like rainwater harvesting and retention basins must be promoted and supported with government subsidies and financial help to capture and manage the rainwater during heavy rainfall events. Regular maintenance and inspection of the existing drainage and sewer system should be done to ensure their functionality and capacity to drain the floodwater volumes with changing climate events and rising urbanization in the city. It is essential to understand that the changing climate is not a myth, and no part of the world is untouched by its adversities, so there is an urgent need to consider the impacts of changing climate in every form of development. It is necessary to induct climate change adaptation techniques in urban planning and flood management to cater to the chances of flooding due to excess runoff due to extreme rainfall events. A vulnerability assessment of the city to identify the flood-prone areas should be carried out to start with the targeted robustness of the infrastructure and other measures. Soft interventions like flood early warning systems, multi-stakeholder collaboration for mobilizing resources, and expertise for implementing effective flood mitigation measures based on scientific research should be ensured. Policy and governance should be used to strengthen the existing flood management policies and regulations and their follow-up to align with the best practices and emerging challenges. Strict land use regulations should be enforced to control encroachment in flood-prone areas and protect the natural drainage system. Other anthropogenic activities, like the dumping of solid waste into the sewer system, which leads to the choking of the drainage system, should be checked and supported with proper solid waste management in the city.

8 Recommendation for Patna

The urban local bodies in Patna need to concentrate on the strict follow-up of the latest master plan proposed for the city to avoid issues like encroachment of the natural drainage and illegal constructions on the flood plains. The central areas, especially the commercial zone near Gandhi Maidan Area, which often faces flooding, need to upgrade and expand the stormwater drainage system to ensure efficient water drainage during heavy rainfall. Since the area is highly commercialized and occupied, the best way to cope is to maintain by regularly cleaning and desilting the existing drains to prevent clogging and obstruction. The city's master plan should integrate sustainable planning techniques emphasizing green spaces, porous surfaces, and water retention areas. The highly dense mix settlements, i.e., residential and commercial, like Patna city, Rajeev Nagar, Buddha Colony, and kankarbagh, face the issue of flooding mainly because of the chocked drainage as a result of poor waste management and lack of green permeable spaces to drain, hold, and absorb rainwater. The local urban bodies should develop green belts, parks, and rain gardens in these areas to absorb excess water and reduce runoff. From Fig. 3b, it is evident that the urban sprawl in Patna is expanding outward now, extending toward Bihta in the west and Fatuha in the east; it is high time for the planners and other concerned authorities to identify the vulnerable areas and prioritize them for intervention in advance. The areas near the secretariat, Anne Marg, and the Zoo are relatively less susceptible to flooding and, hence, could be under the low-priority areas.

9 Conclusion

This study has provided valuable insights into the complex dynamics of urban flooding in Patna. The use of time-series analysis of the rainfall events, intensity, and generated runoff, along with the help of a GIS-based study of the changes in the land use, i.e., built-up and vegetation cover, provides a deeper understanding of the factors influencing the flood events and their interconnections for the Patna city. The study concludes that a combination of natural and anthropogenic factors affects the frequent flood events in Patna. Though the number of rainy days in the southwest monsoon season in Patna has not changed significantly, Patna still received higher rainfall in 2019 than in September 2007. This clearly shows that the rainfall intensity in the region has increased, and the distribution of the rainfall events has changed and is not equally received. The major flood events in Patna occurred in August and September, which means that the intensity of rainfall has shifted or is more prevalent toward the end of the southwest monsoon.

On the one hand, where the rainfall runoff has increased (+12.69%) in the last decade; on the other hand, it is coupled with the rising built-up cover (+23.44%) and decreasing vegetation cover (−12.44%), leading to increased surface runoff (+3.87%). The cause-effect relationship of the urban flood in Patna, established through spatial and statistical analysis of changing land use patterns, rainfall, and runoff volume, has highlighted the importance of sustainable urban planning, as suggested in the recommendation section.

It is evident from the study that urban floods require a long-term intervention (need for robust infrastructure, the addition of GI and BGI, etc.) along with a holistic approach that integrates the policy reformation, multi-stake holder's involvement with scientific forecasting of the urbanization rate and runoff generation in the future under the changing climate. It is apparent that controlling the rainfall events and transforming back in terms of land use is not easily possible, so the best we can afford is to be prepared. As Patna continues to grow, the challenges brought by urbanization in terms of urban floods will be aggravated in the future also, to which this study provides insight that will be crucial in formulating effective flood management strategies, supporting evidence-based policy decisions, and safeguarding the well-being of its residents.

Moreover, this research has significant implications for other cities with the problem of urban flooding and rapid urbanization. The findings of this study present a current trend of changing land use patterns globally, where the built-up cover is exponentially rising at the cost of depleting vegetation cover. Similar studies could be implemented in other cities globally using the mix of spatial investigation of the land use change and relating it with rainfall frequency, intensity, and runoff. Further, the hotspots of flood-prone areas could be analyzed and identified. The areas with intensification of the built-up regions and reduced vegetation cover trends could be analyzed. Such studies will help mitigate the impact of urban floods in the flood-prone areas of cities.

Acknowledgements This study was supported by the Climatology Lab at the Department of Architecture and Planning, Indian Institute of Technology Roorkee.

References

Assennato F, Smiraglia D, Cavalli A, Congedo L, Giuliani C, Riitano N, Strollo A, Munafò M (2022). The impact of urbanization on land: a biophysical-based assessment of ecosystem services loss supported by remote sensed indicators. Land 11(2). https://doi.org/10.3390/land11020236

Bagyaraj M, Senapathi V, Chung SY, Gopalakrishnan G, Xiao Y, Karthikeyan S, Nadiri AA, Barzegar R (2023) A geospatial approach for assessing urban flood risk zones in Chennai, Tamil Nadu, India. Environ Sci Pollut Res 30(45):100562–100575. https://doi.org/10.1007/s11356-023-29132-1

Bhat GK, Karanth A, Dashora L, Rajasekar U (2013) Addressing flooding in the city of Surat beyond its boundaries. Environ Urban 25(2):429–441. https://doi.org/10.1177/0956247813495002

Bhatt CM, Gupta A, Roy A, Dalal P, Chauhan P (2021) Geospatial analysis of September, 2019 floods in the lower gangetic plains of Bihar using multi-temporal satellites and river gauge data. Geomat Nat Haz Risk 12(1):84–102. https://doi.org/10.1080/19475705.2020.1861113

Dewan AM, Yamaguchi Y (2009) Land use and land cover change in Greater Dhaka, Bangladesh: Using remote sensing to promote sustainable urbanization. Appl Geogr 29(3):390–401. https://doi.org/10.1016/j.apgeog.2008.12.005

Hassan BT, Yassine M, Amin D (2022). Comparison of urbanization, climate change, and drainage design impacts on urban flashfloods in an arid region: case study, New Cairo, Egypt. Water (Switzerland) 14(15). https://doi.org/10.3390/w14152430

Islam SN, Winkel J (2017). Climate Change and Social Inequality*. http://www.ejnetindiaresource.org/ejissues/bali.pdfen

Jain S, Jaiswal RK, Lohani AK, Galkate R (2021). Development of cloud-based rainfall-run-off model using Google Earth Engine. Curr Sci 121(11):1433–1440. https://doi.org/10.18520/cs/v121/i11/1433-1440

Kuddus MA, Tynan E, McBryde E (2020) Urbanization: aproblem for the rich and the poor? Public Health Rev 41(1). https://doi.org/10.1186/s40985-019-0116-0

Kumar A, Yadava RS (2018) A study in population growth and characteristics of Patna urban agglomeration. An Int Ref J (NGSI-BHU) 64(4).

Li GF, Xiang XY, Tong YY, Wang HM (2013) Impact assessment of urbanization on flood risk in the Yangtze River Delta. Stoch Env Res Risk Assess 27(7):1683–1693. https://doi.org/10.1007/s00477-013-0706-1

Mohuya F, Walsh C, Fowler H (2023) Urban Flood Risk in Dhaka, Bangladesh. https://doi.org/10.5194/egusphere-egu23-7194

Mzava P, Valimba P, Nobert J (2021) Quantitative analysis of the impacts of climate and land-cover changes on urban flood runoffs: A case of dar es salaam, Tanzania. J Water Clim Chang 12(6):2835–2853. https://doi.org/10.2166/wcc.2021.026

Nath B, Ni-Meister W, Choudhury R (2021) Impact of urbanization on land use and land cover change in Guwahati city, India and its implication on declining groundwater level. Groundwater for Sustainable Development, 12. https://doi.org/10.1016/j.gsd.2020.100500

Neupane B (2018) TigerPrints effectiveness of low impact development practices in reducing urban stormwater runoff under land use and climate change scenarios recommended citation. https://tigerprints.clemson.edu/all_theses

Pettorelli N, Vik JO, Mysterud A, Gaillard J-M, Tucker CJ, Stenseth N Chr (2005) Using the satellite-derived NDVI to assess ecological responses to environmental change. Trends Ecol & Evol 20(9):503–510. https://doi.org/10.1016/j.tree.2005.05.011

Pradhan-Salike I, Raj Pokharel J (2017) Impact of urbanization and climate change on urban flooding: a case of the Kathmandu Valley. Journal of Natural Resources and Development, 56–66. https://doi.org/10.5027/jnrd.v7i0.07

Ramakrishnan D, Bandyopadhyay A, Kusuma KN (2009) SCS-CN and GIS-based approach for identifying potential water harvesting sites in the Kali Watershed, Mahi River Basin, India. J Earth Syst Sci 118(4):355–368. https://doi.org/10.1007/s12040-009-0034-5

Rentschler J, Salhab M, Jafino BA (2022) Flood exposure and poverty in 188 countries. Nat Commun 13(1). https://doi.org/10.1038/s41467-022-30727-4

Rubinato M, Martins R, Kesserwani G, Leandro J, Djordjević S, Shucksmith J (2017) Experimental calibration and validation of sewer/surface flow exchange equations in steady and unsteady flow conditions. J Hydrol 552:421–432. https://doi.org/10.1016/j.jhydrol.2017.06.024

Seemuangngam A, Lin HL (2024) The impact of urbanization on urban flood risk of Nakhon Ratchasima, Thailand. Appl Geogr 162. https://doi.org/10.1016/j.apgeog.2023.103152

Sharma K, Pani P, Delhi N (2017) The river response to urbanization: A study of Patna regional development area Land Degradation, Climatic Variability and Livelihoods in Fragile Ecologies: Linkages and Implications View project River Behaviour and Sensitivity in Arid Environment: An Analysis of Luni Basin, India View project The River Response to Urbanization: A Study

of Patna Regional Development Area. | 271 Trans.Inst.Indian Geographers 39(2). https://www.researchgate.net/publication/322383970
Sugianto S, Deli A, Miswar E, Rusdi M, Irham M (2022) The effect of land use and land cover changes on flood occurrence in Teunom Watershed, Aceh Jaya. Land 11(8). https://doi.org/10.3390/land11081271
Tripathi G, Pandey AC, Parida BR (2022) Flood hazard and risk zonation in north bihar using satellite-derived historical flood events and socio-economic data. Sustainability (Switzerland) 14(3). https://doi.org/10.3390/su14031472
UN DESA (2019) World urbanization prospects the 2018 revision
UN DESA SHPG, JW (2023) Two "population billionaires", China and India, face divergent demographic futures. https://www.un.org/development/desa/pd/sites/www.un.org.development.desa.pd/files/undesa_pd_2023_policy-brief-153.pdf
United Nations (2022) World population prospects 2022 world population prospects 2022 summary of results.
Waghwala RK, Agnihotri PG (2021) Impact assessment of urbanization on flood risk and integrated flood management approach: a case study of Surat city and its surrounding region. ISH J Hydraul Eng 27(S1):577–587. https://doi.org/10.1080/09715010.2019.1658548
Wang J, Hu C, Ma B, Mu X (2020) Rapid urbanization impact on the hydrological processes in Zhengzhou, China. Water (Switzerland) 12(7). https://doi.org/10.3390/W12071870
Wang XC, Fu G (n.d.). Edited by Water-wise cities and sustainable water systems.
Williams DS, Máñez Costa M, Sutherland C, Celliers L, Scheffran J (2019) Vulnerability of informal settlements in the context of rapid urbanization and climate change. Environ Urban 31(1):157–176. https://doi.org/10.1177/0956247818819694
Willner SN, Otto C, Levermann A (2018) Global economic response to river floods. Nat Clim Chang 8(7):594–598. https://doi.org/10.1038/s41558-018-0173-2
Winsemius HC, Aerts JCJH, Van Beek LPH, Bierkens MFP, Bouwman A, Jongman B, Kwadijk JCJ, Ligtvoet W, Lucas PL, Van Vuuren DP, Ward PJ (2015) Global drivers of future river flood risk. Nat Clim Chang 6(4):381–385. https://doi.org/10.1038/nclimate2893
Xu C, Rahman M, Haase D, Wu Y, Su M, Pauleit S (2020) Surface runoff in urban areas: The role of residential cover and urban growth form. J CleaProd 262. https://doi.org/10.1016/j.jclepro.2020.121421
Zang W, Liu S, Huang S, Li J, Fu Y, Sun Y, Zheng J (2019) Impact of urbanization on hydrological processes under different precipitation scenarios. Nat Hazards 99(3):1233–1257. https://doi.org/10.1007/s11069-018-3534-2
Zha Y, Gao J, Ni S (2003) Use of normalized difference built-up index in automatically mapping urban areas from TM imagery. Int J Remote Sens 24(3):583–594. https://doi.org/10.1080/01431160304987
Zhou Q, Leng G, Su J, Ren Y (2019) Comparison of urbanization and climate change impacts on urban flood volumes: Importance of urban planning and drainage adaptation. Sci Total Environ 658:24–33. https://doi.org/10.1016/j.scitotenv.2018.12.184

Sustainable Megacities and Spatial Planning for Urban Centres

Inclusive City Design for Future Megacity: Reclaiming Streets Through Place-Making in an Inner Urban Core of Surat, India

Jinal Boricha, Satyaki Sarkar, and Prashant Prasad

1 Introduction

Megacities have emerged because of urban transformation over time. All indications point to numerous Asian cities speeding towards becoming megacities or cities with ten million or more residents as we usher in the year 2021. Most recent cities with populations over 10 million are found in Asia and Africa. People have been drawn to cities by more excellent employment opportunities, a centralised market, higher wages, and more extraordinary personal riches; for these reasons, urbanisation is occasionally linked to economics. And these pull factors propelled city expansion for a very long time. High fertility rates, migration from rural areas, and the development of the city's borders are a few factors that have contributed to this increase. In the metropolis, where the prospect of food shortages, traffic jams, and inadequate educational facilities has become a stark reality, the population is exceedingly practically all support infrastructures. Megacities, previously regarded as the proudest successes of the modern age, now appear to be the biggest issue of the recent past.

India has also been impacted by the megacity's extraordinary expansion and equally dramatic collapse. In India, rapid urbanisation and population growth have led to increased transportation networks, which have strained road infrastructure and created parking problems in many megacities. The urban core of these cities experiences tremendous economic pressure due to high return value per unit space. These inner cores also share the increased intensity of use of all public places and tremendous financial stress leading to traffic and congestion problems. Inadequate infrastructure in such cores was devoid of the city of social interaction and

J. Boricha · S. Sarkar (✉) · P. Prasad
Department of Architecture & Planning, BIT Mesra, Ranchi, India
e-mail: satyakisarkar@bitmesra.ac.in

P. Prasad
e-mail: pprasad@bitmesra.ac.in

M. Firoz C et al. (eds.), *Urban Planning and Design for Megacities in the Global South*, Advances in 21st Century Human Settlements,
https://doi.org/10.1007/978-981-97-8370-0_5

activities, thereby calling for inclusive city design. Therefore, the focus of public place designs in inner city cores must shift to a people-centric approach, considering all stakeholders, including street users, from sidewalk vendors to wage-earning commuters.

India currently has five megacities with a population of over 10 million, but by 2030 there will be seven. With a whopping 9.6 million new residents, Delhi will continue to have the second-highest population in the world by 2030, the most of any megacity. Through its report, the UN has projected the people of Surat, India, to cross ten million by 2032, which will pave the way for the birth of the eighth megacity in India. Over the past few decades, residents' lifestyles have been severely changed by growing global urbanisation, haphazard development, and population expansion. The need for road transport has increased significantly due to rapid urbanisation and population growth. The transportation network is overloaded due to the rate at which motorised vehicles are being purchased, making the quality of service a significant concern. If there are more cars than parking spots, there will be an inferred right of way, and parking space saturation will be a problem. In contrast, streets only intended for cars and pedestrians have recently begun to succumb to various activities, encroachment, and spillover pollution. Sharing the street is only feasible through absorbing its publicness.

Local and national groups and organisations have attempted to regain control of their environment throughout the past few decades. The idea of "place-making," a committed attempt to strengthen the bond between individuals and their comfortable surroundings, accomplishes this in a collaborative and inspirational manner. Place-making is a methodology that aims to improve the quality of places and lives by enabling and motivating people to reimagine and transform public spaces collaboratively. It has been used since Jane Jacobs and William H. Whyte initially presented the idea of place-making in the 1960s. In India, streets are renowned for being centres of activity and life and for being traffic headaches. They are inextricably tied to place-making on some different levels. Place-making India showcases the nation's naturally comfortable public areas, the adaptability of its urban relationships, and the variety of professions represented in this self-managed organisation of place-makers. Due to the ongoing rebuilding of many of India's streets, Place-making India was established on 7 December 2019 in Pune, Maharashtra, with the help of Place-making X, Smart Citizens, and the International Society of City and Regional Planners. As part of this event, the Indian government introduced the "Streets for People Challenge" in 2020 to inspire more than 100 communities to reconsider their streets as safe, enjoyable, and healthy public spaces, which calls for a paradigm shift away from streets that are focused on automobiles and towards streets that are focused on people.

Using space syntax analysis, this section attempts to establish a methodology for intervention in the Indian streets. With an aim for enhancing public spaces to build places, focusing on converting the street into a public space and improving relationships between people and these places, this chapter makes a noble attempt to intervene in Indian streets with a study of Surat. When redeveloping the street as a public area, the concept of inclusive street design frequently needs to give respect to significant street stakeholders such as walkers, cyclists, sidewalk sellers, and wage auto commuters. Enhancing the street's position as a public space is required to transition from car-centric roads to people-centric streets. This prompts a set of pertinent questions for this study, including:

1. The needs and upcoming roles of reclaiming the street.
2. What are the requirements for pedestrians and cyclists on streets?
3. How happy are the local users with the facilities already in place?
4. What are the crucial factors to consider while creating the street design?
5. What kind of place-making and street reclamation techniques are likely to work in current cities?

This chapter, aiming to research inclusive design for the inner urban core of Surat, will examine the relationship between urban streets, node connections, and individuals using space syntax analysis. Further, it will determine how to promote social interaction and activities and prioritise public space that can help create more inclusive and sustainable communities and cities. Different segments of society can be integrated into public spaces by creating inclusive public spaces.

2 Literature Review

This section reviews the place-making, inclusive cities, and space syntax theory from basic concepts to broad applications. Specific examples of space syntax procedures are shown, including parameters, software implementations, and data analytics. Based on those understandings, the study is conducted to investigate how the CBD area of Surat city developed.

Place-making transforms a "space" into a "place" by giving guests a sense of direction. Researchers in surveying, geography, planning, and architecture have examined this dynamic process's various phases and implications, emphasised its range of dimensions, and identified its immediate and long-term impacts on a local and global scale. The study concludes that highlighting local features in urban environments is essential to place-making and urban renewal. Incorporating local cultural identity and place-making into modern urban development necessitates user participation in the planning and decision-making processes. In addition to enhancing a sense of place, sustainable place-making produces feasible social, economic, and environmental advancements. It is possible to create sustainable place-making solutions by balancing the competing interests of the community, the government, and developers (Dupre 2018).

Tore Omholt addresses techniques for inclusive place-making in metropolitan areas and offers a novel method for examining place-making structures and processes. Comparative analyses have been done on three different place-making situations, showing examples of networking and urban government strategies. Action research interventions are used to create in-person interactions (Omholt 2019). While attempting to remake itself through place-building, the American State of

Michigan applied new informational technologies, models, and online training. Picture Michigan: An informatics project shows the process of democratising data and incorporating it into methodology and tools for scenario planning. The place-making framework developed by the Michigan State University Land Policy Institute comprises an online and offline training programme for municipal planners and citizen planners (Beyea 2009). In recent years, cities like Shanghai and Shenzhen have seen the emergence of participatory urban regeneration practices and alternative development modalities. Urban renewal initiatives have integrated social goals like community development and capacity building. In this approach, new participatory governance models help place-making in urban regeneration by allowing social entrepreneurs, NGOs, and volunteer groups to work together to achieve these social goals while new public spaces are built and maintained (Yawei Chen 2020).

To make Indian SMART cities inclusive, using the concepts of e-Topia and a case study of Hubli-Dharwad City in Karnataka, N Sridharam identified areas of deprivation in the city and how they are left out of the development process by utilising three aspects of spatial inclusion: water supply, governance, and governance variables. Various stakeholders from different territorial levels can interchange and share spatial data more quickly and efficiently when a spatial data infrastructure is in place (Sridharan 2015). Three specialists, Sonali Anusree Patro, Sujaya Rathi, and Srimoyee Bhattacharya, looked at indicators for five significant urban sectors, including housing, that are used both domestically and abroad. The analysis aimed to determine whether the indicators used to measure the sustainability of the infrastructure and services generated were enough. It also established a framework for this analysis based on sustainable development ideas. Well-being, equity, efficiency, and foresight were the main metrics considered (Shrimoyee Bhattacharya 2016). Ar. Preeti Jaiswal and En. Suresh Hatkar identified the urban planning process, which must integrate land use and environmental planning with transportation planning, socioeconomic and financial planning with spatial planning, and so on, to be more responsive to the changing needs and demands of the populace. Moreover, an assessment was conducted on applying the Inclusive City policy framework to guarantee favourable social results concerning housing, neighbourhoods, economic growth, education, accessibility and mobility, shelter preservation, community services, and the provision of cultural significance. The main metrics considered were well-being, equity, efficiency, and foresight (Ar. Preeti Jaiswal 2017). A study conducted in Thailand examined how inclusiveness may be addressed in smart cities. The study unpacked the development of inclusivity in Smart City projects based on involving parties in the project from start to finish by using critical discourse analysis to describe the Smart City issued by the Thai government, particularly in Khon Kaen City municipal. The study considered three factors: thinking adjustment, method adjustment, and practice adjustment (Roengtam 2020). João Soares de Oliveira Neto and Sergio Takeo Kofuji provided a quick overview of the definitions and key components of Smart City projects, highlighting accessibility issues in urban areas and emphasising the significance of ensuring every city dweller has independence and autonomy (Kofuji 2016). In a study, Madhurima Waghmare and Shaleen Singhal looked at the general level of inclusion in smart cities and trends in several different areas in small, medium, and large cities in India. Three dimensions have been considered for the framework: social, spatial, and economic. Additionally, it created a multifaceted framework for monitoring and evaluating by reviewing and analysing 20 Indian smart cities (Singhal 2022). In a Brazilian city, a study was conducted to determine what people thought about their quality of life in smart cities and what criteria were most important in determining how satisfied they were. Curitiba, a city in southern Brazil that is regarded as one of the ten brightest cities in the world and is livable, green, and inclusive, was the subject of the study. Four primary QOL areas were discovered through interviews with 400 residents: community integration, material well-being, environmental well-being, and socio-structural relationships (Janaina Macke 2018). Green Infrastructure has been implemented in Southern and Eastern Africa to make cities more inclusive and sustainable. Based on a thorough literature review, the researcher concluded that expanding the definition of urban green infrastructure (GI) and connecting it to rights-based and governance conceptualisations could open up new avenues for the inclusive, sustainable, and green development of cities in Southern and Eastern Africa (Alexandra Titz 2019). Before releasing a comprehensive framework for evaluating inclusive growth, Xiong conceptualised urban inclusiveness in China. The state of the territory before Xiong'an's formation and its current development status, along with the management structure and policy measures, are examined to identify the barriers to inclusive progress. Variables such as land and housing, transportation, education and training, and ecological services were considered for the evaluation (Zhaowen Liu 2020). Mauro Romanelli determined cities' course to become inclusive, sustainable, and SMART urban communities. Using a qualitative, descriptive, and exploratory technique, the study analyses how cities reconsider their future growth as inclusive, intelligent, and sustainable communities (Romanelli 2022). The main metrics considered were well-being, equity, efficiency, and foresight.

To build lively, sustainable, and thriving communities, place-making and inclusivity are intertwined ideas. Place-making is the cooperative process of creating and modifying public areas to improve the lives of those who utilise them. Contrarily, inclusivity is the belief that every person should feel respected, accepted, and able to engage in and contribute to community life, regardless of their background or unique qualities. Examining the relationships between social cohesion, accessibility, cultural diversity, community involvement, and economic vibrancy can help one understand the relationship between inclusiveness and place-making. In a nutshell, place-making and inclusion are related because they both aim to create courteous, open spaces representing the community's identities and needs. The benefits of inclusive place-making include:

- Strengthening interpersonal relationships.
- Encouraging accessibility.
- Honouring cultural diversity.
- Involving the community.
- Enhancing social and economic well-being.

Reclaiming the streets is essential to inclusive city planning and place-making. It entails reusing or reconstructing streets with the needs of the community, bikers, and pedestrians rather than just automobile traffic. This method helps to make cities more inclusive, livable, and accessible. Street reclamation is one of the most effective tools for inclusive city planning and place-making. Cities may promote community and improve citizens' general living standards by putting people before cars, involving the community, and developing lively, accessible public areas. Numerous studies have examined how streets functioned in recent decades, primarily to determine the variables influencing how streets are used. Many factors influence how people use roads, as evidenced by a review of the studies in this area.

The following parameters (Fig. 1) were used, extracted from the various literature reviews. A brief flow chart is prepared for the surveys and their tools and techniques (Fig. 1). The analysis considers the respondents' location, travel parameters, vehicle ownership, and socioeconomic characteristics. Graphs and charts in a database created using Excel spreadsheets were used to assess this. Data from the part containing user perception metrics must be gathered to examine the aspects of the respondent's happiness. A Likert-type scale was used to collect these replies for every one of the factors. Space syntax analysis was used for Street Connectivity, Bike Ability, Walkability, and Accessibility. An AutoCAD-prepared database with DepthmapX performance characteristics were the analytical tool.

Space Syntax Theory

One specific set of socio-spatial theories has been chosen for this section to put it into practice: space syntax. Space syntax has made a name for itself as a respected tool for built environment research. According to Hillier in 2007, space syntax calls for considering four factors when doing urban research (1) A brief explanation of what constitutes a space, (2) A collection of methods for analysing urban structures

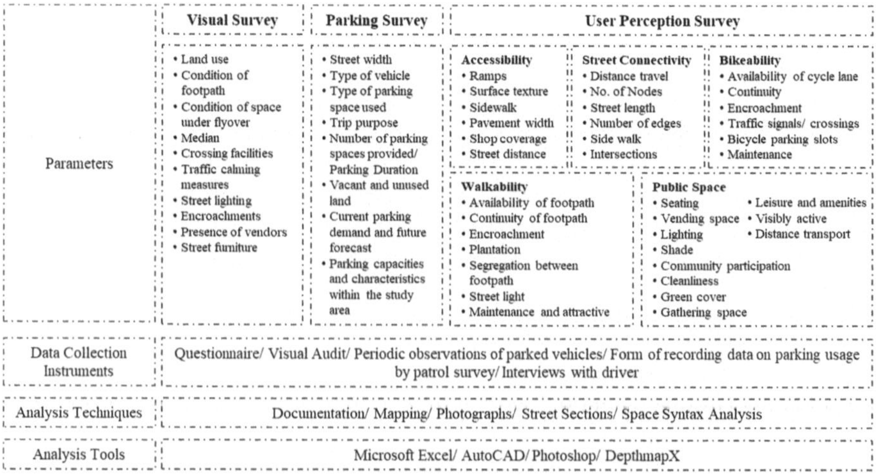

Fig. 1 Flow chart representation of tools & techniques for surveys conducted

is generated by the arrangement, clustering, and orientation of buildings and conceptualised as networks connecting space, (3) The research must consider the social environment by examining how space networks connect in a larger context. In other words, how and why do individuals use space? This is accomplished by examining various social factors, such as pedestrian mobility patterns, land usage, or social well-being. (4) The elements above and the urban network itself are shaped by theories on how an urban network relates to social, economic, and cognitive characteristics.

The urban network's two main components are the focus of space syntax. "To-movement" and "through-movement" are the terms used for them. Through-movement measures the likelihood of people travelling through groups of segments relative to all other segments in the network, whereas to-movement measures the accessibility potential of a specific segment. The social behaviours of a city's people, such as shopping and travel, are connected to the purposes of the city.

Space syntax can be applied at several dimensions and levels when studying the built environment, from local public spaces like town squares to large city-wide roadway networks.

In space syntax studies, the axial model is the primary representation model for urban street networks. The nodes in the model are the axial lines, and the edges are the connections between the axial lines. Then, using the justified graph, it is possible to determine the connectedness, depth, local integration, and global integration of each axial line. The most obvious parameter in a morphological examination is connectivity. It is described as the number of nodes in the justified graph directly connected to each node. A key idea in topological computation is depth, which refers to the shortest paths (or steps) between two nodes. The total minimum distance between a given node and every other node in the network is the global depth of that node. It will be recalled that the choice (betweenness) value describes how frequently a node may be used in trips from any space to any other space inside the city. Finding frequent usage of axial line segments compared to full axial line gives an in-depth picture. These are the former parameters in space syntax. While the next two parameters, synergy and embeddedness, look at the correlation or change rate between various radii to analyse the spatial organisation of the entire system.

Lei Wang et al.(2022) used deep learning and spatial syntactic methods to explore people's perceptions of municipal roadway planning. They employed space syntax and deep learning methods to link street accessibility and urban spatial perception. To automatically integrate the street view photographs, Python software was developed. This work established a human-centred artificial intelligence-based method for quantifying urban spatial quality based on a solid space syntactic theory (Lei Wang 2022). Using spatial syntactic analysis techniques, Girmay Berhle and Saif Haq discussed possible land uses and transportation choices in American cities. Four American cities—Boston, Massachusetts; Pittsburgh, Pennsylvania; Lubbock, Texas; and Salt Lake City, Utah—were the subjects of a case study examination to ascertain whether street layout influences the mode of transportation choice (Girmay Berhie 2017). Wensheng Zheng et al. (2021), have looked at how to comprehend the metropolitan agglomeration's city transportation system through enhanced space syntactic analysis tools. This study employed space syntax to look at the physical integration of cities (Wensheng Zheng 2021). An auditing tool designed explicitly for remote examination of built environment micro-scale characteristics was developed to evaluate the validity of Virtual-STEPS. Madeleine Steinmetz-Wood looked at the virtual and field audits agreement and the inter-rater agreement for the Virtual-STEPS tool (Madeleine Steinmetz-Wood 2019).

Space syntax analysis uses a variety of indicators to quantify and comprehend spatial layouts. Table 1 displays the indicators used in various space syntax analysis tool literature. One of the most well-known software in modern space syntax research is Depthmap, which Turner A (2004) designed. It is a single software platform that allows networks to use the previously discussed space syntax techniques. It functions on various scales, from individual buildings to entire cities or states. To perform space syntax analysis QGIS plug-in was used as it has tools for managing and analysing urban data, including information on land use, frontages, entrances, pedestrian traffic, road centrelines, and service areas. DepthmapX was used for this research analysis. The programme's objectives are to map open space elements at every scale, connect them via connections like interconnectivity or inter-visibility, and then do graph analysis on the resultant network. The comments offered by Depthmap include segment analysis, axial map production and analysis, and visibility analysis. Since visibility (isovist) analysis and automated production of axial lines are primarily used in building or small urban dimensions, this study would entail axial line and segment analysis.

Analysing space syntax is the first step of this study. This is accomplished by creating a graph representation of the studied spaces. These maps will be made using a variety of software programmes. The Axial, Integration, Choice, and Visibility maps are the four main ways the area is depicted. The four maps represent the outcomes of analysis techniques for assessing and forecasting activities.

Table 1 The index used in space syntax analysis

Index used in various literatures Space syntax analysis tool		Research	Reason for applicability	Reference
Axial map Segment map Convex map	Integration Connectivity Visibility Choice	Visibility Graph Axial Graph Integration Graph Choice Graph	To analyse the spatial layout of streets and human activity patterns in urban areas	Dipanjan Nag (2022), Lei Wang (2022), Girmay Berhie (2017, December), Wensheng Zheng (2021)

3 Methodology

This chapter includes the overall methodology of the study. The book chapter title was initiated and motivated by the context of planning to enhance the quality of the place with a place-making approach through a space syntax analysis tool. The aim of the study was derived from the topic. It covers some essential aspects of place-making. The strategies for gathering the necessary data on a specific context and region are presented in the data collection for inclusive development. The methodology is focused on observing, hearing from, and questioning those who use a specific area to live, work, and play to comprehend their needs and goals for that space and their community. The literature review provides information about the global and Indian scenario of research and interventions related to place-making, inclusive cities, and space syntax analysis tools and techniques. An extensive literature review has been conducted before the street reclamation programme. The study aimed to ascertain the key factors and optimal approaches to prosperous street reclamation initiatives. Significant concepts that can direct the successful redesign and repurposing of our streets have been retrieved through a thorough review of published academic works, case studies, and publications on urban planning. The set of parameters that have been defined takes into account many factors, such as the safety of pedestrians, accessibility, community involvement, and the development of lively public areas. These factors have been carefully chosen to ensure they harmoniously mesh with our streets' unique characteristics and needs. The criteria were determined by considering the distinct characteristics of the roads. Parameters from the literature studied promoted the identification of final parameters included in the questionnaire preparation.

This data collection section has been divided into three survey categories: the Visual Survey, Parking Survey, and User Perception Survey, as well as the tools and techniques required for these Surveys. The tools and techniques are selected based on how the data is being processed for the study (Fig. 2).

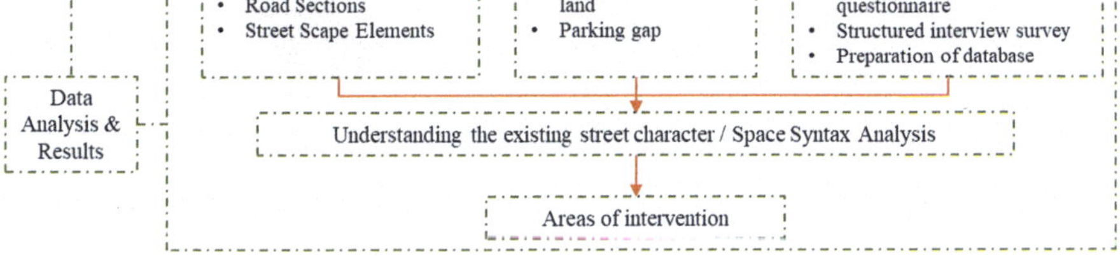

Fig. 2 Methodology flow diagram

4 Results and Discussions

The inner city of Surat, Gujarat, has been chosen as the research study area. In the southern part of Gujarat, India, the coastal city of Surat, governed by the Municipal Corporation of Surat, lies. Surat is a central industrial and commercial hub in Gujarat and the second most populous city in the state. It is also the eighth most populated city in India. The city has a strong and thriving economy, making it a popular destination for job seekers nationwide. It faces significant challenges because it is still growing. The city's heart is home to formal and informal market areas and a sizable neighbourhood population.

A more comprehensive map of Surat's location can be found in Fig. 3. With a close-up view of Surat, Fig. 4 shows a thorough city map that shows the boundaries of the various zones in the city. Each corresponds to distinct functional or administrative areas, and these zones are distinguished on the map by colour coding. This map provides a thorough understanding of the spatial distribution within Surat, which helps visualise the organisation and planning of the city.

This section reviews Surat City's growth, particularly from a spatial perspective. It is a conventionally developed commercial town with distinctively organic street networks. Therefore, the spatial organisation is more organic and spontaneously driven by regional conditions, unlike capital cities with ideological grids. Particularly with the advent of the twenty-first century, the city is growing and integrating with nearby towns and villages. Contemporary roadways have been inserted into the old

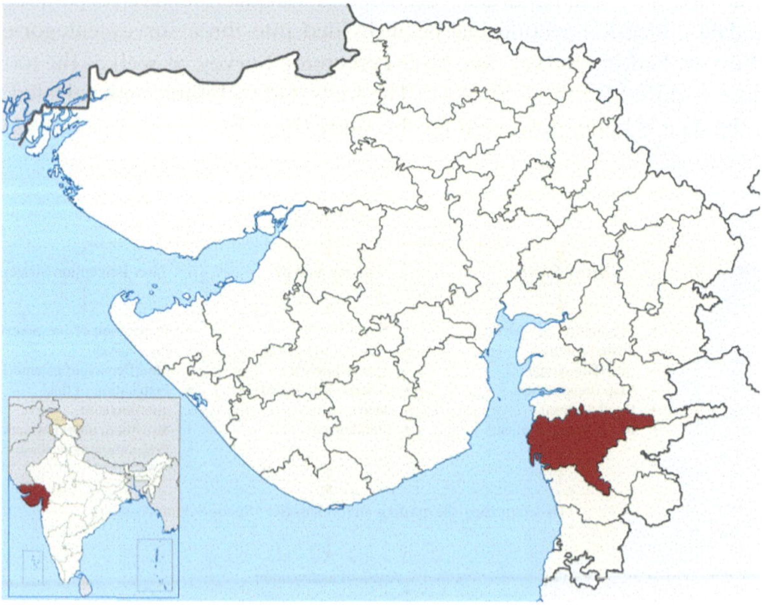

Fig. 3 Location map (*Image Source* https://en.wikipedia.org/wiki/Surat_district)

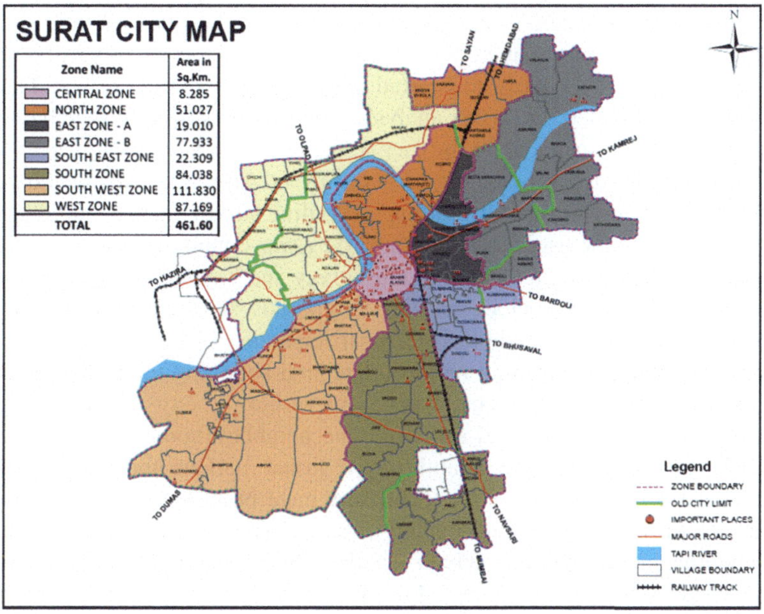

Fig. 4 Surat city map with zone boundary (*Image Source* Surat Municipal Corporation)

areas, creating a new grid structure. Post review, this section uses syntactic space techniques to analyse Surat's street network. The CBD region and its interactions will be the focus of the micro-scale analysis and interpretations. Through place-making, this study aims to promote the understanding of the socio-spatial context of the CBD core as well as the reclamation of the street. The region chosen included a 12-km stretch of pedestrian and semi-pedestrian streets, a flyover, parts of the under-flyover, street intersections, and urban squares that were particularly representative of the local urban geography. Surveys, observations, and questionnaires were distributed to respondents from various societal segments who were primarily local stakeholders at multiple times, including weekdays and weekends, during the day, and in the late afternoon. Urbanists, architects, and government officials were among the specialists and professionals who were brought into the realm of the responders to clarify their answers.

4.1 Street Characteristics of CBD

The research's data collection was separated into sections based on the needs of the study. The literature review and a list of the different parameters taken into consideration in the studies carried out are then included in the papers. A simple procedure for gathering data is used following the parameters found, further narrowing the list of appropriate parameters for the primary survey. The questionnaire preparation using the first set of parameters was considered for the pilot survey. Following the survey, the collected data is transferred onto a spreadsheet to create a database, then organised and sorted for further analysis. The survey was performed based on a visual audit, and the instrument for the user perception survey was designed to record responses regarding various parameters listed (Fig. 1). After preparing the final questionnaire, the survey was done by rolling out online survey forms and conducting intercept surveys by the users on site. The survey was conducted online and offline to overcome the shortcomings of the online survey and get maximum responses for respondents with different socioeconomic backgrounds. A 95% confidence level was used for a user perception survey with 270 participants.

Small-scale shops, substantial retail outlets, and shopping malls are all present in plenty in the area. It draws customers from within the city and outside thanks to its wide variety of goods and services. The neighbourhood is well-known for its bustling street life and the availability of regional markets and bazaars. Central Banks, financial institutions, and stockbrokers have branches there. It centres on financial dealings, investing, and related services. The area's economic health is boosted by financial institutions, which also draw businesses and people looking for financial services. It is well connected to other city areas through road and public transportation networks and is close to significant transportation hubs. Accessibility helps people and commodities flow into and out of the CBD.

This mixed-use development makes the region livelier and more vibrant (Fig. 5). Theatres, galleries, restaurants, and cafes enhance the CBD's cultural landscape and draw locals and tourists.

Figure 6 and 7 show the characteristics of the edge. Its buildings, activities, and public relations strategies characterise the entire edge. The neighbourhood's edge junctions facilitate movement, connect with adjacent areas, and establish social hubs that unite people, culture, and space. Regarding current conditions, traffic, and infrastructure issues, many action areas within the study region are the same, but the actions

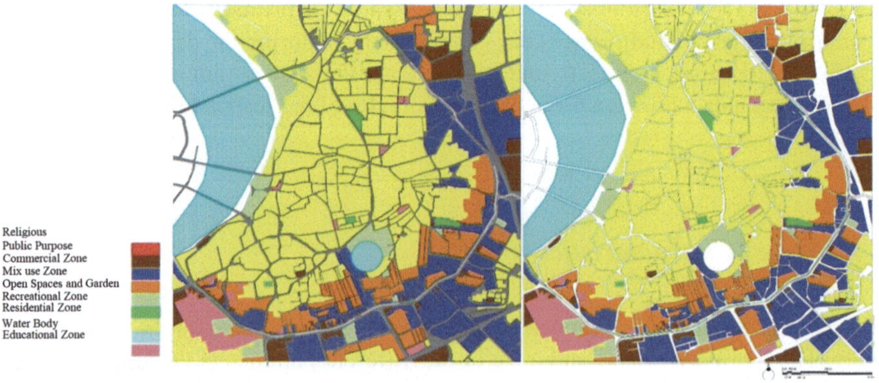

Fig. 5 Land use map—ground floor & first floor

Fig. 6 Unsegregated lane and traffic congestion during the daytime

Fig. 7 Encroach footpath (*Image Source* https://timesofindia.indiatimes.com/city/surat/cops-issue-notices-to-textile-markets-over-parking-forvisitors/articleshow/70299615.cms)

for other areas differ. The study areas are hubs of activity due to the influx of vehicles, the encroachment of auto rental locations, and the presence of food vendor development.

The flyover at this intersection adds extra traffic segregation at two levels. The stretch's parking gap is expected to hold about 1285 automobiles according to the current situation (Table 2).

To determine the quality of parking services, a poll of 330 respondents was conducted with a 95% confidence level (Table 3). The area's unoccupied and unutilised land was discovered, and it may be used to minimise on-street parking in response to anticipated future parking demand. Table 3 displays the parking parameters selected from several literature reviews and rated by respondents based on their level of satisfaction. According to data, surveyed citizens are least interested in travelling via public transport like buses and auto rickshaws. Surveyed citizens were disappointed by the area's terrible parking options.

Table 4 displays the walkability parameters selected from several literature reviews and rated by respondents based on their level of satisfaction. Citizens from different age groups were surveyed; some were students, and some were employed

Table 2 Current parking demand

Land use type	Area (Sq. Mt.)	Permissible ECS as per URDPFI	Required parking (ECS)	Existing parking (ECS)	Total required parking (ECS)
Commercial	42,648	2 ECS per 100 sq. Mt.	852	918	2203-918 1285
Mix use	39,784	2 ECS per 100 sq. Mt..	795		
Public–Semi-public	27,837	2 ECS per 100 sq. Mt	556		

Table 3 Response to parking survey

Parking	1-Excellent	2-Good	3-Average	4-Poor	5-Very Poor
Parking from your destination	–	4 - 4%	14 - 14%	15 - 15%	67 - 67%
Availability of parking	6 - 6%	7 - 7%	17 - 17%	15 - 15%	55 - 55%
Range of parking facilities in the area?	3 - 3%	6 - 6%	13 - 13%	17 - 17%	61 - 61%
Spaces under the flyover for parking	2 - 2%	9 - 9%	21 - 21%	15 - 15%	53 - 53%

Note The first number in the column displays the absolute value, and the second column shows the percentage of those values

and were very dissatisfied with the overall pedestrian facilities in the area. 4% and 5% are satisfied with the availability of footpaths and street benches.

Table 5 displays the bike-ability parameters selected from several literature reviews and rated by respondents based on their level of satisfaction. Surveyed citizens from different age groups are very dissatisfied with the overall facilities for Bicycles in the area. People of all ages who were surveyed expressed extreme dissatisfaction with the neighbourhood's general pedestrian-level amenities. A similar poll on bicycle abilities reveals extreme unhappiness with bicycle infrastructure.

A similar poll clarified comments on open space attributes, green space, street furniture, hardware and sculpture, and urban art (Table 6). More green spaces, accessible, clean entertainment areas, room for walking and bicycling, safety from traffic, and area beautification were also suggested by respondents. As a result, full access will be provided to all places, including those currently vacant, unused, mistreated, or abandoned. To improve the local identity and direct visitors to the region, these accessible places must be supplemented with seats and relaxation areas, flexibility to revitalise life after dusk, and prominence and dominance of landscaping, signs, lighting, and shading.

Table 4 Response to the walkability survey

Parameters (Walkability)	1-Very satisfied	2-Somewhat satisfied	3-Neutral	4-Somewhat dissatisfied	5-Very dissatisfied
Availability of footpath	–	15 - 4%	51 - 14%	55 - 15%	248 - 67%
Continuity of footpath	–	–	–	77 - 21%	292 - 79%
Accessibility of footpath	–	–	–	33 - 9%	336 - 91%
Encroachment-free footpath	–	–	15 - 4%	100 - 27%	255 - 69%
Hygiene and cleanliness	–	–	18 - 5%	122 - 33%	233 - 63%
Green cover/shades	–	–	–	33 - 9%	336 - 91%
Signage	–	–	4 - 1%	22 - 6%	344 - 93%
Street lighting	–	–	7 - 2%	40 - 11%	322 - 87%
Security and surveillance	–	–	11 - 3%	59 - 16%	300 - 81%
The attractiveness of the route	–	–	–	26 - 7%	344 - 93%
Street benches and breakout area	–	18 - 5%	18 - 5%	48 - 13%	284 - 77%

Note The first number in the column displays the absolute value, and the second column shows the percentage of those values

4.2 Representation of Space

This study uses a space syntax analysis based on four significant indicators—the visibility, axial, choice, and integration graphs—found through a literature review. Together, these indicators provide information about decision points, spatial hierarchy, visual connections, and general accessibility in the built environment, creating a thorough framework for the research.

The Visibility Graph analysis is used to examine the connections and connections between network nodes. The concept of visibility between nodes is particularly pertinent when analysing spatial networks, such as road, social, or communication networks. In a visibility graph analysis, nodes stand in for objects or places of interest, and edges between nodes are drawn based on their visibility. If there is an unobstructed, direct line of sight between two nodes, then they are said to be visible to one another.

The project's axis has the longest line of visibility. The precinct's central zones, defined by the building blocks, are where the regions with moderately high values are located. Figure 8 displays a representation of the area's visibility indices. The outputs display link values in a colour-coded schematic from red (high, warm colour) to blue

Table 5 Response to the bike ability survey

Parameters (Bike ability)	1-Very satisfied	2-Somewhat satisfied	3-Neutral	4-Somewhat dissatisfied	5-Very dissatisfied
Availability of cycle track lanes	–	–	7 - 2%	40 - 11%	322 - 87%
Continuity of cycle track lanes	–	–	–	33 - 9%	336 - 91%
Accessibility of cycle track lanes	–	–	4 - 1%	22 - 6%	344 - 93%
Encroachment-free cycle track	–	–	11 - 3%	59 - 16%	300 - 81%
Hygiene and cleanliness	–	18 - 5%	18 - 5%	48 - 13%	284 - 77%
Green cover/shades	–	–	–	77 - 21%	292 - 79%
Signage	–	–	37 - 10%	62 - 17%	307 - 83%
Street lighting	–	15 - 4%	51 - 14%	55 - 15%	248 - 67%
Security and surveillance	–	–	18 - 5%	122 –33%	233 - 63%
The attractiveness of the route	–	–	15 - 4%	100 - 27%	255 - 69%
Cycle parking and breakout area	–	–	–	26 - 7%	344 - 93%

Note The first number in the column displays the absolute value, and the second column shows the percentage of those values

Table 6 Response to park and open spaces

Parameters	1-Excellent	2-Good	3-Average	4-Poor	5-Very Poor
Overall, how satisfied or dissatisfied are you with the parks and open spaces in the area?	–	15 - 4%	51 - 14%	55 - 15%	248 - 67%
How easily can you access and get around the park or open space?	22 - 6%	26 - 7%	63 - 17%	55 - 15%	203 - 55%
Range of facilities in the area?	11 - 3%	22 - 6%	48 - 13%	63 - 17%	225 - 61%
Are the spaces under the flyover clean and well-maintained?	7 - 2%	33 - 9%	77 - 21%	55 - 15%	196 - 53%

Note The first number in the column displays the absolute value, and the second column shows the percentage of those values

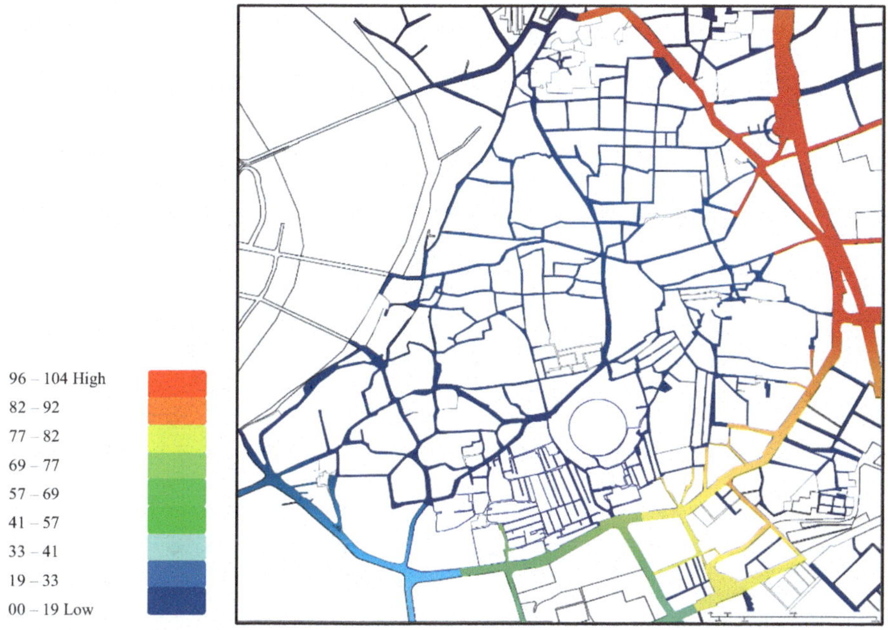

Fig. 8 Visibility graph

(low, cold colour). The larger (red/orange) values were seen in the main thoroughfares with substantial pedestrian traffic and commercial activity.

Streets and other spatial segments are shown as lines in the Axial Graph, and their crossings are shown as nodes. The lengths of the lines typically correspond to the sizes of the streets or segments. The area's axial parameter is shown in Fig. 9. An area's integration value and connectedness to other lines increase with the length of its axial line and vice versa. In contrast to the back street, which requires more direction changes to reach all streets, the main street may reach all streets with a minimal total number of direction changes.

The Integration Graph is an essential measurement of the axial graph. The integration value represents the depth of a segment and all segments that are topologically closer to a cul-de-sac together. It assesses how well a section is integrated and accessible into the more extensive network. While segments with lower integration values are frequently remote or more complicated to access, segments with greater integration values are more central and well connected.

The red and orange axial lines show the most highly integrated streets in Fig. 10. The ring road is where the streets and roadways are the most integrated. The different local shopping streets are identified in this research by yellow and orange lines.

A Choice Graph comprises nodes and edges, where nodes stand for places where people can choose between various paths, and edges stand for the connections or ties that connect these choice points. To comprehend the patterns of movement and decision-making within the geographical network, analysis of the choice graph

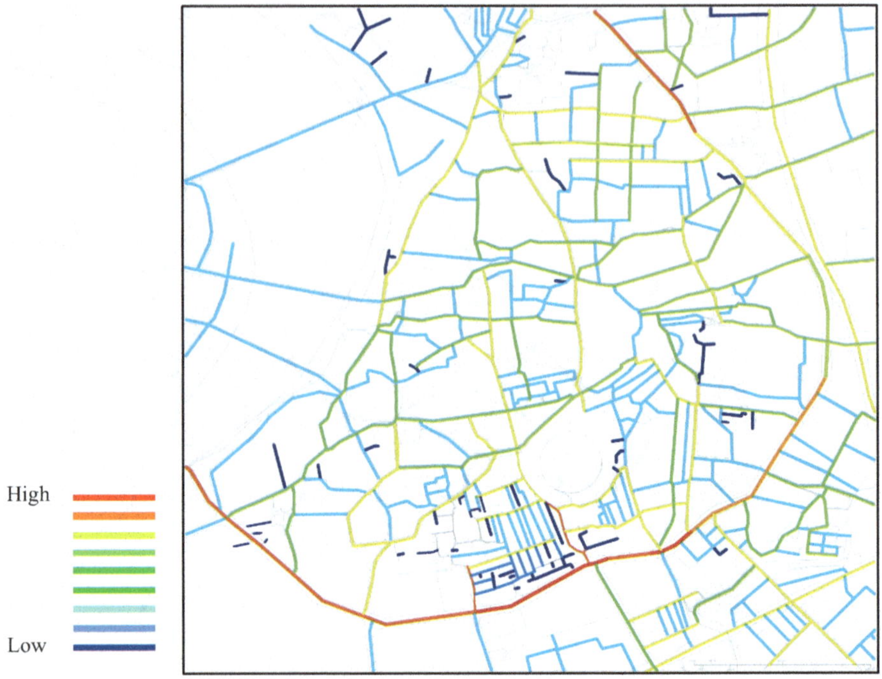

Fig. 9 Axial graph

entails a variety of measurements and metrics. Analyses frequently utilised include Route Choice Analysis, Accessibility Analysis, and Decision Points Analysis.

Through choice graph analysis (Fig. 11), it is possible to enhance the structure of a street network for several types of accessibility, such as motorist traffic, bicycle traffic, and pedestrian traffic, by looking at the network's connectivity. The blue-coloured highlight shows the walkable distance near the highly desired destination, and then it gradually goes down to the light colour for vehicular movement. A radius between 400 and 600 m was used for pedestrians and bicyclists, and a radius between 1200 and 200 m was used for motorists.

Space syntactic analysis measurements and survey responses were used to identify intervention areas. Three intersections were chosen from the periphery. The first parallels the flyover and runs from Athwagate to MajuraGate Junction. The intersection beneath the flyover, Sub-jail Junction, is the second, and the intersection where the flyover begins and ends, Mandarwaja Junction, is the third.

The space syntax analysis in this study was consistent with what was predicted by the theory. From the space syntax analysis maps, the visibility graph shows how one of the junctions, called Athwagate to Majuragate junction (Fig. 12[1]), has the lowest visibility in the entire stretch. In contrast, the axial graph shows that one of the junctions, Sub-jail to Udhna Darwaja Junction (Fig. 12[2]), has the highest street connection network. The integration graph shows another junction of the stretch: the

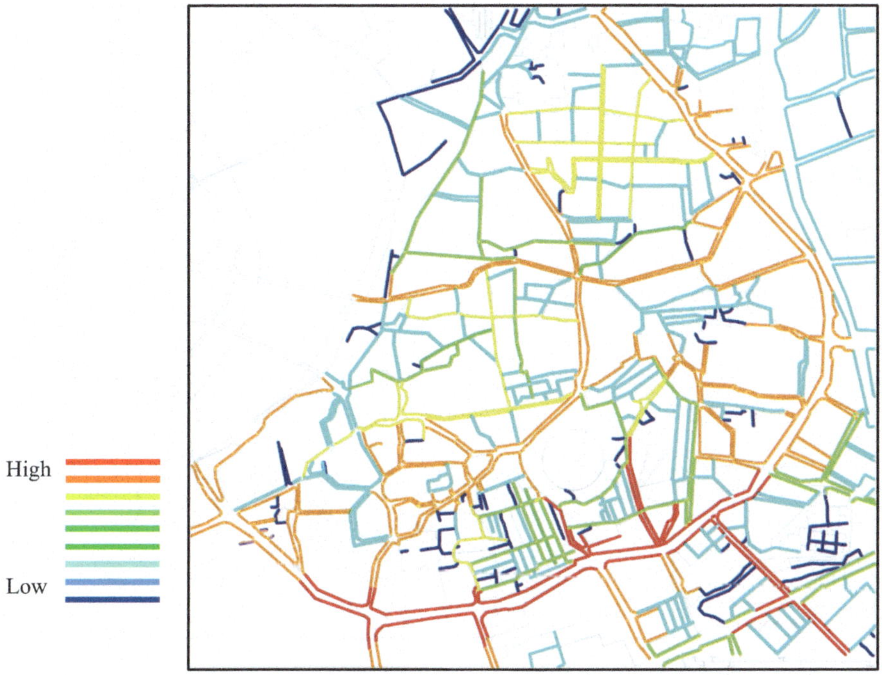

Fig. 10 Integration graph

Ring Road to Jay Prakash Narayan Marg (Fig. 12[3]), which is marked as a highly desired destination because of various activities, i.e., commercial and educational activities. The choice graph shows how accessible the entire stretch is on foot or a bicycle. In the whole stretch, there is no continuity of the footpath, the surface of the footpath is unmaintained, and no continuation in the footpath and an unhygienic and unclean experience is found along the route. The dry dump, vendors, hawkers, and haphazardly parked vehicles obstruct the footpath with no proper segregation of the carriageway.

The stretch has no bicycle lane, unmaintained public bicycle service, or bicycle stand. Citizens use the carriageway for bicycle rides. The entire stretch is very active during the different times of the day because of the variety of area usage. In the area, poor street signages and markings lead to commotion. High business activity leading to colossal traffic footfall creates an urban heat island effect. Haphazard plantations, low waste management, and areas below flyover turning into dump yards create street pollution. There is a limited amount of open space available. The area highlights the conflict between commercial and public uses of parks, plazas, and other shared spaces. The lack of interactive spaces and gathering points in the area diminishes the stakeholders' sense of place. Spillover pollution is caused by hawkers on the road and footpaths, causing congestion. The parking potential of the interior streets connecting to Ring Road must be utilised. More parking spaces are required as the

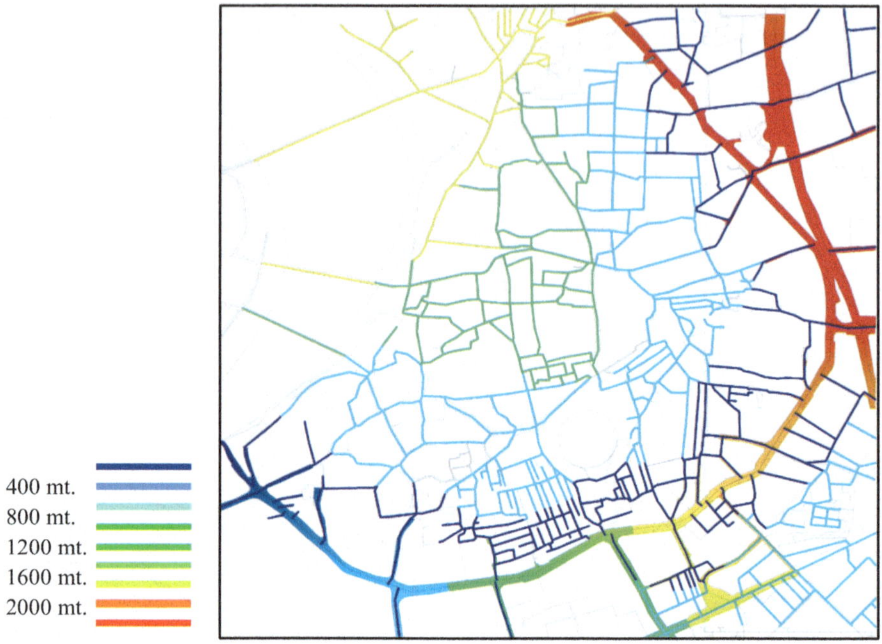

Fig. 11 Choice graph

Fig. 12 Identified action area

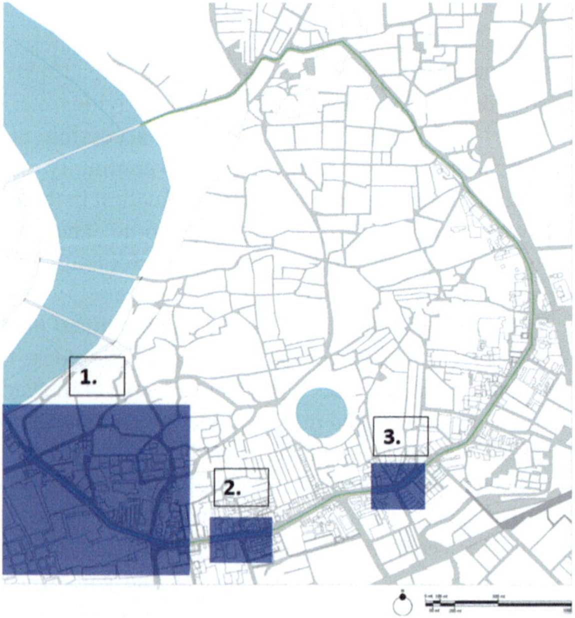

existing parking space is insufficient for the present load. Due to various commercial activities, the area needs enough visibility through adequate lighting to create a workable atmosphere. Impregnated with problems, the CBD area of Surat stands juxtaposed between revival and obsoletion. Hence, intervention in every aspect of place-making through street reclamation is of ardent necessity.

4.3 Potential Planning Guidelines and Management Strategies for "Inclusive Street Design Through Place-Making"

Based on the inferences from the space syntax analysis, the following planning guidelines are suggested:

Visibility: High-contrast colours, well-lit areas, and clear sightlines improve visibility. The visually handicapped benefit from tactile paving, while safety is increased with designated crosswalks and accessible transit stops. Pedestrian areas and universal design ideas encourage diversity. Sustained effectiveness requires frequent maintenance and community involvement.

Linkages: Create multi-modal connectivity, prioritise safe crossings, and incorporate accessible routes to improve city links. For diverse mobility, use green corridors and "complete streets" ideas. Universal design and well-considered placement of street furniture contribute to accessibility. Create an inclusive urban environment with public art, clear signs, safety measures, innovative navigation technology, and community engagement for local insights.

Accessibility: Designing inclusive streets: Make accessibility a top priority by implementing tactile elements, universal signage, and apparent pathways. Make sure parking, buildings, and public transportation are accessible. For continued inclusivity, include community feedback and intelligent technology.

Nodes & Edges: Multipurpose nodes with universal accessibility that include green areas, benches, and shelters should be the critical objective of inclusive city design. Efficient navigation, aesthetically pleasing elements, and public art contribute to the overall experience.

In addition to creating inclusive streets, the planning criteria listed below should be incorporated into the city's overall inclusion.

1. Creating detailed plans for traffic management to reduce congestion and enhance traffic flow in CBD locations.
2. Expanding parking infrastructure, such as multi-level parking structures and effective parking management systems.
3. Improving pedestrian infrastructure, including wider sidewalks, designated pedestrian areas, and secure crossings.
4. Putting into practice ways to reduce noise and air pollution, like encouraging the use of electric vehicles, green building techniques, and noise reduction techniques.

5. The development and improvement of green spaces within the CBD area to enhance environmental quality and offer leisure possibilities for locals and workers.
6. Infrastructure needs to be improved to keep up with demand, including utilities, trash disposal, and water supply.
7. Promoting public transportation, biking, walking, and other sustainable modes of transport as practical substitutes for private vehicles.
8. Collaborating with all interested parties to create and execute solutions that address the unique requirements and difficulties of the CBD region, including local people, companies, and community organisations.

4.4 Recommendations

By prioritising people's needs, encouraging social interaction, and ensuring fair access and pleasure for all community members, urban areas are designed and transformed to create inclusive streets.

For reclaiming the streets of the CBD of Surat, some crucial concepts and strategies for creating inclusive streets using a place-making strategy are mentioned in Fig. 13.

Community involvement: Including the stakeholders in the planning and design processes paves the way for participatory planning, which is the need for the hour. This may be further reinforced through consultation with locals, companies, advocacy groups, and community organisations. This ensures that the community's needs, goals, and cultural identities are reflected in the street design.

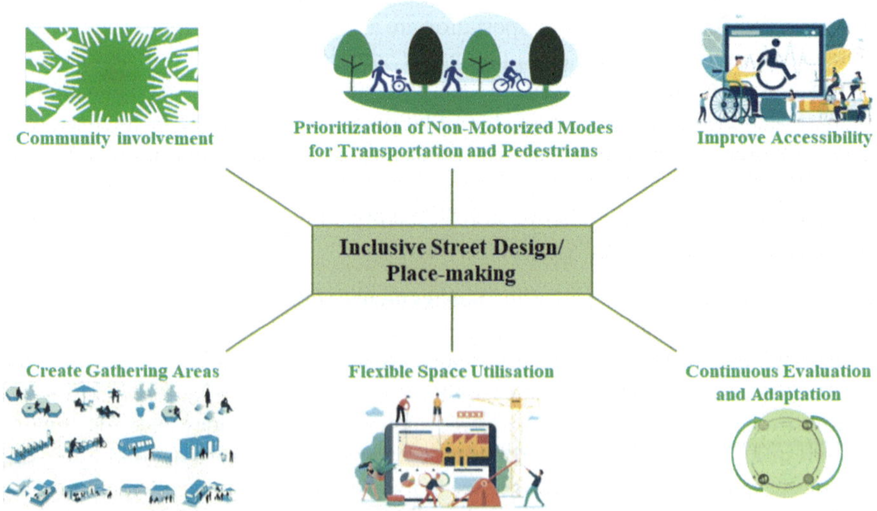

Fig. 13 Inclusive street design elements/place-making

Prioritisation of Non-motorised Modes for Transportation and Pedestrians: The construction of secure, cosy, and easily accessible walking environments puts pedestrians at the centre of street design. Wide sidewalks, crosswalks for pedestrians, and amenities like seats, awnings, and public art enhance the area's usability. Including dedicated lanes and bike parking facilities to accommodate and promote non-motorised means of transportation, such as cycling and micro-mobility choices, will further streamline the modal choice for the CBD of Surat.

Improve Accessibility: Everyone can use the streets regardless of age, ability, or mobility. The inclusion of curb cuts, ramps, tactile paving, and strategically located crosswalks in the identified CBD area of Surat would make it convenient for individuals with impairments. Further, the area needs to be facilitated with design interventions ensuring clear signage, appropriate lighting, and good acoustics for people with sensory impairments.

Create Gathering Areas: Surat CBD is deficient in gathering spaces. By including gathering areas along the roadway through the relocation of some less preferred activities, road pockets can be reclaimed for social interaction, creating a sense of community. The spillover pollution occurring below flyovers in the CBD area needs to be organised, and new park-o-mat created public plazas, rest places, mini-parks, and community gardens. These reclaimed streets may be aesthetically improved by introducing public art displays and community notice boards.

Flexible Space Utilisation: Reclaimed streets can support various activities and events. Through transit-oriented and flexible street designs, traffic may be closed for special events like farmers' markets, festivals, or street fairs. This will encourage the participation of the community in the designed public space.

Continuous Evaluation and Adaptation: A cyclic evaluation matrix may be designed to analyse the success of the street design and how it affects inclusivity regularly. Through community feedback, appropriate changes may be incorporated in phases to address any new problems and prospects.

Based on analyses and proposals, the CBD of Surat is destined to have reclaimed streets that are lively, welcoming, and attentive to the needs of its citizens by using a place-making strategy that prioritises inclusivity. This strategy promotes a sense of pride, ownership, and social cohesiveness, ultimately making streets more liveable, fair, and entertaining for everyone.

5 Conclusion

In this chapter, we have comprehensively reviewed the existing protocols, auditing tools, and frameworks, examining their approaches, target audiences, and scope of functions. A standout example of the impact of place-making can be observed in the Surat Central Business District (CBD), renowned for its bustling street life, soaring high-rise structures, and vibrant mix of business offices and retail outlets. This dynamic setting attracts numerous visitors, contributing to the city's rich cultural diversity. Our literature studies have identified the crucial elements that empower place-making initiatives to reclaim the streets as public spaces for their users. The policies and proposals in this chapter are set to pave the way for inclusive street design through street reclamation and effective place-making interventions that will benefit all users. With the implementation of these forward-looking strategies, the future streets of Surat are poised to showcase a more thoughtfully managed, aesthetically pleasing, and conducive urban environment, fostering growth and interaction among stakeholders. It is envisaged that the city's streets will not merely be helpful passages but will evolve into vibrant and welcoming communal spaces. Surat can enhance its unique identity by embracing place-making while accommodating its inhabitants' diverse needs and aspirations.

The success of place-making initiatives lies in their inclusivity, drawing on the collective wisdom and perspectives of the community. Active engagement with residents, business owners, urban planners, and policymakers will be critical in creating spaces that reflect the city's character while providing opportunities for shared experiences and interactions. Additionally, ensuring accessibility for all individuals, including those with disabilities, will be instrumental in fostering an environment that prioritises equity and social cohesion.

However, our study analysis has limitations: Based on current and projected demand and the development scope, an Action Area was chosen for the study. The data's availability and correctness may impact the conclusions of the spatial analysis. It overlooks non-linear paths, shifting street widths, and irregular street shapes that may impact how people move through and interact with space by failing to monitor and measure human behaviour patterns. Technological advancements can help to increase the availability and accuracy of data related to urban street systems. Any space syntax option does not consider the changes in human movement patterns, usage, and connections that may occur in various time zones. It's difficult to concentrate on long-term effects. The study area abides by Surat's existing Zonal Plan and Development Plan.

In conclusion, the utilisation of place-making holds immense promise in addressing the myriad challenges megacities face. Place-making for streets and inclusive city design are related because they both aim to create urban environments that are welcoming, engaging, and focused on the community. Applying inclusive city design to streets means that public places are developed with a focus on universal accessibility, considering a range of requirements, ages, and abilities. Place-making is a complementary approach that centres on converting streets into lively, sociable spaces that promote a feeling of community. When streets are designed as dynamic and inclusive with amenities like seating, accessible walkways, and cultural aspects promoting social interaction, they become more than just thoroughfares and interconnected. To successfully integrate streets as essential elements of a city where everyone is accepted and able to contribute to the urban fabric, inclusive city planning, and place-making must work together. The pursuit of place-making offers a transformative path for megacities like Surat to navigate the complexities of urbanisation. Through collaborative efforts, thoughtful planning, and an unwavering commitment to inclusivity, cities can create spaces that celebrate their diversity, promote sustainable practices, and foster a sense of community and belonging among their inhabitants. By embracing place-making and implementing the proposed strategies, Surat can aspire to become a thriving urban ecosystem that exemplifies harmonious coexistence, where the streets become more than mere transportation arteries but flourishing hubs of human interaction and expression. Through collective efforts and a commitment to place-making, we can build cities that tackle their unique challenges and thrive and evolve into vibrant, inclusive, and harmonious spaces for all.

Acknowledgements We want to thank the Department of Architecture and Planning, Birla Institute of Technology, Mesra, for providing us with various facilities needed to complete the work. We would also like to thank Surat Municipal Corporation for their help in collecting data. Finally, we also acknowledge that AICTE MODROB facilitated 3D City Engine for facilitating the project.

References

Alexandra Titz SS (2019). Pathways for sustainable and inclusive cities in southern and eastern Africa through urban green infrastructure? Sustainability—MDPI, 27.

Ar. Bhaghyalaxmi MAS (2014) Urban streets and nodes: a case of Bangalore. Int J Sci & Eng Res, 8.

Ar. Preeti Jaiswal ES (2017) Inclusive planning: smart solution for city design. Int J Emerg Technol 4.

Ashish Pandey AG (2020). Assessment of on-street parking issues in CBD. Int Res J Eng Technol (IRJET) 5.

Ayse Ozbil TG (2019). Comparative associations of street network design, streetscape attributes and land-use characteristics on pedestrian flows in peripheral neighbourhoods. Int J Environ Res Public Health 23.

Beyea W (2009) Place-making through participatory planning. Handbook of Research on Urban Informatics: The Practice and Promise of the Real-Time City, 13.

Bhopal Smart City LD (2017) Smart city Bhopal—Placemaking project guidelines. NIUA,Bhopal

Das D (2016) Non-motorised transportation for revitalising the city centeres of South Africa. National Research Foundation, 15.

Debasish Das MA (2018). Level of service for on-street parking. KSCE J Civ Eng 11.

DepthmapX Development Team, (2017) The Bartlett school of architecture. Retrieved from DepthmapX. https://www.ucl.ac.uk/bartlett/architecture/research/space-syntax/depthmapx#:~:text=depthmapX%20is%20an%20open%2Dsource,source%20and%20available%20as%20depthmapX

Dipanjan Nag JS (2022) Measuring connectivity of pedestrian street networks in the built environment for walking: a space-syntax approach. Transportation in Developing Economies 19.

Dupre K (2018) Trends and gaps in place-making in the context of urban development and tourism. J Place Manag Dev, 20.

Eng. Soha AbuBakr DM (2021). Public spaces under flyovers: Qualitative data analysis of users' interests in Heliopolis. A+ArchDesign, 8.

Girmay Berhie SH (2017). Land Use and transport mode choices: space syntax analysis of American cities. The ARCC J Arch Res, 23.

Huanmei Qin FZ (2022) Analysis of the effect of demand-driven dynamic parking pricing on on-street parking demand. IEEE Access, 12.

Indian Roads Congress (2015) Recommended practice for the design and layout of cycle tracks. Indian Roads Congress, New Delhi

Janaina Macke RM (2018) Smart city and quality of life: citizens' perception in a Brazilian case study. J Clean Prod, 24.
Kaur R (2019) Decoding advanced streetscapes: Case study of Delhi. ResearchGate, 4.
Kofuji JS (2016) Inclusive smart city: an exploratory study. Springer International Publishing Switzerland, 10.
Lei Wang XH (2022) Measuring residents' perceptions of city streets to inform better street planning through deep learning and space syntax. ISPRS J Photogramm Remote Sens, 16.
M. M. Rahman MR (2015). Parking demand and supply analysis of major commercial spaces. In: International conference on recent innovation in civil engineering for sustainable development (IICSD-2015). ResearchGate, Gazipur, Bangladesh, p 7
Madeleine Steinmetz-Wood KV (2019) Assessing the micro-scale environment using Google Street view: the virtual systematic tool for evaluating pedestrian streetscapes (Virtual-STEPS). BMC Public Health, 11.
Md. Asaduzzaman Noor, SA (2021) Assessment of traffic congestion scenario at the CBD areas in a developing city: In the context of Khulna City, Bangladesh. Transportation Research Interdisciplinary Perspectives, 11.
Mwangi Josephine GA (2021) The impact of street layout design on non-motorized activities with nairobi city. Scientific Research Publishing, Kenya, p 27
Nehad S. Abd El Gawad KS-H (2019) Place making as an approach to revitalize Neglected Urban Open Spaces (NUOS): A case study on Rod El Farag Flyover in Shoubra, Cairo. Alex Eng J, 10
Omholt T (2019) Strategies for inclusive place-making. J Place Manag Dev, 19.
Placemakingindia. (2021, 07 28). Coimbatore, streets for people. https://placemakingindia.org/: https://placemakingindia.org/blog/coimbatore-creates-case-for-tactical-urbanism/
Place-makingX I (2019, 12 7). ISOCARP knowledge for better cities. Retrieved from Placemaking India, Pune. https://isocarp.org/activities/international-collaborations/placemaking-india-pune-india-7-9-december-2019/#:~:text=On%20December%207%2C%202019%2C%20Placemaking,ISOCARP%2C%20PlacemakingX%20and%20Smart%20Citizen
Project for Public Spaces (2007) What is placemaking. https://www.pps.org/article/what-is-placemaking#:~:text=Placemaking%20shows%20people%20just%20how,begins%20at%20the%20smallest%20scale
Rezaei M (2022) Urban "Place-Making" through "Walkability"1. Urban Planning Knowledge, 21.
Roengtam S (2020) Inclusive smart city development: a lesson learned from Thailand. Public Administration Program, Faculty of Humanities and Social Sciences, 16.
Romanelli M (2022) Towards Smart Inclusive Cities. Puntoorg Int J, 19.
Shrimoyee Bhattacharya SA (2016) Creating inclusive cities: a review of indicators for measuring sustainability for urban infrastructure in India. Environment and Urbanization Asia. Sage Publications, 20.
Singhal MW (2022) Monitoring and evaluation framework for inclusive smart cities in India. Development in Practice. Routledge, 20.
Sridharan N (2015) Can smart city be an inclusive city?—Spatial Targeting (ST) and Spatial Data Infrastructure (SDI). Springer Science+Business Media Singapore, 12.
The better India, 2. (2023, 03 28). The better India. Retrieved from https://www.thebetterindia.com/313823/viral-video-of-playground-under-navi-mumbai-flyover-inspires-other-cities/
Wensheng Zheng ND (2021) Understanding the citytransport system of urban agglomeration through improved space syntax analysis. journals.sagepub.com, 27.
Wyckoff MA (2016) Definition of placemaking: four different types. MSU Land Policy Institute's, Michigan
Yawei Chen LQ (2020). Emerging participative approaches for urban regeneration in Chinese megacities. American Society of Civil Engineers, 13.
Zhaowen Liu, MD (2020) towards developing a new model for inclusive cities in China—the case of Xiong'an new area. Sustainability—MDPI, 25.

Sustainable Urbanism-Making Indian Religious Cities Accessible

Kakoli Saha, Rajrani Kalra, and Rachna Khare

1 Introduction

While there are varying definitions of the notion of sustainability, it is a consensus that sustainability deals with eliminating problems for future generations (Ferdous et al. 2020). Although the endeavor of sustainability itself began in the 1800's, a truly sustainable approach to development that meets the three pillars of sustainability—environmental, economic, and social—is now a global concern that reflects a shift in collective societal paradigms and scientific understanding. In September 2015, the General Assembly of the United Nations adopted2030 Agenda for Sustainable Development Goals, which included 17 Goals in which the protection of the world's natural and cultural heritage was considered an important step under the 'Goal 11: Make cities and human settlements inclusive, safe, resilient and sustainable' (United Nations 2015). To achieve the goal, one of the targets is to provide safe mobility for persons with disabilities (United Nations 2015). This chapter aims to assess the universal accessibility of an Indian religious city from the perspective of mobility of a visitor with disability.

K. Saha (✉) · R. Khare
Department of Architecture, School of Planning and Architecture, Bhopal, India
e-mail: kakolisaha@spabhopal.ac.in

R. Kalra
Department of Geography and Environmental Studies, California State University, San Bernardino, CA, USA

M. Firoz C et al. (eds.), *Urban Planning and Design for Megacities in the Global South*, Advances in 21st Century Human Settlements,
https://doi.org/10.1007/978-981-97-8370-0_6

1.1 Religious Cities in India

Religious cities (or pilgrim-towns as conventionally known) are special places where urbanization processes are driven by visitor influxes that visit these places for cultural and religious reasons (Shinde 2016). A person travels to a sacred place for paying homage or being healed by blessings is known as Pilgrim and the journey undertaken by him or her as, "Pilgrimage" (Sivakumar et al. 2019). United Nation World Tourism Organization (2011) estimates around 170 million visits to more than 2,000 pilgrim-towns in India (NCAER 2003; UNWTO 2011). Such influx drives not only the economy but also urban change in pilgrim-towns (Woodward 2004; for Mecca, see Henderson 2010; for sites in Italy, see Presti and Petrillo 2010). Shackley observes that many such sites 'have grown from villages to cities as a result of the pilgrim trade' (Author 2001, p. 9) and many more have experienced rapid urban growth and transformations with burgeoning religious tourism.

Madurai is an ancient city in South India with an administrative and religious basis. The structure of the ancient city follows both a traditional and modern city structure. The traditional city structure is based on the Mandala concept. This notion establishes the separation of settlements by caste and status. The lower castes neighborhood are situated in the most unfavorable direction and the upper caste neighborhood are located in the most favorable direction (Dutt et al. 2016). Further, the city's economic structure is based on activities like retail trade, manufacturing, education, and administration and Meenakshi temple-related activities are the central basis of the city's economic activities (Chatterjee and Dutt 2012). The city has been losing population and is shrinking although in 2010 the city expanded to envelope the entire urban agglomeration (Kundu et al. 2019).

Madurai city contains Hindu temples famous for architecture and heritage, other religious shrines, heritage buildings like museums, palaces, and theme parks. Pilgrimage in Hindu tradition is seen as a form of penance and indeed even today imposed as punishment or relief for offenses. Among the pilgrims, persons with disability and aged are a significant part of the visitors to the Hindu temples in Madurai city. Visitors to these temples may wish to visit other tourist attractions as well. In this context, this study examines sustainable urbanism of Madurai city in the context of safe mobility of visitors with disability. So far lot of works have been published on temple architecture, divinity, sacred space, festivals of Madurai city (Stein 1977; Lewandowski and King 1980; Reynolds 1987; Hudson 1993; Branfoot 2000; Branfoot 2013; Fuller 1980; Bharne 2019), but none is available on accessibility status of Madurai city from the perspective of mobility of a visitor with disability.

1.2 Disability Status in India

The two main official sources of nation-wide disability statistics are the National Sample Survey (NSS) and the Census of India. The census is an enumeration of the entire population of India while the NSS uses a nationally representative stratified sample. Launched in 1969 by the Ministry of Statistics and Programme Implementation, Government of India, the purpose of National Sample Survey (NSS) programme was to collect micro data. In the NSS, a disability schedule is administered every 11 years, with the most recent one collated in 2018. In the census, a disability question was included during selected years, the most recently was in 2011 (Mitra et al. 2006). Since both census and NSS collect disability data in the administrative hierarchy of national, state, and district; this research uses district level data.

The Convention on the Rights of Persons with Disabilities (2006) defined accessibility as a cross-cutting issue that enables persons with disabilities to live independently and participate fully in all aspects of life. The Agenda for Sustainable Development (2030) called for actions to ensure accessibility, following the universal design approach, by removing barriers to the physical environment, and transportation. The concepts of accessibility and universal design originated in western nations, whereas in India disability is primarily ingrained as an attribute of poverty, rural life, and social indifference (Khare et al. 2014). Currently, in India, majority of people with disabilities live in a physical environment which is either unplanned or underdeveloped. According to census 2011, there were 26,814,994 persons with disability living in India which is 2.21% of total population. Figure 1 shows the percent increase of persons with disability in India during 2001–2011 both in rural and urban areas.

According to Fig. 1A, nation-wide increase in total persons with disability during 2001–2011 was 22.41%. While the increase in urban population with disability was 48.21%, the rural increase was only 13.72%. The fact emphasizes the need to make Indian cities accessible for persons with disability. Figure 1B further details percent of people in each disability type in urban India according to NSS (2018). According to Fig. 1B, persons with locomotor disability constitutes 54.45% among all disability types in urban India. According to NSS (2018), 85.5% and 61.70% persons with

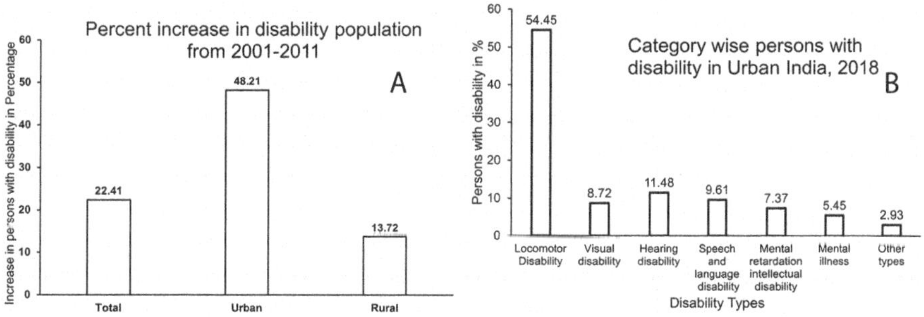

Fig. 1 **A** Percentage increase in disability population from 2001 to 2011 in India, **B** Distribution of persons with disability in different categories of disability defined by NSS

disability in urban areas faced difficulty in accessing public transportation and public buildings, respectively. The above facts emphasize the need of making built environment of urban India accessible which is also one of the targets of sustainable urbanism. In this research, accessibility of built environment of Madurai city was assessed using geospatial technique. This research addresses an overarching research question: Can GIS technology make Indian cities sustainable from the perspective of accessibility?

2 Selection of Madurai District in Tamil Nadu State

Tamil Nadu is India's southernmost state (Fig. 2A). It has several religious cities and because of that receives significant number of pilgrims to total tourists (Fig. 2B). In Tamil Nadu, domestic tourists constitute the majority proportion of pilgrims (Fig. 2C).

According to Fig. 2B, there was a steady increase in pilgrims from 1997 to 1999 and then 1.8% decrease was observed between 1999 and 2000. Figure 2C shows that more than 98% of the pilgrims are domestic tourists. Due to the reason mentioned above (Sect. 1), we can safely assume that these domestic pilgrims constitute significant number of persons with disability which leads to the need of making religious cities accessible. The assumption is important in this research and used in subsequent analysis as data for tourists with disability are not available in India. Since religious towns are distributed all over Tamil Nadu state, study district was selected based on total tourist footfall and total domestic tourist footfall. Figure 5 shows district-wise total tourist and domestic tourist footfall in Tamil Nadu.

Figure 3 shows total and domestic tourists footfall in thirty-seven districts of Tamil Nadu in 2020. It is observed that Madurai district experienced highest number of both total and domestic tourist visitors. Also, among all districts, Madurai contributed maximum to states' total tourist footfall (9.84%) and to states' total domestic tourist footfall (9.87%) as well. Further inference can be drawn that among other districts in Tamil Nadu, Madurai receives maximum tourists with disability since pilgrims are

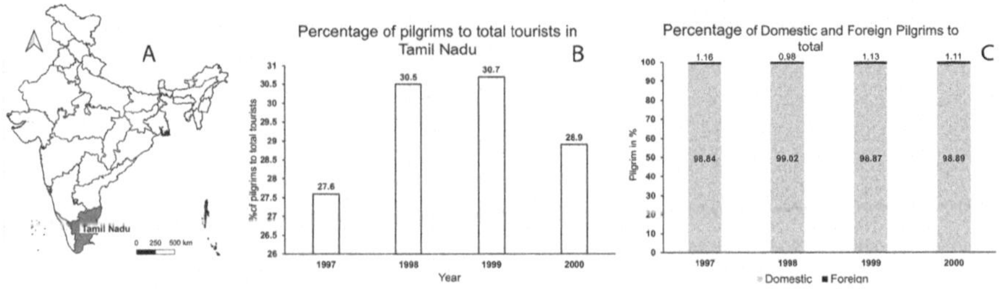

Fig. 2 **A** Location of Tamil Nadu in India, **B** Percentage distribution of domestic and foreign pilgrims to total pilgrims in Tamil Nadu

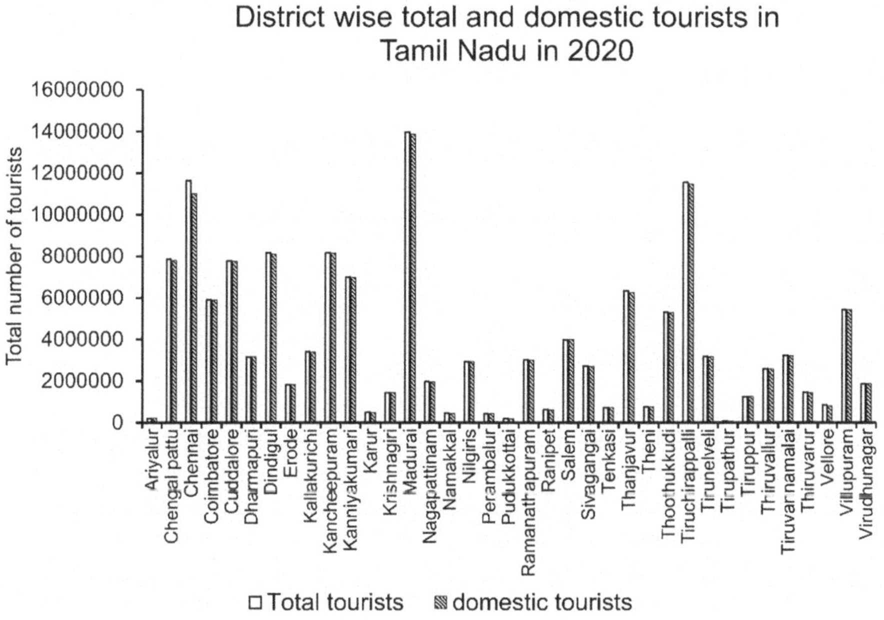

Fig. 3 District wise total and domestic tourist footfall in Tamil Nadu

mostly domestic tourists and pilgrims mostly constitute of persons with disability (refer Sect. 1.1). Thus, Madurai is chosen for further study and analysis.

3 Tourist Attractions in Madurai City

Madurai district had a population of 30,38,252 and 60.78% of its population was urban according to the 2011 census of India. Out of total urban population in Madurai district, 79.36% resides in Madurai city. Madurai city serves as the prime administrative and religious center for the Madurai District. The city is located on the banks of the Vaigai River and has been a major settlement for over two millennia. Further, it has been one of the world's oldest and continuously inhabited cities. The religious attractions of Madurai can be classified into tangible and intangible. The tangible religious landscape is comprised of natural, built, and art heritage. On the other hand, intangible religious landscape include festivals and socio-religious customs (Table 1).

a. **Temples in Madurai**

The **Meenakshi temple** (Fig. 4) forms the genesis point around which the entire city of Madurai is structured. The city was originally planned as per the mandala principles with concentric streets around the Meenakshi Temple precinct. The streets

Table 1 List of tangible and intangible religious attractions of Madurai

Tangible			Intangible
Natural	Built	Art	
River Vaigai, natural and tanks like Vandiyur Thepakkulam, Mariamman, Tepakulam, etc.	Religious structure (temple, churches and mosques), townscape with Mandala planning, palaces, isolated structure like fort wall. Parks, Gopurams	Mural painting found on the Meenakshi temple, Mahal and heritage spots. Various coins and epigraphs found in the region	Festivals like Chithrai festival, Alagar Festival

Source Heritage Development Plan (2008)

namely Adhi, Chithirai, Avanimoola, and Masi streets were laid in the form of four expanding concentric squares with Meenakshi temple in the center. Enormous Gopurams or gateway towers with intricate sculptures were constructed on all four cardinal directions. The **Thirumalai Nayak Palace** (Fig. 4) near the temple has considerable importance in the history of the city. There was a sacred tunnel connecting the Meenakshi temple to the palace which no longer exist. The **Vaigai River**, **Alagar Koil,** and **Mariamman Thepakkulam** (Fig. 4) having mythological references are regarded as significant part of history. Located at about eight kilometers away from Madurai city core is the **Thirupurankundram temple** (Fig. 4) considered as one of the six celestial Murugan temples. It has a scenic location beside a hill and is one of the favorite tourist spots of the city. The Jain caves (Fig. 4) in the hills surrounding the city add another layer to the religious context of the place.

b. **Festivals in Madurai**

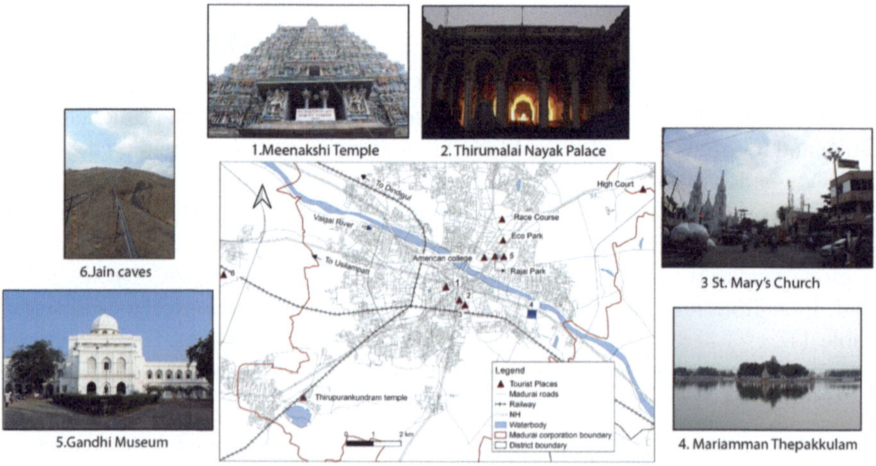

Fig. 4 Tourist attraction in and around Madurai city

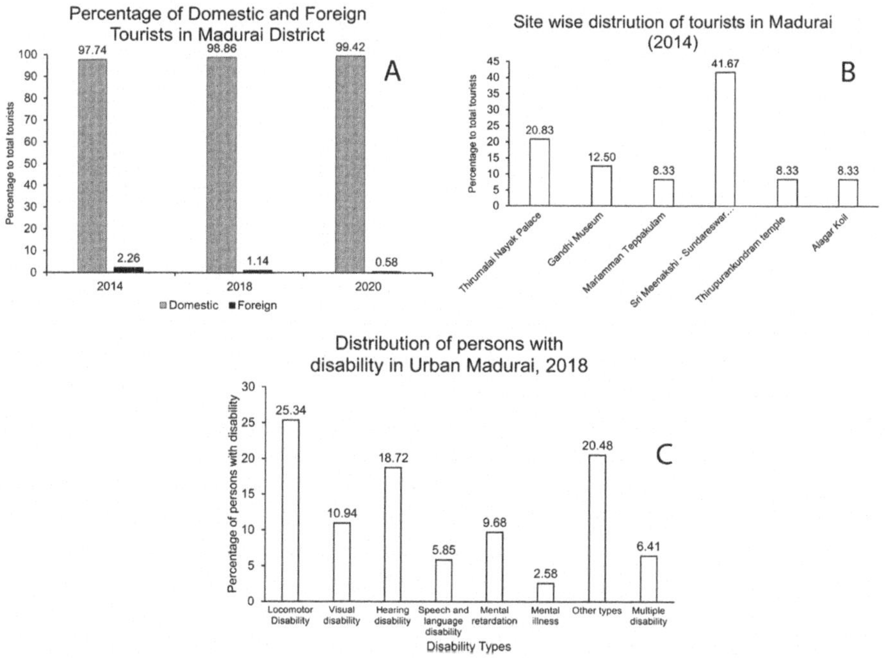

Fig. 5 **A** Percentage distribution of domestic and foreign tourists in Madurai district, **B** Site-wise distribution of tourists in Madurai city, **C** Distribution of person with disability in different disability types defined by NSS

In the festive city of Madurai, festivals are celebrated throughout the twelve months of Tamil calendar. Chithirai festival in the first month of the Tamil Calendar (April/May) is a combination of two festivals. The first fifteen days is a Shaivite (followers of Lord Shiva) festival, the next fifteen days is a Vaishnavite (followers of Lord Vishnu) festival which dramatically re-enacts the visitation of Lord Alagar (a form of Lord Vishnu and brother of Goddess Meenakshi) to Madurai from Alagar Koil who comes to attend the celestial marriage. Processions are integral part of these festivals. Since processions are taken out on streets/roads around temples it is important that they are inclusive so that visitors with disability can also participate in those religious events.

c. **Other Tourist Attractions in Madurai**

Madurai played a very important role in India's freedom fighting movement as the father of the nation Mahatma Gandhi denounced his upper cloth at this location. The Gandhi museum (Fig. 4) with extensive exhibits and narratives on the life of Mahatma Gandhi and the role of Madurai in the Indian freedom struggle is one of the most coveted sites in the Madurai city. In addition, American college, Madras High Court, St. Mary's Church, and the Town Hall (Fig. 4) are the important colonial buildings of the city. Several cultural, recreational, and sports activities are also organized in Rajai Park, Eco Park, Race Course, etc., (Fig. 4).

3.1 *Religious Tourism and Universal Accessibility Status in Madurai*

Figure 5A summarizes the tourist data collected by Ministry of Tourism, Government of India (GoI) over the years. According to Fig. 5A, domestic tourists constitute more than 97% of total tourist footfall in the district from 2014 to 2020 and the percentage of domestic tourist shows a rising trend. During 2014, the Ministry of Tourism has performed a tourist site wise survey which reveals that Sri Meenakshi—Sundareswar Temple received the highest number of visitors (41.67%) followed by Thirumalai Nayak Palace (20.83%) (Fig. 5B).

Gandhi Museum receives the third highest percentage (12.50%) of the total tourists. Since major tourist attractions are concentrated within city limit, universal accessibility assessment was done within Madurai city limit. In this research, Madurai Municipal Corporation boundary was considered as city limit (Fig. 4).

According to the 2011 census, Madurai district had 46,868 persons with disability among which 57.31% lived in urban areas. The disability types include locomotor disability, visual disability, hearing disability, speech and language disability, mental retardation (Fig. 5C). Like urban India, locomotor disability dominates disability categories in urban Madurai as well. Among the sample surveyed, 77.50% of surveyed persons with disability in urban Madurai confirms that they faced difficulties while accessing public buildings. 83.33% said they faced difficulties while accessing public transport (NSS 2018). The difficulties they faced were: unavailability of ramp/ lift, difficulty in opening the doors, lack of seating arrangement in the waiting area, difficulty at the point of receiving service, lacked special toilet seats, unavailability of Signage for direction/instruction/public announcement were mentioned (NSS 2018). The data reveals that there are accessibility issues in urban Madurai. To study these issues in detail, universal accessibility assessment was performed within the city limits of Madurai.

4 Methodology for Universal Accessibility Assessment

The methodology applied in this research for assessing universal accessibility of Madurai City comprises the following three steps—(1) Selection of Road Stretches: Considering the mobility pattern of religious visitor with disability, roads of Madurai City are categorized and sample road stretches are selected from each category; (2) Performing the Universal Accessibility Audit: Accessibility audit was performed on selected road stretches using purpose made checklists (the Audit Checklist prepared by authors from secondary sources); (3) GIS analysis—GIS analysis was performed to prepare a map showing the accessibility of religious and other tourist attractions within and around Madurai city. Details of the methodology described in a flow chart (Fig. 6).

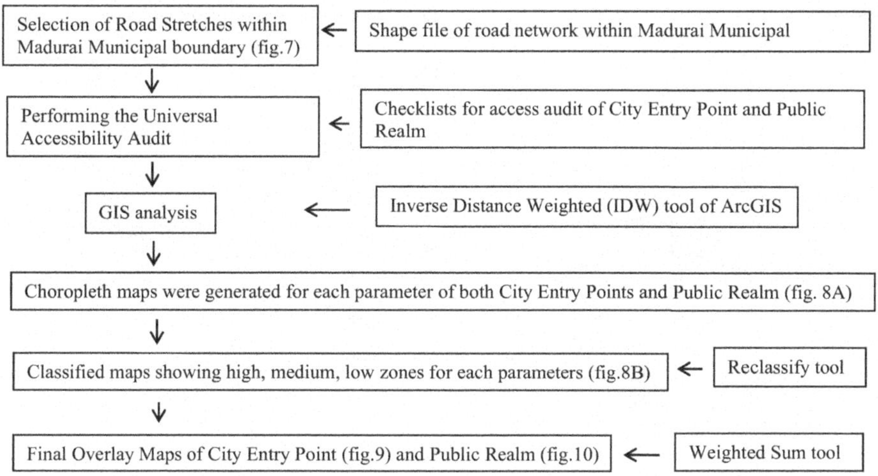

Fig. 6 Flow chart and corresponding figures showing methodology of the research

4.1 Selection of Road Stretches

Since this study was aimed to assess universal accessibility from the perspective of a visitor with disability, roads of Madurai city played a very important role. The total length of road network is 1572 km within Madurai Municipal Corporation. To identify road stretches, the activities of visitors were monitored.

Based on activities and footfalls of tourists, the major roads stretches were divided into three categories such as, (1) Road stretches along city entry points (bus stand, railway stations, airport), (2) Road stretches along city entry points to accommodation points, (3) Road stretches along heritage sites. Six road stretches were selected for the first category (Fig. 7B), two were selected for the second category (Fig. 7C) and three were selected for the third category (Fig. 7D).

4.2 Preparation of Audit Checklists

The method to prepare a universal accessibility audit checklist is explained in detail in Saha et al. (2020). In addition a report prepared by Indian Institute of Travel Management (Problems and Prospects of Accessible Tourism in India, 2010) and other relevant guidelines and checklists (Access-improving the accessibility of historic buildings and places 2011; Code of Practice-on Accessibility of Public Services and Information provided by Public Bodies—National Disability Authority, Dublin, Ireland; Improving Access to Heritage Buildings; Accessible Route in Historical Cities; Harmonized Guidelines, 2016) were consulted to prepare the audit checklist. The finding of NSS (2018) specific to Madurai urban areas (Sect. 3.1) was

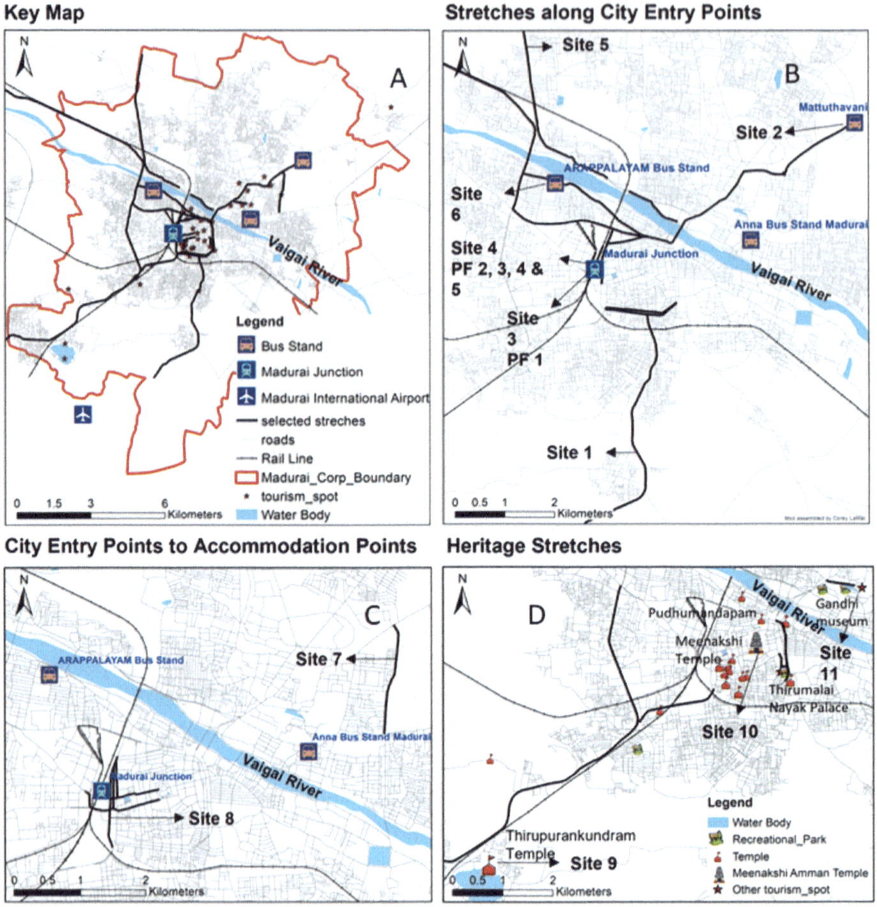

Fig. 7 **A** Road stretches with Madurai Municipal Corporation Boundary, **B** Road stretches along city entry points, **C** Road stretches along city entry points to accommodation points, **D** Road stretches along heritage sites

also referred to prepare the audit checklist to assess accessibility of religious tourist destination like Madurai city. The checklists considered disabilities such as Cognitive Impaired, Complete Blindness, Elderly, Hearing Impaired, Limb, Partial Blind, Speech Impaired, persons in Wheel Chair.

The existing checklist has two types such as:

1. Checklist for city entry points—category1 road stretches were audited using this checklist.
2. Checklist for public realms—category 2 & 3 road stretches were audited using this checklist.

Parameters like Signages, Kerb Ramps, Tactile Guiding, and Warning Blocks are common for both checklists. Parameters like Reservation and Information Counters,

Toilet Facility, Platforms (train), Seating Area, Ramps, Walks and Paths, Stairs, Lifts were included in checklist for city entry points. On the other hand, parameters like Traffic Signals, Parking Space, Approach to Building, Planned Pedestrian Routes, Subways and Foot Over Bridges, Sidewalks, and Footpaths are added in the checklist for Public Realms. Each parameter has a sub-parameter ranging from five to thirty. These existing checklists are used to map the universal accessibility status of Madurai city.

4.3 Performing the Universal Accessibility Survey Using the Checklists

It is very difficult for a non-disabled person to have the perspective for the inner nature of the mobility support needed for a person with disability. A group of students who helped to conduct the survey were first given a short training on universal accessibility and universal access audit. Students were introduced to Epicollect5 application which is a mobile & web application for free and easy data collection. It provides both the web and mobile application for the generation of forms (questionnaires) and freely hosted project websites for data collection. Projects are created by using the web application at five.epicollect.net, then downloaded to the device to perform the data collection. Data are collected (including GPS and media) using multiple devices and all data can be viewed on a central server (via map, tables and charts). Separate projects for city entry points and public realm are created in the application and using 'form builder' tool of the app, the audit checklists are incorporated within the application.

The students installed Epicollect5 app in their smart phone and collected data at the selected road stretches. As mentioned above, every parameter in the checklist has several sub-parameters. Students put numeric values against each parameter. For example, if a particular sub-parameter exists then 1, if it is partially exists, then 0.5 and in case of non-existence they put the value of 0. After collecting data for consecutive days, students came back and accessed the project in the desktop/laptop computer. In the window of the project, students had an option of 'view data' which has tabs like—download, table and map. On the map, location of data points can be viewed. Download tab can be used to download the data in.csv format. The.csv file is then opened in excel format. Subsequently, they calculated the total value of each of ten main indicators using the sum function of excel. They copied and pasted the ten indicators and their total value in separate excel sheet and calculated the total score for each indicator. Since each parameter has a different number of sub-parameters, they calculated the relative percentage of the scores. The formula for calculating the relative percentage was (individual score of a location point/ total score of that location point)*100 (Table 2) and further saved in csv format. Using the 'Display XY data" tool available in GIS software like ArcMap, data points in csv can be displayed by setting up the coordinates. The data points were then converted to shape file and saved.

4.4 GIS Analysis

The relative percentages were then added to the attribute table of the point shape file. Using Inverse Distance Weighted (IDW) tool of ArcGIS, choropleth maps was generated for each indicator. IDW tool interpolates a surface from points using an IDW technique (ArcGIS Desktop Help). The weight is a function of inverse distance. Thus, nearby data will have the most influence, and the surface will have more detail (be less smooth). In IDW maps, higher values indicate that the accessibility status of the parameter is also high. The IDW map was then reclassified using 'Reclassify' tool in ArcMap to generate three zones such as high, medium, low accessibility zones. An example of such map is shown in Fig. 8.

Since there were eleven parameters to assess universal accessibility, eleven choropleth maps were generated. Eleven classified maps showing high, medium, low zones for eleven accessibility parameter were further added to Weighted Sum tool. It is an ArcGIS geoprocessing tool that overlays several raster layers, multiplying each by their given weight and summing them together. Weightages were assigned to individual layers of accessibility parameter on the basis of expert's opinion. The group of experts consisted of academicians and stakeholders. As there were eleven parameters, weights were given in the order of 11 to 1. 'Platforms' got maximum weightage that was 11 and 'Stairs' got the minimum weightage that was 1. The Weighted Sum tool multiplied each raster layer by their given weight and summed them together.

The same processes described above was adopted to analyze accessibility data of category 2 and 3 road stretches. Figures 9 and 10 show the final output for category 1, 2, and 3 road stretches respectively.

5 Results and Discussion

The aim of this chapter was to apply GIS based method to map universal accessibility of Madurai city from the perspective of mobility of a visitor with disability. We began this by performing universal access audit of the identified road stretches within the city. The information gathered through the audit was analyzed and represented through cartographic technique.

From Fig. 9, it can be concluded that road stretches around bus stands are more accessible compare to railway stations and the road stretch toward the airport. Zones of high accessibility include Site 2 (MGR Bus stands). Presence of relative score in all ten parameters except lift (see Table 2) contributed to this high accessibility. Medium accessibility zone includes site 3, 4 (platforms of Madurai junction), and site 6 (Arapalaya bus stand). All these three sites score zero in seating

Table 2 Relative accessibility—Category 1 road stretches (city entry points)

	Toilet facility (%)	Platforms (%)	Seating area (%)	Kerb ramps (%)	Ramps (%)	Walks and paths (%)	Stairs (%)	Lifts (%)	Tactile guiding & warning blocks (%)	Signages (%)	Reservation and information counters (%)	Total score (%)
Site 1 (Airport)	0.0	0.0	0.0	0.0	0.0	24.2	0.0	0.0	0.0	75.7	0.0	100
Site 2 (MGR Bus Stand)	9.4	11.1	8.1	14.9	0.3	18.6	6.5	0.0	2.1	16.2	12.8	100
Site 3 (Platform 1)	12.9	22.7	0.0	0.0	0.0	18.1	0.0	0.0	8.9	14.8	22.7	100
Site 4 (Platform 2, 3, 4, & 5)	12.9	22.7	0.0	0.0	0.0	18.1	0.0	0.0	8.9	14.8	22.7	100
Site 5 (Sellue Station)	0.0	0.0	0.0	0.0	0.0	0.0	0.0	0.0	0.0	100.0	0.0	100
Site 6 (Arapalayam Bus Station)	15	23	0	0	0	15	0	0	7	17	23	100

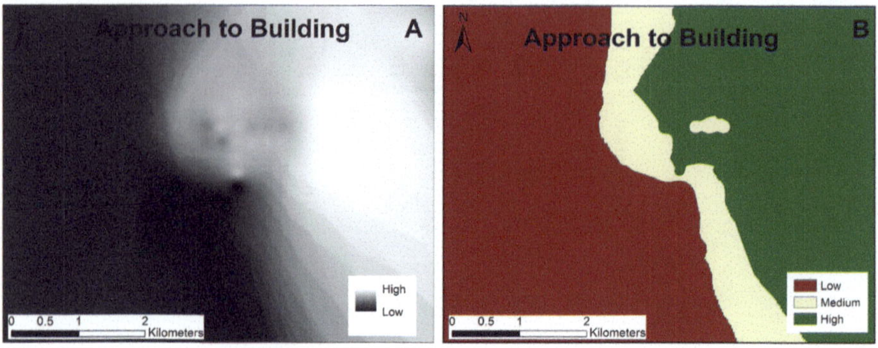

Fig. 8 **A** IDW map generated for the parameter Approach to Building, **B** Reclassified map of the same parameter

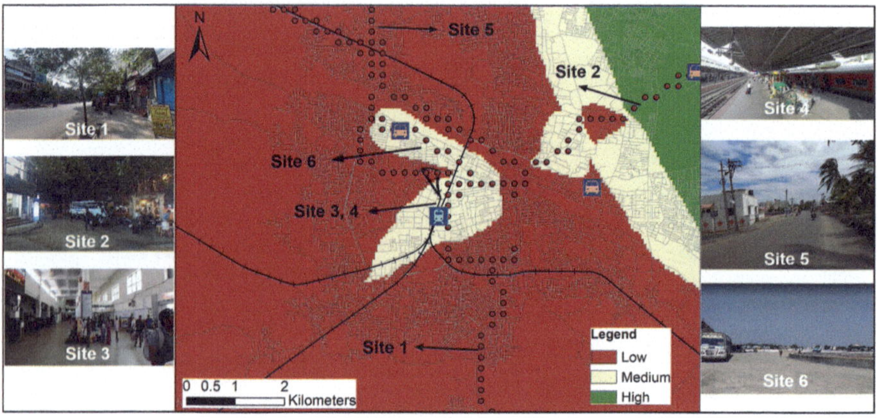

Fig. 9 Final output map and corresponding images showing location of category 1 road stretches in different accessibility zones

arrangement, Kerb Ramp, Ramp, Stairs, and Lifts (Table 3). Since railways connect Madurai City with different parts of India, it is the most favorable transportation mode of Indian tourists. Authorities can enhance Madurai's' scope as a religious tourist destination for visitor with disability by installing the missing parameters. The funding for accessible infrastructure development can be secured through National Mission on 'Pilgrimage Rejuvenation and Spiritual, Heritage Augmentation Drive' (PRASAD) scheme and Swadesh Darshan Scheme launched in 2014–2015. Both of these schemes emphasizes upon infrastructural development of religious and spiritual cities.

Site 1 (road stretch to and from the airport) and site 5 (Sellue station) fall under the zone of lowest accessibility, while Sellue station scored zero in all parameters except Signages. Site 1 scores zero in all parameters except Walks & Path and Signages. As explained in Saha et al. (2020), each of these accessible parameters has psychological

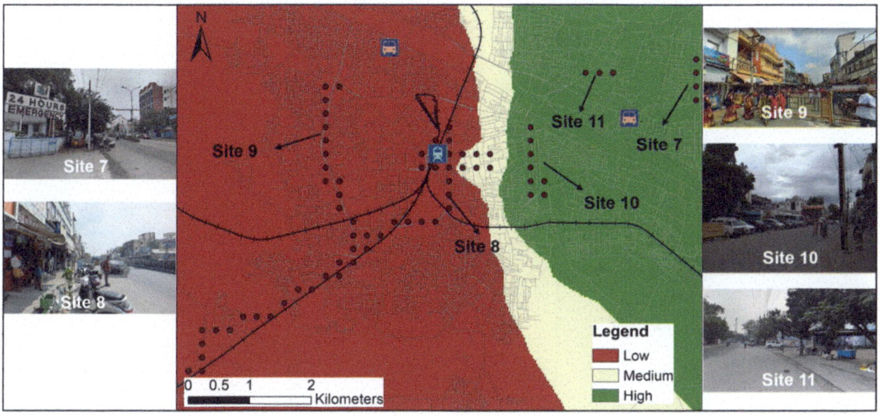

Fig. 10 Final output map and corresponding images showing location of category 2 & 3 road stretches in different accessibility zones

impact on persons with disability. For example, existence of Kerb Ramp and Ramp will not only help visitors with locomotor disability, wheel chair and crutch users to access the sidewalks; it also provides them a sense of safety as they are away from busy streets. Absence of it may cause sense of insecurity among them. Persons with mobility restrictions or sight loss spent much energy on orientation and on details compared to able bodied persons which makes them exhausted easily (Lid and Solvang 2016). Adequate seating areas within railway station will help crutch users, aged, or blind visitors to relax during transition resulting more enjoyable travel experience.

On the other hand, for public realm, zone of high accessibility includes site 7, 10, and 11 (Fig. 10). While site 7 is road stretches along high-end accommodation, site 10 and 11 include road stretches from accommodation points to religious and other tourist sites. Site 10 is the Meenakshi Amman temple where footfall of religious tourists is highest. Site 11 contains Gandhi museum which is also a major landmark for tourists. High accessibility in these sites will definitely benefit tourism of the city and pull more tourists. However, according to Table 3, site 10 (temple precinct) scored zero in Subways and Foot Over Bridges, Kerb Ramps, Tactile Guiding and Warning Blocks, Planned Pedestrian Routes, Sidewalks, and Footpaths. As articulated in Saha et al. (2020), maximum spatial interaction with built environment and social interaction with people happens in the road stretches around the temple. Visitors with disability have to pay greater attention to details and negotiate with hindrances large in both number and complexity to access the temples. Lack of above-mentioned parameters may cause feeling of annoyance leading to bad tourism experience. On the other hand, site 11 score zero in Subways and Foot Over Bridges, Kerb Ramps, Tactile Guiding and Warning Blocks, and Signages. Tactile Guiding and Warning Blocks help people with visual impairment (both complete and partial) to orient themselves and relate to space (Saha et al. 2020). Gandhi museum is the

Table 3 Relative accessibility—Category 2 & 3 road stretch (Public Realm)

	Subways and foot over bridges (%)	Traffic Signals (%)	Kerb ramps (%)	Tactile guiding & warning blocks (%)	Barriers and Hazards (%)	Planned pedestrian routes (%)	Signages (%)	Sidewalk/ footpaths (%)	Parking space (%)	Approach to building (%)	Total score (%)
Site 7 (High End Accommodation)	0.0	0.0	0.0	0.0	42.8	10.0	14.5	12.5	5.5	14.7	100.0
Site 8 (Railway Station to Accommodation & temple)	0.0	27.7	0.0	0.0	14.9	6.7	0.0	19.1	0.0	31.7	100.0
Site 9 (Tirupurankundram Temple to Mappillai Vinayaga Complex)	0.0	0.0	0.0	0.0	24.9	0.0	14.4	37.0	11.0	12.7	100.0
Site 10 (Nehru Nagar to Meenakshi Temple)	0.0	32.5	0.0	0.0	18.5	0.0	8.1	0.0	21.8	19.1	100.0
Site 11 (Hotel Northgate to Gandhi Museum)	0.0	11.7	0.0	0.0	5.0	28.1	0.0	17.6	18.5	19.1	100.0

major cultural attraction of Madurai city (see Sect. 3.1). Lack of Tactile Guiding and Warning Blocks will make the site inaccessible for the tourists with total or partial visual impairment.

The above discussion shows that, visitors with disability willing to visit religious and cultural places in Madurai city can face many difficulties as only 4 out of 7 sites fall within the zone of high accessibility. The sites within public realm such as religious and other tourist attractions have better accessibility compare to city entry points. But sites with complete absence of parameters like Subways and Foot over bridge, Kerb Ramp, and Tactile Guiding and Warning Blocks are of major concern. Since accessibility is interwoven with psychosocial level of the visitors (Saha et al.2020), lack of inclusivity will not only prevent them from enjoying tourism experience, but also lower their confidence level.

It can be inferred from above discussion that accessibility challenges can affect disabled tourists at the social, psychological, and physical levels. Enhancing the accessibility of the built environment involves more than just eliminating disabling obstacles; it serves as a tactic to foster an engaged and self-directed lifestyle for individuals living with impairments (Lid and Solvang 2016). Geographic Information Systems (GIS) can contribute to this endeavor. The proposed GIS framework can assist local administrations in establishing an environment conducive to participation by visitors with disabilities in religious tourist destinations.

6 Conclusion

Sustainable urbanism is the SDG 11 adopted by UN in their 2030 agenda. To achieve the goal, one of the targets was to provide safe mobility for persons with disabilities. According to Census of India (2011), person with disability constitutes 2.21% of total population and significant proportion of that resides in urban areas. The same trend is followed in states as well as in districts. It reflects the fact that to make urbanism sustainable, cities need to be universally accessible. According to NSS (2018) data, there are accessibility issues in urban areas of Madurai district (refer Sect. 3.1). In India, religious cities get significant number of visitors with disability as pilgrimage in Hindu tradition is always seen as a form of penance and indeed even today it is imposed as punishment or relief for offenses. This research attempts to assess sustainable urbanism of an Indian religious city from the perspective of physical mobility of a visitor with disability taking Madurai as a case study city. The methodology proposed in this research for assessing universal accessibility in Madurai City comprises three steps—(1) Selection of Road Stretches, (2) Performing the Universal Accessibility Audit, and (3) GIS analysis. Road stretches were selected based on mobility pattern of religious visitor with disability. Universal access audit was performed on selected road stretches using purpose made checklists. The checklists were of two types such as checklist for city entry points and public realms. The audit results are used to generate accessibility zone maps in GIS. The analysis shows that road stretches around MGR bus stand are more accessible compared to both

Madurai Junction and Sellue railway stations. On the other hand, for public realm, the Meenakshi Amman temple falls in the zone of high accessibility but scored zero in major universal accessibility parameters. Improving accessibility of built environment at temple precinct is necessary as maximum spatial and social interaction with happens in the road stretches around the temple.

RPwD Act (2016) states that appropriate government and the local authorities are responsible for providing information and accessible infrastructure to persons with disability. The success of the local administration is based on an excellent information management. The effectiveness of management requires a quick handling of great volumes of spatial (territorial) data related to different local tasks. Through the course of this study, it has been understood that due to its flexibility and robustness GIS has the potential to represent administrative data in spatial context. Once data is fed into the system it can be shared and updated in no time ultimately helping in easy coordination between different service departments. If Madurai city becomes more accessible for visitors with disability, it will generate more revenue from religious tourism. Greater accessibility will not only help the visitors with disability, Madurai city residents will also be benefitted. By achieving the target of safe mobility for persons with disability, Madurai will ultimately attain sustainable urbanism.

The authors have conducted religious tourism in the context of accessibility for Ujjain city in Madhya Pradesh (Saha et al. 2020) and Madurai city in Tamil Nadu. More studies need to be conducted in different cities in India so that such micro studies can lead to a macro spatial generalization. Such planning studies will aid in understanding how cities can be improved to make them more disability friendly and attain the goals of sustainable urbanism.

References

Access-improving the accessibility of historic buildings and places (2011) Ireland: National Disability Authority

ArcGIS Desktop Help (V. 10.6). Environmental systems research institute (ESRI), Retrieved October 21, 2021, from https://desktop.arcgis.com/en/arcmap/10.6/tools/3danalyst-toolbox/idw.htm

Branfoot, C (2000) Approaching the temple in nayaka-period Madurai: the kūṭal aḻakar temple. Artibus Asiae, 197–221

Branfoot, C (2013) Remaking the past: tamil sacred landscape and temple renovations. Bull Sch Orient Afr Stud 76(1):21–47

Census of India (2011) Data on disability. Registrar general and census commissioner of India, Ministry of home affairs, New Delhi, India. Retrieved July 18, 2023, from https://censusindia.gov.in/2011

Chatterjee M, Dutt AK (2012) Madurai: a holy city of pilgrimage. In: Dutt AK, Wadhwa V, Thakur B, Costa FJ (eds) Facets of social geography: international and indian perspectives. Cambridge University Press, Delhi, pp 526–540

Code of Practice-on Accessibility of Public Services and Information (2006) Ireland: Public Bodies–National Disability Authority

Dutt A, Pomeroy G, Islam I, Chatterjee I (2016) Cities of South Asia. In: Brunn S, Hays-Michell M, Zeigler DJ, Graybill JK (eds) Cities of the world: regional patterns and urban environment, 6th ed. Rowman & Littlefield, New York, pp 369–412

Ferdous F, Lawless J, Silva KD (2020) Sustainable urbanism and urban heritage conservation. In Silva KD (ed) The Routledge handbook on historic urban landscapes in the Asia-Pacific: Routledge, London, pp 363–376

Harmonised Guidelines and Space Standards for Barrier Free Built Environment for Person with Disability and Elderly Person (February, 2016). India: Ministry of Urban Development

Henderson JC (2010) Religious tourism and its management: The Hajj in Saudi Arabia. Int J Tour Res 13(6):541–552

Heritage Development Plan (2008) Ministry of Urban development, Government of India https://www.scribd.com/document/499034526/City-Heritage-Development-Plan. Accessed 31 July 2023

Hudson, DD (1993) Madurai: The city as goddess. Stud Hist Art 31:125–142

Indian Institute of Tourism and Travel Management (2010) A report on ‘Problems and Prospects of Accessible Tourism in India.’ https://tourism.gov.in/sites/default/files/2020-04/PPA.pdf. Accessed 11 July 2023

Khare R, Khare A (2014) Uniting differences—universal design for exploring the world heritage sites in India. SPA Press, Bhopal, India

Kundu D, Lahiri B, Pandey A, Sharma P (2019) City profile: Madurai. Environ Urban ASIA 10(2):308–330

Lewandowski S, King AD (1980) The hindu temple in south India. Build soc: Essays soc Dev Built Environ, 67–82

Lid IM, Solvang PK (2016) (Dis) ability and the experience of accessibility in the urban environment. Alter 10(2):181–194

Mitra S, Sambamoorthi U (2006) Disability estimates in India: what the census and NSS tell us. EPW: 4022–4026.

NCAER (2003) Domestic Tourism Survey: 2002–03.National Council of Applied Economic Research and Ministry of Tourism and Culture, Government of India, New Delhi, p 45

NSS (2018) Organisation, ministry of statistics and programme implementation. National Sample Survey (NSS). Report no 485

Presti OL, Petrillo CS (2010) Co-management of religious heritage: an Italian case-study. Tourism 58(3):301–311

Reynolds, HB (1987) Madurai: kōyil nakar. In The City as a Sacred Center Brill, pp 12–44

Saha K, Kalra R, Khare R (2020) A geospatial approach to enhance religious tourism in India—a case of Ujjain city, Madhya Pradesh. Geojournal 87(3):1793–1810

Shackley M (2001) Managing sacred sites: service provision and visitor experience. Continuum, London

Shinde KA (2016) Planning for urbanization in religious tourism destinations: insights from Shirdi, India. Plan Pract Res 32(2):132–151

Sivakumar V, Saranya PR, Iyappan A, Leelapriyadharsini S (2019) A preliminary review of the cultural heritage and emerging pilgrimage tourism in Tamil Nadu. JTHCA 11(2):94–105

Stein, B (1977) Temples in tamil country, 1300–1750 AD. The Indian Econ Soc Hist Rev 14(1):11–45

United Nations (2015) Transforming our world: The 2030 agenda for sustainable development, resolution adopted by the General Assembly on 25 September 2015, A/RES 70/1. The United Nations, New York

UNWTO (2011) Religious tourism in Asia and the pacific. United Nations World Tourism Organisation (UNWTO), Madrid

Woodward SC (2004) Faith and tourism: planning tourism in relation to places of worship. Tour Plan Dev 2(2):173–186

Role of Development Control Regulations in Achieving Compact Urban Development: A Case of Ahmedabad, Gujarat, India

Apurv Bhogibhai Patel and Paulose N. Kuriakose

1 Introduction

Rapid urbanization and migration contribute to the urban area's expansion and increase the housing demand and other basic infrastructure, thereby contributing to the city's overall density (Bahadure and Kotharkar 2015). The urban fabric and infrastructure of cities are impacted by variations in population density (Banerjee et al. 2022). The distribution of people in a place and the sustainable use of a variety of resources are significantly influenced by population density. Thus, density serves as a crucial indicator of sustainable growth (Burton et al. 1996; Christofer Robert 2012; Neuman 2005).

Development has a big impact on urbanization. According to the 2011 India Census, 377 million people call metropolitan areas home. By 2031, 600 million people (an increase of 59% from 2011) are anticipated to reside in urban India (Census of India 2011). Urban housing is becoming increasingly scarce, with 66 million people living in slums in India in 2011 (Mckinsey 2010).

Urban areas have some cultural value, which depends on various factors & these are directly affecting the individual communities (regime shifts), which is a result of migrated population, which increases day by day and increases the poverty level (Millennium Challenge Corporation compact development guideline) (Bertaud 2013). Consequently, the city's size expands, necessitating robust connectivity and efficient public transportation systems. However, this urban growth also leads to increased energy consumption and heightened environmental issues, thereby

A. B. Patel (✉) · P. N. Kuriakose
Department of Urban & Regional Planning, School of Planning and Architecture Bhopal, Bhopal, Madhya Pradesh, India
e-mail: apurvab1112@gmail.com

P. N. Kuriakose
e-mail: paulosenk@spabhopal.ac.in

M. Firoz C et al. (eds.), *Urban Planning and Design for Megacities in the Global South*, Advances in 21st Century Human Settlements,
https://doi.org/10.1007/978-981-97-8370-0_7

adversely affecting the overall quality of life (Bertaud 2018; Burton et al. 1996; OECD 2010).

By prioritizing residential densities, transit-oriented development, diversified land use, and social equity in compactness, economic viability can be attained, public transportation alternatives can be expanded, uncontrolled city sprawl can be curbed, and agricultural land can be preserved (Singh Pratibha 2016). The shape of cities has drawn increased attention during the past 15 to 20 years. With the constant discussion of compact development and the sustainability of cities in developing nations, it is important to look at a sustainable form that is relevant to the context (Kumavat 2013).

As a result, compact development within mixed land use has emerged as an appealing strategy for growing cities, particularly in communities with the highest availability of amenities and transportation (Adhvaryu 2010). To meet future needs and appropriately segregate resources at the lowest cost, it is now important to enhance the use of the current property due to the scarcity of accessible land for development (Burton et al. 1996; Conticelli 2019; Jabareen 2006; M M, n.d.; OECD 2010). The compact city concept, adopted by a number of developed countries in their urban planning strategies, aims to efficiently use the land while reducing sprawl, decreasing the transport network and relying on mass transportation, providing infrastructure, creating a socially supportive atmosphere with the vibrancy of actions, and ensuring financial viability (Kjaeras 2021). It is still unclear, however, whether dense cities in developing nations like India will benefit from the compact urban fabric. Assessing these cities' urban morphology and compactness is becoming essential for comprehending the spatial urban structure and intervening appropriately for sustainable urban development (Yao et al. 2018). Urban morphology and compactness are led by different development control tools like DP, TPs, LAP, etc. (Beatley Timothy 2000; Fan et al. 2019). This approach seeks to enhance the understanding of the land allocated for open space and open spaces, the land used by structures, the land covered by private open spaces, and the overall compactness of the city. It looks at Ahmedabad's five highly developed neighborhoods' urban forms.

2 Theoretical, Conceptual, and Discursive Foundations

In the lexicon of urban research, urban sprawl is a term used to describe certain types of urban form, such as commercial strip development, low-density, discontinuity, and leapfrog that are generally considered as "sprawl" (Bibri et al. 2020). Urban sprawl can have several negative impacts on the environment, such as loss of natural habitats, increased water and air pollution, and increased greenhouse gas emissions from transportation. Urban planning strategies such as zoning and land use regulations can be used to limit sprawling and promote more forms that are more sustainable for development (Arbury 2019; Mouratidis 2019). The urban form may be quantified in several sizes, including the scale of a single structure, a street, an urban block, a neighborhood, or an entire city (Burton et al. 1996; Kamble and Bahadure 2021).

Various researchers use different trends and identify indicators, and methodologies to measure urban form some of them are mentioned in Table 1.

Table 1 Various Approaches to Measure Urban Form

Researcher	Measured trends	Indicators for research
Jabareen, Y.R.	Sustainable urban forms and their types	Evaluating various urban components or concepts across different urban forms, such as mixed-use, density, diversity, sustainable transportation, compactness, and ecological design, facilitates the appraisal of urban forms concerning their sustainability
(Glaster et al. 2012)	Urban sprawl index	Land use patterns that may be measured include clustering, proximity, centrality, density, concentration, continuity, and nuclearity
(Ewing et al. 2013)	Sprawl indices	Urban form indicators: (a) neighborhood mix of homes, jobs, and services; (b) residential density; (c) accessibility of the street network; (d) strength of activity centers and downtowns
(Bertaud 2013)	Spatial organization of a city	Two components: (a) census statistics indicate the spatial distribution of the population; (b) the frequency with which individuals travel from their homes to other areas for purposes such as jobs, school, shopping, socializing, etc
(Song and Knaap 2004)	Development patterns	A neighborhood's composition is primarily defined by five essential components: street layout, circulation infrastructure, land use variety, accessibility, and pedestrian pathways
(Hess et al. 1999)	The correlation between site design and pedestrian travel is significant	Using indications such as block size, length, and the availability of all sidewalks
(Fulton and Pendall 2001)	Trends in urban form and land usage	The amount of urbanized land and the density of cities
(Knaap et al. 2013)	Multidisciplinary Measures of urban sprawl	Measures of form, accessibility, transportation networks, community design, landscape ecology, and other factors are used to determine density, distribution, and variability
(Tsai 2016)	Measures at the urban level to distinguish between sprawl and compactness	In measuring a neighborhood's form, size, population density, equal distribution (Gini coefficient), and clustering (Moran coefficient) are utilized

Sources Bertaud (2013), Bibri (2020), Ewing et al. (2013), Fan et al. (2019), Glaster et al. (2012), Hess et al. (1999), and Knaap et al. (2013)

2.1 Compactness

In 1997 Richardson and Gordon characterized Compactness as monocentric development or high density. Anderson (1996) defined compact development as monocentric and polycentric, but Ewing's (1997) definition asked for some concentration of housing and a mix of land uses and employment. In 1999, Bertaud and Malpezzi created a compactness measure that compares the average travel time from a residence to the CBD to its equivalent in a hypothetical cylindrical metropolis with an equal dispersion of development (Bertaud 2004). Compactness, according to Galster (2001), is the degree to which development is concentrated and reduces the quantity of area developed per square mile.

Compact cities and compact development reveal distinct differences in terms of development size, open space, privacy/amenity space, density, growth considerations, and market value. A compact city pertains to the overall size of a city, encompassing its urban form and population density, while compact development operates at a smaller scale, focusing on specific areas within the city such as clusters or neighborhoods. Compact cities often have a low open space ratio, resulting in limited green spaces, whereas compact development allows for more flexibility in open space ratios. In terms of privacy and amenity space, compact cities may experience a loss of amenity space and reduced privacy due to the dense urban environment, whereas compact development emphasizes maintaining privacy and providing sufficient amenity spaces. Density differs between the two, with compact cities often featuring high-rise, high-density developments, while compact development encompasses varying densities tailored to the specific needs of each cluster or neighborhood. In terms of market value, compact cities can lead to increased social segregation due to expensive housing in the city center and wealthier suburbs, whereas compact development aims to mitigate segregation by creating different clusters with their central business districts, offering a range of housing options.

2.2 A Conundrum Involving Compact Development

Compact development offers benefits such as efficient land use, sustainable transportation, vibrant communities, and infrastructure efficiency. However, it also presents challenges related to limited open spaces, housing affordability, infrastructure upgrades, high levels of noise pollution, unfavorable living conditions for low-income groups, and the need to carefully balance density and livability, overcrowding, lack of privacy, or diminished livability. Successful implementation of compact development requires thoughtful planning, effective urban design, and consideration of the specific context and needs of the urban area in question.

2.3 Need for Context-Specific Evaluation

Compact city strategies in developing countries require careful consideration, as the high population density in developed nations may equate to low-density living in some developing regions. Furthermore, developing countries face unique challenges, including rapid population growth rates, which necessitate tailored approaches to implementing compact city models (Li, Yang, Ren, Li, & Wang 2018). It is well understood that regions exhibit unique characteristics that require context-specific evaluation. Even within a large country, regions with distinct spatial and historical factors can influence density distribution and facility allocation (Bibri 2020; Kjaeras 2021; Kotharkar et al. 2012).

In certain areas, urban planning policies that focus on a single center result in high-density concentration in the CBD, attracting residents and facilities. However, this concentration can have negative impacts on those living outside the CBD as they must travel further for work, leisure, and shopping. On the other hand, polycentric cities distribute facilities and populations across multiple CBDs, promoting even density distribution. Studies that evaluate context-specific density can provide insight into the unique problems and preferences of a particular area, such as the desirability of low density versus high density. These evaluations can help identify how much land is needed and what features make an urban region appealing to its residents (Kjaeras 2021; Kotharkar et al. 2012; Neuman 2005; Rawat Kuhoo 2008; Wang 2022). Table 2 demonstrates principles and strategies that are the most prevalent underlying the compact city model.

The analysis of various studies on sustainable urban planning (Table 2) reveals a range of dimensions that contribute to creating environmentally and economically sustainable cities. These dimensions include characteristics of the built environment, land use mix, density, compactness, sustainable transport, and the integration of diverse land uses. Understanding and implementing these dimensions can guide policymakers and urban planners in developing strategies to foster sustainable urban environments.

3 Methodology

The research focused on evaluating development control tools concerning compactness and pursued three main objectives. Firstly, it aimed to comprehend the concept of compact urban development and identify parameters for assessing compactness. Secondly, it examined the implementation rate of development control tools employed in the last two decades and analyzed their impact on compactness and built form. Lastly, the study proposed mechanisms and recommendations to enhance compactness and steer development toward smart urbanism. The study adopts a three-stage methodology aligned with objectives. The first stage aims to elucidate the compact urban development concept and identifies parameters for assessing

Table 2 Dimensions of the compact urban form

Scholars, Theorists, and Organizations	Study focus	Dimensions
(Jones and Macdonald 2004)	Economic sustainability and components of urban form	• Characteristics of the built environment • Land uses a mix • Transport infrastructure • Density
(Neuman 2005)	Dynamic planning against static design. Such as forms and processes	• Mixture of land uses • High residential and employment density • Land use with fine granularity • Contiguous development, including surface parking and derelict or abandoned plots or buildings • More interactions in the social and economic realms • Urban infrastructure, including water and sewage lines • Limited urban development with readable boundaries • Multimodal transportation
(Jabareen and Rafeq Yosef, 2006)	Concepts and contribution of sustainable urban forms	• Land use mix • Density • Compactness • Sustainable transport
(OECD 2012)	Policy Evaluation of Compact Cities	• Compactness • Impact of policies
(Bertaud 2013)	Analysis of how major cities are laid up spatially	• Spatial Distribution of trips and Population • Density profile • Average land usage and density • Population based on distance from the center of gravity and land consumption
(Bahadur et al. 2014)	Calculate the density of the urban form	• Accessibility • Density • Land uses a mix • Transportation network

(continued)

Table 2 (continued)

Scholars, Theorists, and Organizations	Study focus	Dimensions
(UN Habitat 2015)	The neighborhood sustainability planning approach	• Limited land use specialization • Adequate space for streets • High density • Social mix • Efficient street network • Mixed land uses
(Bibri 2020)	Use urban planning techniques to create sustainable development	• Density • Compactness • Diversity • Mixed land use • Green space • Sustainable Transportation

Sources Bertaud (2013), Bibri (2020), Ewing et al. (2013), Fan et al. (2019), Glaster et al. (2012), Hess et al. (1999), and Knaap et al. (2013)

compactness (Jabareen 2006; Jayasinghe et al. 2021; Mouratidis 2022). This stage is subdivided into three sub-tasks, which aim to identify the variables for assessing the nature of compactness in the urban area, which include density, diversity, floor use mix, green accessibility, centrality, street connectivity, and transit stops, with the help of this parameter chapter able to derive comprehensive compactness index of the city (Bibri et al. 2020; Lan et al. 2021; Mouratidis 2019; Tsai 2016). Within the AUDA limit, there are various neighborhood typologies, each with a unique mix of land uses. Population, Density (PPH), Major land use (Development plan-2021), Distribution of FSI, Connectivity, Public transport, Walkability, and Distance from the core city, based on this Parameter five neighborhood are selected which contains unique characters and diverse land use with different density (Bahadure and Kotharkar 2015; Kotharkar et al. 2014).

The second stage investigates the development control tools applied in the past two decades and demonstrates their effects on the density, compactness, and built form of Ahmedabad. The study includes an examination of the development plan, town planning schemes, local area plans, and development control regulations (DCR), aiming to understand the inter-relationship of FSI and density with different parameters. Parameters of DCR such as zoning, permissible use, margin & setbacks, FSI, road width, building height, and ground coverage are crucial factors that have a direct or indirect impact on the compactness and density of a city (Adhvaryu 2010; Parekh and Makwana 2020; Patel et al. 2018; Shah and Patel 2021). By analyzing development control tools, this research tries to establish the proposed scenario for the city.

To understand the ground reality of development, a primary survey was conducted. The survey included a reconnaissance survey of the whole of Ahmedabad and a detailed study of a selected neighborhood. Based on this literature background, data collation was divided into three parts. The study involves collecting qualitative data through interviews with academicians and town planners to gain insight into the city's growth and development direction. Focus group discussions with developers and contractors provide stakeholders' perspectives on development, and land use analysis is conducted (Bibri et al. 2020; Jia et al. 2019; Rahman et al. 2022; Tsai 2016). The land use analysis considered land use incompatibility, land use indicators such as density, land use diversity, destination accessibility, and land use inventory, which led to the derivation of neighborhood compactness (Cyril et al. 2019; Fan et al. 2019; Hong et al. 2022; Li and Zhang 2022; Zhi Cai et al. 2022).

The second part involved quantitative data collection, including surveys to understand travel behavior and mode of choice for travel. Regarding the neighborhood population, a total of 273 surveys were conducted with a confidence rate of 10%. The surveys were carried out in both commercial and residential areas to gain insights into travel behavior. This provided insights into developing and assessing the use of public transport (Cyril et al. 2019; Karteek 2015; Nallathiga 2005; Singh 2020; Thara 2013).

The third stage of primary data collection is dedicated to spatial analysis, which involves the assessment of various parameters such as building use, height, age, ground coverage, FSI, and green spaces to understand the urban morphology (Adolphson 2022; Hess et al. 1999; Kjaeras 2021; Kotharkar et al. 2014; Mckinsey 2010; Wenjian Pan and Juan Du 2020). Additionally, data related to the connectivity, road width, public transport, employment center, and travel pattern are also collected from the selected neighborhood to gain insights into the transportation system (Bibri 2020; Boyko and Cooper 2011; Kamble and Bahadure 2021; OECD 2017; Parekh and Makwana 2020; Stratigea et al. 2015; Y.-H. Tsai 2016). This analysis helps to comprehend the spatial layout and connectivity of the neighborhood, which further assists in understanding the compactness and density of the urban area (Byahut et al. 2020; Cyril et al. 2019; Dadras et al. 2014; Rahman et al. 2022).

After data collection, the study aggregated the data using different methods derived from the literature and compared the ground situation with the proposal in development control tools to identify gaps in implementation. The third part of the study identifies various reasons for the implementation gaps and suggests mechanisms/ recommendations for better compactness, leading development of smart urbanism. Different land use modeling scenarios based on density, diversity, and destination accessibility are used in the study (Abdullahi et al. 2015; Rahman et al. 2022; Rawat Kuhoo 2008).

4 Study Area Description

Gujarat is one of India's most developed states. The state has extensive mineral riches and is distinguished by significant industrialization. It produces over 20% of India's total industrial production and has a strong GDP growth rate. Gujarat has eight municipal corporations and nine urban development authorities. Ahmedabad is the largest city in terms of both population and area in Gujarat. Due to rapid growth,

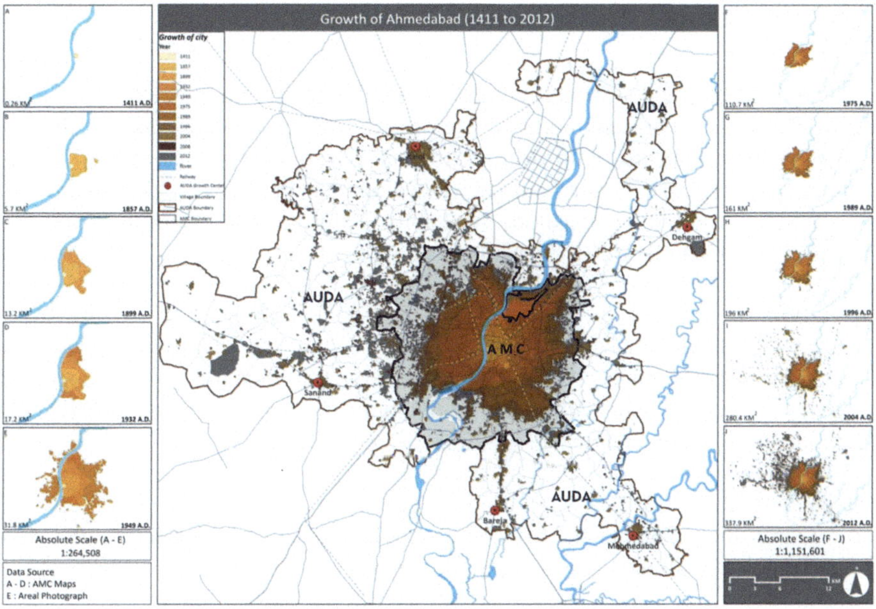

Fig. 1 Growth of Ahmedabad 1411 to 2012 (*Source* AUDA 2014a)

it is imperative to assess and reorient the developmental and growth trajectories in the forthcoming decade (AMC 2022; AUDA 2014a) (Fig. 1).

Located in west-central India, Ahmedabad spans across a vast plain encompassing both sides of the Sabarmati River (Shah and Patel 2021). As an older primate city, its strategic positioning has allowed for unrestricted growth and the emergence of a polycentric urban pattern (AUDA 2021). However, this expansive geography has presented challenges in terms of policy coordination and the integration of the metropolitan area's burgeoning population (Patel et al. 2018). Despite these obstacles, Ahmedabad has reaped significant developmental advantages from its favorable location (Mahadevia et al. 2014). This research presents an analysis of India's urbanization process, focusing on the Ahmedabad city region. Ahmedabad has experienced a remarkable surge in population growth over the last three decades, positioning it among the world's 19 fastest-growing cities (Bharti et al. 2020). AUDA and AMC have utilized a town planning mechanism to execute the development plan.

A shifting pattern of land use mix has been observed in Ahmedabad, notably inside the highly mixed ecosystem of the center. Given the high land and rental values in these locations, it is impossible to use the property exclusively for residential purposes, forcing mixed land use (Mahadevia et al. 2014).

5 Site Selection

To select neighborhoods various parameters are identified from the literature. Based on this Parameter five neighborhood are selected which contains unique characters and diverse land use with different density (Nargunde Aadish 2011). Population, Density (PPH), Major land use (Development plan-2021), Distribution of FSI, Connectivity, Public transport, Walkability, and Distance from the core city (Bahadure and Kotharkar 2015; Kotharkar et al. 2014).

Population
The study area of Ahmedabad was divided into a 1X1 km grid, which overlapped with the population data from the 2011 census (Fig. 2[1]). Based on these populations, the research selected the grids with higher populations to identify compact urban areas (Abdullahi et al. 2015; Rahman et al. 2022).

Density
The research then overlaid the grid with population density per hectare data and selected the grid with the highest population density for further analysis (Fig. 2[2]).

Connectivity
In Ahmedabad, the street network is highly developed, but as distance increases, travel time also increases (Shivanand Swamy 2017). To address this, the authorities have developed several highways, including S.P.Ring Road, S.G. Highway, Ashram Road, C.G. Road, and other major corridors with mixed land use and high density, as promoted by the DP2021 plan. Well-connected neighborhoods give priority to the selection (Cyril et al. 2019; Wilbur smith associate 2012) (Fig. 2[3]).

Major Land Use
Analysis of various land usage patterns in distinct areas of the city and identified neighborhoods with varied land uses to be examined by DP2021 (Fig. 2[4]) (AMC 2022; AUDA 2014a).

Distribution of FSI
Specifically, the proposed development plan for 2021 was used to identify areas with varying FSI values. The neighborhoods with the highest FSI values were given priority for analysis, and the research attempted to cover different neighborhoods with varying FSI values for comparison purposes (Abdullahi et al. 2015; AUDA 2014b). (Fig. 2[5]).

Public Transport
Although Ahmedabad's public transport system has improved, there is still space for development in terms of frequency, reliability, and price (Shivanand Swamy 2017; Tejendra Nagabhoina 2022). For research purposes, areas with higher connectivity have been selected for study (Fig. 2[6]).

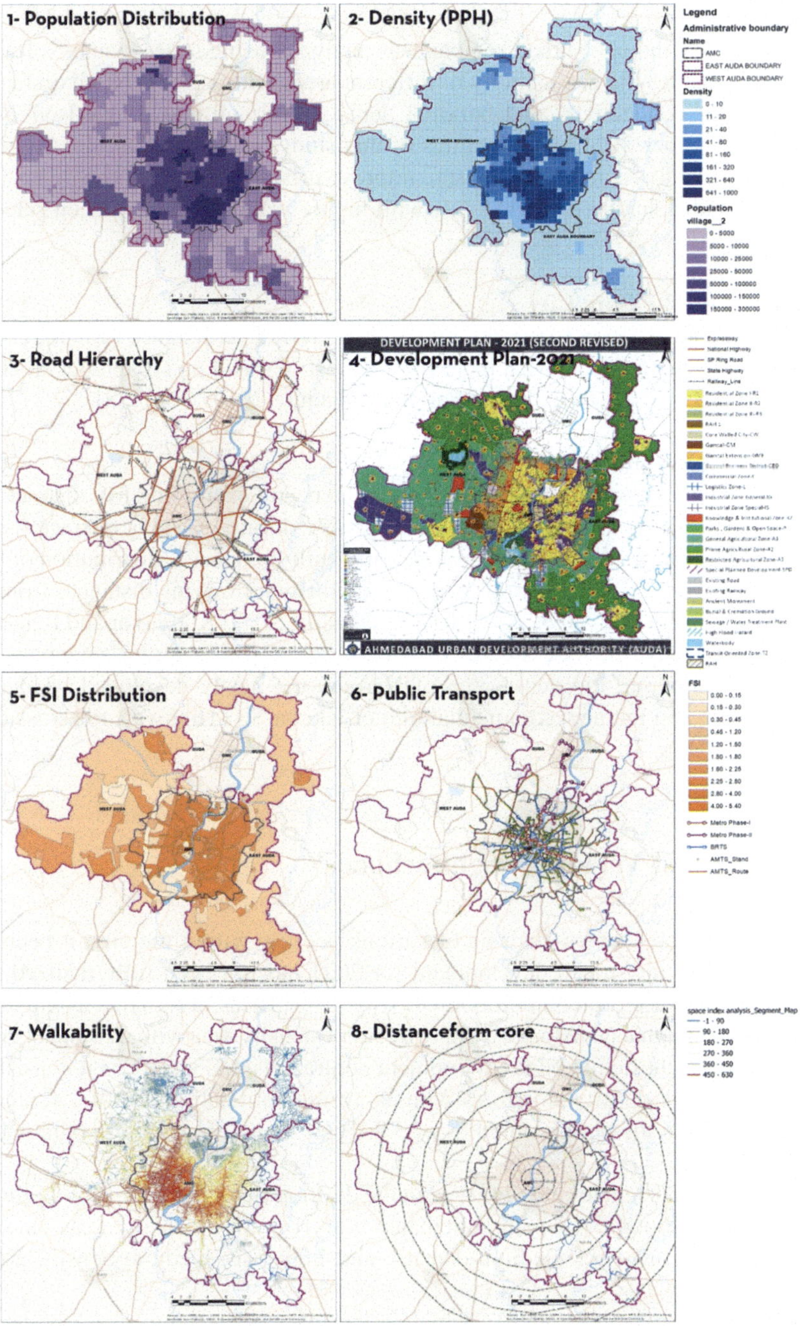

Fig. 2 Site Selection Parameters (*Source* Generated by Author; Ahmedabad Urban Development Authority, 2014)

Walkability

The walkability index of urban regions was analyzed using space syntax software. This tool measures the accessibility of different areas within a city, based on factors such as street connectivity and building density (Bertaud and Richardson 2014; Boyko and Cooper 2011). The findings of this study can be used to identify more pedestrian-friendly regions and adjust to improve urban walkability (Fan et al. 2019; Kotharkar and Bahadure 2019). Areas with higher walkability have been selected for the study (Fig. 2[7]).

Distance from the Core City

For the study, areas closer to the city center are preferred, as distance from the core is considered a negative factor that increases the cost of traveling (Byahut et al. 2020; Kamble and Bahadure 2021; Kotharkar and Bahadure 2019) (Fig. 2[8]).

Selected Neighborhoods

The study encompassed overlapping all parameters and examining five distinct 2 × 2 km neighborhoods as case studies. These areas exhibited varying characteristics, including land use patterns, population density, built-up density, distribution of FSI, high connectivity, good public transport, walkability, and varying distances from the city center (Byahut et al. 2020; Kamble and Bahadure 2021; Kotharkar and Bahadure 2019). Specifically, the case studies focused on the CBD area of Usmanpura, the old city of Jamalpur, the dense area of Naroda, the newly developed residential area of Vadaj, and the new commercial hub of Iskcon-S.G.Highway, to assess their compactness (Fig. 3).

6 Analysis of Various Neighborhood

Through a comprehensive analysis of various pockets within the city, it becomes evident that the envisioned development for 2021 has not been fully realized. For instance, in the case of neighborhood-01, (Fig. 6) which is designated as the CBD, the development plan envisioned a high-rise, high-density area with a compact urban form. However, the actual built environment contradicts this vision, as it consists of medium-rise, low-density structures with an average utilization FSI of 1.4, significantly lower than the permissible FSI of 3.8 (Figs. 4 and 5) and the maximum FSI of 5.4.

Similar inconsistencies are observed in the development plans for neighborhood-03 and neighborhood-04, which anticipated high-rise, high-density areas but have materialized as low-rise, medium-density, and high-rise, medium-density zones, respectively.

In the case of neighborhood-02, which was envisioned as a low-rise, low-density zone, the actual urban fabric reveals a medium-rise, low-density environment that has evolved into an employment hub for the city. Developers have taken advantage of TDR and TOZ mechanisms, constructing medium-rise buildings (averaging 12

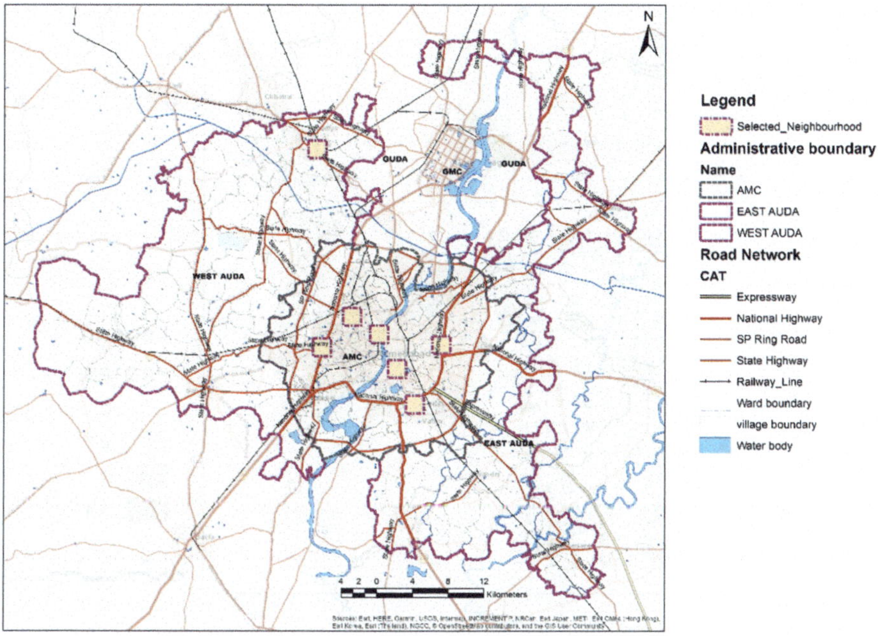

Fig. 3 Selected neighborhoods (*Source* Generated by Author)

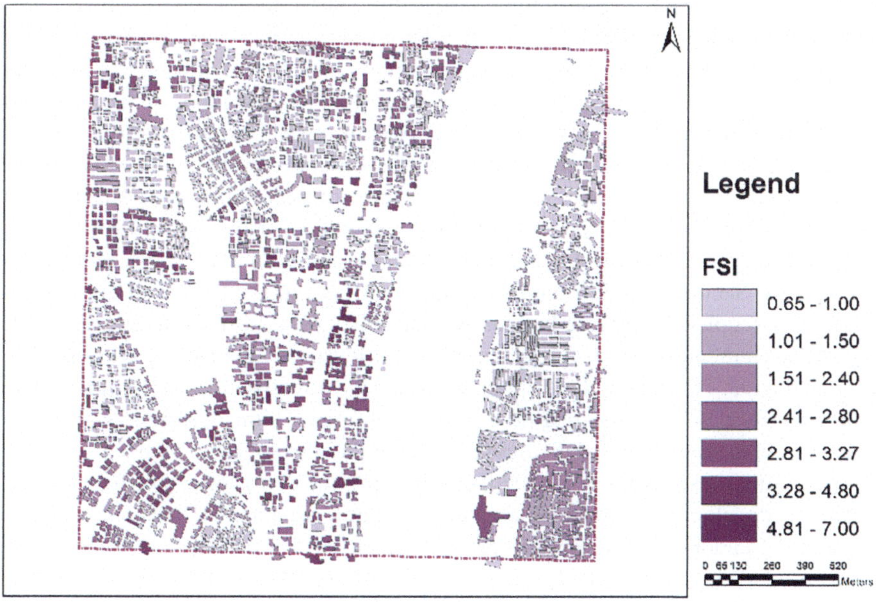

Fig. 4 Utilized FSI distribution of NH-01

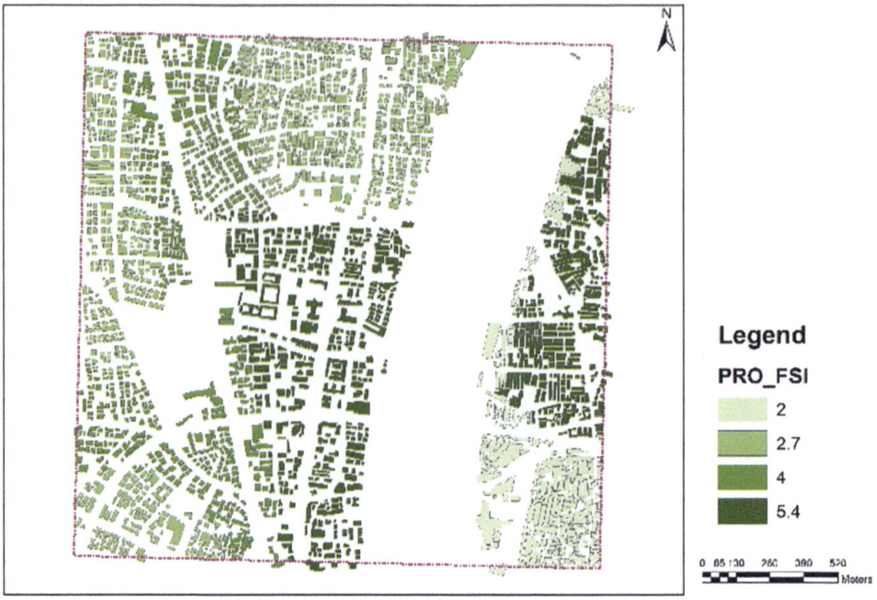

Fig. 5 Proposed FSI of NH-01

floors) and several high-rise structures (averaging 15–18 floors). This analysis highlights a clear disparity between the city's-built environment and the development plan for 2021, attributable to several factors. Firstly, the exorbitant land prices in the CBD and core city areas make it financially unfeasible for developers to acquire land there, leading them to prefer peripheral areas with lower land costs. Additionally, the selling price of units in high-rise buildings is often unaffordable for the middle-income and upper-middle-income groups, who constitute a significant portion of the workforce in these areas. As a result, the higher-income group tends to seek housing on the city's outskirts, away from the urban chaos and with convenient transportation options. Secondly, the limited availability of land in the CBD and core areas, which are occupied by private bungalows and other uses, poses a significant challenge to implementing the development plan. Furthermore, the lack of coordination between the DP, TP, and local area plans hampers the effective utilization of FSI. The complicated and time-consuming process of obtaining building permissions from the authorities also contributes to the deviation from the development plan.

Addressing these issues is crucial for improving the compactness of the city and steering its development toward smart and sustainable urbanism. Coordinated efforts are needed to address the high land prices, streamline the permission process, and ensure proper alignment between the DP, TP, and local area plans. By addressing these challenges, we can strive toward achieving a more coherent and efficient urban form, in line with the envisioned development plan (Fig. 6).

Fig. 6 Existing and proposed scenario of building distribution (*Source* Generated by Author [Primary survey])

7 Data Interpretation

The research looks at Ahmedabad's urban setting and then uses the findings to assess if compact development is an effective growth approach for the city. The study provides a detailed analysis of the urban form characteristics of Ahmedabad.

Density

Population density—the population density was computed through an examination of the number of individuals accommodated per unit area of floor space. A higher population density is indicative of a more compact environment, while a lower population density is reflective of a more sprawling setting (Abdullahi et al. 2015; Kotharkar et al. 2014; Rahman et al. 2022).

$$\text{Population Density} = \text{Pn}/\text{An}$$

where Pn = selected neighborhood's population; An = selected neighborhood's total floor area.

Built-up land density—the built-up density has been derived by dividing the neighborhood's entire built-up area by its overall area.

$$\text{Built-up land density} = \text{Bn}/\text{An}$$

where Bn = Built-up area of the selected neighborhood; An = Total area of the selected neighborhood.

Diversity-Land Use Mix

To assess the diversity of land uses, the entropy index was employed to gauge the range of land uses within a specific area and to reveal the degree of integration or segregation among different land uses (Abdullahi et al. 2015; Kotharkar et al. 2014; Rahman et al. 2022). The entropy index, ranging from 0 to 1, serves as a metric for land use diversity. A value of 0 signifies a uniform distribution of land uses, whereas a value of 1 indicates the highest level of heterogeneity in land use distribution.

Floor Use Mix

The analysis of floor use mix was conducted to determine the vertical land use mix scenarios. Each structure in each cell was categorized into one of three usage categories (Bahadure and Kotharkar 2015; Kotharkar et al. 2014; Li and Zhang 2022; Rahman et al. 2022).

i. Single use
ii. Double types of floor use
iii. Three types of floor use.

Centrality

The distance to the nearest city center or subcenter. As distance increases, travel time will increase which is negative for the city's sustainability (Kotharkar et al. 2014; Li and Zhang 2022; Rahman et al. 2022).

Street Connectivity

Furthermore, the number of real junctions in a 400 × 400 mt grid is used to measure street connectivity. The street connectivity of each grid cell is determined by counting the number of intersections with three or more legs within a 400-m radius of the center of each grid cell. Higher intersections are considered as highly assessable streets, they help to reduce the distance to travel and increase compactness (Bahadure and Kotharkar 2015; Cyril et al. 2019; Kotharkar et al. 2014; Rahman et al. 2022).

Transit Stops

The number of transit stops per grid cell was calculated with a 400 m radius (URDPFI Vol-1) (Abdullahi et al. 2015; Cyril et al. 2019; Kotharkar et al. 2014; Rahman et al. 2022) (Fig. 7).

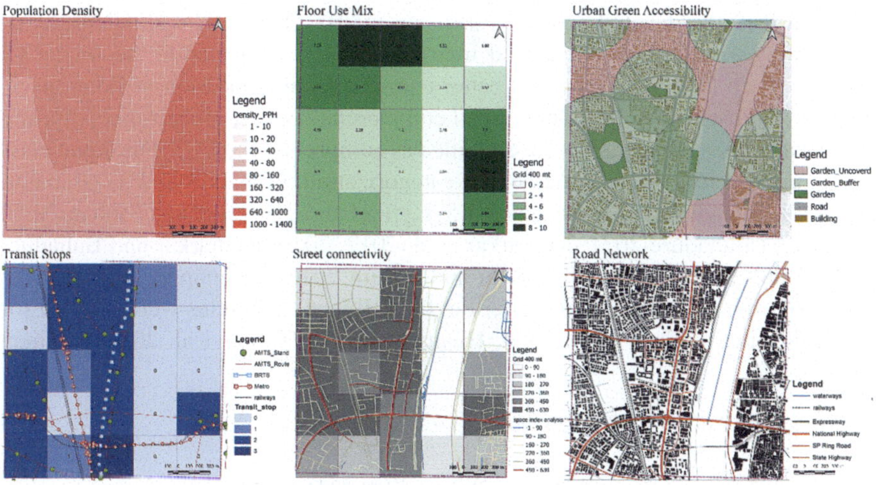

Fig. 7 Parameter of the Compactness (*Source* Generated by Author [Primary survey])

Urban Green Accessibility
The research assessed the per capita public green space and coverage of publicly accessible green spaces within each neighborhood. Parks and other publicly accessible green spaces in urban settings are referred to as public urban green spaces (Abdullahi et al. 2015; Kotharkar et al. 2014; Rahman et al. 2022).

Compactness Index Derivation Method
After calculating measures for compactness, the research employed the Simple Additive Weighting (SAW) technique in a Multiple-criteria analysis (MCA) to generate the Compactness Composite Index (CCI) for the examined neighborhoods (Abdullahi et al. 2015; Kotharkar et al. 2014; Rahman et al. 2022). The SAW method was chosen because it is a straightforward Multi-criteria Decision-Making (MCDM) approach that allows for the comprehensive evaluation of multiple spatial indicators with different units of measurement (Abdullahi et al. 2015; Cyril et al. 2019; Kotharkar et al. 2014). Each criterion in the MCDM evaluation was classified as a benefit or cost criterion based on how its value affected compactness (Cyril et al. 2019).

Data Normalization
To ensure the comparability of data that were measured using different scales, data normalization was performed in this research (Cyril et al. 2019). The proportion-based normalization method, also known as the Max–Min Method, was chosen due to its capability to construct cost and benefit, and criteria (Huang et al. 2021). The study created a decision matrix with urban form indicators and neighborhoods as the two dimensions (In this study, all the variables except centrality were considered as benefit criteria, as they had a positive impact on the compactness level. The

SAW involved the development of a decision matrix (Table 3), normalizing data, and building a weighted normalized matrix (Table 3). to generate the final CCI scores for each NH. When there is better equality in the allocation of floor space, closer proximity between residential and commercial sectors, greater diversity, and functional connectedness, sites are classified as compact by the CCI. The equal class interval approach is used to obtain this categorization. (Tables 3 and 4) (Bahadur et al. 2014; Bahadure and Kotharkar 2015; Cyril et al. 2019; Goel and Mohan 2020; Kamble and Bahadure 2021; Kotharkar et al. 2012, 2014; Kotharkar and Bahadure 2019; Li and Zhang 2022; Nurmalini and Rahim 2018; Rahman et al. 2022). To normalize each benefit and cost criterion, the following formulas were utilized for beneficial criteria (Cyril et al. 2019; Huang et al. 2021; Lan et al. 2021; Rahman et al. 2022).

Equation 1- Equation for non-beneficial or cost criteria

$$R_{ij} = \begin{cases} \frac{x_{ij}}{Max\, x_{ij}} & \text{(When attribute benefit)} \\ \frac{Min\, x_{ij}}{x_{ij}} & \text{(When attribute cost)} \end{cases}$$

Table 3 Decision Matrix—Urban form variables are measured at the neighborhood level

Urban form variables are measured at the neighborhood level							
Parameters	CBD (1)	Iskcon (2)	Naroda (3)	Vadaj (4)	Maninagar (5)	Criteria (Cost/ Benefit)	Weightage
Population Density	783	73	997	317	1278	B	0.125
Built-up density	0.64	0.98	1.02	1.11	0.77	B	0.125
Entropy index	0.6960	0.6318	0.6143	0.4177	0.7592	B	0.125
Floor use mix	4.08	2.93	11.78	6.92	9.26	B	0.125
Urban green accessibility	67	97.9	75.5	60.4	79	B	0.125
Street connectivity	315.18	239.36	229.12	217.12	391.28	B	0.125
Transit stops	1.08	0.72	1.56	1.56	2.2	B	0.125
Centrality	2.8	8.7	7.6	5.6	2.4	C	0.125
Score	0.6629	0.5295	0.7362	0.5869	0.8100		
Rank	3	4	2	5	1		
Status	Moderate Compact	Low Compact	High Compact	Low Compact	High Compact		

Source Generatedby Author

Table 4 Comparison of the neighborhood with development control tools 01

Neighborhood	Existing CCI		Proposed CCI (DP2021)		Built-up area (Ha)			
	Existing score	Existing rank	Proposed score	Proposed rank	2001	2014	2023	DP 2021
CBD	0.66	3	0.89	2	229.9	245.5	257.8	543.4
Iskcon	0.52	5	0.64	4	179.18	287.05	394.05	412.95
Naroda	0.73	2	0.77	3	370.01	401.68	408.94	683.33
Vadaj	0.58	4	0.62	5	326.32	365.45	445.15	644.1
Maninagar	0.81	1	0.90	1	253.58	277.85	308.24	734.61

Source Generatedby Author

- R_{ij}: Performance rating;
- X_{ij}: Rows and columns of the matrix;
- Max_{ij}: Each row and column's maximum value;
- Min_{ij}: Each row and column's minimum value;

Equation 2- Equation for Composite Compactness Index (CCI)

$$\mathrm{CCI}_i = \sum_{j=1}^{M} w_j r_{ij}$$

- Where CCI_i = ith neighborhood's total score;
- M = urban form indicators;
- N = Number of neighborhoods;
- W_j = jth criterion's weight;
- r_{ij} = ith neighborhood's normalized value for the jth criteria.

Normalized Decision Matrix

After the decision matrix was normalized, a weighted normalized matrix was created by multiplying the dimensionless indicator values by their respective weights (Kjaeras 2021). The equation used for calculating the overall score of each neighborhood was determined (Abdullahi et al. 2015; Cyril et al. 2019; Kamble and Bahadure 2021; Kotharkar and Bahadure 2019; Kumavat 2013; Rahman et al. 2022).

8 Result

8.1 Composite Compactness Index (CCI)

The SAW-based MCDM technique was used to assess the general compactness of each neighborhood (Cyril et al. 2019; Kotharkar et al. 2014; Li and Zhang 2022; Nurmalini and Rahim 2018; Rahman et al. 2022).

In this study, all the variables except centrality were considered as benefit criteria, as they had a positive impact on the compactness level. The SAW involved the development of a decision matrix (Table 3), normalizing data, and building a weighted normalized matrix (Table 3) to generate the final CCI scores for each NH. When there is better equality in the allocation of floor space, closer proximity between residential and commercial sectors, greater diversity, and functional connectedness, sites are classified as compact by the CCI. The equal class interval approach is used to obtain this categorization (Tables 3 and 4) (Bahadur et al. 2014; Bahadure and Kotharkar 2015; Cyril et al. 2019; Goel and Mohan 2020; Kamble and Bahadure 2021; Kotharkar et al. 2012, 2014; Kotharkar and Bahadure 2019; Li and Zhang 2022; Nurmalini and Rahim 2018; Rahman et al. 2022).

The data shows the neighborhood categorization based on the CCI score. The CCI score is used to measure the compactness of a neighborhood, with higher scores indicating higher compactness. The neighborhood of Maninagar (Old City) has the highest CCI score and is ranked first, indicating that it is the most compact neighborhood in terms of built-up area, FSI, and building height. Naroda (Industrial area) is ranked second, indicating that it is also highly compact. CBD is ranked third indicating moderate compactness, Vadaj (residential area) and Iskcon (commercial area) are ranked fourth and fifth, indicating low compactness.

This categorization provides useful information to understand the level of compactness and potential for urban development in each neighborhood. This is associated with a higher population density, polycentric clustering of spatial distribution, diverse land use, and a well-connected road network (Table 3).

An analysis of five distinct neighborhoods with varying characteristics was conducted to calculate the overall compactness of the AMC area. MCDM methods with the same parameters were employed, where distance from the core was considered a negative factor. Neighborhoods within a 3 km distance were assigned a weightage of 0.25, while those within 3–6 km and 6–9 km distances were given a weightage of 0.20 and 0.15, respectively. A composite compactness index was derived from this analysis, resulting in a score of 0.67 for the existing condition and 0.78 as per the development plan for 2021. This information provides insights into the city's compactness and can aid in determining the effectiveness of urban planning strategies in achieving sustainable development goals. Further analysis and evaluation are required to fully understand the implications of the compactness index and its potential impact on urban development and sustainability.

8.2 Study of Development Control Tools of the Neighborhood

Planning for land use and zoning, building specifications, and site planning are all covered under development regulations (Patel et al. 2018; Singh 2020). Health, greater levels of safety, and sustainability; to ease traffic and crowding; to prevent the development of different land uses; to avoid minor violations of building control regulations during construction, etc.

Building control regulations are most potent and effective in shaping the built environment (Sikarwar and Chattopadhyay 2016). They affect the environment, foster or discourage socialization, and establish the physical characteristics of the urban area (Banerjee et al. 2022; Vaghani et al. 2010). The maximum FSI, minimum lot sizes, maximum ground coverage, minimum road widths, number of floors or maximum building height, minimum open spaces, margins, and minimum building setbacks, parking, and other requirements mandated by planning regulations also mandate a certain level of land consumption (Kumar and Pushplata 2017; Patel et al. 2018). The regulations facilitate the transition toward compactness in the built environment. Table 4 illustrates the relationship between compactness and development control tools by examining the built-up area of each neighborhood through primary data collection. The analysis shows that Old City has the highest score of 0.81 with a built-up area of 308.24 hectares in 2023, followed by Naroda with 0.7362, indicating a higher level of urbanization in these areas.

8.3 Probable Factors Affecting Density & Compactness

Ahmedabad has not been able to fully leverage its potential to be developed as a compact, mixed-use city, as envisioned in the development plan. Various hurdles frequently obstruct the execution of compact urban development and growing density. The absence of cooperation between the DP, TPs, and LAPs is one of the major impediments. After the decade of the Development Plan, only 47% of the plan has been implemented, and there is a lack of coordination between TPs and LAPs, resulting in poor implementation of proposed plans.

Another barrier is the land market, where the value of land in the core city is higher than in the periphery, making it difficult to acquire land for new developments. Additionally, dedicated funds are absent for improving the density that orchestrates and stage development. Funding for TOD also tends to be low on the priority list of many projects, as the "soft costs" for TOD features such as public-space enhancements and streetscape are not prioritized.

Institutional barriers play a significant role in hindering compact urban development and increasing density. Lack of coordination between various planning agencies, and among authorities regarding the effectiveness of compactness in solving local problems is some of the critical institutional factors. These factors contribute to the ineffective implementation of compact urban development policies. Apart from institutional coordination, several other institutional factors contribute to the challenges faced in implementing compact urban development. The lack of a supportive institutional setting, weak market demand, skepticism among developers, excessive parking facilities, weak lender/investor support, zoning restrictions, and inadequate public transport services are significant barriers to the successful implementation of compact urban development policies (Fig. 8).

Table 5 presents an evaluation of the CCI score for various neighborhoods based on their built environment and the corresponding implications and actions required to enhance compact urban development and increase density.

9 Conclusion

The study looks at the elements that influence urban density and compactness in Ahmedabad, as well as the interaction between development control tools and urban morphology. It is evident that Ahmedabad has the potential to be developed as a compact, mixed-use city, but various barriers hinder its implementation. The study identified five areas of 2 km $\times$ 2 km as case study areas and examined the built environment and compactness score of each area. The CCI score was used to determine the corresponding implications and actions required to enhance compact urban development and increase density. The Old City, Naroda, CBD, and Iskcon areas were evaluated and found to have different compactness levels and development needs.

One of the significant barriers to the implementation of compact urban development and increasing density is the lack of coordination between the DP, TPs, and LAP. The land market is another obstacle, where the value of land in the core city is higher than in the periphery, making it difficult to acquire land for new developments. Institutional barriers such as a lack of coordination between various planning agencies, limited local expertise with Transit-Oriented Development, weak political support, and skepticism among authorities also contribute to the ineffective implementation of compact urban development policies (Cervero and Dai 2014; Li and Zhang 2022).

Polycentric cities are considered a preferable urban development model to improve density and compactness. Ahmedabad is already implementing policies to promote higher density and compactness in the city, such as TOD, TDR, and CBD areas that permit an FSI of up to 5.4. However, in practice, Ahmedabad has not been effective in capitalizing on the advantages of high-density high-rise structures. Achieving a compact policy requires the collaboration of the authorities and the community.

The effectiveness of density in urban development requires certain measures to be undertaken (Goel and Mohan 2020). One such step is promoting mixed land use, which entails assessing and providing for essential land uses like schools, clinics, and recreational spaces in required areas (Sankhe et al. 2010). To encourage

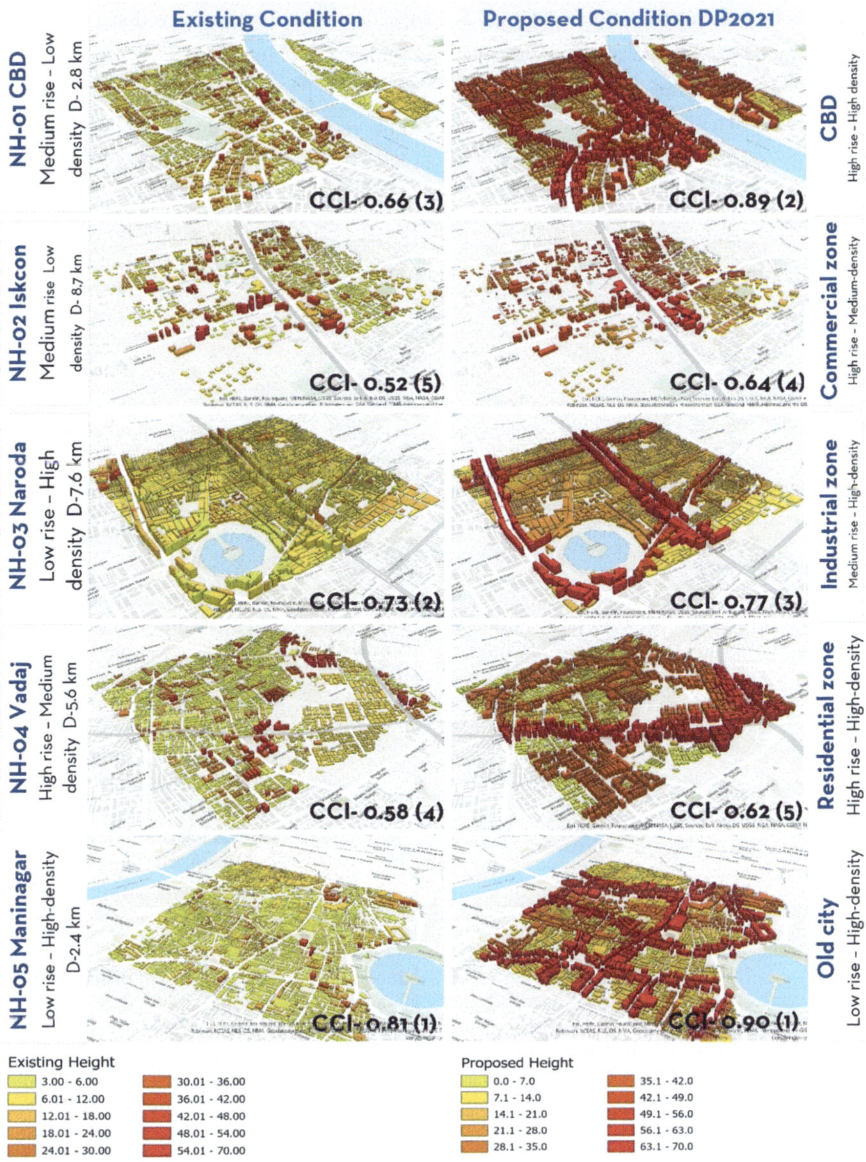

Fig. 8 Volumetric analysis (*Source* Generated by Author)

walking, walkable paths should be provided to essential land use sites, which can also strengthen the bond within the community (Pallares Barbera et al. 2011).

To ensure that infill or brownfield development is systematic and effective, policies must support the construction of newly built space and the increased density

Table 5 - Probable factors affecting density & compactness with the required action

Neighborhood	CCI score		Evaluation	Built environment	Implication	Required Actions
	Existing	Proposed				
Old City	0.81	0.9	High Compact	Low-rise – High density	A densely built environment with low-rise development	Need to reduce traffic congestion and restrict development and encroachment surrounding heritage structures
Naroda (Industrial area)	0.73	0.77	High Compact	Low-rise – High density	Nearby industrial areas increase density and major low-rise built areas due to LIG class	need slum rehabilitation schemes, need to improve infrastructure
CBD	0.66	0.89	Moderate Compact	High-rise – Low density	Unaffordable to buy, Land market—land price is higher, Lack of Dedicated Funding for TOD, lack of public open space	Regulate land market, introduce joint venture projects, improve institutional Setting, increase Investor Support
Iskcon (Commercial area)	0.52	0.64	Low Compact	High-rise – Low density	Restricted development by R3 zone, developed as a commercial hub due to low land price	Revised land use distribution, and improve PT and amenities, increase walkability
Vadaj (Residential area)	0.58	0.62	Low Compact	High-rise – Medium density	Land use changes restrict development (Green belt), lack of public open space & PT	Revised land use distribution, improve PT, green space, and amenities

Source Generatedby Author (Primary survey)

of existing areas (Hong and Fath 2009; Tsai 2005). The success of such policies is contingent on infrastructure upgrades, which must keep pace with the new development (Jadhav Rhujuta et al. 2015; Lan et al. 2021).

Policies and strategies can be implemented to regulate the distribution of population density in urban areas and promote compactness. High travel costs and weak public transport have contributed to the increase in density (Burton et al. 1996; Cervero and Dai 2014). As cities grow, multiple nodes emerge, resulting in polycentric cities. Therefore, policies should be implemented to ensure that necessary infrastructure and amenities are available in each node, such as providing adequate housing, public transportation, and job opportunities to reduce the need for long commutes (Kjaeras 2021; Mouratidis 2019).

In conclusion, several strategies can be implemented to improve density and compactness in urban areas, including urban renewal projects, creating real estate demand in the CBD, promoting mixed land use and walkable neighborhoods, providing adequate infrastructure and amenities, and regulating the distribution of population density (Kotharkar et al. 2014; Rogatka and Ramos Ribeiro 2015). These strategies require the collaboration of both the authorities and the community and must be supported by policies that keep pace with new developments (Bibri et al. 2020; Fan et al. 2019). Achieving compactness in urban development can lead to reduced traffic congestion, decreased pollution levels, and improved health benefits, making it imperative for cities like Ahmedabad to implement such policies effectively.

References

Abdullahi S, Pradhan B, Jebur MN (2015) GIS-based sustainable city compactness assessment using integration of MCDM, Bayes theorem and RADAR technology. Geocarto Int 30(4):365–387. https://doi.org/10.1080/10106049.2014.911967

Adhvaryu B (2010) Enhancing urban planning using simplified models: SIMPLAN for Ahmedabad, India. Prog Plan 73(3):113–207. https://doi.org/10.1016/j.progress.2009.11.001

Adolphson M (2022) Urban morphology, lifestyles and work-related travel behaviour: Evidence from the Stockholm region. Transp Res Interdiscip Perspect, 16. https://doi.org/10.1016/j.trip.2022.100706

AMC (2022) Ahmedabad administrative wing

Arbury J (2019) From urban sprawl to compact city-an analysis of urban growth management in Auckland acknowledgements

AUDA (2014a) Comprehensive development plan 2021 (Second Revised)

AUDA (2014b) General development control regulations

AUDA (2021) Urban planning and development practices of Ahmedabad development authority

Bahadur R, Vyas PN, Kothakar (2014) Compact City concept: its relevance and applicability for planning of Indian cities. 28th International PLEA Conference

Bahadure S, Kotharkar R (2015) Assessing sustainability of mixed use neighbourhoods through residents' travel behaviour and perception: the case of Nagpur, India. Sustainability (Switzerland) 7(9):12164–12189. https://doi.org/10.3390/su70912164

Banerjee S, Ching NYG, Yik SK, Dzyuban Y, Crank PJ, Pek Xin Yi R, Chow, WTL (2022) Analysing impacts of urban morphological variables and density on outdoor microclimate for tropical cities:

a review and a framework proposal for future research directions. In: Building and Environment, vol 225. Elsevier Ltd.. https://doi.org/10.1016/j.buildenv.2022.109646
Beatley Timothy (2000) Green urbanism, 1st ed. Island Press
Bertaud A (2004) The spatial organization of cities: deliberate outcome or unforeseen consequence?
Bertaud A (2013) Metropolis: a measure of the spatial organization of 7 large cities
Bertaud A (2018) Order without design: how markets shape cities. The MIT Press
Bertaud A, Richardson HW (2014) Transit and density: Atlanta, the United States and Western Europe
Bharti M, Mehrotra S, Lewis L, Orloff M, Olberding B (2020) Ahmedabad, India scaling up with contiguous replication of town planning schemes (eds: Mehrotra S, Lewis L, Orloff M, Olberding B). The World Bank, 1st ed. www.worldbank.org
Bhavsar N (2018) Land value capture and value capture financing—case of Ahmedabad. CEPT University
Bibri SE (2020) Compact urbanism and the synergic potential of its integration with data-driven smart urbanism: an extensive interdisciplinary literature review. Land Use Policy, 97. https://doi.org/10.1016/j.landusepol.2020.104703
Bibri SE, Krogstie J, Kärrholm M (2020) Compact city planning and development: Emerging practices and strategies for achieving the goals of sustainability. Dev Built Environ, 4. https://doi.org/10.1016/j.dibe.2020.100021
Boyko CT, Cooper R (2011) Clarifying and re-conceptualising density. Progress in Planning 76(1):1–61. https://doi.org/10.1016/j.progress.2011.07.001
Burton E, Jenks M, Williams K (1996) The compact city a sustainable urban form? Routledge
Byahut S, Patel B, Mehta J (2020) Emergence of sub-optimal land utilization patterns in Indian cities. J Urban Des 25(6):758–777. https://doi.org/10.1080/13574809.2020.1752646
Census of India (2011) District census handbook, Ahmedabad district 2011
Cervero R, Dai D (2014) BRT TOD: Leveraging transit oriented development with bus rapid transit investments. Transp Policy 36:127–138. https://doi.org/10.1016/j.tranpol.2014.08.001
Christofer Robert (2012) Compact city (High density with mixed uses)
Conticelli E (2019) Compact city as a model achieving sustainable development, pp 1–10. https://doi.org/10.1007/978-3-319-71061-7_35-1
Cyril A, Mulangi RH, George V (2019) Development of a GIS-based composite public transport accessibility index. J Urban Environ Eng 13(2):235–245. https://doi.org/10.4090/juee.2019.v13n2.235245
Dadras M, Shafri HZM, Ahmad N, Pradhan B, Safarpour S (2014) A combined fuzzy MCDM approach for identifying the suitable lands for Urban development: An example from bandar ABBS, Iran. J Urban Environ Eng 8(1):11–27. https://doi.org/10.4090/juee.2014.v8n1.011027
Ewing R, Pendall R, Chen D (2013) Smart growth America
Fan P, Lee YC, Ouyang Z, Huang, S. L. (2019a). Compact and green urban development - Towards a framework to assess urban development for a high-density metropolis. Environ Res Lett 14(11). https://doi.org/10.1088/1748-9326/ab4635
Fulton WB, Pendall R (2001) Who sprawls most?: how growth patterns differ across the US
Glaster G, Hanson R, Ratcliffe MR, Wolman H, Coleman S, Freihage J (2012) Wrestling Sprawl to the ground: defining and measuring an elusive. Housing Policy Debate, 681–717
Goel R, Mohan D (2020) Investigating the association between population density and travel patterns in Indian cities—an analysis of 2011 census data. Cities, 100. https://doi.org/10.1016/j.cities.2020.102656
Government of Gujarat (2022) Gujarat GDP Report
Haghani M, Sabri S, De Gruyter C, Ardeshiri A, Shahhoseini Z, Sanchez TW, Acuto M (2023) The landscape and evolution of urban planning science. Cities, 136. https://doi.org/10.1016/j.cities.2023.104261
Hess PM, Moudon AV, Snyder MC, Stanilov K (1999) Site Design and Pedestrian Travel. Transp. Res. Record, 9–19

Hong, Fath M (2009) Measurement and spatial distribution of urban land use compactness in Chaoyang District of Beijing, China. http://pure.iiasa.ac.at/9107/
Hong S, Hui EC-M, Lin Y (2022) Relationship between urban spatial structure and carbon emissions: a literature review. In: Ecological indicators, vol 144. Elsevier B.V. https://doi.org/10.1016/j.ecolind.2022.109456
Huang Q, Xu C, Jiang W, Yue W, Rong Q, Gu Z, Su M (2021) Urban compactness and patch complexity influence PM2.5 concentrations in contrasting ways: Evidence from the Guangdong-Hong Kong-Macao Greater Bay Area of China. Ecological Indicators, 133. https://doi.org/10.1016/j.ecolind.2021.108407
Jabareen, Rafeq Yosef (2006) Sustainable urban forms: Their typologies, models, and concepts. J Plan Educ Res 26(1):38–52. https://doi.org/10.1177/0739456X05285119
Jabareen YR (2006) Sustainable urban forms: their typologies, models, and concepts. J Plan Educ Res 26(1):38–52. https://doi.org/10.1177/0739456X05285119
Jadhav Rhujuta, Kiran Pratyusha, Motwani Rhea (2015) Compact city sustainability concepts. COEP
Jayasinghe A, Madusanka NBS, Abenayake C, Mahanama PKS (2021) A modeling framework: To analyze the relationship between accessibility, land use and densities in urban areas. Sustainability (Switzerland) 13(2):1–18. https://doi.org/10.3390/su13020467
Jenks M, Burgess R (2001) Compact cities: sustainable urban forms for developing countries (compact city), 1st ed. Routledge
Jenks M, Burton E, Williams K (1996) Debating urban compaction THE Compact city: a sustainable urban form?
Jia Y, Tang L, Xu M, Yang X (2019) Landscape pattern indices for evaluating urban spatial morphology—A case study of Chinese cities. Ecol Ind 99:27–37. https://doi.org/10.1016/j.ecolind.2018.12.007
Jones C, Macdonald C (2004) Sustainable urban form and real estate markets, 2–5
Kamble T, Bahadure S (2021) Investigating application of compact urban form in central Indian cities. Land Use Policy, 109. https://doi.org/10.1016/j.landusepol.2021.105694
Karteek G (2015) Is FSI dependent on land availability and densities? a comparative review of FSI in Indian cities. Eur J Sustain Dev 4(2). https://doi.org/10.14207/ejsd.2015.v4n2p27
Kjaeras K (2021) Towards a relational conception of the compact city. Urban Studies 58(6):1176–1192. https://doi.org/10.1177/0042098020907281
Knaap GJ, Song Y, Ewing R, Clifton K (2013) Seeing the elephant: Multi-disciplinary measures of urban sprawl
Kotharkar R, Bahadure P (2019) Achieving compact city form through density distribution: case of Indian cities. Am Soc Civil Eng. https://doi.org/10.1061/(ASCE)
Kotharkar R, Bahadure PN, Vyas A (2012) Compact City concept: it's relevance and applicability for planning of Indian cities
Kotharkar R, Bahadure P, Sarda N (2014) Measuring compact urban form: a case of Nagpur city, India. Sustainability (Switzerland) 6(7):4246–4272. https://doi.org/10.3390/su6074246
Kraff NJ, Wurm M, Taubenböck H (2020) The dynamics of poor urban areas - analyzing morphologic transformations across the globe using Earth observation data. Cities, 107. https://doi.org/10.1016/j.cities.2020.102905
Kumar A, Pushplata (2017) Problems and prospects of building regulations in Shimla, India—a step towards achieving sustainable development. Int J Sustain Built Environ 6(1):207–215). Elsevier B.V. https://doi.org/10.1016/j.ijsbe.2017.03.009
Kumavat R (2013) Compact development frame work for Indian cities case of Pune City. CEPT
Lan T, Shao G, Xu Z, Tang L, Sun L (2021) Measuring urban compactness based on functional characterization and human activity intensity by integrating multiple geospatial data sources. Ecol Ind 121. https://doi.org/10.1016/j.ecolind.2020.107177
Li Y, Zhang S (2022) Applied research methods in urban and regional planning. In: Applied research methods in urban and regional planning. Springer International Publishing. https://doi.org/10.1007/978-3-030-93574-0

M M. (n.d.). Sustainable urban form
Mahadevia D, Desai R, Vyas S (2014) City profile: Ahmedabad centre for urban equity (CUE). www.cept.ac.in/cue
Mckinsey (2010) India's urban awaking building inclusive cities sustainable economic growth. Urban Studies
Mouratidis K (2019) Compact city, urban sprawl, and subjective well-being. Cities 92:261–272. https://doi.org/10.1016/j.cities.2019.04.013
Mouratidis K (2022) COVID-19 and the compact city: implications for well-being and sustainable urban planning. Sci Total Environ 811. https://doi.org/10.1016/j.scitotenv.2021.152332
Nallathiga R (2005) The impacts of density regulation on cities and markets: Evidence from Mumbai. Int J Regul Gov 5(1):13–39. https://doi.org/10.3233/IJR-120043
Nargunde Aadish (2011) Urban growth patterns in landuse planning of Ahmedabad—is compact city concept achievable? CEPT
Neuman M (2005) The compact city fallacy. J Plan Educ Res 25(1):11–26. Sage Publications Inc. https://doi.org/10.1177/0739456X04270466
Nurmalini, Rahim R (2018) Study approach of simple additive weighting for decision support system
OECD (2010) Regulatory policy and the road to sustainable growth regulatory policy and the road to sustainable growth. OECD. www.oecd.org
OECD (2012) Compact city policies: a comparative assessment. ECD Green Growth Studies, 123–158
OECD (2017) The governance of land use. OECD
Pallares Barbera, Badia A, Duch J (2011) The need for a new city and service provision. Urbani Izziv, 79–92
Parekh V, Makwana H (2020) Transformation of urban structure and emergence of higher-order commercial areas—case of Ahmedabad, India. Proceedings Article, 189–199. https://doi.org/10.38027/N182020ICCAUA316369
Patel B, Byahut S, Bhatha B (2018) Building regulations are a barrier to affordable housing in Indian cities: the case of Ahmedabad. J Housing Built Environ 33(1):175–195. https://doi.org/10.1007/s10901-017-9552-7
Rahman MH, Islam MH, Neema MN (2022) GIS-based compactness measurement of urban form at neighborhood scale: The case of Dhaka, Bangladesh. J Urban Manag 11(1):6–22. https://doi.org/10.1016/j.jum.2021.08.005
Rawat Kuhoo (2008) Envisaging residential zone conversion in Ahmedabad development plan by 2011 to ensure appropriate compactness of city form. CEPT
Rogatka K, Ramos Ribeiro RR (2015) A compact city and its social perception: a case study. Urbani Izziv 26(1):121–131. https://doi.org/10.5379/urbani-izziv-en-2015-26-01-005
Sankhe S, Vittal I, Dobbs R, Mohan A, Gulati A, Ablett J, Gupta S, Kim A (2010) India's urban awakening: building inclusive cities, sustaining economic growth. Mumbai
Shah DP, Patel DP (2021) A comparison between national air quality index, india and composite air quality index for Ahmedabad, India. Environ Challenges, 5. https://doi.org/10.1016/j.envc.2021.100356
Shivanand Swamy M (2017) Transit oriented development—a process oriented approach centre of excellence in urban transport
Sikarwar A, Chattopadhyay A (2016) Change in land use-land cover and population dynamics: A town-level Study of Ahmedabad city sub-District of Gujarat. Int J Geomat Geosci 7(2). http://glovis.usgs.gov
Singh Madhu (2011) What is a compact city? how could it be measured? CEPT
Singh Pratibha (2016) Assessment of TOD policies and implementation process cases of Delhi-Ahmedabad
Singh R (2020) Land utilization in Indian cities and the role of development control regulation: a case study of Ahmedabad

Song Y, Knaap G (2004) Measuring urban Sprawl: Is Portland winning the war on Sprawl? J Am Plan Assoc, 210–225

Stratigea A, Papadopoulou CA, Panagiotopoulou M (2015) Tools and technologies for planning the development of smart cities. J Urban Technol 22(2):43–62. https://doi.org/10.1080/10630732.2015.1018725

Nagabhoina T (2022) Impact of Development control regulations on social life in urban transport hubs—a case of Secunderabad. J Plan, Arch Des 2(1):1–8

Thara D (2013) Draft development plan review

Tsai YH (2005) Quantifying urban form: compactness versus "sprawl." Urban Stud 42(1):141–161. https://doi.org/10.1080/0042098042000309748

Tsai Y-H (2016) Quantifying urban form: compactness versus "Sprawl." Sage J

UN Habitat (2015) Issue paper on urban and spatial planning and design

Vaghani KB, Shah NC, Chauhan KA (2010) Impact of building byelaws on housing: a comparative study of development control rules of Jaipur and Pune. India J 7(2):34–40

Wang M (2022) Building a compact city—spatial planning in Yinchuan City, Western China, 1st ed. Springer Cham

Wenjian Pan, Juan Du (2020) Towards sustainable urban redevelopment: urban design informed by morphological patterns and ecologies of informal settlements. Urban Ecology, Chapter 20, pp 377–411

Wilbur Smith Associate (2012) Study on traffic and transportation policies and strategies in urban. Areas in India

Yao J, Zhang X, Murray AT (2018) Spatial optimization for land-use allocation: accounting for sustainability concerns. Int Reg Sci Rev 41(6):579–600. Sage Publications Inc. https://doi.org/10.1177/0160017617728551

Zhi Cai, Matthias Demuzere, Yan Tang, Yongman Wan (2022) The characteristic and transformation of 3D urban morphology in three Chinese mega-cities. Cities, 131

A Comparative Analysis of Commercial Centres at Neighbourhood Level in Traditional and Planned Indian Cities

Archiman Biswas, Subrata Chattopadhyay, and Haimanti Banerji

1 Introduction

In the context of urban planning, neighbourhoods are defined as spatial units characterized by an array of spatially based attributes associated with clusters of residences, and social cohesion, often in conjunction with other land uses (Galster 2001). The social, economic, and physical characteristics differentiate one neighbourhood from another (Rohe 2009). Neighbourhoods have evolved from self-contained units with essential amenities to collectives fostering social interaction (Un-Habitat 2016). They provide shelter, security, and access to infrastructure, emphasizing safety, cleanliness, and sustainability (UN 1992; Choguill 2008). In the Indian context, traditional neighbourhoods are known as "*mahallas*" or "*paras*", which are compact, small spatial units characterized by narrow streets and mixed land uses. *Mahallas* and *paras* usually grew organically and densely around commercial, religious, or cultural centres (Narayanan 2015). Unlike many planned neighbourhoods, *mahallas* have a strong sense of community and social cohesion fostered by close-knit relationships among residents. The interconnectedness of residences, public and semi-public spaces, and commercial centres promote community participation, social interaction, and social cohesiveness (Dhingra and Chattopadhyay 2016). The narrow streets and compact layouts of mahallas make them pedestrian-friendly (Dhingra et al. 2016). Several studies explained that traditional Indian neighbourhoods could outperform newly planned neighbourhoods in terms of sustainability on various aspects (Lara et al. 2016; Dhingra and Chattopadhyay 2016; Sen et al. 2017; Dhingra et al. 2017; Balducci and Ferrara 2018; Axelsson and Granath 2018).

The neighbourhood commercial centre (NCC) is a crucial element in achieving neighbourhood sustainability, as highlighted by various theories and researchers

A. Biswas (✉) · S. Chattopadhyay · H. Banerji
Indian Institute of Technology Kharagpur, Kharagpur, West Bengal 721302, India
e-mail: biswasarchiman9@kgpian.iitkgp.ac.in

M. Firoz C et al. (eds.), *Urban Planning and Design for Megacities in the Global South*, Advances in 21st Century Human Settlements,
https://doi.org/10.1007/978-981-97-8370-0_8

(Qureshi et al. 2002; Perry 1929; UN 1992; Choguill 2008; Larsen 2008; Rohe 2009; Johnson 2010; Aulia et al. 2017). Perry's neighbourhood model proposed a NCC for each neighbourhood to provide residents with daily necessities and essential goods within a convenient travel distance (Perry 1929). These NCCs played a vital role in promoting a sense of community and social interaction within the neighbourhood (Rohe 2009). In the plan for Radburn, Clarence Stein and Henry Wright recognized the significance of neighbourhood commercial centres in providing essential needs and fostering social cohesion among residents (Larsen 2008). The United Nations (UN) highlighted the importance of these commercial centres in economic growth, poverty reduction, and employment generation at the neighbourhood level (UN 1992). The UN further emphasized that these commercial centres should be accessible, inclusive, and vibrant. Different types of NCCs are present in the Indian context, including informal open-air markets, weekly markets, *mandis*, hawker markets, organically grown street markets in traditional neighbourhood, and multi-storeyed market buildings in planned neighbourhoods (Lakhani 2021).

In many Indian traditional neighbourhoods, NCCs grew organically along the streets. However, in the present-day context, several planned neighbourhoods incorporate multistorey buildings as NCCs. The literature review indicates negligible studies have compared NCCs between planned and traditional neighbourhoods based on sustainability indicators. This comparison can provide a foundation for designing future markets that adhere to principles of sustainable urban development. To bridge this gap, this chapter aims to compare the street markets in traditional neighbourhoods (SMTN) and market buildings in planned neighbourhoods (MBPN) with the help of sustainability indicators through case studies in the Indian context. The specific objectives of this study are as follows:

i. Identify the factors associated with the NCC that contribute to the sustainability of the neighbourhood.
ii. Compare the SMTNs and MBPNs based on the identified sustainability indicators and provide recommendations for enhancing the overall sustainability of NCCs in future developments.

2 Literature Review

Over time, the concept and definition of neighbourhoods have evolved. In the early twentieth century, Perry conceptualized neighbourhoods as self contained units occupying less than 500 acres and accommodating populations ranging from 5,000 to 9,000 (Perry 1929). Perry's neighbourhood model aimed to provide diverse essential amenities, such as schools, commercial centres, and green spaces, while fostering a strong sense of community and citizen participation (Choguill 2008). Another planner, Clarence Stein, described neighbourhoods as collectives of individuals sharing a common bond, emphasizing the importance of social interaction and diversity in housing and land use mix for neighbourhood planning (Larsen 2008). The United Nations defined neighbourhoods as small identifiable spaces within urban areas that provide shelter, security, privacy, and access to basic infrastructure such as educational institutions, healthcare facilities, and commercial centres, with a focus on safety, cleanliness, and sustainability (UN 1992). Neighbourhood Commercial Centres (NCCs) are essential for neighbourhood sustainability, providing convenient access to necessities and fostering community interaction (Lara et al. 2016; Dhingra and Chattopadhyay 2016; Sen et al. 2017; Dhingra et al. 2017; Balducci and Ferrara 2018; Axelsson and Granath 2018). The UN highlights their role in economic growth, poverty reduction, and employment generation at the local level, emphasizing the importance of accessibility, inclusivity, and vibrancy (UN 1992). The objective of this literature review is to identify the indicators related to sustainability of the NCCs.

Existing literature mainly discusses four major dimensions of sustainability of NCC: economic sustainability (Dillard et al. 2008; Rydin 2013; Vaughan 2018), accessibility (Johnson 2010; Roger Tym and Partners 2010; AlWaer et al. 2011; Rydin 2019), environmental sustainability (Dillard et al. 2008; Rydin 2019), and social sustainability (Dillard et al. 2008; Hall 2012; Bowen McBeath et al. 2013; Zukin et al. 2015).

NCCs play a crucial role in promoting economic sustainability within neighbourhoods (Rydin 2019). These centres support local businesses, encourage entrepreneurship, and create job opportunities, contributing to the overall economic vitality of the community (Wells and Watson 2006). Thriving NCCs have been shown to attract customers and generate revenue, thereby sustaining the economic well-being of the neighbourhood and benefiting the local community (Johnson 2010). Moreover, NCCs serve as entry points for small and medium-sized enterprises (SMEs) in affordable locations (Vaughan 2018). This aspect is particularly beneficial for promoting entrepreneurship and inclusivity, empowering individuals, and supporting the growth of local residents (Vaughan 2018; Rydin 2019). Research conducted by the London Assembly (London Assembly Planning and Housing Committee 2010) has revealed compelling evidence that the profits generated within neighbourhood areas tend to circulate within the community. Specifically, the study found that approximately 50% of the turnover of local shops is reinvested back into the neighbourhood, in stark contrast to a mere 5% for large supermarket chains. This stark disparity highlights the significant role that neighbourhood businesses play in ensuring that economic benefits remain within the local community. Furthermore, NCCs offer a variety of goods and services, ensuring consumer choice and preventing over-reliance on larger commercial establishments (Nathan et al. 2012). This diversity fosters a robust and resilient local economy, enhancing its long-term sustainability (Moser 2010).

NCCs prioritize accessibility, ensuring that residents have convenient access to goods and services (Rydin 2019). These centres are strategically located within or near residential neighbourhoods, allowing residents to easily reach them on foot or by bicycle (Johnson 2010). The physical proximity of NCCs reduces barriers to access, making it easier for individuals without cars or with limited mobility, such as the elderly and disabled, to meet their daily needs (AlWaer et al. 2011). The availability of local shops within walking distance reduces barriers to access, allowing these individuals to meet their daily needs more easily (Ministry of Housing 2018). Roger Tym and Partners (2010) explained that neighbourhood shopping areas often offer a range of goods at affordable prices, catering to a diverse population with different income levels. This affordability enhances accessibility for lower-income households and promotes social equity (Mehta 2009; Nathan et al. 2012). In summary, neighbourhood shopping areas prioritize accessibility by providing proximity, affordability, and inclusivity, ensuring that residents can conveniently access the goods and services they require for their daily lives.

NCCs actively contribute to environmental sustainability by promoting sustainable modes of transportation and effectively reducing carbon emissions (Zukin et al. 2015). Their convenient accessibility within neighbourhoods encourage residents to opt for walking or cycling as means to access goods and services, thereby decreasing reliance on cars and minimizing carbon-based emissions (Barton et al. 2012; Sharifi and Murayama 2015; Wolday 2023). This emphasis on active transportation aligns with the principles of sustainable urban development and fosters an environmentally friendly lifestyle (Verovšek et al. 2022). Moreover, NCCs demonstrate land use efficiency by concentrating amenities within a compact and walkable environment, effectively discouraging urban sprawl and mitigating the need for long commutes, which in turn reduces associated environmental impacts (Barton 2009). In addition to transportation-related measures, NCCs also serve as catalysts for implementing sustainable practices within the community. They can integrate energy-efficient buildings, capitalize on natural light and ventilation, and promote responsible electricity use (Rydin 2019). Furthermore, Pitt (2005) highlighted that NCCs can play an important role in solid waste management by implementing effective waste reduction initiatives and encouraging recycling and proper disposal practices. These comprehensive efforts collectively contribute to the creation of environmentally sustainable neighbourhoods that prioritize resource efficiency and environmental well-being.

Neighbourhood commercial centres (NCCs) play a pivotal role in fostering social sustainability within urban communities. By providing accessible goods and services within walking distance, these centres serve as vital gathering places that promote social interaction, community engagement, and a sense of belonging (Bowen McBeath et al. 2013). Additionally, NCCs contribute to the availability of public gathering spaces, which further enhances the social fabric of the neighbourhood (Dillard et al. 2008). NCCs frequently host a wide range of community events, creating opportunities for social interaction and nurturing a strong sense of community (Dillard et al. 2008; Hall 2012; Zukin et al. 2015). They provide opportunities for residents to meet, socialize, and engage in various activities, that induce social cohesion and community pride (Adams et al. 2001). NCCs that reflect the cultural diversity and unique character of the neighbourhood contribute to its identity and create a strong sense of place (Mehta 2009).

The literature review identified 20 indicators specifically associated with NCCs that can effectively gauge their sustainability (Table 1).

The literature review shows that a considerable amount of study has been done on these domains encompassing various aspects. This includes comparative analyses of sustainability between traditional and planned neighbourhoods (Lara et al. 2016; Dhingra and Chattopadhyay 2016; Sen et al. 2017; Dhingra et al. 2017; Balducci and

Table 1 Indicators identified from the literature review

Indicator	Source
Economic sustainability	
Revenue generated by shopkeepers	(Johnson 2010; London Assembly Planning and Housing Committee 2010)
Adequacy of number of shops in the NCC	(Vaughan 2007, 2018)
Presence of businesses that have been in operation for more than five years	(Rydin 2019)
Average salaries paid by businesses	(Rydin 2019)
Percentage of vacant stores	(Moser 2010)
Range and diversity of products and services	(Nathan et al. 2012; Rydin 2019)
Tax and rental revenue generated by the local government from NCC	(London Assembly Planning and Housing Committee 2010)
Number of jobs created at NCC	(Reimers and Clulow 2004; Wells and Watson 2006; Vaughan 2018)
Accessibility	
Ease of navigation in and around the NCC	(Johnson 2010)
Walkability to the NCC	(Barton 2009; Roger Tym and Partners 2010; Ministry of Housing 2018; Farahani et al. 2022; Verovšek et al. 2022; Dong et al. 2023)
Pedestrian safety measures on the way to NCC	(Nathan et al. 2012; Zukin et al. 2015)
Accessible for persons with disabilities	(AlWaer et al. 2011; Zhang et al. 2023)
Availability of the parking space	(Barton et al. 2012; Sharifi and Murayama 2015; Rydin 2019)
Shopping trip experience to NCC	(Mehta 2009; Wolday 2023)
Environmental sustainability	
Monthly electricity consumption by the shops in NCC	(Rydin 2019)
Utilization of natural light in NCC	(Rydin 2019)
Solid waste management practices in NCC	(Pitt 2005)
Social sustainability	
Availability of public gathering spaces	(Bowen McBeath et al. 2013; Cao et al. 2022)
Community events or activities hosted in the NCC premises	(Dillard et al. 2008; Zukin et al. 2015)
NCC enhance safety and security by providing natural surveillance	(Hall 2012; Farahani et al. 2022)

Ferrara 2018; Axelsson and Granath 2018), exploration of the multifaceted dimensions related to the sustainability of NCCs (Table 1), and assessing sustainability of SMTN (Vural-Arslan and Cahantimur 2011; Bradley and Sparling 2016; Tlemsani Bozdağ et al. 2022). However, we found only one article that conducted a comparative analysis between NCCs from traditional and planned neighbourhood (Scheer and Petkov 2007). Conducting such a comparative analysis can offer valuable insights into the strengths and weaknesses of each type of NCCs. This analysis can inform future neighbourhood planning efforts by identifying best practices and design principles that can be incorporated into sustainable neighbourhood development.

3 Materials and Methods

3.1 Study Area

For this study, two areas within the Kolkata urban agglomeration in India were chosen: North Kolkata and Salt Lake. Kolkata is among the world's 30 largest megacities, with a population of more than 10 million (Majumdar and Chatterjee 2021). Eight SMTNs located in the traditional neighbourhoods of North Kolkata, India, and fourteen MBPNs located in Salt Lake are selected. North Kolkata's neighbourhoods are characterized by high population density and accommodates residents from diverse income groups (Mandal et al. 2019). This area was the part of the capital city of British India (Bose 2012). It is approximately 8.8 sq. km in area which is demarcated by Baranagar at the north, Dum Dum and the Circular Canal at the east, Mahatma Gandhi Road at the south and the Hooghly River at the west. The NCCs in North Kolkata have evolved organically and serve as vibrant hubs for commerce and social interactions along the roads (Fig. 1). During our primary survey, the authors observed that these street markets exhibit a horizontal spread of mixed-use development. The majority of the buildings in these areas feature ground-floor commercial spaces, while the upper floors are commonly utilized for residential and various other purposes (Fig. 1d, e and f). These NCCs have a rich history of specialized trades and traditional craftsmanship, with local artisans and traders playing a significant role in the local economy (Sinha et al. 2023). Presently, these NCCs have a mix of both old and new shops (Chaudhuri 2014).

In North Kolkata, a total of eight closely located SMTNs are selected for this study, namely Raja Bazar Market, Bagmari Market, Telengabagan Market, Maniktala Market, Hatibagan Market, Garpar Market, Uttara Market, and Madhav Market (Fig. 2).

The second study area, Salt Lake is a planned satellite city of Kolkata designed to accommodate the city's growing population (Tošković 2008). It is divided into 71 blocks planned based on concept of the neighbourhood model by Perry and garden city concept by Howard (Banerjee et al. 2022). The selected study area (excludes Sector V of Salt Lake) is approximately 11 sq. km. Salt Lake houses

Fig. 1 SMTN from North Kolkata. **a** Activities on the street, **b** Narrow inner lanes, **c** Social events like Durga Puja inside the market premises, **d** Schematic section of typical SMTN, **e** Schematic section of narrow inner lanes, **f** Encroachment by informal retailers

residents belonging to the middle-income to high-middle-income group (Rumbach 2014). Within the study area, there are well-planned and evenly distributed NCCs at the block level, totalling fourteen in number (Fig. 4). These 14 MBPNs of Salt Lake were studied. These MBPNs in Salt Lake adopt a modular design with similar units, following a consistent approach with minor variations (Fig. 3f). The commercial centres offer a diverse mix of shops and feature a small courtyard at their centre (Fig. 3).

4 Data Collection

From the literature review, 20 sustainability indicators that are relevant to NCCs are identified. Among these indicators, 11 are focused on customers, 5 pertain to shopkeepers, and 4 are related to NCC-level observations. To gather the necessary data, a combination of survey and observation methods was employed. Both customers and shopkeepers from both types of NCCs were surveyed using structured questionnaires. The customer survey utilized a five-point Likert scale-based questionnaire (questions mentioned in Table 3), while shopkeepers were directly asked about various aspects such as the number of employees, average salaries, monthly income, rent payments, electricity bills, and years of establishment (Appendix 1). In the observational survey,

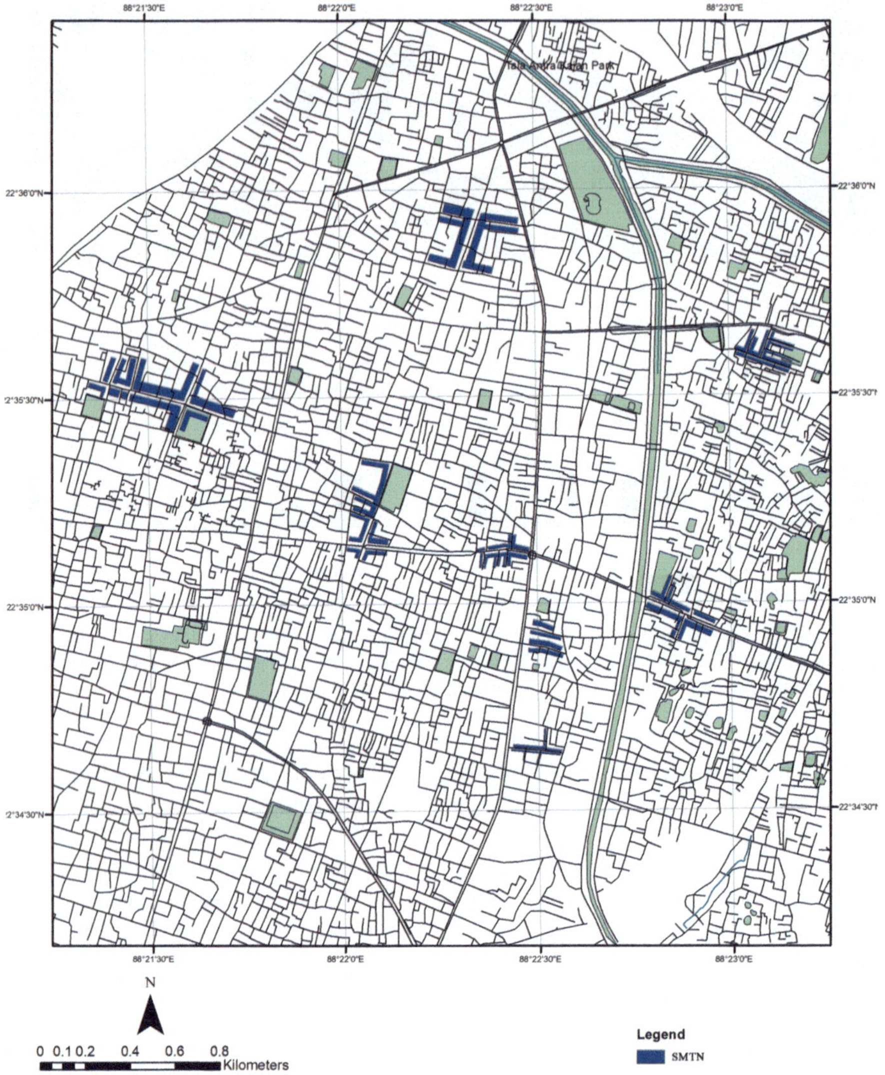

Fig. 2 Location of SMTNs in North Kolkata, India

NCC-level information was documented, including the number of vacant stores, the solid waste collection system, the basic planning and layout of the NCC, and the quality and character of both the NCC premises and the surrounding areas.

Prior to conducting the surveys, the minimum sample size was determined using Cochran's formula. A total of 216 customers from SMTNs (27 samples from each NCC using systematic random sampling) and 224 customers from MBPNs (16 samples from each NCC using systematic random sampling) participated in the survey. After removing outliers, responses with missing values, and errors, a total of

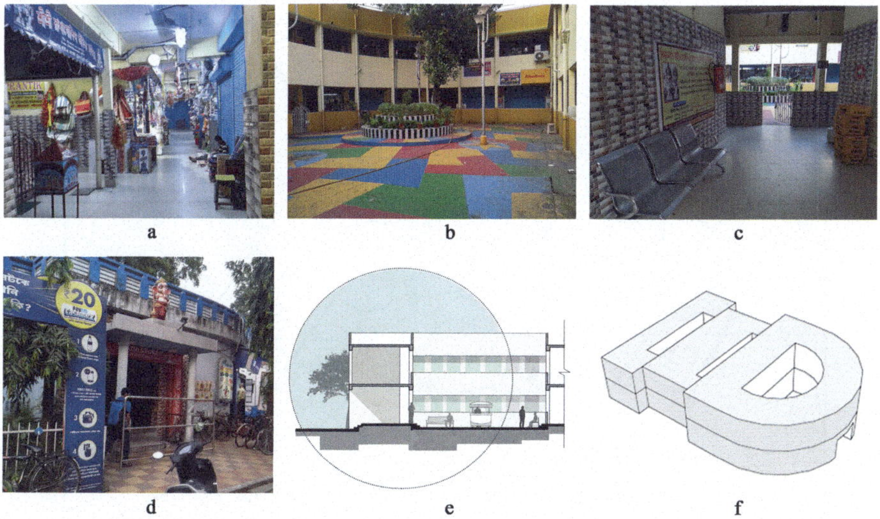

Fig. 3 MBPNs of Salt Lake, **a** temple inside the market building, **b** courtyard of IA Market, **c** sitting space inside the market building, **d** entrance: inadequate entrances width for wheelchairs and absence of ramp, **e** schematic section showing the MBPN courtyard, **f** basic form of the MBPNs in Salt Lake

400 samples (199 from SMTNs and 201 from MBPNs) were considered for further analysis. In the shop-owner survey, a total of 432 shopkeepers participated (222 from SMTNs and 210 from MBPNs) (survey questionnaire in Appendix 1). After eliminating outliers, responses with missing values, and errors, a total of 400 samples were used for analysis (Fig. 3).

5 Tools and Techniques

To compare SMTNs and MBPNs based on the identified sustainability indicators, a pairwise t-test was conducted with IBM SPSS 22. The pairwise t-test is a statistical technique used to compare the means of two groups or conditions. By analysing the data collected from surveys and observations, the pairwise t-test provided insights into the distinct contributions of these two types of NCCs. In accordance with the assumptions of the pairwise t-test, random sampling techniques were employed to collect a substantial sample size of over 384 participants. Additionally, measures were taken to ensure the integrity of the data by identifying and removing outliers (Fig. 3).

This statistical analysis facilitated a comprehensive comparison between SMTNs and MBPNs, allowing for a better understanding of their respective strengths and weaknesses in promoting neighbourhood sustainability. The findings of the pairwise t-test served as the basis for formulating proposals and recommendations for

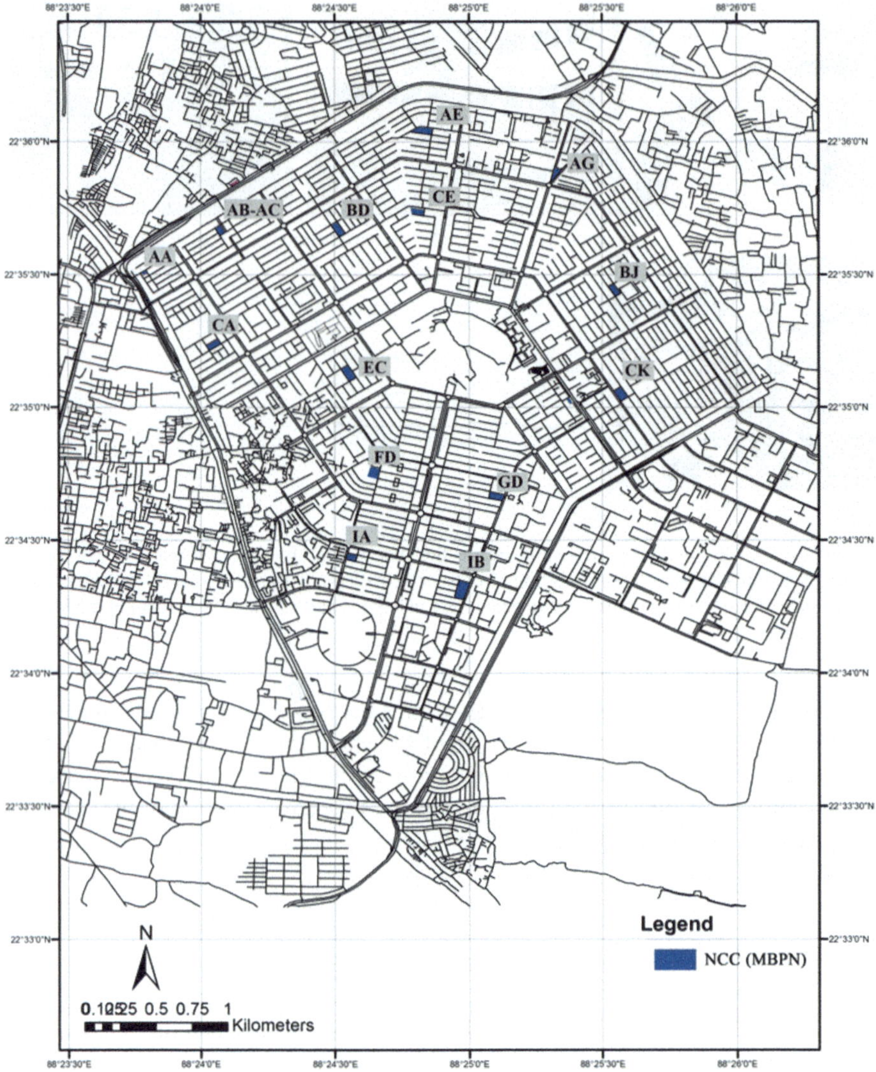

Fig. 4 Location of MBPNs in Salt Lake

future NCC planning. By identifying the areas where SMTNs excel or lag behind compared to MBPNs, the study aimed to provide actionable insights to improve the overall sustainability of neighbourhoods through targeted interventions and planning strategies (Fig. 5).

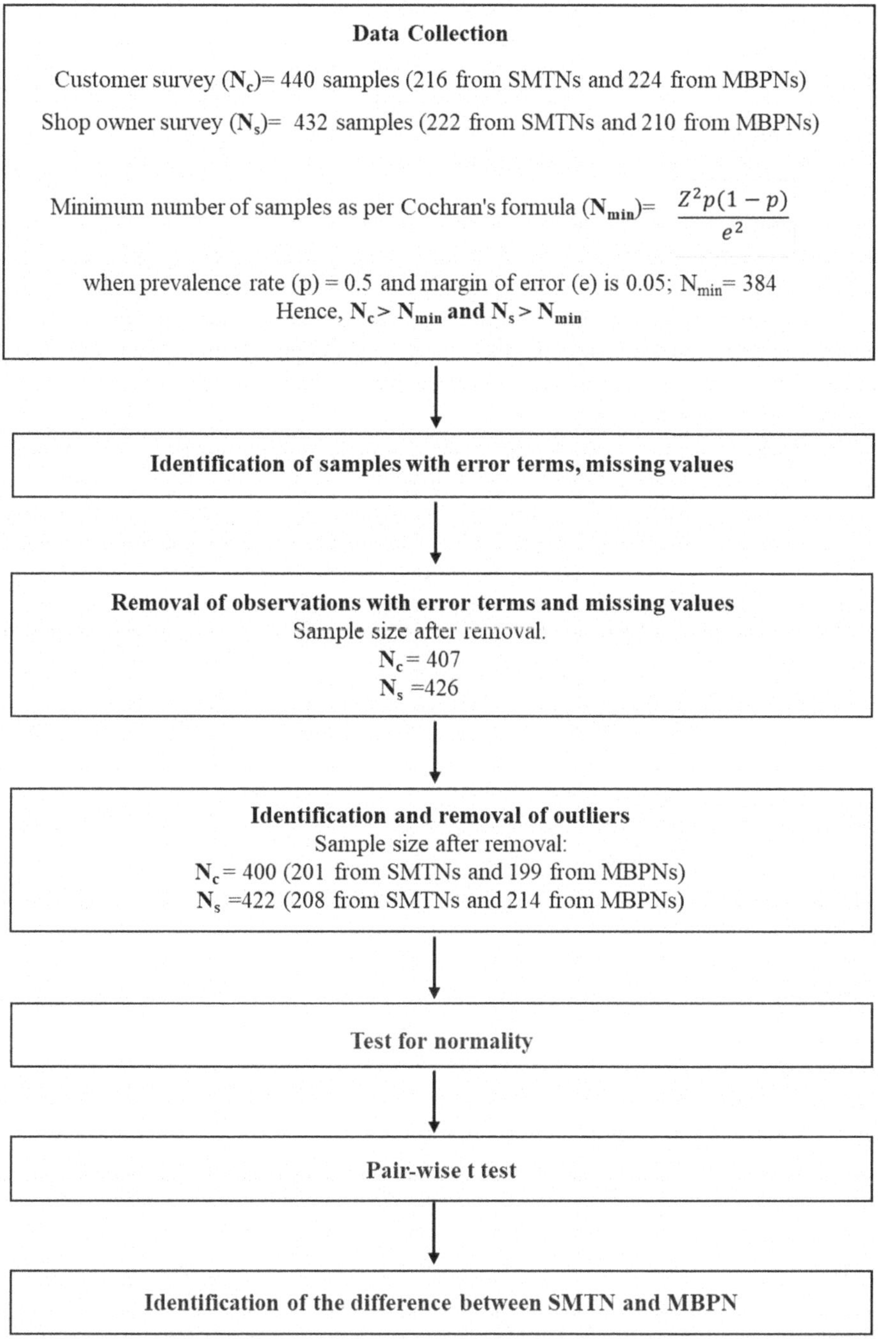

Fig. 5 Data analysis methodology

6 Results

To compare SMTNs and MBPNs, a pairwise t-test was conducted using data collected from customers of the NCCs, shopkeepers, and NCC-level information from 22 NCCs. The grouping variable used in the analysis was the NCC type (SMTN and MBPN).

To determine statistical differences between the two types of NCCs, the t-values and critical t-values were examined. The t-test indicated that the groups were not statistically different if the resulting t-value was lower than the critical t-value. Conversely, if the t-value exceeded the critical t-value, it meant that the groups were statistically different. Additionally, the 2-tailed significance value greater than 0.05 indicated no significant differences between the groups, and values less than 0.05 signified significant differences.

The results (Tables 2, 3 and 4 , 7th column) revealed that indicators such as rent or lease paid to the local government per month by shopkeepers, monthly electricity bills for shops, walkability to the NCC, availability of parking, accessibility for differently abled populations, level of pedestrian safety measures, presence of adequate signage and wayfinding systems, availability of public gathering spaces, NCC's contribution to enhancing safety and security through natural surveillance, utilization of natural light, and solid waste management exhibited a significance value lower than 0.05 and t-values higher than the critical t-values (Appendix 2). Hence, it can be concluded that SMTN and MBPN significantly differed in terms of these parameters.

Table 2 Pairwise t-test result of the data collected from the shopkeepers

Question	SMTN ($n = 208$)		MBPN ($n = 214$)		t	Sig. (2-tailed)	Difference
	Mean	S.D.	Mean	S.D.			
Number of employees	1.3558	1.33636	1.3318	1.33118	0.185	0.854	00.02399
Average monthly salary paid to your employees	4269.231	3110.232	4285.047	3140.387	−0.052	0.959	−15.81596
Monthly income as a shop owner	30,273.56	13,205.61	30,009.35	13,416.67	0.204	0.839	264.21190
Pay for rent or lease to the government per month	0	0.00	1	0.00	−6.616	0.000	0.28370
Monthly electricity bill for the shop	1125.769	658.9014	937.1963	523.4713	3.260	0.001	188.57297

Table 3 Pairwise t-test result of the data collected from the customers

Question	SMTN ($n = 201$)		MBPN ($n = 199$)		t	Sig. (2-tailed)	Difference
	Mean	S.D.	Mean	S.D.			
Do you think your neighbourhood's commercial centre has an adequate number of shops?	4.0151	0.80702	4.0299	0.81799	−0.182	0.856	−0.01478
Rate the walkability to your neighbourhood's commercial centre	3.9749	0.78777	3.0547	1.43596	7.934	0	0.92015
Rate the availability of parking at your neighbourhood's commercial centre	2.4623	1.13143	4.0199	0.80598	−15.87	0	−1.55759
Do you think your neighbourhood's commercial centre is well accessible for differently abled populations?	2.9246	0.78454	2.4478	1.11737	4.935	0	0.47686
Rate the level of pedestrian safety measures implemented for road crossings on the way to the neighbourhood commercial centre	2.5729	1.16925	3.5622	1.12132	−8.637	0	−0.98932
Evaluate the presence of adequate signage and wayfinding systems for easy navigation	1.9598	0.85776	2.9851	0.82145	−12.21	0	−1.02528

(continued)

Table 3 (continued)

Question	SMTN ($n = 201$)		MBPN ($n = 199$)		t	Sig. (2-tailed)	Difference
	Mean	S.D.	Mean	S.D.			
How satisfied are you with the range and diversity of shops available at the neighbourhood commercial centre?	3.4523	1.09935	3.4527	1.11311	−0.004	0.997	−0.00048
Rate the shopping trip experience to the NCC	1.9196	0.81251	1.9403	0.81634	−0.254	0.799	−0.0207
How satisfied are you with the availability of public gathering spaces in or around the neighbourhood commercial centre?	4.0553	0.85396	3.0000	0.84853	12.397	0	1.05528
To what extent does the neighbourhood commercial centre enhance safety and security by providing "eyes on the street" or natural surveillance?	3.9648	0.82497	2.4726	1.07726	15.542	0	1.49219
To what extent do the shops in the neighbourhood commercial centre utilize natural light effectively?	2.9648	0.81263	4.0896	0.81972	−13.78	0	−1.12473

On the other hand, the indicators (Tables 2, 3 and 4 , 7th column), including the number of employees, average monthly salary of employees, monthly income of shopkeepers, adequacy and variety of shops, percentage of vacant stores, percentage of businesses operating for more than five years, and percentage of new businesses opened in the last one year showed a significance value higher than 0.05 and resulted

Table 4 Pairwise t-test result of the data collected from the NCC

Indicators	SMTN ($n = 8$)		MBPN ($n = 14$)		t	Sig. (2-tailed)	Difference
	Mean	S.D.	Mean	S.D.			
Percentage of vacant stores	6.6583	1.19602	6.6049	1.20764	0.100	0.921	0.05342
Percentage of businesses that have been in operation for 5+ years	88.6958	2.50642	90.5132	2.71775	−1.550	0.137	−1.81740
Number of community events in NCC premises in last one year	9.2500	1.90863	2.3571	0.74495	12.160	0.000	6.89286
Practice segregation at the source in case of solid waste collection	0	0	0.6429	0.49725	−3.618	0.002	−0.64286

t-values lower than the critical t-values (Appendix 2). Thus, it is determined that SMTN and MBPN were not significantly different in relation to these parameters.

For further insights into the performance of both NCC types, mean values for MBPN and SMTN on all indicators are provided in Tables 2, 3, and 4 (2nd and 4th column). These tables offer a comprehensive overview of how each NCC type performs across different parameters.

7 Discussion

Results from both case studies helped us to compare the two different types of NCC: SMTN and MBPN. The sustainability of a NCC heavily relies on its economic factors. Result of the pairwise comparison using t-test shows that there is not significant difference between the MBPN and SMTN in terms of revenue generated by shopkeepers, adequacy of number of shops in the NCC, presence of businesses that have been in operation for more than five years, average salaries paid by businesses, percentage of vacant stores and range and diversity of products and services (Table 5). In case of MBPNs, the shopkeepers pay rent to the local government, which not only generates revenue for the government but also helps prevent encroachment by informal retailers. On the other hand, in SMTNs, shopkeepers do not pay any rent to the government. In SMTNs, it was found that a significant proportion of shops

(17.18%) were located in encroached areas such as footpaths and open spaces, occupied by informal retailers (Appendix 3). One notable difference between MBPNs and SMTNs is the provision for growth over time. SMTNs have the ability to expand along the street as the demand increases. In contrast, MBPNs have a fixed number of shops and total shop space. Despite these differences, both types of commercial centres contribute to economic sustainability by offering a wide variety of stores and providing a range and diversity of products and services. Additionally, they allow shopkeepers to adapt their businesses based on the demand of the local population.

Accessible NCCs offer numerous benefits, including reduced travel time, travel costs, and carbon footprint by minimizing reliance on motorized vehicles. Pairwise t-tests revealed that SMTNs have superior walkability and accessibility for individuals with disabilities compared to MBPNs (Table 6). Conversely, MBPNs provide better parking spaces, signage, and safety measures. Our primary survey showed that SMTNs, spread horizontally along the street, offer better perceived accessibility due to fewer level changes. In contrast, MBPNs, typically multi-storied buildings lack lifts, escalators, ramps, and have limited universal accessibility (Fig. 3d). However, the mean value of the "accessibility for differently-abled populations" score is significantly low for both types of NCCs (less than 3). In the primary survey, it was observed that both SMTN and MBPN have inadequate entrances width for wheelchairs, ramps, elevators, tactile pavers, and street furniture for resting of elderly people.

Pairwise T-test results show that SMTNs are more walkable compared to the MBPN. Better walkability in SMTNs reduces reliance on motorized vehicles for shopping trips. On the other hand, in Salt Lake, most residents prefer private vehicles due to the availability of parking spaces near MBPNs. Pairwise t-test result shows that MBPNs are significantly better than SMTN in "ease of navigation in and around the NCC". MBPNs feature well-designed market buildings with shop lists and wayfinding plans, simplifying the process of locating shops within the NCC. However, it is important to note that both types of NCCs performed poorly in terms of ease of navigation within and around the NCC, with mean scores below 3. The

Table 5 Comparison of NCCs based on economic sustainability (where red indicates worse performance, green indicates better performance, and grey indicates no significant difference)

Indicators	SMTN	MBPN
1. Revenue generated by shopkeepers		
2. Adequacy of number of shops in the NCC		
3. Presence of businesses that have been in operation for more than five years		
4. Average salaries paid by businesses		
5. Percentage of vacant stores		
6. Range and diversity of products and services		
7. Tax and rental revenue generated by the local government from NCC		
8. Number of jobs created at NCC		

Table 6 Comparison of NCCs based on accessibility (where red indicates worse performance, green indicates better performance, and grey indicates no significant difference)

Indicators	SMTN	MBPN
1. Ease of navigation in and around the NCC	[red]	[green]
2. Walkability to the NCC	[green]	[red]
3. Pedestrian safety measures on the way to NCC	[red]	[green]
4. Accessible for persons with disabilities	[green]	[red]
5. Availability of the parking space	[red]	[green]
6. Shopping trip experience to NCC	[grey]	[grey]

primary survey revealed insufficient signage systems in both types of NCCs, lacking braille-based signage or voice-activated controls for visually impaired individuals.

Regarding the shopping trip experience, the pairwise t-tests found no significant difference between MBPNs and SMTNs, and the mean shopping trip experience score for both NCC types was significantly low. This could be attributed to poor pedestrian infrastructure and poorly maintained footpaths, as observed during the field study.

SMTNs and MBPNs demonstrate contrasting characteristics in terms of environmental sustainability. The t-test analysis shows that MBPNs have lower average electricity usage and make better use of natural light compared to SMTNs. The well-planned nature of MBPNs allows for enhanced natural light penetration within shops, while SMTNs face limitations due to their organic growth. However, it is worth noting that the average monthly electricity consumption in both types of NCCs remains high, possibly due to the use of air conditioners and artificial lights during the daytime (Table 7).

The results of the t-test indicate that SMTNs outperform MBPNs in terms of social sustainability. A pairwise t-test analysis reveals that SMTNs offer a higher number of public gathering spaces and host a greater number of community events compared to MBPNs. Moreover, SMTNs contribute to neighbourhood sustainability by providing natural surveillance. The horizontal spread of the shops along the street in SMTNs allows for a diverse range of activities to take place at the ground-floor level. This configuration creates a sense of "eyes on the street", enhancing the perceived safety and security of the neighbourhood.

Table 7 Comparison of NCCs based on environmental sustainability (where red indicates worse performance, green indicates better performance, and grey indicates no significant difference)

Indicators	SMTN	MBPN
1. Monthly electricity consumption by the shops in NCC	[red]	[green]
2. Utilization of natural light in NCC	[red]	[green]
3. Solid waste management practices in NCC	[red]	[green]

Our results reveal that MBPNs and SMTNs are performing similarly in terms of economic sustainability parameters. This is one of the reasons behind the existence of both these types of NCCs in the present-day context.

Results suggest that due to the development context, the NCCs in the traditional neighbourhood are significantly different from the NCCs in the planned neighbourhood. Although both have positives and negatives on various factors, it is clear that street markets are a better fit for traditional neighbourhoods for their seamless integration, pedestrian-friendly design, and historical significance. These NCCs also cater to the preference for walking, offering accessible pathways for residents to interact and engage socially (Table 6). SMTNs serve as vital public gathering spaces, hosting numerous community events like social gatherings and *Durga puja* (Fig. 1c), a significant festival in West Bengal. Furthermore, SMTNs contribute to safety and security by providing "eyes on the street" or natural surveillance, creating a lively and secure environment for both residents and visitors (Table 8) (Amiri and Crain 2020). Another study comparing commercial centres based on morphology supports these findings (Scheer and Petkov 2007).

On the other hand, this study shows that market building offers several advantages in planned cities like Salt Lake. Their strategic design enhances urban layout, providing familiarity and convenience to residents (Table 6). The modular design (Fig. 3f) of MBPNs allows for efficient infrastructure planning and construction. The key learnings from these case studies and recommendations for the NCCs are listed in Table 9.

Table 8 Comparison of NCCs based on social sustainability (where red indicates worse performance, green indicates better performance, and grey indicates no significant difference)

Indicators	SMTN	MBPN
1. Availability of public gathering spaces		
2. Community events or activities hosted in the NCC premises		
3. NCC enhance safety and security by providing "eyes on the street" or natural surveillance		

Table 9 Key learnings and recommendations

Aspect	SMTNs	MBPNs
Site Selection	• Growth type: Organic • Location selection: Driven by the market demand and not pre-planned	• Growth type: Planned • Location selection: Planned to ensure equitable access to goods and services
Shop mix and shop type	• Shop mix and number driven by market forces • Number of shops increased with the increase in the population and demand	• At the initial stage, the number of shops and shop types in MBPNs was determined based on various planning guidelines • As the needs of the community change, shop mix of MBPNs changed to better meet the demands of the residents
Contextuality	• Blends harmoniously with traditional neighbourhoods	• Provides convenience and an identity in planned cities
Advantages	• Integrates well with local heritage • Enhances walkability, reduces reliance on vehicles • Adaptable to changing demands • Provides natural surveillance • Efficient utilization of land with the help of mixed used development	• Strategically designed for efficiency • Ability to generate revenue for local government • Efficient infrastructure and waste management
Disadvantages	• Expansion may encourage encroachments and informal commercial activities • Lacks proper waste disposal systems and public restrooms • Navigational challenges making it challenging to locate specific shops or services • Absence of barrier-free design considerations	• Limited capacity for expansions and modifications • Absence of barrier-free design considerations
Recommendations	• Implement formalization and regulation to prevent disorderly expansion and encroachments • Prioritize braille-based signage, voice-activated controls, wider entrances, ramps, and elevators to enhance accessibility for all • Install clear shop lists and wayfinding maps to ease navigation • Promote eco-friendly practices, waste segregation, and maximize natural light and ventilation • Planning authorities should focus on upgrading and renovating existing SMTNs with adequate consideration of their socio-economic and cultural context	• Incorporate braille-based signage, voice-activated controls, wider entrances, and ramps to create a barrier-free shopping experience • Future MBPNs can have a greater number of floors (3 or more) with office spaces on upper floors, encouraging mixed-use development • Provision for conversion of upper floors as per future commercial needs • Provision for designated vending zones along the street, to enhance natural surveillance and promote organized vending activities

8 Conclusion

This chapter presents a comparative analysis between SMTNs and MBPNs. The literature review identified four significant dimensions related to neighbourhood sustainability: economic, accessibility, environmental, and social, along with their corresponding indicators related to NCCs. This study used these sustainability indicators to compare SMTNs from North Kolkata with MBPNs from Salt Lake.

Further, the analysis found no significant difference in economic sustainability between SMTNs and MBPNs, except for MBPNs in Salt Lake generating higher rental revenue. SMTNs exhibited higher social sustainability with public spaces and community events, while MBPNs showed better environmental sustainability with lower electricity usage. This research also highlights the importance of considering context-specific factors for sustainable NCC development.

Due to the contextuality of development, the NCCs in the traditional neighbourhood significantly differ from the NCCs in the planned neighbourhood. Although both have positives and negatives in various factors, street markets are a better fit for traditional neighbourhoods than the market buildings. On the other hand, the market buildings offer several advantages in planned neighbourhoods like Salt Lake. In this study, the comparison between the two types of NCCs based on the identified four sustainability factors. It is important to note that these factors may be interconnected, and improvements in one factor could influence others. However, the study did not attempt to statistically assess the relative importance of different factors and the correlation between these factors, which can be addressed in future studies. Hence, while preparing development proposals for NCCs, planners should aim for holistic development keeping the context in mind. Comparative studies in various contexts in developing and developed countries can provide broader insights. Further research can also examine how NCCs influence the overall sustainability of neighbourhoods.

This chapter offers a comprehensive understanding of their respective performances and potential areas for improvement, contributing significantly to the advancement of sustainable urban development practices.

Acknowledgements This study is funded through the institute scholarship provided by the Ministry of Education (MoE), Government of India. The authors thank the Indian Institute of Technology Kharagpur for providing the work facilities.

Appendix 1: Questions Asked to Shopkeepers During Primary Survey

Sl. No.	Questions
1	How many employees do you have working in your shop?
2	What is the average salary you pay to your employees?
3	What is the approximate monthly income generated by your shop?
4	How much do you typically spend on rent payments for your shop?
5	What is the average amount you pay for electricity bills related to your shop?
6	How many years has your shop been established in its current location?

Appendix 2: Comparison of Calculated and Critical t-values

	DF	Calculated *t*-value	Critical *t*-value
Revenue generated by shopkeepers	420	0.204	± 1.965628
Adequacy of number of shops in the NCC	398	−0.182	± 1.965942
Presence of businesses that have been in operation for more than five years	20	−1.550	± 2.085963
Average salaries paid by businesses	420	0.052	± 1.965628
Percentage of vacant stores	20	0.100	± 2.085963
Range and diversity of products and services	398		± 1.965942
Tax and rental revenue generated by the local government from NCC	420	−6.616	± 1.965628
Number of jobs created at NCC	420	0.185	± 1.965628
Ease of navigation in and around the NCC	398	−12.210	± 1.965942
Walkability to the NCC	398	7.934	± 1.965942
Pedestrian safety measures on the way to NCC	398	−8.637	± 1.965942
Accessible for persons with disabilities	398	4.935	± 1.965942
Availability of the parking space	398	15.870	± 1.965942
Shopping trip experience to NCC	398	−0.254	± 1.965942
Monthly electricity consumption by the shops in NCC	420	3.260	± 1.965628
Utilization of natural light in NCC	398	−13.780	± 1.965942

(continued)

(continued)

	DF	Calculated t-value	Critical t-value
Solid waste management practices in NCC	20	−3.618	± 2.085963
Availability of public gathering spaces	398	12.397	± 1.965942
Community events or activities hosted in the NCC premises	20	12.160	± 2.085963
NCC enhance safety and security by providing "eyes on the street" or natural surveillance	398	15.542	± 1.965942

Appendix 3: Details Related to the NCCs

Attributes	SMTN	MBPN
Average number of shops	178	64
Average area of the shop	10.84	9.16
Percentage of shops in the encroached areas	1.82	17.18
Shopping trip mode choice of the shoppers (Walk: cycle: paratransit: motorized two wheelers: car)	48:14:6: 24:8	33:11:13: 23:20
Building type	G + 1 to G + 7 buildings with ground-floor commercial	G + 1 market building

References

Adams PC, Hoelscher SD, Till KE (2001) Textures of place: exploring humanist geographies, 461

AlWaer H, Sibley M, Lewis J (2011) Factors and Priorities for Assessing Sustainability of Regional Shopping Centres in the UK. Arch Sci Rev 51:391–402. https://doi.org/10.3763/ASRE.2008.5143

Amiri S, Crain DR (2020) Quantifying Jacobs ' notion of ' eyes upon the street ' in 3-dimensions. J Urban Des 25:467–485. https://doi.org/10.1080/13574809.2019.1691440

Aulia DN, Marpaung BO, Zahrah W (2017) Place shaping in residential area as a process of increasing livable neighbourhood

Axelsson K, Granath M (2018) Stakeholders' stake and relation to smartness in smart city development: Insights from a Swedish city planning project. Gov Inf Q 35:693–702. https://doi.org/10.1016/J.GIQ.2018.09.001

Balducci F, Ferrara A (2018) Using urban environmental policy data to understand the domains of smartness: An analysis of spatial autocorrelation for all the Italian chief towns. Ecol Indic 89:386–396. https://doi.org/10.1016/J.ECOLIND.2017.12.064

Banerjee S, Sengupta A, Kar S, et al (2022) Salt lake archives: a report

Barton H (2009) Land use planning and health and well-being. Land Use Policy 26:S115–S123. https://doi.org/10.1016/J.LANDUSEPOL.2009.09.008

Barton H, Horswell M, Millar P (2012) Neighbourhood Accessibility and Active Travel. Plan Pract Res 27:177–201. https://doi.org/10.1080/02697459.2012.661636

Bose S (2012) Historic buildings at Chitpur in Kolkata: problems and prospects through urban conservation and planning. Disp-Plan Rev 48:68–82. https://doi.org/10.1080/02513625.2012.702987

Bowen McBeath, Laura Nissen, Charlotte Goodluck EMB (2013) The intersection of disparity reduction and social sustainability: crafting a stronger foundation and alliance. Routledge

Bradley Q, Sparling W (2016) The impact of neighbourhood planning and localism on house-building in England. Hous, Theory Soc 34:106–118. https://doi.org/10.1080/14036096.2016.1197852

Cao Y, Heng CK, Fung JC (2022) Towards human-centric planning: Older people's everyday life in neighbourhood spaces in Singapore's public housing. Habitat Int 127:102649. https://doi.org/10.1016/J.HABITATINT.2022.102649

Chaudhuri D (2014) Analysis of the awareness of green products in the city of Kolkata. J Glob Mark 27:207–212. https://doi.org/10.1080/08911762.2014.880769

Choguill CL (2008) Developing sustainable neighbourhoods. Habitat Int 32:41–48. https://doi.org/10.1016/J.HABITATINT.2007.06.007

Dhingra M, Chattopadhyay S (2016) Advancing smartness of traditional settlements-case analysis of Indian and Arab old cities. Int J Sustain Built Environ 5:549–563. https://doi.org/10.1016/J.IJSBE.2016.08.004

Dhingra M, Singh MK, Chattopadhyay S (2016) Rapid assessment tool for traditional Indian neighbourhoods: a case study of Alwar Walled City in Rajasthan. Sustain Cities Soc 26:364–382. https://doi.org/10.1016/J.SCS.2016.06.015

Dhingra M, Singh MK, Chattopadhyay S (2017) Macro level characterization of historic urban landscape: case study of Alwar walled city. City, Cult Soc 9:39–53. https://doi.org/10.1016/J.CCS.2016.10.001

Dillard J, Dujon V, King MC (2008) Understanding the social dimension of sustainability. Underst Soc Dimens Sustain 1–300. https://doi.org/10.4324/9780203892978/UNDERSTANDING-SOCIAL-DIMENSION-SUSTAINABILITY-MARY-KING-VERONICA-DUJON-JESSE-DILLARD

Dong Y, Li F, Cao J, Dong W (2023) What neighborhood factors are critical to resident satisfaction with old neighborhoods? An integration of ground theory and impact asymmetry analysis. Cities 141:104460. https://doi.org/10.1016/J.CITIES.2023.104460

Farahani LM, Izadpanahi P, Tucker R (2022) The death and life of Australian suburbs: Relationships between social activity and the physical qualities of Australian suburban neighbourhood centres. City, Cult Soc 28:100426. https://doi.org/10.1016/J.CCS.2021.100426

Galster G (2001) On the nature of neighbourhood. Urban Studies 38:2111–2124. https://doi.org/10.1080/00420980120087072

Hall S (2012) City, Street and Citizen: The Measure of the Ordinary. City, Str Citiz Meas Ordinary 1–156. https://doi.org/10.4324/9780203118597/CITY-STREET-CITIZEN-SUZANNE-HALL

Johnson DL (2010) Origin of the Neighbourhood Unit. Plan Perspect 17:227–245. https://doi.org/10.1080/02665430210129306

Lakhani AR (2021) Informal in formal - street vending in neighbourhood markets of Chandigarh. Descriptio 3:

Lara AP, Da Costa EM, Furlani TZ, Yigitcanlar T (2016) Smartness that matters: towards a comprehensive and human-centred characterisation of smart cities. J Open Innov Technol Mark Complex 2:1–13. https://doi.org/10.1186/S40852-016-0034-Z

Larsen K (2008) Research in progress: the Radburn idea as an emergent concept: Henry Wright's regional city. Plan Perspect 23:381–395. https://doi.org/10.1080/02665430802102849

London Assembly Planning and Housing Committee (2010) Cornered shops: London's small shops and the planning system. GLA, London

Majumdar S, Chatterjee U (2021) Modelling urban growth using Urban growth deterministic model in Kolkata metropolitan area: a geo-statistical approach. Model Earth Syst Environ 7:2241–2249. https://doi.org/10.1007/S40808-020-00985-6/FIGURES/5

Mandal J, Ghosh N, Mukhopadhyay A (2019) Urban growth dynamics and changing land-use land-cover of megacity kolkata and its environs. J Indian Soc Remote Sens 47:1707–1725. https://doi.org/10.1007/S12524-019-01020-7/TABLES/13

Mehta V (2009) Look Closely and You Will See, Listen Carefully and You Will Hear: Urban Design and Social Interaction on Streets. J Urban Des 14:29–64. https://doi.org/10.1080/13574800802452658

Ministry of Housing C and LG (2018) National planning policy framework. MHCLG, London

Moser S (2010) Putrajaya: Malaysia's new federal administrative capital. Cities 27:285–297. https://doi.org/10.1016/J.CITIES.2009.11.002

Narayanan Y (2015) Religion, heritage and the sustainable city: Hinduism and urbanisation in Jaipur. Routledge, Abingdon, Oxon, New York, NY

Nathan A, Pereira G, Foster S et al (2012) Access to neighbourhood commercial destinations and associations with walking in older Australians. J Sci Med Sport 15:S46. https://doi.org/10.1016/j.jsams.2012.11.111

Perry C (1929) The Neighborhood Unit: A scheme of arrangement for the family-life community. Reg Plan New York Its Environs New York VII:636–637

Pitt M (2005) Trends in shopping centre waste management. Facilities 23:522–533. https://doi.org/10.1108/02632770510618480/FULL/PDF

Qureshi S, Memon IA, Aftab M, Talpur H (2002) Association between Objectively Measured Neighbourhood Built Environment and Walkability. Mehran Univ Res J Eng Technol 41:157–168. https://doi.org/10.3316/INFORMIT.263594976826073

Reimers V, Clulow V (2004) Retail concentration: a comparison of spatial convenience in shopping strips and shopping centres. J Retail Consum Serv 11:207–221. https://doi.org/10.1016/S0969-6989(03)00038-9

Roger Tym & Partners (2010) Greater London authority London SMALL shops study 2010

Rohe WM (2009) From local to global: One hundred years of neighborhood planning. J Am Plan Assoc 75:209–230. https://doi.org/10.1080/01944360902751077

Rumbach A (2014) Do new towns increase disaster risk? Evidence from Kolkata, India. Habitat Int 43:117–124. https://doi.org/10.1016/J.HABITATINT.2014.03.005

Rydin Y (2013) The future of planning. https://doi.org/10.2307/J.CTT9QGWG6

Rydin Y (2019) Planning for Sustainability: Lessons from Studying Neighbourhood Shopping Areas. Plan Pract & Res 34:522–536. https://doi.org/10.1080/02697459.2019.1679428

Scheer BC, Petkov M (2007) Edge city morphology: a comparison of commercial centers. J Am Plan Assoc 64:298–310. https://doi.org/10.1080/01944369808975987

Sen J, Majumdar M, Saha D, Chaudhuri A (2017) Smart Economy in smart cities Varanasi India: case of a smart traditional economy of knowledge-based institutional services and creative-cultural products. Adv 21st Century Hum Settlements, 553–576. https://doi.org/10.1007/978-981-10-1610-3_21/FIGURES/12

Sharifi A, Murayama A (2015) Viability of using global standards for neighbourhood sustainability assessment: insights from a comparative case study. J Environ Plan Manag 58:1–23. https://doi.org/10.1080/09640568.2013.866077/SUPPL_FILE/CJEP_A_866077_SM4950.DOCX

Sinha C, Mazumdar T, Dutta S, Saha S (2023) The industry of idol-making in Kolkata: a survey in the light of UNESCO recognition. J Indian Asian Stud. https://doi.org/10.1142/S271754132350002X

Tlemsani Bozdağ H, Benabbou R, Vural Arslan T (2022) A framework proposal for resilience assessment in traditional commercial centres: case of the historical bazaar of bursa as a resilient world heritage site. Herit Sci 10:1–21. https://doi.org/10.1186/S40494-022-00792-5/TABLES/10

Tošković D (2008) A review on salt lake city, Kolkata, India: Master planning and realization. Spatium 98–105. https://doi.org/10.2298/SPAT0818098T

Un-Habitat (2016) Planning Sustainable Cities. Nairobi

UN (1992) Global Strategy for Shelter to the Year 2000. 3 p

Vaughan L (2018) High street diversity. Suburb Urbanities 153–174. https://doi.org/10.2307/J.CTT1G69Z0M.15

Vaughan L (2007) The spatial syntax of urban segregation. Prog Plann 67:205–294. https://doi.org/10.1016/J.PROGRESS.2007.03.001

Verovšek Š, Juvančič M, Petrovčič S et al (2022) An integrative approach to neighbourhood sustainability assessments using publicly available traffic data. Comput Environ Urban Syst 95:101805. https://doi.org/10.1016/J.COMPENVURBSYS.2022.101805

Vural-Arslan T, Cahantimur A (2011) Revival of a traditional community engagement model for the sustainable future of a historical commercial district: Bursa/Turkey as a case. Futures 43:361–373. https://doi.org/10.1016/J.FUTURES.2011.01.003

Wells K, Watson S (2006) A politics of resentment: Shopkeepers in a London neighbourhood. Ethn Racial Stud 28:261–277. https://doi.org/10.1080/0141987042000315843

Wolday F (2023) The effect of neighbourhood and urban center structures on active travel in small cities. Cities 132:104050. https://doi.org/10.1016/J.CITIES.2022.104050

Zhang S, Wu W, Xiao Z et al (2023) Creating livable cities for healthy ageing: Cognitive health in older adults and their 15-minute walkable neighbourhoods. Cities 137:104312. https://doi.org/10.1016/J.CITIES.2023.104312

Zukin S, Kasinitz P, Chen X (2015) Global cities, local streets: Everyday diversity from New York to Shanghai. Glob Cities, Local Streets Everyday Divers from New York to Shanghai, 1–230. https://doi.org/10.4324/9781315776194

Sustainable Megacities and Planning for Urban Expansion and Regeneration

Enhancing Resilience Through Sustainable Management of Brownfield Industrial Heritage: A Case of Kolkata

Abantika Mukherjee, Haimanti Banerji, and Subrata Chattopadhyay

1 Introduction

The chapter aims to study brownfields, which may classify as industrial heritage and located in an urban context, in this case, the megacity of Kolkata. According to the United Nations, a megacity should have 10 million or more population, including the urban sprawl (United Nations 2019). Megacities have gained attention of researchers and decision-makers due to the growing issues of deindustrialization, urban sprawl and environmental contamination (Yeung 2009). Integrating sustainability and urban resilience principles into the urban interventions of these megacities are taken as a priority (Kumar et al. 2021). This study identifies the qualitative potential of the urban brownfields through establishment of the interlinkage between industrial heritage, sustainable development, and urban resilience in the context of the megacity of Kolkata.

1.1 Formation of Industrial Brownfields in Indian Megacities

Most of the Indian megacities have a history of development as urban industrial areas. In the present context, these historic industrial hubs have evolved into megacities with prominent commercial activities and vast cultural diversity (Kumar et al. 2021). The

A. Mukherjee (✉) · H. Banerji · S. Chattopadhyay
Indian Institute of Technology Kharagpur, Kharagpur, West Bengal 721302, India
e-mail: abantika0m@gmail.com

H. Banerji
e-mail: haimanti@arp.iitkgp.ac.in

S. Chattopadhyay
e-mail: schat@arp.iitkgp.ac.in

M. Firoz C et al. (eds.), *Urban Planning and Design for Megacities in the Global South*, Advances in 21st Century Human Settlements,
https://doi.org/10.1007/978-981-97-8370-0_9

Indian industrial revolution began with the advent of the British colonial rule (Logan 2022). One major factor for industrial decline in India is the independence which resulted in change in ownership status and loss of capital. This led to abandoned "sick" industrial units. The term "sick industrial unit" was coined as part of the Sick Industrial Companies Act (SICA), 1985, defined through the incurred loss of the industrial unit (Navulla and Sunitha 2016). This phenomenon of industrial sickness led to abandonment of sites generating industrial brownfields. Although no formal definition of brownfields exists in India, the concept of brownfield has been widely established in researches on such derelict industrial sites (Majumdar and Sen 2018). Simultaneously, the need to act towards urban resilience in Indian megacities have been pointed out (TERI 2011). In this context, rejuvenation of industrial brownfields might be considered as an attempt to achieve resilience.

Like the other megacities, industrialization of Kolkata flourished during the British colonial rule (Kumar et al. 2021). As the nation's capital at the time, Kolkata developed into the hub of economic and trading operations because of its advantageous location along the Hooghly River. In order to facilitate the trade and commerce, the Kolkata port was established (Kolkata Port Trust 2014). The change in trading policies after the Indian independence resulted in decline of trading of activities for goods like tea, salt, raw jute, etc. (Rai 1978). This phenomenon of deindustrialization resulted in abandonment of industrial sites, which may be categorized as brownfield. This study elaborates the observations from one such industrial brownfield site, which, because of its historical and architectural significance, may also be considered as a potential industrial heritage site.

1.2 Potentials of Brownfield Industrial Heritage

The potential of industrial brownfields to portray the historical evolution of industrial culture and technological advancement of an urban area has been mentioned in many studies (Cenci 2018; Canevaro et al. 2019). Industrial heritage identification is often linked as an opportunity for brownfield regeneration (Mostadi and Biara 2019). Interventions in industrial brownfields generally involve environmental remediation and demolition of existing built structure to convert into vacant buildable lands (Navratil et al. 2018a). Identifying the heritage potential and utilizing the principles of heritage conservation and interpretation enhances both the property value, as well as the image of the site and its surroundings (CABARNET 2006; Navratil et al. 2018a). The main challenges of brownfield sites include cost of environmental remediation (Han and Zhang 2022), balancing of needs of multiple stakeholders (Mirzakhani et al. 2021) and overcoming legal constraints (Matković and Jakovčić 2020). The regeneration approaches should overcome these challenges by utilizing the potentials of industrial brownfields as industrial heritage, which include opportunities for improving urban image, creating of sense of identity, prospects for economic diversity, and ecological sustainability.

2 Literature Review

2.1 Urban Brownfields and Sustainable Development

A wide variety of sites are recognized as urban brownfields depending on the level of contamination and the previous land use. These may include industrial complexes, derelict residential areas, vacant transportation related facilities, abandoned military bases, etc. (Gregorová et al. 2020). Urban brownfields have an inherent potential for relevant densification strategies, economic revival, and remediation of built environment (Navratil et al. 2018a; Rey et al. 2022a) which aligns with the social, economic, and environmental principles of sustainable development (Matković and Jakovčić 2020). Relevance of linking sustainable development of urban brownfields and heritage conservation has been established in previous studies (Harfst et al. 2021; Nedučin and Krklješ 2022). This study considers the three pillars of sustainability, namely the social, economic, and environmental pillars (Purvis et al. 2019) as a theoretical framework for identifying factors of brownfield industrial heritage regeneration. The concepts of sustainable development had been reinforced through the 17 Sustainable Development Goals (SDGs) adopted in the United Nations 2030 Development Agenda. The SDG Goal 11 "Make cities and human settlements inclusive, safe, resilient and sustainable" is relevant to regeneration of such brownfield industrial heritage sites (Mallick 2021). The results of this study have been attempted to be positioned within the context of the SDGs.

2.2 Brownfield Industrial Heritage and Urban Resilience

The environmental and social vulnerability are high in developing countries, and thus the concept of urban resilience are most studied in these regions (Walisser et al. 2005; Mallick 2021). Studies have defined urban resilience as the capacity of tolerance to alterations of an urban area (TERI 2011), the ability to adapt to socio-ecological changes (da Silva et al. 2020) or the ability of modern cities to respond to current social, economic, and environmental challenges (Rogatka et al. 2021). Brownfields can be considered as regenerating resource (Rogatka et al. 2021) and the heritage potential can respond to the socio-economic resilience needs. Thus, brownfield industrial heritage regeneration, as an urban element, may be utilized as one of the tools for enhancing resilience. Comparative studies on available city resilience planning tools (UN-Habitat 2017; Spaliviero et al. 2020) have indicated the lack of emphasis on community participation or stakeholder consultation process. For the case of brownfield industrial heritage sites with multiple levels of stakeholders, a tool which involves a bottom-up approach should be considered. Hence, this study draws from the five pillars of urban resilience from the City Resilience Action Planning Tool (CityRAP), developed by UN-Habitat (UN-Habitat 2019), which are as follows:

1. Urban governance
2. Urban planning and environment

3. Resilient infrastructure and basic services
4. Urban society and economy
5. Urban disaster risk management

This tool has been referred to in recent studies on resilience planning in fast-growing cities (Spaliviero et al. 2020).

3 Materials and Methods

The aim of this study is to highlight how a regeneration strategy for brownfield industrial heritage embodies significant potential to improve the urban resilience and connect to sustainable development principles. For the purpose of the study, a case study of brownfield industrial heritage in Kolkata, West Bengal was selected. Based on the observations in the site, a scenario of proposed regeneration strategy was developed. Then a comparative analysis of the impact on the urban resilience principles between existing status and proposed scenario was conducted. The three stages of the study are elaborated as follows.

3.1 Systematic Literature Search

The first stage of the study was to develop a theoretical framework on the factors of brownfield industrial heritage sites which contribute to sustainable development. For this, a content analysis was conducted on the existing literature from the SCOPUS database, based on a systematic literature search. The search inclusion criteria were as follows:

- Search terms: "brownfield*" AND "heritage*" AND "sustainable" AND "development"
- Source type: Journal, book, book series and conference proceeding
- Year: 2015–2023
- Accessibility: All open access
- Language: English
- Publishing status: Published

After sorting of the final literature according to relevance and removing repetition, a final list of 109 literature sources were read in full for the content analysis.

3.2 Cluster Analysis Using NVIVO Word Query

The second stage of the study aimed to categorize the factors of brownfield industrial heritage regeneration under the three pillars of sustainable development, namely, the social, economic, and environmental pillars. For this a keyword cluster analysis was conducted on the 109 literature sources using the word frequency query tool in NVivo software (Viola et al. 2019; Vecchio and Arku 2020). This cluster analysis also helped in combining the factors with similarity of meaning. The query criteria were as follows:

- Minimum word length: 3
- Total words displayed: 100 most frequent words
- Grouping: With stemmed words
- Similarity metrics: Pearson's correlation coefficient

Additionally, certain verbs and nouns, years and other numerical values and conjunctions were added to the "Stop Words" list, so that it was not shown in the cluster diagram.

The results of the cluster analysis (refer Appendix) were displayed as a dendrogram, which is generated using the hierarchical clustering technique that is used to form the clusters. After analysing the complete dendrogram, individual clusters were identified and sorted under the three pillars of sustainable development. Then the full texts were read and statements related to each word cluster were sorted. A final list of 5 environmental sustainability factors, 4 economic sustainability factors, and 3 social sustainability factors had been identified. The aim was to connect these factors to urban resilience theories using observations from a case study. The observation parameters required for the case study, for each of the factors, were identified based on the literature review along with urban resilience theories as mentioned in the following sections.

3.3 Inventorying the Identified Factors from Case Study and Proposed Use

After the establishment of the factors and observation parameters from literature survey, the selected brownfield industrial heritage site in Kolkata was surveyed by the authors along with a team of post graduate students from the Department of Architecture and Regional Planning, Indian Institute of Technology, Kharagpur in March 2023. Data collection was conducted using inventory formats for condition assessment of each site, semi-structured interviews of stakeholders, photo and video documentation of the site and surrounding urban area and focus-group discussion with experts. A proposed regeneration scenario was developed after studying the current situation and the stakeholders' needs, and the study was then undertaken by comparing the differences between the existing condition and the proposed scenario. Therefore, the results of this research had been synthesized from three different sources: literature, field survey, and proposal by researchers. The observation parameters were then compared against the five resilience pillars (refer Sect. 2.2) to determine which ones are improved or enhanced by the regeneration scenario.

4 Results and Discussion

4.1 Sustainability Factors of Brownfield Industrial Heritage Management

The dendrogram (refer Appendix) indicates proximity of environmental and economic factors, supporting studies that include the cost of remediation as one of the challenges facing brownfield regeneration projects. Aspects related to industrial heritage like architectural conservation, preservation of historicity, and adaptive reuse are clustered both under environmental and social factors. In this section, the individual identified clusters of environmental, economic, and social factors have been identified. A content analysis of each of the individual clusters have led to the identification of factors, which are as follows.

4.2 Factors of Environmental Sustainability

The primary topic of discussion in any brownfield regeneration includes decontamination (Gregorová et al. 2020; Rey et al. 2022a) and designs that address environmental remediation (refer Fig. 1, part b) (Botezan et al. 2020; Lai et al. 2020). Adaptive reuse or regeneration of brownfield sites are accompanied with decontamination from the past industrial activities (Navratil et al. 2018a) which leads to reduction in health hazards of the community (Rey et al. 2022a) and sustainable urban renewal (Jašarović et al. 2021). Moreover, improved accessibility and connectivity due to any revitalization and reuse (Križan et al. 2022; Tzortzi et al. 2022; Zhang et al. 2022) leads to reduction in transportation (Ţoc and Alexandrescu 2022), spacing of activities for community support (refer Fig. 1, part a). Regeneration or infill development of industrial brownfields into recreational open spaces (refer Fig. 1, part a and c) provides a scope for creating a green infrastructure network (Włodarczyk-Marciniak et al. 2020; Tzortzi et al. 2022). Moreover, conservation of built environment in brownfields (refer Fig. 1, part d) may be considered as an approach for optimizing urban land resources, leading to an approach for regenerative design and circular economy (Chowdhury et al. 2020; Scolaro and De Medici 2021; Anna 2022), thus contributing to climate mitigation and sustainable development (Sala Benites et al. 2022; Pintossi et al. 2023). In fast-growing cities, pressure of urban sprawl leads to fragmentation and ecological degradation of the urban periphery (Elfadaly and Lasaponara 2019; Natividade-Jesus et al. 2019; Tzortzi et al. 2022). In response to this, reuse or infill development of industrial brownfields provide opportunity to increase urban density and mitigate or limit urban sprawl (Navratil et al. 2018a; Rey et al. 2022b).

Thus, the factors of environmental sustainability for regeneration of brownfield industrial heritage are identified as follows:

1. Improved connectivity
2. Decontamination

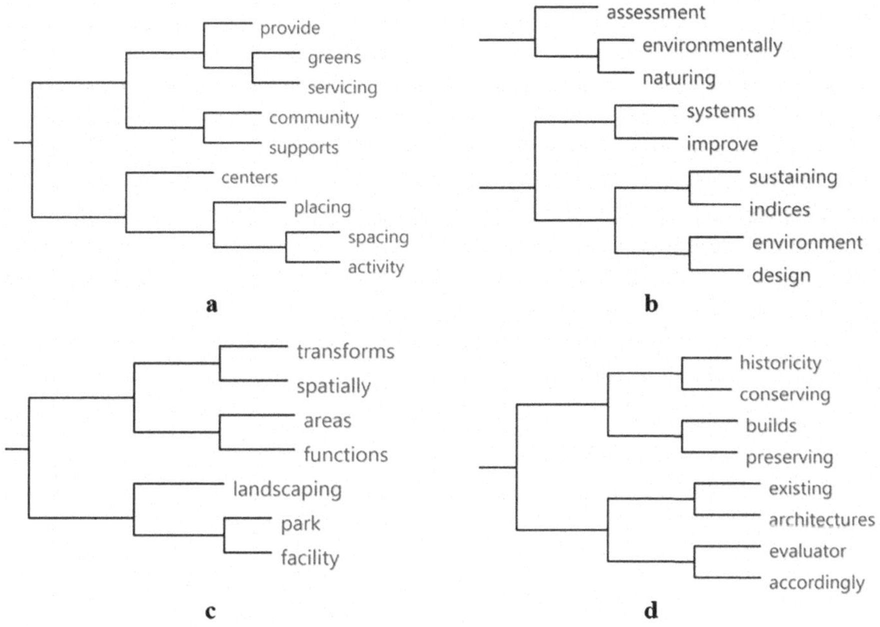

Fig. 1 (a, b, c, d) Four clusters of environmental sustainability factors

3. Green infrastructure
4. Circular economy through reuse
5. Limiting urban sprawl

4.3 Factors of Economic Sustainability

Even though industrial brownfields are a consequence of economic shifts in megacities, studies have highlighted the capacity of regeneration of such sites to enhance and diversify local economic activity (Natividade-Jesus et al. 2019). The regeneration strategies are also indicated to generate local tax income (Lai et al. 2020; Xian et al. 2021), increase employment levels (refer Fig. 2, part b) (Ţoc and Alexandrescu 2022; Jucu 2023), as well as revitalize communities and property values (Lai et al. 2020; Ingaramo et al. 2022). In most countries, brownfield industrial heritage redevelopment is driven by real estate market (refer Fig. 2, part a and c) (Sroka 2019; Wang et al. 2020; Jašarović et al. 2021; Nedučin et al. 2021; Xian et al. 2021; Lorens and Bugalski 2021; Rey et al. 2022a; Simic et al. 2022; Bianco 2023), which may be considered as an important economic factor. In this aspect, the kind of proposed use has an impact, where the commercial usage or mixed use is looked upon as advantageous by real estate investors, while tourism-related uses are more of interest for urban local bodies (refer Fig. 2, part d) (Seo and Lee 2019; Lorens and Bugalski

2021; Križan et al. 2022). The real estate value is also increased by creating more public green spaces or riverside access (Simic et al. 2022; He and Reith 2022). As industrial sites, these were historically a major source of economic activity (Matković and Jakovčić 2020; Zhang et al. 2022), and hence a reuse strategy will require equal or more diverse options of economic activities (refer Fig. 2, part a and b) (Lee et al. 2022; Sadovnikova et al. 2023). This may include tourism (Jóźwiak and Sieg 2021; Giannakopoulos et al. 2022) and other commercial ecosystems (Križan et al. 2022; Benites and Osmond 2023). Expansion of tax bases (Wang et al. 2020; Nedučin et al. 2021; Ingaramo et al. 2022; Rey et al. 2022a; Zhang et al. 2023) is considered as an opportunity derived from brownfield redevelopment.

Thus, the factors of economic sustainability for regeneration of brownfield industrial heritage are identified as follows:

1. Real estate value
2. Diversified economic activities
3. Local tax income
4. Tourism potential

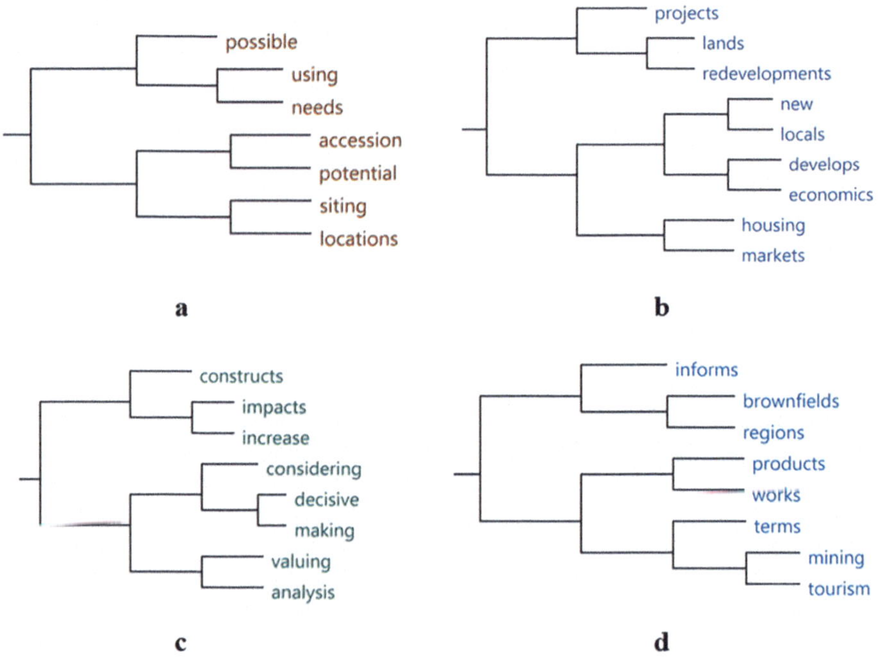

Fig. 2 (a, b, c, d) Four clusters of economic sustainability factors

4.4 Factors of Social Sustainability

Most industrial brownfields in urban areas carry evidences of the historic industrial culture (refer Fig. 3, part a and d) of the area and thus regeneration of such sites may support cultural heritage valorization (Oevermann and Mieg 2021). Valorization of such brownfield sites as industrial heritage improves the image and perception of the urban area (Ginzarly and Teller 2018; Anna 2022; Lee et al. 2022; Pinto and Dos Santos 2022; Zhang et al. 2022). Formation of new vistas and visual ambiences, thus enhancing the quality of public place, is one of the key advantages of revitalization (refer Fig. 3, part b) (Jung et al. 2015; Navratil et al. 2018b; Simic et al. 2022). Revitalization also provides an opportunity to consolidate social cohesion through creation of a new entrepreneurial community (Marian-Potra et al. 2020; Han and Zhang 2022) and inclusion of diverse communities and uses (refer Fig. 3, part c and d) (Anna 2022; Ingaramo et al. 2022). A revitalization strategy of brownfield industrial sites that involves mixed-use facilities have often been linked to sustainable development (Canevaro et al. 2019; Seo and Lee 2019). In comparison to single ownership, the combined ones like public private partnerships or co-operatives lead to more access and sharing of skill sets and funding and better addresses needs of the community (refer Fig. 3, part b and d) (Klusáček et al. 2022; Zhang et al. 2023). Moreover, industrial heritage tourism or refurbishment of brownfield into recreational spaces leads to successful place-making or place value (Rostański 2021; Banerjee 2023) and facilitates optimized use of social infrastructure facilities (Gregorová et al. 2020; Wu et al. 2022).

Thus, the factors of social sustainability for regeneration of brownfield industrial heritage are identified as follows:

1. Revitalized urban image
2. Social cohesion
3. Scope for tourism

4.5 Case Study: Tea Trading (TT) Shed and Chief Mechanical Engineer (CME) Workshop at Kolkata

The Kolkata Port Trust, currently called the Syama Prasad Mookerjee Port Trust is a statutory body under the Ministry of Shipping, Government of India (Kolkata Port Trust 2014). It has around 2300 acres of land under active possession of lessees and licensees in Kolkata. During the period of colonial rule some of the lands along the edge of river Hooghly were utilized to build several warehouses for trading of goods, including tea. The sites, located at the industrial core of the city, also housed the offices, industrial workshops, dispensaries, and relevant infrastructural services. During the phase of post-independence deindustrialization, the trading activities had declined and these sites were left abandoned in the urban core. One such site is the Tea Trading (TT) shed and Chief Mechanical Engineer (CME) workshop at Garden Reach area of Kolkata, which has been chosen as the case study. Apart from

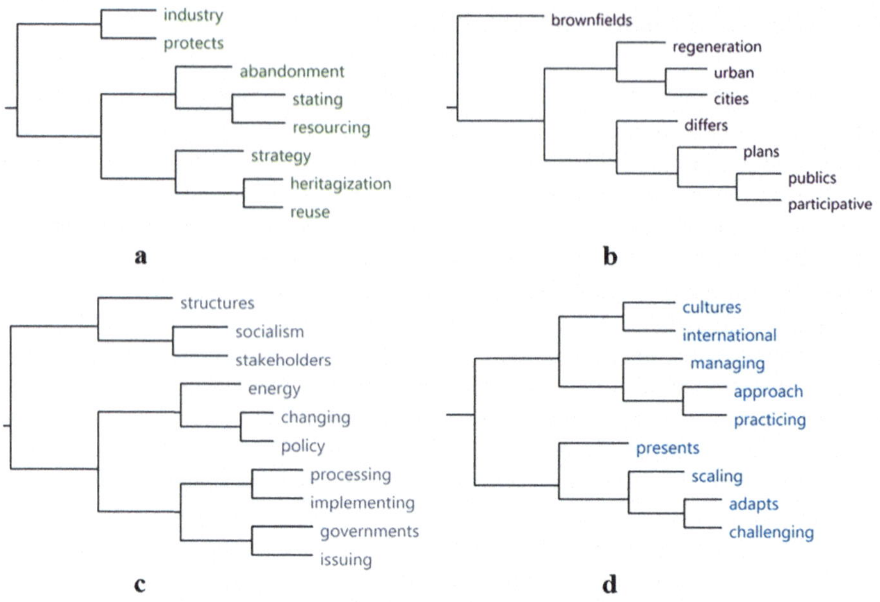

Fig. 3 (a, b, c, d) Four clusters of social sustainability factors

being a prominent parcel of brownfield land, the site is a remnant of the industrial architectural marvel and technological advancements built during the British colonial rule. The site has also been proposed for an adaptive reuse strategy, by the port trust authorities. Hence, due to its potential as an industrial heritage, the site is suitable for this exploration.

4.5.1 Location and History

The TT shed is located on the eastern bank of the river Hooghly, in Kolkata (refer Fig. 4). It was formerly a warehouse for the Tea Trading Company, which was liquidated in 2004–2005. The Ministry of Shipping's Central Inland Water Transport Corporation Ltd. historically used the dock close to the shed. Currently, it is utilized for cargo transportation along rivers. Previously, the largest workshop run by the port was the CME workshop, which was utilized for the production and assembly of parts for shipbuilding and related equipment. The TT shed and this are both on the same campus. The workshop has been abandoned for years and is no longer in operation.

4.5.2 Current Condition

At present, roughly 46% of the built space within the site is abandoned. Currently the ground floors of most of the TT shed buildings (refer Image 1) are being used as

Fig. 4 Location map of TT shed and CME workshop at Kolkata

warehouses for mica, light bulb, incense sticks, fly ash, etc., while the mechanical workshop buildings are left completely abandoned (refer Image 2). At the north-eastern periphery of the site, there is a historic clock tower built around 1899 (refer Image 3). Along this tower, a memorial podium, called the Indenture memorial, was established by the Global Organization of People of Indian Origin (GOPIO). Public access to the memorial requires permission from the port trust.

Recently, this north-eastern section of the site, has been converted into a recreational area with a dock for private cruise ships (refer Image 4). This conversion is part of the mixed-use adaptive reuse strategy formulated by the port trust. For this study, a scenario has been created for a proposed mixed-use development of the entire site, based on the requirements and existing strategy of the port trust.

Image 1 TT shed warehouses

Image 2 Abandoned CME workshop

Image 3 Indenture Memorial and clocktower

Image 4 Dock for private cruise ship

Fig. 5 Plan showing proposed regeneration of TT shed and CME workshop site

4.5.3 Proposed Scenario

Based on the present status, a mixed-use regeneration scenario (refer Fig. 5) containing a waterfront commercial facility, a workshop cum exhibition centre, a hotel, and a café was suggested. A recreational park was designed for the open area along the river edge. The following sections elaborate the comparative analysis of the present status to proposed scenario.

According to the proposal, the observation parameters are compared in the following section.

4.5.4 Impact on Resilience by Environmental Factors

According to the environmental factor findings (refer Table 1), the resilience pillars 4 and 5, which address environmental contamination and the distribution of disease-prone locations, have been significantly impacted. The issue of contamination from the underutilized and derelict industrial brownfields along the banks of the river Hooghly is commonly discussed among urban decision-makers. This proposed regeneration strategy provides scope for decontamination of the site and the river edge (refer Table 1). Currently, the site is accessible through bus routes and private cars. The dock inside the site is used by industrial cargos. The proposed scenario provides additional transportation modes like a formal pedestrian and bicycle access, improving the overall connectivity and reducing traffic-related contaminations. Resilience pillars are positively impacted by the green infrastructure characteristics, which include preserving trees and including public open spaces in the scenario that has been provided. Additionally, because it is suggested that every abandoned structure be re-utilized, it directly promotes circular economy.

To reduce risks like river contamination, the loss of greenery, limited public space utilization, along the industrial belts of Hooghly riverfront, it is necessary to improve environmental factors through sustainable regeneration or reuse strategies of brownfield industrial heritage in the context of Kolkata.

4.5.5 Impact on Resilience by Economic Factors

Resilience pillar 4, which deals with diversity and growth in economic activities, is positively impacted by the suggested scenario (refer Table 2) of adaptive reuse into a recreational cum commercial mixed-use centre. By adopting a new use, maximum of the allowable FAR can be consumed, more public access can be included in the site and the real estate value may be increased. The building use distribution and availability of physical infrastructure in the existing condition had been documented, and the regeneration strategy helps in maximum utility of site and building, creating further opportunity for sources of income. At present the site is under ownership of Syama Prasad Mookerjee Port Trust, which being autonomous, is not liable to much of the local area taxes. With the proposal, a change in the ownership status and introduction of commercial and recreational uses will allow increased opportunity for tax income by the Kolkata Municipal Corporation. Introducing tourism and commercial activities will allow a range of employment opportunities, increase the source of revenue generation impacting the local area employment ratio and quality of life.

Thus, along with enhancing the resilience pillar of urban society and economy, the economic factors also influence the urban governance largely, where we see a shift in approach for operations and maintenance of the brownfield site, having limited private access, to a more public hold. This can in turn increase participatory decision-making process, improving the relationship between citizens and local government.

4.5.6 Impact on Resilience by Social Factors

Recognition of the heritage potential of the site is very limited at present (refer Table 3). The buildings, which are maintained are used only for their utilitarian purposes. The urban scape with views and vistas towards the river Hooghly is not highlighted except for the Indenture Memorial area. The proposal recognizes the potential of such sites to improve the public perception and the overall urban image by conserving and interpreting the heritage structures. The cultural importance of such sites is indicated by how most of the resilience pillars are affected by the parameters of social cohesion and scope for tourism (refer Table 3). The proposal puts special emphasis on the riverfront accessibility and interpretation facilities.

At present the area around Garden Reach is known for its trading activities, surrounded by solid high walls of industrial complexes. There is limited scope for walkability in the neighbourhood and movement of women and children is almost

Table 1 Observations on site for environmental factors

Environmental factors	Observation parameters	Existing status	Proposed scenario	Resilience pillar*				
				1	2	3	4	5
Improved connectivity	Accessible transportation modes	Bus	Additional pedestrian and bicycle access		✓	✓		
		Jetty						
		Car						
	Parking spaces	Informal parking of goods vehicles	50–80 cars and 2-wheelers, formal			✓		
Decontamination	Waste generation sources	Old machineries and industrial structures	Food waste from users		✓		✓	✓
		Godown wastes	Sanitation wastes					
		Food waste from users						
		Sanitation wastes						
	Soil pollution source	Fly ash	None				✓	✓
	Sound pollution source	Heavy goods vehicle, loading and unloading	Vehicle noises				✓	✓
Decontamination	Air pollution source	Heavy goods vehicle	Vehicular access				✓	✓
Green infrastructure	Public access to open space (in %)	0%	30%		✓	✓		✓
	Trees conserved in and around the site (in %)	10%	90%					✓
Circular economy through reuse	Abandoned built spaces (in %)	46%	0%		✓		✓	✓
Limiting urban sprawl	Low density single use/ high density single use/ high density multiple use	Low density single use	High density multiple use		✓		✓	✓
	Loss of wetlands/ core habitats in neighbourhood	Yes	No		✓			✓

*Resilience pillar as indicated:
1 Urban governance, *2* Urban planning and environment, *3* Resilient infrastructure and basic services, *4* Urban society and economy, *5* Urban disaster risk management

Table 2 Observations on site for economic factors

Economic factors	Observation parameters	Existing status	Proposed scenario	Resilience pillar*				
				1	2	3	4	5
Real estate value	Floor Area Ration (FAR) consumed	1.5	1.8–2	✓			✓	
	Property type for valuation	Industrial	Mixed-use, commercial				✓	
	Public access to riverside	Limited	Completely accessible		✓		✓	✓
Diversified economic activities	Building use distribution (in %)	Unoccupied: 46%	Commercial: 30%		✓		✓	
		Commercial: 43%	Public/ semi-public: 15%					
		Public/ semi-public: 5%	Recreational: 45%					
		Residential: 3%	Mixed use: 10%					
		Industrial: 2%						
		Mixed use: 1%						
	Access to physical infrastructure (in %)	Electricity, water supply, and drainage: 2%	Electricity, water supply, drainage: 100%			✓		
		Electricity and water supply: 20%						
		Electricity: 49%						
		No infrastructure: 24%						
		Inaccessible: 5%						
	Distance to social infrastructure (in km)	Education facilities: 1.2km	Additional recreational facilities added within the complex			✓	✓	
		Health facilities: 0.6km						

(continued)

Table 2 (continued)

Economic factors	Observation parameters	Existing status	Proposed scenario	Resilience pillar*				
				1	2	3	4	5
		Recreational facilities: 1.5km						
Local tax income	Ownership	Syama Prasad Mookerjee Port Trust	Co-operative/ Public–private partnership	✓	✓		✓	
	Potential tax sources to urban local bodies	Water supply charges to Kolkata Municipal Corporation	Commercial property tax	✓			✓	
			Municipality taxes if any for the infrastructures like road and park					
Tourism potential	Site Accessibility (in km)	Bus: 0.5 km	Additional pedestrian and bicycle access			✓		
		Ferry: 0.2 km						
		Railway station: 0.8 km						
		Airport: 18 km						
	Employment sources	Goods carrier	Salesperson				✓	
		Vehicle driver	Museum guide					
		Tea stall owner	Café staff					
		Manager	Hotel staff					
		Accountant	Maintenance staff					
		Security	Gardener					
			Manager					
			Accountant					
			Security					
	Revenue source for owner	Tea, Mica, Light, Incense sticks, Mixed Godown rents	Commercial space rent				✓	
			Advertising and sponsorships					
			Parking fees					
			Exhibition or event space rent					

(continued)

Table 2 (continued)

Economic factors	Observation parameters	Existing status	Proposed scenario	Resilience pillar*				
				1	2	3	4	5
			Hotel room rent					
			Food and beverage sales					

*Resilience pillar as indicated:
1 Urban governance, *2* Urban planning and environment, *3* Resilient infrastructure and basic services, *4* Urban society and economy, *5* Urban disaster risk management
km: Kilometres

nil at all times of the day. With the regeneration strategy, the sense of security is also enhanced by provision of illumination, street furniture, structural consolidation of all abandoned buildings, provision of signages and security kiosks. Also, the mixed-use development strategy enables evenly distributed access to the site, throughout the day.

5 Conclusions

In the process of urban redevelopment, brownfield regeneration is acknowledged as a critical issue in cities all over the world (Jašarović et al. 2021; Mirzakhani et al. 2021). Meanwhile, it has been determined that adaptive reuse of industrial heritage positively impacts socio-cultural sustainability (Ţoc and Alexandrescu 2022). Promoting sustainable development approaches will be made less challenging by recognizing brownfield industrial heritage as a significantly impacting element of urban resilience.

5.1 Linking Sustainable Development to Urban Resilience

Both the concepts of sustainability and resilience recognize the need for regeneration and reuse of available resources (da Silva et al. 2020) and urban brownfield regeneration can thus directly be linked to these concepts. From this study, a linkage has been established (refer Fig. 6) which identifies the relationship and impacts of the pillars of urban resilience to concepts of sustainability for a regeneration of brownfield industrial heritage sites in the context of Indian megacities. The outcomes of the regeneration scenario analysis offer guidance for overcoming the problems posed by expanding megacities. The study also emphasizes how recognizing heritage potential enables socio-cultural sustainability, leading to a resilient future.

Although brownfield regeneration as an urban asset and a potential heritage resource is oriented towards SDG Goal 11, the individual factors identified in this study also have impacts on the other SDGs. Access and upgradation of physical infrastructure helps to achieve SDG 7. Environmental impact reduction, urban landscape improvement and resilient infrastructure development can be used to support

Table 3 Observations on site for social factors

Social factors	Observation parameters	Existing status	Proposed scenario	Resilience pillar*				
				1	2	3	4	5
Revitalized urban image	Significant industrial heritage which are maintained/ conserved/ reused	CME Canteen	CME Canteen and office		✓			
		CME store yard	CME store yard					
		CME office	Diesel workshop					
		Workshop dispensary	Blacksmith workshop					
			Buoy shed					
			Fabrication workshop					
			CI smith workshop					
			Workshop dispensary					
			Electric Converter house					
			Tea trading sheds					
	Number of accessible views/ vistas	2	8		✓			✓
	Sources of recognition and pride	Employment	Employment		✓			
			Interpretation of industrial heritage					
			Views to the river					
Social cohesion	Sources of security	Security kiosk	Security kiosk		✓	✓		
		Boundary wall	Boundary wall					
			Illumination					
			Signage					
			Visual access					
			Structural safety of buildings and objects					
	Diversity of communities	Industrial workers	Industrial workers		✓		✓	
			Residents					
			Students					
			Tourists					

(continued)

Table 3 (continued)

Social factors	Observation parameters	Existing status	Proposed scenario	Resilience pillar*				
				1	2	3	4	5
	Opportunity for creative-cultural events	No	Yes			✓	✓	
	% of total built-up area having mixed-use	1%	10%		✓	✓	✓	
	Ownership	Syama Prasad Mookerjee Port Trust	Public–private partnership	✓			✓	
Scope for tourism	Integrity of historic built fabric (in %)	2%	90%				✓	✓
	Access to recreational spaces/ open spaces/ waterbodies	No public access	Complete public access to riverside and park			✓	✓	✓
	Accessible social infrastructure	None	Park		✓	✓		✓
			Ferry point					
			Cafe					
			Hotel					
			Museum					
	% of use in public, semi-public spaces	5%	15%		✓		✓	

* Resilience pillar as indicated:
1 Urban governance, *2* Urban planning and environment, *3* Resilient infrastructure and basic services, *4* Urban society and economy, *5* Urban disaster risk management

achieving SDG 13 and SDG 17. As the proposal is aimed at promoting economic diversity and enhanced quality of life and social cohesion, it is expected that the results can also promote SDG 8 and SDG 16.

5.2 *Enhancing Resilience in Indian Megacities*

If all the riverside industrial brownfields are identified in Kolkata, a connecting network of green infrastructure can be created which will impact a larger urban area. Due to the large scale of the brownfield sites, any regeneration represents regeneration on the neighbourhood or a larger urban area (Laprise et al. 2015). The method of scenario generation is a way to explore strategies for improving the efficiency of development plans and to measure vulnerabilities. From the findings it has been determined that addressing the needs of individual factors may not enhance all the

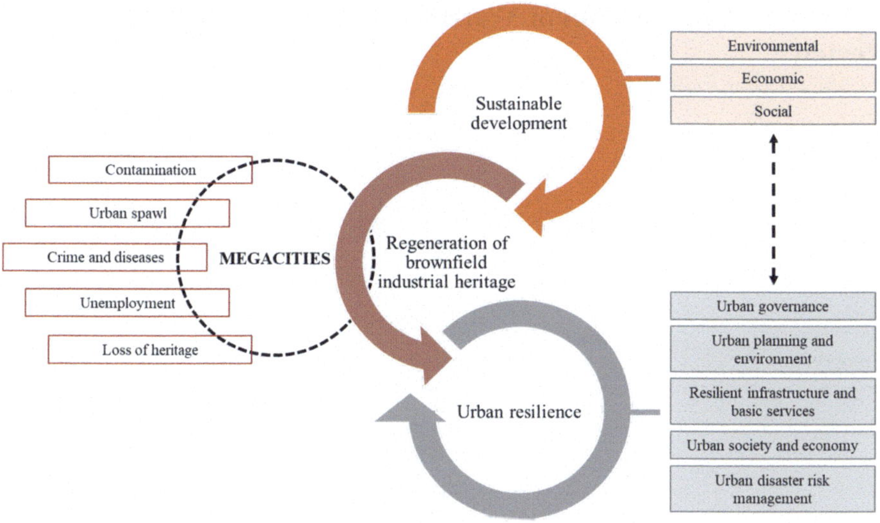

Fig. 6 Sustainable development and urban resilience linkage diagram for brownfield industrial heritage

resilience pillars. Rather, the regeneration strategies should be such that there is a balance between all the economic, social, and environmental factors. In the Indian context, city resilience may be mainstreamed (TERI 2011) through master plans and city development plans. Using this study as a theoretical framework, future studies may develop a quantitative measurement tool for the factors enhancing urban resilience. Adaptive reuse, introduction of green infrastructure, employment diversification, promotion of circular economy are all actions to enhance overall resilience of Indian megacities.

The findings establish how urban resilience pillars are enhanced by indicators of sustainable development in a proposed scenario of brownfield industrial heritage regeneration. The study discusses the contribution of the regeneration strategy on several Sustainable Development Goal targets and proposes a way forward for measuring changes in urban resilience through brownfield regeneration in Indian megacities.

Acknowledgements The authors express gratitude to the post graduate students of the Department of Architecture and Regional Planning, IIT Kharagpur, involved in the data collection process in Kolkata, India. The authors also express gratitude to the Syama Prasad Mookerjee Port Trust for providing data and allowing access to the site. This study is part of an ongoing Ph.D. research conducted at the Department of Architecture and Regional Planning, Indian Institute of Technology Kharagpur and is funded by the Institute Research Assistantship under Ministry of Education, Government of India.

Disclosure Statement The authors declare that they have no conflict of interest.

Appendix: Dendrogram of Word Clusters from Literature Review

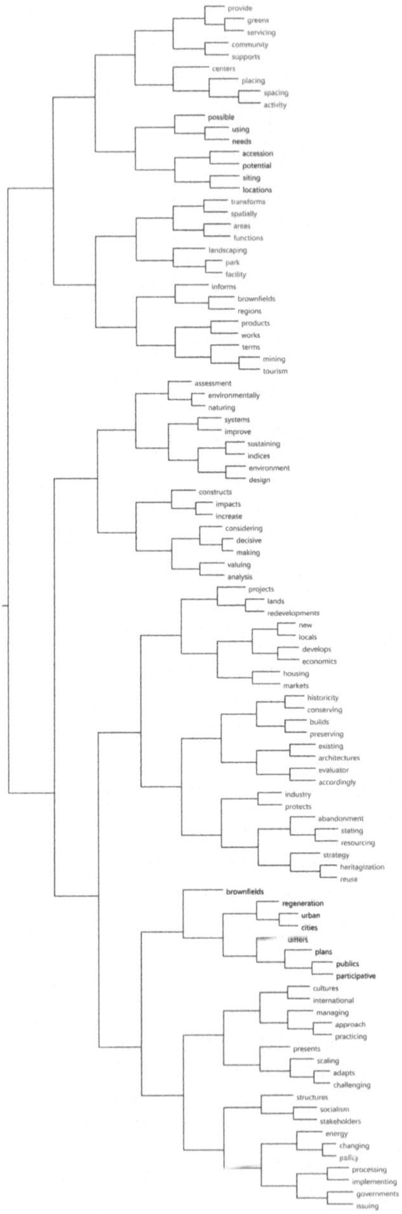

References

Anna FD (2022) What advantages do adaptive industrial heritage reuse processes provide ? An econometric model for estimating the impact on the surrounding residential housing market. Heritage 5:1572–1592. https://doi.org/10.3390/heritage5030082

Banerjee I (2023) Production of railwayscape in urban environment: Analysing railway heritage tourism potential in Siliguri City, India. Reg Sustain 4:1–12. https://doi.org/10.1016/j.regsus.2023.02.001

Benites HS, Osmond P (2023) A future-proof built environment through regenerative and circular lenses—Delphi approach for criteria selection. Sustainability 15. https://doi.org/10.3390/su15010616

Bianco L (2023) Adaptive re-use of historic covered markets: A review of selected cases in European capital cities. Heritage 6:1089–1102. https://doi.org/10.3390/heritage6020060

Botezan C, Constantin V, Meltzer M et al (2020) Is there sustainable development after mining? A case study of three mining areas in the Apuseni Region (Romania). Sustain 12:1–20. https://doi.org/10.3390/su12239791

CABARNET (2006) Sustainable Brownfield Regeneration: CABERNET Network Report

Canevaro E, Ingaramo R, Lami IM, et al (2019) Strategies for the sustainable reindustrialization of brownfields. IOP Conf Ser Earth Environ Sci 296:. https://doi.org/10.1088/1755-1315/296/1/012010

Cenci J (2018) From factory to symbol: Identity and resilience in the reuse of abandoned industrial sites of Belgium. Hist Environ Policy Pract. https://doi.org/10.1080/17567505.2018.1506017

Chowdhury S, Kain J-H, Adelfio M, et al (2020) Greening the browns: A bio-based land use framework for analysing the potential of urban brownfields in an urban circular economy. Sustain 12. https://doi.org/10.3390/SU12156278

da Silva CA, dos Santos EA, Maier SM, da Rosa FS (2020) Urban resilience and sustainable development policies: An analysis of smart cities in the state of São Paulo. Rev Gest 27:61–78. https://doi.org/10.1108/REGE-12-2018-0117

Elfadaly A, Lasaponara R (2019) On the use of satellite imagery and GIS tools to detect and characterize the urbanization around heritage sites: The case studies of the Catacombs of Mustafa Kamel in Alexandria, Egypt and the Aragonese Castle in Baia, Italy. Sustain 11. https://doi.org/10.3390/SU11072110

Giannakopoulos D, Karekou Z, Menegaki E, et al (2022) Reuse of historic buildings in the medieval city of Rhodes to comply with the needs of sustainable urban development. Land 11. https://doi.org/10.3390/land11081214

Ginzarly M, Teller J (2018) Eliciting cultural heritage values: Landscape preferences vs representative images of the city. J Cult Herit Manag Sustain Dev 8:257–275. https://doi.org/10.1108/JCHMSD-06-2017-0031

Gregorová B, Hronček P, Tometzová D et al (2020) Transforming brownfields as tourism destinations and their sustainability on the example of Slovakia. Sustain 12:1–28. https://doi.org/10.3390/su122410569

Han SH, Zhang H (2022) Progress and prospects in industrial heritage reconstruction and reuse research during the past five years: Review and outlook. Land 11:2119. https://doi.org/10.3390/land11122119

Harfst J, Sandriester J, Fischer W (2021) Industrial heritage tourism as a driver of sustainable development? A case study of Steirische Eisenstrasse (Austria). Sustain 13. https://doi.org/10.3390/su13073857

He Q, Reith A (2022) (Re)defining restorative and regenerative urban design and their relation to UNSDGs—A systematic review. Sustain 14:16715. https://doi.org/10.3390/su142416715

Ingaramo R, Lami IM, Robiglio M (2022) How to activate the value in existing stocks through adaptive reuse: An incremental architecture strategy. Sustain 14. https://doi.org/10.3390/su14095514

Jašarović EA, Perović S, Žarić SP (2021) Impacts of arsenal brownfield regeneration on urban development of Tivat in Montenegro: From industrial settlement to center of nautical tourism. Sustain 13. https://doi.org/10.3390/su13158549

Jóźwiak M, Sieg P (2021) Tourism development in post-industrial facilities as a regional business model. Sustain 13:1–16. https://doi.org/10.3390/su13042028

Jucu IS (2023) Replace old industries and local brownfields. Land 12. https://doi.org/10.3390/land12010037

Jung TH, Lee J, Yap MHT, Ineson EM (2015) The role of stakeholder collaboration in culture-led urban regeneration: A case study of the Gwangju project, Korea. Cities 44:29–39. https://doi.org/10.1016/j.cities.2014.12.003

Klusáček P, Charvátová K, Navrátil J, et al (2022) Regeneration of post-agricultural brownfield for social care needs in rural community: Is there any transferable experience? Int J Environ Res Public Health 19. https://doi.org/10.3390/ijerph19010240

Kolkata Port Trust (2014) Brief History of KoPT-Kolkata Port Trust. https://smportkolkata.shipping.gov.in/index1.php?layout=1&lang=1&level=1&sublinkid=1429&lid=563. Accessed 19 Jul 2023

Križan F, Kunc J, Bilková K, Novotná M (2022) Transformation and sustainable development of shopping centers: Case of Czech and Slovak cities. Sustain 14. https://doi.org/10.3390/su14010062

Kumar S, Bhaumik S, Banerji H (2021) Methodology for framing indicators for assessing economic-socio-cultural sustainability of the neighbourhood level urban communities in Indian megacities: Evidence from Kolkata. Soc Indic Res 154:511–544. https://doi.org/10.1007/s11205-020-02559-6

Lai Y, Chen K, Zhang J, Liu F (2020) Transformation of industrial land in urban renewal in Shenzhen, China. Land 9:1–21. https://doi.org/10.3390/land9100371

Laprise M, Lufkin S, Rey E (2015) An indicator system for the assessment of sustainability integrated into the project dynamics of regeneration of disused urban areas. Build Environ 86:29–38. https://doi.org/10.1016/j.buildenv.2014.12.002

Lee W-Y, Shin S-H, Jang S-H (2022) Sustainable urban regeneration strategies in Korea's abandoned mine area using industrial heritage. Adv Civ Eng. https://doi.org/10.1155/2022/7401027

Logan WAT (2022) The industrial revolution in India. In: A technological history of cold-war India, 1947–1969. Springer International Publishing, Cham, pp 15–42

Lorens P, Bugalski Ł (2021) Reshaping the Gdańsk shipyard—the birthplace of the solidarity movement. The complexity of adaptive reuse in the heritage context. Sustainability 13:7183

Majumdar M, Sen J (2018) A spatio-temporal assessment of brownfield transformation in a metropolis: Case of Kolkata India. Proc Annu Int Conf Archit Civ Eng 307–312. https://doi.org/10.5176/2301-394X_ACE18.80

Mallick SK (2021) Prediction-Adaptation-Resilience (PAR) approach—a new pathway towards future resilience and sustainable development of urban landscape. Geogr Sustain 2:127–133. https://doi.org/10.1016/j.geosus.2021.06.002

Marian-Potra AC, Işfănescu-Ivan R, Pavel S, Ancuţa C (2020) Temporary uses of urban brownfields for creative activities in a post-socialist city. Case study: Timişoara (Romania). Sustain 12:1–20. https://doi.org/10.3390/su12198095

Matković I, Jakovčić M (2020) Conversion and sustainable use of abandoned military sites in the Zagreb urban agglomeration. Hrvat Geogr Glas 82:81–106. https://doi.org/10.21861/HGG.2020.82.02.03

Mirzakhani A, Turró M, Jalilisadrabad S (2021) Key stakeholders and operation processes in the regeneration of historical urban fabrics in Iran. Cities 118:103362. https://doi.org/10.1016/j.cities.2021.103362

Mostadi A, Biara RW (2019) Sustainable development of brownfield site for a new landscape perception of an industrial heritage in the city of Kenadsa. Int J Conserv Sci 10:165–176

Natividade-Jesus E, Almeida A, Sousa N, Coutinho-Rodrigues J (2019) A case study driven integrated methodology to support sustainable urban regeneration planning and management. Sustain 11. https://doi.org/10.3390/su11154129
Navratil J, Krejci T, Martinat S et al (2018a) Brownfields do not “only live twice”: the possibilities for heritage preservation and the enlargement of leisure time activities in Brno, the Czech Republic. Cities 74:52–63. https://doi.org/10.1016/j.cities.2017.11.003
Navratil J, Picha K, Martinat S et al (2018b) Resident’s preferences for urban brownfield revitalization: Insights from two Czech cities. Land Use Policy 76:224–234
Navulla D, Sunitha G (2016) A study on industrial sickness in India. Int J Sci Technol Manag 5:83–96
Nedučin D, Krklješ M (2022) Culture-led regeneration of industrial brownfield hosting temporary uses: A post-socialist context–case study from Novi Sad, Serbia. Sustain 14. https://doi.org/10.3390/su142316150
Nedučin D, Krklješ M, Perović SK (2021) Demolition-based urban regeneration from a post-socialist perspective: case study of a neighborhood in Novi Sad, Serbia. Sustain 13. https://doi.org/10.3390/su131810430
Oevermann H, Mieg HA (2021) Urban development planning and world cultural heritage: Two perspectives of planning practice dealing with industrial heritage. Plan Pract Res 36:430–441. https://doi.org/10.1080/02697459.2021.1874639
Pinto PJ, Dos Santos GL (2022) Olympic waterfronts: an evaluation of wasted opportunities and lasting legacies. Sustain 14:. https://doi.org/10.3390/su14041968
Pintossi N, Ikiz Kaya D, Pereira Roders A (2023) Cultural heritage adaptive reuse in Salerno: challenges and solutions. City, Cult Soc 33:100505. https://doi.org/10.1016/j.ccs.2023.100505
Purvis B, Mao Y, Robinson D (2019) Three pillars of sustainability: in search of conceptual origins. Sustain Sci 14:681–695. https://doi.org/10.1007/s11625-018-0627-5
Rai A (1978) Trends in the jute industry since independence. Soc Sci 6:83. https://doi.org/10.2307/3516545
Rey E, Laprise M, Lufkin S (2022a) The Multiple Potentials of Urban Brownfields. In: The Urban Book Series. Springer, pp 47–63
Rey E, Laprise M, Lufkin S (2022b) Sustainability Issues at the Neighbourhood Scale. In: Urban Book Series. Springer, pp 77–93
Rogatka K, Starczewski T, Kowalski M (2021) Urban resilience in spatial planning of Polish cities—true or false? Transformational Perspective. Land Use Policy 101:105172. https://doi.org/10.1016/J.LANDUSEPOL.2020.105172
Rostański KM (2021) Contaminated areas as recreational places-exploring the validity of the decisions taken in the development of Antonia Hill in Ruda Śląska, Poland. Land 10. https://doi.org/10.3390/land10111165
Sadovnikova N, Savina O, Parygin D et al (2023) Application of scenario forecasting methods and Fuzzy Multi-Criteria Modeling in substantiation of urban area development strategies. Information 14:241. https://doi.org/10.3390/info14040241
Sala Benites H, Osmond P, Prasad D (2022) A future-proof built environment through regenerative and circular lenses—Delphi approach for criteria selection. Sustainability 15:616. https://doi.org/10.3390/su15010616
Scolaro AM, De Medici S (2021) Downcycling and upcycling in rehabilitation and adaptive reuse of pre-existing buildings: re-designing technological performances in an environmental perspective. Energies 14. https://doi.org/10.3390/en14216863
Seo J-W, Lee J-L (2019) Characteristics and retrofit constraints of grayfield in Korean cities. Sustain 11. https://doi.org/10.3390/su11082381
Simic I, Stupar A, Grujicic A, et al (2022) The transformation of Dorćol Power Plant: Triggering a sustainable urban regeneration or selling the heritage? Sustain 14. https://doi.org/10.3390/su14010523

Spaliviero M, Pelling M, Lopes LF et al (2020) Resilience planning under information scarcity in fast growing African cities and towns: the CityRAP approach. Int J Disaster Risk Reduct 44:101419. https://doi.org/10.1016/j.ijdrr.2019.101419

Sroka B (2019) Specificity of brownfield's revitalisation in Polish legal framework: Discussion on current legislature problems based on case study. IOP Conf Ser Mater Sci Eng 471. https://doi.org/10.1088/1757-899X/471/7/072049

TERI (2011) Mainstreaming urban resilience planning in Indian Cities: a Policy Perspective

Ţoc S, Alexandrescu FM (2022) Post-coal fantasies: An actor-network theory-inspired critique of post-coal development strategies in the Jiu Valley, Romania. Land 11. https://doi.org/10.3390/land11071022

Tzortzi JN, Guaita L, Kouzoupi A (2022) Sustainable strategies for urban and landscape regeneration related to agri-cultural heritage in the urban-periphery of South Milan. Sustain 14. https://doi.org/10.3390/su14116581

UN-Habitat (2017) Trends in Urban Resilience 2017. Nairobi UN Habitat

UN-Habitat (2019) City Resilience Action Planning (CityRAP) Tool. In: UN-Habitat (United Nations Hum. Settlements Program. Sub-Regional Tech. Cent. Disaster Risk Manag. Sustain. Urban Resil. https://dimsur.org/3-cityrap-tool/. Accessed 17 Jul 2023

United Nations (2019) World urbanization prospects: The 2018 Revision (ST/ESA/SER.A/420). New York

Vecchio M, Arku G (2020) Promoting adaptive reuse in Ontario: a planning policy tool for making the best of manufacturing decline. Urban Plan 5:338–350. https://doi.org/10.17645/UP.V5I3.3188

Viola DD, Viola S, Diano D (2019) Repurposing the built environment: emerging challenges and key entry points for future research. Sustainability 11:4669. https://doi.org/10.3390/su11174669

Walisser B, Mueller B, McLean C (2005) The resilient city. Prepared for the World Urban Forum 2006 by the Vancouver Working Group

Wang N, Golubchikov O, Chen W, Liu Z (2020) The hybrid spatialities of post-industrial Beijing: Communism, neoliberalism, and brownfield redevelopment. Sustain 12. https://doi.org/10.3390/su12125029

Włodarczyk-Marciniak R, Sikorska D, Krauze K (2020) Residents' awareness of the role of informal green spaces in a post-industrial city, with a focus on regulating services and urban adaptation potential. Sustain Cities Soc 59. https://doi.org/10.1016/j.scs.2020.102236

Wu Y, Pottgiesser U, Quist W, Zhou Q (2022) The guidance and control of urban planning for reuse of industrial heritage: A study of Nanjing. Land 11. https://doi.org/10.3390/land11060852

Xian BWX, Rahmawati Y, Al-Aidrous A-HMH, et al (2021) Value-based decision to redevelop transportation facilities: A case study of an abandoned airport. Sustain 13. https://doi.org/10.3390/su13094959

Yeung YM (2009) Mega-cities. Int Encycl Hum Geogr 40–47. https://doi.org/10.1016/B978-008044910-4.01061-0

Zhang H, Liu G, Han Q, Chen G (2023) Mapping the barriers of utilizing public private partnership into brownfield remediation projects in the public land ownership. Land 12. https://doi.org/10.3390/land12010073

Zhang Q, Lee J, Jiang B, Kim G (2022) Revitalization of the waterfront park based on industrial heritage using post-occupancy evaluation—A case study of Shanghai (China). Int J Environ Res Public Health 19. https://doi.org/10.3390/ijerph19159107

Re-opening the Halic Shipyard (Istanbul, TK) to the Local Community: Risks and Opportunities of the Regeneration of an Industrial Site in a Megacity

Mario Paris and Alara Bilgic

1 Introduction

Nowadays, the shift from an industrial-based to a service-based economy has impacted the Global South megacities. This phenomenon affects the urban patterns, as well as the cultural, social, and economic conditions of their inhabitants, together with relevant environmental consequences.

The process is even more evident when the transformation affects traditional manufacturing locations and industrial spaces that are part of the inner areas of the city, where current transformations erase the traces of the first wave of globalisation. The image of these factories belongs to the consolidated imaginary of a neighbourhood, and their presence is part of the everyday experience of the inhabitants, even though they are often configured as fenced spaces that leave the urban life and flow outside.

Nowadays, megacities in the Global South experience an increasing and combined pressure, in which several former factories and industrial spaces became palatable opportunities for real estate investments due to their location, their connection to dense neighbourhoods and public transport networks, the availability of spaces, and, in some cases, the presence of heritage buildings.

Indeed, the valorisation of the former industrial spaces often became an opportunity for the simplistic re-use of these areas as luxury housing compounds, gated communities, or detached campuses with a limited impact on the population and the close contexts. Together with this consolidated approach, a new interest in the potential role of those spaces in their urban and metropolitan context has risen. In

M. Paris (✉)
DISA – Università degli Studi di Bergamo, Bergamo, Italy
e-mail: mario.paris@unibg.it

A. Bilgic
Istanbul, Turkey

M. Firoz C et al. (eds.), *Urban Planning and Design for Megacities in the Global South*, Advances in 21st Century Human Settlements,
https://doi.org/10.1007/978-981-97-8370-0_10

urban studies, there is a need for new design tools and strategies for the integration of these fragments within the urban system, reconciling the megacity and its local/ neighbourhood scale.

Therefore, a reframing of current approaches involved in the design of these spaces and their latent potential is needed. However, what are the problems of the current approaches, and what are the underholed features of a space that the designers could include in their proposals? In this light, urban designers seek innovative design forms based on unconventional readings of the condition of the place and its historical role within the city.

This contribution aims to explore the opportunities of an alternative method based on a research-by-design methodology for a decommissioned site in Istanbul. As recognised by R. Roggema (2017: 1), the complexity due to the existing needs and the persistent problems 'cannot be approached with regular thinking. They require counterintuitive thinking, and the development of new knowledge', and the most effective and creative approach that architects and planners can experiment is the research-by-design method.

Through this approach, designers suggested alternatives to the development of the area, and these hypotheses could animate a public debate about the site's future, involving the local population, future users, and stakeholders in a co-design process.

Often, the operations that involve decommissioned industrial areas in megacities of the Global South repropose exogenous and neoliberal models, collecting high-rise buildings and/or hyper-dense typologies, starting from a *tabula rasa* of existing presences and relationships. The proposed and tested design methodology represents a different approach instrumental to the discussed topic. It suggests a configuration based on the recognition of the existing geo-historical values of an area, the opportunities related to a new urban role of the heritage buildings, and the potential of design proposals based on the reconnection of these spaces in the local networks.[1]

After a brief review of the status of the question (Sect. 2), the paper proposes a case study exploration and a discussion of the results of a research-by-design experience (Sect. 3). In the conclusions (Sect. 4), we propose a reflection on the approach that designers should adopt for the development of future projects.

[1] The research has been developed by Alara Bilgic in the context of her MS thesis in Politecnico di Milano, entitled *Halic Shipyard. Urban regeneration of a former industrial site* (A.Y. 2019/ 2020), and Mario Paris supported this study as a supervisor. Taking advantage of the produced materials, the authors reworked part of them, and the present paper is one of the first opportunities to disseminate the results of this ongoing research. Although the authors cooperated in writing the paper, Mario Paris took the main responsibility for Sects. 1 and 2 and Alara Bilgic for Sects. 3 and 4. Section 5 is the output of the shared work of the two authors.

2 Urban Regeneration for the Megacities of the Global South

After the 1944 Bretton Woods Conference, the international arrangement devoted to economic openness and the consequent involvement of African, Asian, and Latin-American countries in a new global economic order (Fioretos and Heldt 2019) generated a new territorial asset at local and international scales. The trends of financial economies and the progressive localisation of production and transformation factories in countries and cities heretofore excluded by global networks of capital (Conway 2017), as well as the establishment of new tourism destinations, generated a complex and still-ongoing process of urban transformation for the recipient cities. Those changes emerged in the allocation of resources according to market forces and trends (Soederberg 2005) and in the radical enlargement of the geography of global cities (Sassen 1991) that became the engine for the world's neoliberal mutation (Callison and Manfredi 2020). Starting in the late 1990s, cities in the Global South began to undergo deep metamorphosis, surpassing any previous prediction.

The rapid growth of urban populations, the demand for inclusive urban spaces equipped with equal infrastructures of public spaces and services, and other claims produced by rising citizenship of newcomers, often with a low income and a relative dishabituation to high-density living, are still a challenge. In addition, the issues connected with the new normality developed after the COVID-19 outbreak (climate change, health and access to open and public services, proximity, happiness) influenced even more the transformation of the spaces of the regional urbanisation process (Soja 2011) in the Global South, which often still lacks adequate planning and experience, increasing inequity (Arku and Marais 2021; Hardoy et al. 2019).

This transformation is extraordinary both in its speed and results, and we have no intention of portraying a simplified version of it in this paper (Morandi and Paris 2015). At the same time, it was related to a technical, financial, and cultural framework that, today, is still under transformation, and cities, together with the issues connected to the growth, face the challenges of their transformation and the recovery of existing parts that are degraded, underused, or abandoned. Therefore, more than an impossible analysis of the adaptation of neoliberal urbanism (Peck et al. 2009) on Global South cities, we want to underline that the approach to the urban regeneration of these urban systems is still under-considered in the field of urban studies. The main objective of this section is to understand whether and how concepts and ideas already employed in the current approach to urban regeneration of cities are involved in the emerging practices of urban design—or redesign—of existing parts of Global South megacities.

2.1 Regeneration as a Research Field

The initial purpose of this section is to understand how urban planning and design intervene in the current transformation of Global South megacities and which kind of approach today is more effective in supporting the green transition and the improvement of the quality of life of their inhabitants. Academic debates and political rhetoric share a continuous, reassuring, but often blurry, reference to urban regeneration as an adequate tool for acting within the existing city.

This expression, used as a *portmanteau*, should be declined according to the context of the application, as the Global South city. However, the idea of providing a set of policies aimed at improving the quality of the city and its inclusiveness according to changing conditions could be a starting point to face off the rising problems and exceed the conflicts of the urban systems that we study within this paper.

Indeed, planners defined 'urban regeneration' as those interventions that cover a larger field than 'urban renewal'. Working on degraded buildings and blocks, cities adopted actions in the field of urban regeneration to boost social inclusion and ecological aspects of the areas involved, even including actions oriented to the revitalisation of the local economy and its cultural aspects (Paris 2021; Paris and Pezzoni 2020; Angiulli 2017; Couch and Fraser 2003). Therefore, urban regeneration, as noted by M. Bianchi (2018: 4), echoes the definition coined by Roberts, Skyes and Granger (2016)—a 'set of policies aimed at modifying the physical structure of the city in order to respond to the role assigned to it by the specific historical moment and by the interconnection between political will and economic order'.

This approach refers to a well-established community-led interpretation (Sendra and Fitzpatrick 2020; Krüger 1993) of the concept of urban regeneration. These insights have enabled the consolidation of an approach in which interventions are oriented towards the regeneration of the buildings and open spaces of specific neighbourhoods of the consolidated city (containers) and supported by actions to support and reactivate the local economy, cultural animation, and social innovation (contents). This willingness to include both physical/material aspects and socioeconomic and cultural dimensions in projects is the element that marks the radical difference between urban revitalisation and renewal policies and those dedicated to regeneration. In this light, the regeneration needs a set of cross-sectoral interventions and a cooperative and multidisciplinary working method that involves citizens and public and private stakeholders.

2.2 Regenerating a Global South Megacity

According to several studies in the field of urban regeneration of Global South megacities (Rukmana 2020), planners and urban designers must consider that large parts of the cities are composed of a juxtaposition of urban materials and farmlands, with a complex pattern of different densities. Therefore, the mix of inhabited spaces involves gated communities and informal settlements, large transit infrastructures and metropolitan facilities, central business districts, shopping malls and strips, factories and abandoned grey fields, motorways, and railways—in use or not—producing spatial segregation and new poverties (Beard et al. 2016; Ravallion et al. 2007). The fragmentation also often defines the distribution of the marginal population, which produces a geography of settlement opportunities based on parasite—or even historical—settlements and consolidates presences. At the same time, these fragile spaces often produced a strong sense of community and cultural characterisation of entire parts of the city.

Therefore, the examples collected within a larger study we conducted with young colleagues over the last 10 years (mainly in Istanbul and other Turkish cities but also in China, India, and Latin America) confirm that the regeneration of megacities in the Global South is a matter of proposing actions that consider participatory processes and the role of collaborative planning approaches oriented to revitalising existing communities together with their living environments (i.e. Pourzakarya and Bahramjerdi 2021). In this light, the planning approach sensible to citizens' role and urbanity (Boone et al. 2014) can contrast real estate-oriented or authoritarian urban development investments that 'exacerbate existing patterns of inequality, segregation, and isolation' (Mahendra et al. 2021: 172).

The proposed actions should consider cultural perspectives and the presence of market demands that are radically different from those in Western countries, as well as the emerging needs of societies that have experienced significant economic growth.

The answer to these issues cannot be design proposals that aim only to the valorisation intended only as exchange value in H. Lefebvre's terms (Lefebvre 1968). Regeneration projects must be a multiscalar process of socio-spatial transformation (Lefebvre 1970), where the proposals provide functional programmes, community engagement (Anthony 2023), and social innovation activities aimed at generating use values (Hazelkorn 2010), which can benefit the local communities and not only the potential target of real estate projects.

Therefore, exceeding urban regeneration based on flagship mega-projects (Karachalis and Sarantakou 2023) that pair urban renewal with gentrification needs new approaches based on placemaking practices and sensitive design proposals in which the redefinition of spatial assets and the refurbishment of buildings is accompanied by job creation, social redistribution, equity, and participation (Schmid 2012; Misoczky and Misoczky 2018).

In 1976, in his interpretative study of human experience (Seamon and Sowers 2008: 45), R. Relph (p. 141) introduced the definition of 'places' as a portion of space that involves the fusion of human and natural order because both are the significant centres of people's immediate experiences of the world.

Nevertheless, the place is an area where the practices of the use of the space and its physical form merge (Canter 1977) to form a 'socio-physical unit of environmental experience that forms an ecobehavioral model of the space' (Castello 2010: 4). Taking place in a specific space, the everyday experiences of the inhabitants produce the process of territorialisation (Turco 2003). However, what happens when inhabitants have been excluded for a long time from these spaces, and what are the tools and approaches used by the designers in defining the future configurations of these abandoned places?

2.3 Avoiding Ready-Made Proposals

According to Relph (1976), the experiences and the intentions of the communities that live inside them when they interact with their settings (Paris and Fang 2018) define the places. Therefore, the space, filled by material objects, activities, relationships, meanings, and identities, transforms—makes—a space in a place. This placemaking process influences the everyday life of the inhabitants. Indeed, the influence and the stimulation of the experiences and interactions between inhabitants and space have a technical component. Therefore, designers, investors, public actors, and other stakeholders have a key role in the creation of experiences when they shape places, and they have the capacity to shape them by applying tools and using diapositives. This is the first key aspect of this contribution because, in this light, it emerges the opportunity to define a strong link between the technical background of the design proposals oriented to urban regeneration and their impact on the liveability and the urbanity of the results.

In addition, the possibility of transforming the conditions of Global South megacities assigns a significant role to urban designers and planners to fight the existing urban ills such as pollution, congested traffic, and ageing infrastructures. At the same time, designers can deal with existing global challenges such as climate change, improve the quality of life of fragile populations, and seek sustainable urban growth paired with just social costs for everyone. As pointed out by J. Bodnar (2015: 2093), 'Social segregation works against mixing, often aided by urban design which operates on conscious policies of reducing encounters with difference (e.g. Ellin 2001; Smithsimon 2010)'.

Therefore, when planners and designers deal with the regeneration of an existing portion of a Global South megacity, they attempt to reconnect it to the urban system with its networks, communities, and functional successions. In other words, they transform fragments into lived, recognised, and inhabited places. This process generates a long-term value for these spaces and is a process of endogenous development based on specific features and characteristics of the place, its *genius loci*, in Norberg-Schulz's terms (1980), or its 'aura', according to L. Castello (2010). To achieve this result, designers should avoid those protocols often involving grafting ready-made solutions into complex environments without 'paying attention to local structures of meaning and experiences and their different levels of meanings' (Relph 1976: 146). Therefore, the challenge that we analyse here is how to 'make better places', as P. Healey wrote in 2010, and produce new urbanity by designing and regenerating underused districts/areas in Global South megacities, also considering the new needs and lifestyles that have emerged since the COVID-19 pandemic.

Due to the novelty of the topic and the difficulties in producing this re-patching of urban issues and relationships, there is relatively a lack of academic debates about the issue and only a few examples of projects that avoid the repetition of ready-made formats and interacting with local contexts, considering their needs, and taking advantage of their potentials. We look for proposals able to become drivers for the insertion of alternative rhythms and living practices in consolidated parts of the city so that they can host multiple spatialities (Ascher 2001) and re-activate parts of metropolitan landscapes. This feature—which is often cumulative rather than serial, inverting the relationship proposed by R. Sennett (1999)—relies on the density of activities, inhabitants—although temporary—and meanings, which mark these spaces, and on the intensity of goods, data, energies, investments, and information that flow through them (Paris and Fang 2018).

3 Methods and Materials

Within the present study and thanks to previous research,[2] we explored a set of contemporary practices in Turkish (Kasimpasa Salt Warehouse, Istanbul Bilgi University, Bomonti Beer Factory) and international contexts (Shanghai Shipyard, the Raahgiri Day neighbourhood in the city of Gurgaon, India, Rus Playground in Lima, Peru, and the Kwai Chai Hong Alley in Kuala Lumpur, Malaysia). These examples have shown limitations in their ability to understand the true values of the place and their potential role at the urban scale and in their ability to generate urbanity and public life. In addition, we studied two examples, the Camialti and Taskizak shipyards, that have recently been partly renewed (partly ongoing), with interventions that had certain degrees of success in the perspective of the real estate investment, but they had a very weak impact on the urban pattern.

Few examples, especially the ones developed in smaller areas, have the capacity to unleash the local potential, and the regeneration processes that have generated them are often the result of excellent design insights rather than an established approach. Therefore, the intention is to retrace the design experience and give it a testing character for the approaches, tools, and techniques currently used. The definition of a design proposal then becomes an opportunity to identify an innovative approach to design, in which the valorisation of place is measured not only in economic and 'exchange' terms—driven by real estate logic—but also in terms of use value, which could produce a relevant increase in the real estate market and an improvement of the urban quality of the proposal.

[2] The set of case studies has been explored by Yuan Kang and Chua Kim Kiat in their master's degree thesis in Architecture and Urban Design entitled *Making Space a Place*, in which Mario Paris collaborated as a supervisor in the A.Y. 2019/2020.

3.1 The Method: A Research-By-Design

We used a research-by-design methodology for this study because we want to explore an alternative approach that incorporates and develops 'the working method of architects—who searching spatial sketching in a specific material—in the environment of academic research and development' (Hauberg 2011: 52).

Introduced by J. Corner (2006), the 'research-by-design' is 'a type of academic investigation through which design is explored as a method of inquiry, by the development of a project and also exploring the different materials by which a design is carried out—sketches and mapping, among others' (De Queiroz Barbosa et al. 2014: 243). Thanks to these original products, designers can explore people's behaviours and choices when they inhabit a space, according to the landscape urbanism sensitiveness and recent declinations in the field of urban planning (Paris et al. 2023). According to T.M. De Jong (1992), this approach is also useful for the creation of alternative scenarios aimed to 'generate desirable, maybe unexpected, urban perspectives in place of probable, but less desirable, urban developments'.

In this light, the 'project' intended is the formalisation of a way to live and perceive a space and an opportunity to read and influence the living practices of its inhabitants. This operation starts from a succession of composition and decomposition actions and, through them, the exploration of the crossroads between the need for improving the quality of urban spaces, the reactions, and the new need for urban populations focusing on the role of the designer. In addition, the outputs of this methodology are technical products, but thanks to their strong visual impact, they became readable and accessible to peers and others, and they allow a debate that mixes several public actors (practitioners, experts, administrators, inhabitants).

In our opinion, this alternative approach is an opportunity to explore new opportunities for the transformation of existing and underused areas in Istanbul and other underused or abandoned production sites in Global South megacities. In a context where the architectural practice is often ancillary to the real estate business and the proposals tend to erase the existing aura of the spaces, the research-by-design 'can add a creative, experimental and expressive element to research and contribute to theories, rationalisation and generalisation about the object' (Hauberg 2011: 53).

3.2 Criteria of Selection and Interest of the Case Study

The case study for the experience of research-by-design is the Halic Shipyard in Istanbul. Nowadays, the shipyard is an underused, degraded, and neglected area within the Beyoglu district, one of the most historic areas of Istanbul located on the east side of the *Corne d'Or* (Golden Horn), the estuary shared by the Alibeyköy Deresi and Kağıthane Deresi rivers, which flow into the Bosphorus.

The Halic Shipyard is a historic industrial site, raised on the foot of the former Genoese colony of Galata, just in front of the ancient Constantinople.

As other factories and production sites are located along the two riverbanks, the shipyard forms part of a productive cluster connected to the inner port due to the easy access to maritime transport and the proximity to the most populated neighbourhoods of the cities that represent both the labour reserve and the consumer market for the goods produced. After the shipyard, other military factories settled, such as Bakirkoy Baruthane-i Amire (the Bakirkoy gunpowder factory) and Lengerhane-I Amire (chain manufacturing factory), and between the nineteenth century and the beginning of the twentieth century, other industrial productions, such as flour, steel, fabric, bread, wood, leather, tile, and brick, followed the same settlement principles (Fig. 1).

The reasons for selecting this space are many and affect different aspects:

- Its key location is close to the most recognised and historical core of the city, it is connected to several lines of underground and public transport (bus and ferries), and it is surrounded by relevant public facilities (among others, Tersane-i Amire Anatolian High School, the İbb Beyoğlu Yüzme Havuzu swimming pool, and the İbb karaköy coastal park).
- The historical role of this space, together with the industrial memory of the city, represents a privileged perspective of the city and, historically, was also a 'contact point' between the city and the water.
- The dimension of the property (more than 200,000 sqm) and its morphology is composed of a system of buildings/warehouses, an antique mosque, and squares, with the presence of three basins and greenery in the leftover. This low density and the opportunity to rethink this sequence of mineral squares and gardens is a big opportunity due to the scarcity of open spaces and the large plot of the context.
- The succession of design competitions promoted in the area, even without being put into practice, shows the ongoing pressures produced by private investors and the wills expressed by the municipal authority and the port authority. Public institutions advocate for a new future for the area, and the proposals show how introverted and indifferent to the context they can be. Especially, if the projects intend the 'valorisation' of space only as a tool to multiply urban rent and maximise private earnings.
- The composition of our research team, which includes a double perspective: the internal one, connected to the presence of an inhabitant, and an external one. We both tried to examine this space from the perspective of urban designers and planners, as well as from the perspectives of Istanbul citizens and tourists (Figs. 2 and 3).

Due to these reasons stated above, we believe that the Halic Shipyard could be a valuable case study to explore the risks connected to the current design approach oriented to renewal forms based on the analysis of the area full of cliché, which ended up in proposals quite banal. Even proposing a rhetoric on the conservation of the 'aura' of the space, the existing project produced a *tabula rasa*, where the complete substitution of the buildings erases the legacy and the valuable traces present in the area.

Fig. 1 Industrial buildings/sites from late Ottoman period (Elaborated by A. Bilgic, 2020)

3.3 *The Halic Shipyard as a Key Node in the Historical and Current Istanbul*

The Halic Shipyard occupies an area marked by a sloping section that connects the water with the Galata neighbourhood, which used to be a popular excursion and

Fig. 2 Halic Shipyard and, on the background, the Süleymaniye Cami Mosque (Picture by A. Bilgic, 2020)

Fig. 3 The major basin and the open spaces of the shipyard (Picture by A. Bilgic, 2020)

leisure area marked by the presence of mansions, waterfront residences, and pavilions owned by the most affluent families of the city.

Over the last centuries, the pressures connected to the enlargement of the productive sites and informal urbanisation degraded the existing landscape, transforming the space into a commercial and industrial centre. Although the area still shows a lower density in comparison to the peninsula in front, the space is completely urban, with very few gardens. A succession of urban motorways cuts off the area of the shipyard from the neighbourhood staring from the Ataturk Koprusu bridge—which, in fact, links the two riverbanks—and generates a sense of detachment for a part of the city that occupies the strategical position for both metropolitan and local scales. This condition of double centrality (urban and territorial) imposes a reflection on designers, who should consider that the proposals should have different, larger, and deeper impacts on the city (Fig. 4 and Table 1).

On the one hand, the shipyard area is the intersection of two metropolitan axes, which we call the 'cultural/touristic corridor' and the 'open space and urban memory corridor'. The first one starts in Besiktas and, from the Maritime Museum, crosses along the waterfront up to the Rahmi Koç Museum in a continuous succession of more than 4 km, interrupted only in the shipyard and by the major roads that frame it. It connects 12 prominent cultural venues and institutions through a pedestrian and bike lane. This corridor also includes historic tourist attractions like Dolmabahce Palace, Kilic Ali Pasa Mosque and Hammam, and Yeralti Mosque. These kinds of

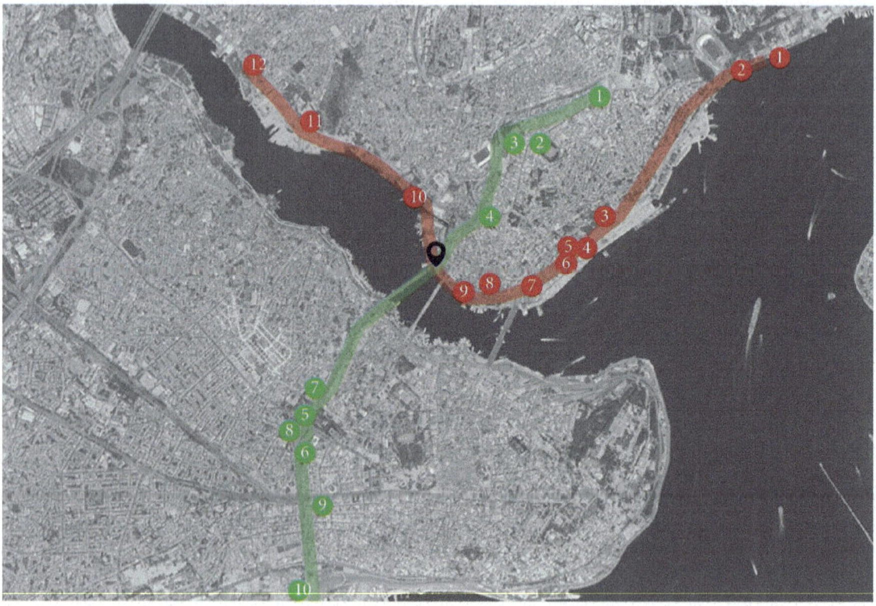

Fig. 4 Two metropolitan axes (Elaborated by A. Bilgic, 2020)

Table 1 Elements of the cultural axes

	Axis 1 The 'cultural/touristic corridor'		Axis 2 The 'open space and urban memory corridor'
1	Maritime Museum	1	Taksim Square
2	Dolmabahçe palace	2	Çiçek Pasajı (Cite de Pera)
3	Mimar Sinan Fine Arts University Museum	3	Nevizade Street
4	Istanbul Moderm Museum	4	Asmalı Mescit Neighbourhood
5	Kiliç Ali Pasa Mosque and Hamam	5	Aqueduct of Valens
6	Galataport Project	6	Saraçhane Arkeological Park
7	Yeralti Mosque	7	Women's Bazaar
8	Salt Galata Gallery (Former Ottoman bank)	8	Horhor Antique Bazaar
9	Arab Mosque	9	Laleli Street
10	Navy Command Building	10	Yenikapı Showground
11	Aynalikavak Pavilion		
12	Rahmi koç Museum		

Elaborated by A. Bilgic, 2020

places are often visited by locals, perhaps once a year, for special and/or religious events, but they are 'must-see' tourist attractions for visitors.

On the other hand, a transect that links the major open and green public space of the city passes from Taksim Square, the Aqueduct of Valens, and the Sarachade Archaeological Park and connects the two sides of the *Corne d'Or*. This corridor also includes historical entertainment and dining places, such as *Çiçek pasajı* (Cite de Pera), Nevizade Street, and the Asmalı Mescit Neighbourhood. These dining areas have both historical and local significance to Istanbul citizens. On this axis, there are also local bazaars and markets (Women's Bazaar, Horhor Antique Bazaar, and Laleli Street) that are mostly used by the locals living close by. Finally, at one of the ends of the axis is the big Yenikapı showground. This event place is used for concerts, events, holidays, and political rallies.

The people who use and inhabit the two axes are different: tourists and city users in the first case and locals in the second one. The differences also impact the offer of economic activities along the two corridors, as well as the housing market and the real estate value. The Halic Shipyard could be the hinge between these two diverse realities, and we believe that its design cannot fail this opportunity.

At the local scale, the shipyard site represents a sort of black hole. Today, it is underused, and only a few people stay there. However, there are many local functions around it, and the neighbourhoods show a certain vitality. Parallel to this, except for the spaces that belong to the open spaces' backbone, there is a lack of green and public spaces. As in other districts of the city, most of the open spaces are cemeteries, and the public ones lack identity and variety. This absence creates social segregation, but, as pointed out by J. Bodnar (2015: 2095), the 'public space thrives on diversity and the lack thereof can kill it'.

A design that does not consider the territorial role of this space could improve the overall metropolitan pattern and, on the other hand, lose the opportunity to improve the quality of life of the inhabitants of the close neighbourhoods and the urban experience of users that could cross the area.

The potential opening of this area to the local communities is an opportunity that designers and public authorities cannot waste proposing a banal project. Therefore, we attempted to define a master plan based on several principles that challenge the current practices (Fig. 5).

4 Results and Discussion

Through a research-by-design experience, we proposed an innovative approach for the re-use of the area of Halic Shipyard, in which the goal is to find an original balance within the overly crowded, overly built, and overly populated neighbourhoods around this heritage industrial space. The master plan, based on the reconnection of this space to two identified urban corridors and aimed at creating a relationship between their different populations, proposes several pedestrian spaces and focuses on the proximity and the space as a value *per sé*.

5 The Proposed Master Plan

The key principle of the design is to take advantage of the presence of a 'void' inside of the overly crowded and overly built Istanbul metropolis and work with subtraction. Thus, the approach to this project was to subtract the buildings and create new relationships among the open spaces inside and outside the area instead of adding even more structures to the site. There are both buildings with historical and/or industrial heritage value in the site, as well as buildings that were recently added in the last few decades. We removed the recently added multi-storey buildings in our simulation. Only two main historical ateliers, two warehouses, the historical carpentry building, and the mosque will be left (Fig. 6).

In parallel with the remotion of the buildings, we worked on the functional programme according to the needs of the municipality expressed in the tenders of the design competition proposed and the opportunities ongoing at the time. For example, we included in Buildings 1 and 2 (former warehouses) the additional campus of Mimar Sinan Fine Arts University—one of the elements of the cultural/touristic corridor—which sought an additional campus dedicated to the conservatory and performance classes. In addition, we re-used the mosque as a small public library to host a permanent exhibition about the shipyard and the neglected industrial memory of the *Corn d'Or*. We also introduced private workspaces in other warehouses and

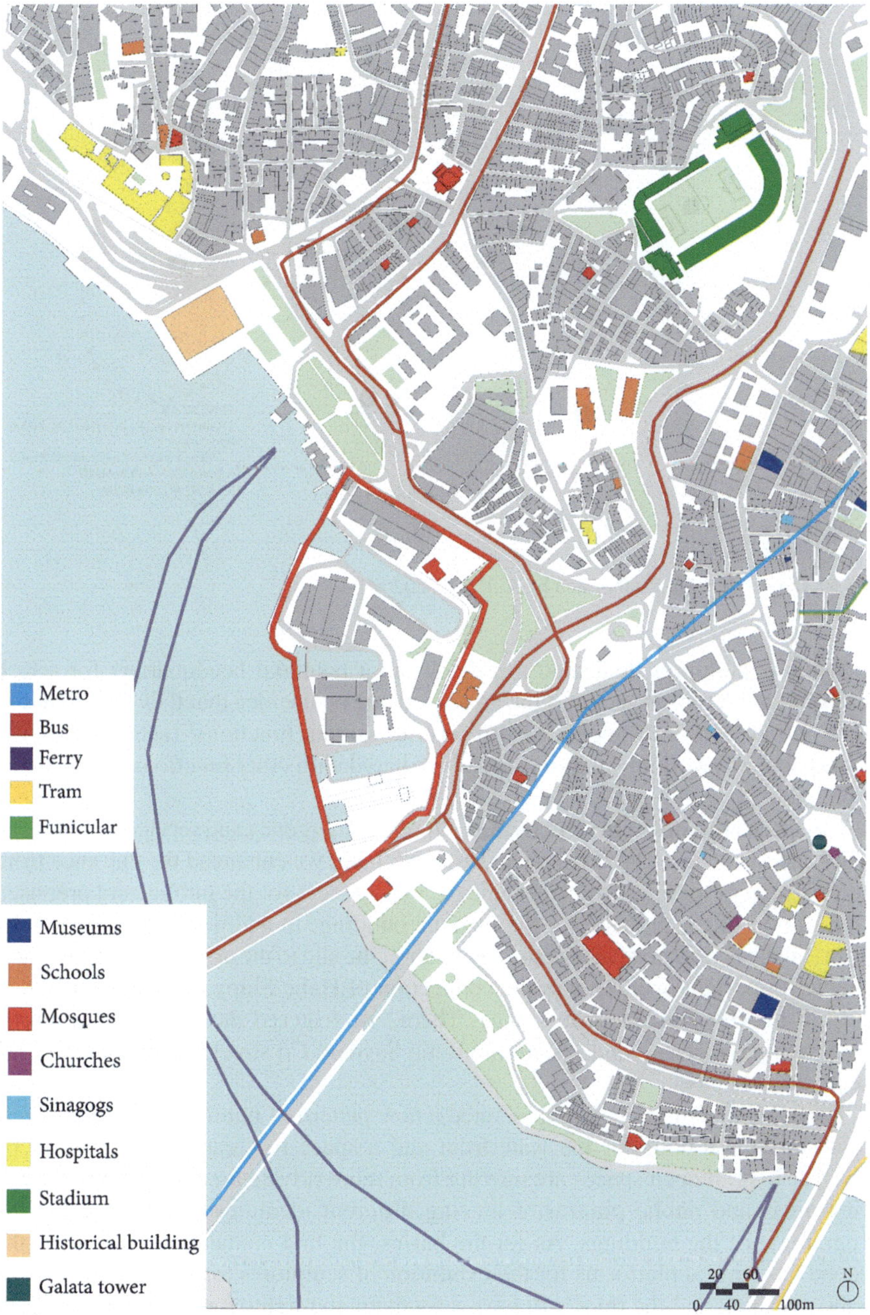

Fig. 5 Land uses and transportation networks in the context of the Halic Shipyard (Elaborated by A. Bilgic, 2020)

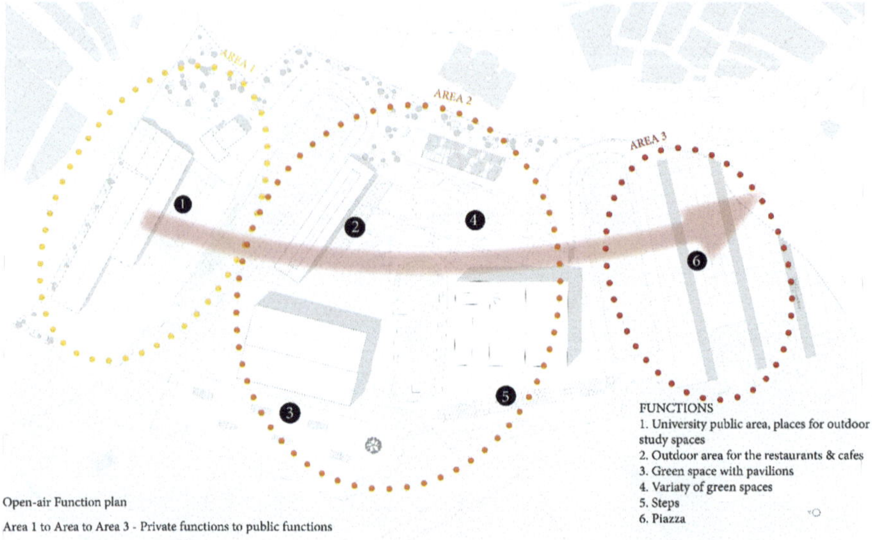

Fig. 6 The master plan (Elaborated by A. Bilgic, 2020)

in two new buildings dedicated to offices and a potential headquarters for a tech company in the central sector of the area, focusing on the idea that they should maintain a certain degree of openness with several 'servant functions' (such as the café/canteen and a coworking area) that could be shared with other functions and the local population.

The space is completely vehicle-free, although emergency and service vehicles can easily access the area. We kept the existing wall, but we enhanced the entrance from the west area of the site, which became the main access for the public, and proposed a renovation of the passage for the mosque building. In addition, we reopened the historical gates located at the northeast corner of the site to the urban pedestrian flows. Another important opening was the connection of Halic Shipyard to the waterfront park on the other side of the bridge. There, we fostered the existing but closed pedestrian passageway underneath the Refic Saydam Cd street, which connects the area to the old city.

These architectural choices generated a new pattern of public spaces in the area, which became a door for the waterfront and a space to create an original social mix. The overall open spaces are moving from more private functions to small gathering areas and public piazzas, achieving different identities and reinforcing the programme of the buildings. As for the basins, the first contains its own flora, the second has several platforms for the exhibition of sculptures to attract more visitors to the museum, and the third will have a fountain and lights for visitors to see from the piazza. The upper parts of the topography will remain a greenery and a space for trees (Fig. 7).

Fig. 7 A new public use for the open spaces and the heritage buildings and structures of the shipyard (Elaborated by A. Bilgic, 2020)

6 Expected Impacts

The consolidated protocol used by real estate developers is always the same, and there is a repetition of the functional programme oriented to a luxury target. The formula included a residential gated community with a high street or a shopping mall and one or more attractive functions dedicated to affluent customers or tourists. The master plans developed for many other decommissioned industrial areas take advantage of the existing walls or fences that usually cut off the spaces from the city, and the design does not invert this feeling, creating a functional—and quite rentable—fragment.

This approach creates closed spaces that exclude the local communities and cut off all forms of interaction and relationship with them, chopping off those urban flows that could link these spaces and the city.

This process is incremental and has been repeated on many occasions, both in Istanbul and in other Global South megacities. The result is a progressive fragmentation of fabrics. The different parts of the city are chopped, and this kind of layout often prevents users—not only the owners of the included houses but also the citizens, city users, and tourists—from generating original forms of use of the space and those innovative practices of inhabiting the space. This is a myopic strategy that disrupts the continuity of the systems of public open spaces that have proven important in making cities more liveable, inclusive, and attractive. With this approach, there is an increasing privatisation of the benefit for real estate investors but a negative impact on the nature of the city as a shared social product that public entities should look after. In the medium term, as argued by S. Yavo-Ayalon (2022), the local communities increase the urban displacement (Davidson 2009) and the inner migration patterns that transform the demographic profile of cities. In the long term, the inhabitants of large parts of the city could suffer a sense of a loss of social ties and community. Therefore, the cumulative effect of these transformations produces gentrification and makes life in the cities unaffordable for large parts of the population, not only the most fragile ones, causing a loss of urbanity, vibrancy, and attractiveness, which could also affect the interest of affluent investors in the city.

Throughout the research-by-design experience, we proposed a different strategy based on the partial re-opening of the area to the city and the connection of the system formed by mineral squares, water basins, and green spaces to the existing network of pedestrian pathways. In addition, we proposed the re-use of a part of the existing buildings as private but attractive functions to complete the two successions of metropolitan/touristic attractors identified during the analysis.

This is a way to 'return to a more authentic urbanity' (Sorkin 1992: xv) and to avoid the risks connected to speculative, marked-based projects promoted as a financial driver. Through the consolidate approach, the project proposed easily become an outdated real estate product that cannot attract investors and customers/inhabitants in a few years. We experimented with a sustainable design in terms of budget but were also able to give back vibrancy and attractiveness to the Halic Shipyard and make it a part of the city. Functions and users could change over time, but its backbone of public and accessible open spaces will connect the area to the Istanbul urban pattern.

7 Conclusions: Towards an Innovative Design Approach

In the present contribution, we attempted to bridge the theoretical approach in the field of urban design that tries to identify what makes a place and how a designer can be a 'placemaker'. The current practices lack this sensitiveness, but the research can identify a set of critical nodes, and ideas can be tested thanks to a research-by-design experience. During this process, we used a former industrial area, the Halic Shipyard, in the city of Istanbul. Today, it is underused and decayed, which led us to use it as a context to imagine a proposal for an urban project. This master plan redefines the arrangement of the area and a new programme for the existing buildings, but it is also a testing ground to show how a focus on the existing interfaces between the area and the city produces a better link that can provide a richer and more interesting habitat for citizens, tourists, and city users, as well as a larger biodiversity and a better integration of the existing spaces and the ones included in the proposal. In this process, architects, urban designers, and urban planners will play a relevant role in these transformations, and their actions will reverberate both on the local scale and the city's global image.

To achieve these goals, we focused on public spaces and the role of the 'public city' as a tool to foster the liveability and the quality of life for its inhabitants. This perspective is relevant for all the cities. Nowadays, the pressures of new urban questions (affordable and inclusive housing, access to quality services and infrastructures, the right to live a fair, healthy, and happy life) affect all the urban systems worldwide, particularly the Global South megacities that host the larger part of the urban populations. These cities are the testing ground for the existing practices connected to urban regeneration and oriented to make better places (Healey 2010), producing diversity, improving the public life of their neighbourhoods and communities, and supporting all the different lifestyles that they host.

Thanks to our research-by-design experience, we attempted to reflect on a proposal for a mixed-use complex that could become the critical infrastructure of the city's symbolic economy (Zukin 2010), considering the authenticity of the conditions of the site and its memory as value that can increase the urban interest of the project and its role in the city. Working in Global South megacities is a challenge, but there are many opportunities to propose projects that can have a positive impact on society and the lives of their inhabitants. This dimension is not rhetoric, not colonialist, and is often neglected as a key part of the social role of architecture and architects.

References

Angiulli A (2017) Rigenerazione identitaria e semplificazione nel governo del territorio. In: Stella Richter P (ed) Governo del territorio e patrimonio culturale. Giuffrè, Milano, pp 29–47

Anthony B (2023) The role of community engagement in urban innovation towards the co-creation of smart sustainable cities. J Knowl Econ. https://doi.org/10.1007/s13132-023-01176-1

Arku G, Marais L (2021) Global South Urbanisms and Urban Sustainability—Challenges and the Way Forward. Frotiers in Sutainable Cities 3 https://doi.org/10.3389/frsc.2021.692799

Ascher F (2001) Les nouveaux principles de l'urbanisme. Éditions de l'Aube, La Tour d'Aiguës

Beard VA, Mahendra, A, Westphal MI (2016) Towards a more equal city: framing the challenges and opportunities—Working Paper. World Resources Institute. Available via WRI. https://www.wri.org/research/towards-more-equal-city-framing-challenges-and-opportunities. Accessed 20 Dec 2023

Bianchi M (2018) Rigenerazione urbana ed innovazione sociale: Il caso di Gillet Square, progetto di cultura urbana guidato dalla comunità. Euricse Working Papers 105. https://doi.org/10.2139/ssrn.3297508 Accessed 20 Dec 2023

Bilgic A (2020) Halic Shipyard. Urban regeneration of a former industrial site. Dissertation, Politecnico di Milano

Bodnar J (2015) Reclaming public space. Urban Stud 52(12):2090–2104. https://doi.org/10.1177/0042098015583626

Boone C, Redman CL, Blanco H et al (2014) Reconceptualizing land for sustainable urbanity. In: Seto K, Reenberg A (eds) Rethinking global land use in an urban era. Strüngmann Forum Reports, vol 14. MIT Press, Cambridge, pp 313–330

Callison W, Manfredi Z (eds) (2020) Mutant neoliberalism: market rule and political rupture. Fordham University Press, New York. https://doi.org/10.5422/fordham/9780823285716.001.0001

Canter D (1977) The psychology of place. Architectural Press, London
Castello L (2010) Rethinking the concept of place in architecture-urbanism. Ashgate, Farnham
Conway D (2017) Globalization of labor: increasing complexity, more unruly. In: Conway D Heynen N (eds) Globalization's Contradictions. Routledge, London and New York, pp 79–94
Couch C, Fraser C (2003) Introduction: The European context and theoretical framework. In: Couch C, Fraser C, Percy S (eds) Urban Regeneration in Europe. Blackwell, Oxford, pp 1–16
Davidson M (2009) Displacement, space and dwelling: placing gentrification debate. Ethics, Place Environ 12(2):219–234. https://doi.org/10.1080/13668790902863465
De Jong TM (1992) Kleine Methodologie voor Ontwerpend Onderzoek. BOOM Uitgevers, Amsterdam
De Queiroz Barbosa ER, DeMeulder B, Gerrits Y (2014) Design Studio as a Process of Inquiry: The case of Studio Sao Paulo. AE Revista Lusófona Revista Lusófona De Arquitectura e Educação 11:241–254
Ellin N (2001) Thresholds of fear: embracing the urban shadow. Urban Stud 38(5–6):869–883
Fioretos O, Heldt EC (2019) Legacies and innovations in global economic governance since Bretton Woods. Rev Int Polit Econ: RIPE 26(6):1089–1111. https://doi.org/10.1080/09692290.2019.1635513
Hardoy J, Gencer E, Winograd M (2019) Participatory planning for climate resilient and inclusive urban development in Dosquebradas, Santa Ana and Santa Tomé. Environ Urban 31(1):33–52. https://doi.org/10.1177/0956247819825539
Hauberg J (2011) Research by design—a research strategy. Revista Lusófona de Arquitectura e Educação 5:46–56
Hazelkorn E (2010) Community engagement as social innovation. In: Weber L, Duderstadt J (eds) The role of the Research University in an innovation-driven GlobalSociety, Economica, London and Geneva, pp 63–76
Healey P (2010) Making better places: the planning project in the twenty-first century. Palgrave Macmillan, New York
Kang Y, Kiat CK (2021) Make space a place. Dissertation, Politecnico di Milano
Karachalis, N, Sarantakou E (2023). Urban mega-projects, cultural flagships and their effect on the destination image and visitor economy: the case of the waterfront of the metropolitan area of Athens. In: Katsoni V (eds) Tourism, travel, and hospitality in a smart and sustainable world. IACuDiT 2022. Springer, Cham. https://doi.org/10.1007/978-3-031-29426-6_31
Krüger A (1993) Local Communities and Urban Regeneration: The Contribution of Community Education. Community Development Journal, 28(4), 342–354. http://www.jstor.org/stable/44257112
Lefebvre H (1968) Le droit a la ville. Athropos, Paris
Lefebvre H (1970) La révolution urbaine. Gallimard, Paris
Mahendra A, King R, Du J, et al (2021) Seven transformations for more equitable and sustainable cities. World Resources Report, Towards a More Equal City. World Resources Institute, Washington, DC. https://doi.org/10.46830/wrirpt.19.00124 .
Misoczky MC, Misoczky de Oliveira C (2018) The city and the urban as spaces of capital and social struggle: notes on Henri Lefebvre's enduring contributions. Revista de Administração Pública 52(6):1015–1031. https://doi.org/10.1590/0034-761220170122
Morandi C, Paris M (2015) A territorial role for superplaces? Archivio di Studi Urbani e Regionali 112:101–125. https://doi.org/10.3280/ASUR2015-112006
Norberg-Schulz Ch (1980) Genius loci: Towards a phenomenology of architecture. Rizzoli, New York
Paris M (2021) La rigenerazione urbana nelle agende strategiche delle città metropolitane: uno studio comparativo. Archivio di Studi Urbani e Regionali 132:153–176. https://doi.org/10.3280/ASUR2021-132007
Paris M, Fang L (2018) From luxury to prestigious place-making: an overview. In: Paris M (ed) Making prestigious places: how luxury influences the transformation of the city. Routledge, Oxton, pp 1–20

Paris M, Dezio C, Caricati I, Vettore D (2023) How to regenerate the ordinary public spaces after Covid-19 crisis? Recommendations for planners and urban designers. Contesti 1-2023.
Paris M, Pezzoni N (2020) Rigenerare il territorio, disseminare le progettualità. La ricognizione delle opportunità di trasformazione della periferia Milanese. Archivio di Studi Urbani e Regionali 127:48–71. https://doi.org/10.3280/ASUR2020-127003
Peck J, Theodore N, Brenner N (2009) Neoliberal urbanism: models, moments, mutations. The SAIS Rev Int Aff 29(1):49–66. https://doi.org/10.1353/sais.0.0028
Pourzakarya M, Bahramjerdi SFM (2021) Community-led regeneration practice in Ghalam Gudeh District, Bandar Anzali, Iran: A participatory action research (PAR) Project. Land Use Policy 105:105416. https://doi.org/10.1016/j.landusepol.2021.105416
Ravallion M, Chen S, Sangraula P (2007) The urbanization of global poverty. World Bank Research Digest 1(4):1–8
Relph E (1976) Place and placelessness. Pion, London
Roberts P, Sykes H, Granger R (eds) (2016) Urban regeneration. Sage Publications, London
Roggema R (2017) Research by design: proposition for a methodological approach. Urban Sci 1(1):1–19. https://doi.org/10.3390/urbansci1010002
Rukmana D (2020) The Routledge handbook of planning megacities in the Global South. Routledge, Oxton
Sassen S (1991) The global city. Princeton University Press, Princeton. https://doi.org/10.1177/027046769301300240
Schmid C (2012) Henri Lefebvre, the right to the city, and the new metropolitan mainstream. In: Brenner N Marcuse P Mayer M (eds) Cities for people, not for profit: critical urban theory and the right to the city. Routledge, Oxton, pp 42–62
Seamon D Sowers J (2008) Place and placelessness, Edward Relph. In: Hubbard P Kitchen R Vallentine G (eds) Key texts in human geography. Sage, London, pp 43–51
Sendra P, Fitzpatrick D (2020) Community-led regeneration: a toolkit for residents and planners. UCL Press. https://doi.org/10.2307/j.ctv13xpsj7
Sennett R (1999) Capitalism and the city. In: Echenique M, Saint A (eds) Cities for the new millennium. Spoon Press, London and New York, pp 19–24
Smithsimon G (2010) Inside the empire: ethnography of a global citadel in New York. Urban Stud 47(4):699–724
Soederberg S (2005) Recasting neoliberal dominance in the Global South? A critique of the Monterrey consensus. Alternatives 30(3):325–364. https://doi.org/10.1177/030437540503000304
Soja EW (2011). Regional urbanization and the end of the metropolis era. In: Bridge G, Watson S (eds) The new Blackwell companion to the city. Wiley-Blackwell, Malden, pp 679–689
Sorkin M (ed) (1992) Variations on a theme park: the New American City and the end of public space. Hill and Wang, New York
Turco A (2003) Abitare l'avvenire. Configurazioni territoriali e dinamiche identitarie nell'etàdella globalizzazione. Bollettino della Società Geografica 1:3–20
Yavo-Ayalon S (2022) Privatization and its aftermath: are we facing a new displacement force? J Urban Manag 11:285–297. https://doi.org/10.1016/j.jum.2022.02.002
Zukin S (2010) Naked city: the death and life on authentic urban places. Oxford University Press, Oxford, New York

Measuring Utility of Vertical Expansion in Kolkata Region and Its Future Prospect

Dipanwita Ghosh, Surajit Kar, Subham Mukherjee, and Sumona Bandyopadhyay

1 Introduction

The term vertical expansion of urban zones combines with increasing density of metropolitan built-up regions and high-rise building's growth for residential, commercial, and administrative determinations. Sometimes central city region has fixed demarcated areas; hence it is experiencing vertical expansion through "high-rise buildings." In most of the cases, urban areas are experiencing two city growths: one is a horizontal expansion and another one is a vertical expansion. The horizontal expansion of metropolitan areas is sometimes known as urban agglomeration, where the city captures the surrounding suburban and rural sectors (He et al. 2019). At the same time, vertical expansion is another form of urban growth through the augmentation of commercial, residential, and administrational sectors. The vertical expansion in the city is likely concentrated in some of the zone of the urban sector where the commercial and administrative sectors are concentrated (Abdulghafour and Al-Rawe 2021; Kavilkar and Patil 2014). Since the concentration of higher buildings is the primary integer to define urban vertical expansion, there is not a single explicit definition of vertical expansion in the city region, as the concept of high-rise is relative.

In most of the cases, a tall flat or building, generally a skyscraper, is defined as high-rise building with continuous habitat, and this height varies from region to region (Gan et al. 2017). For example, "ASHRAE (American Society of Heating, Refrigerating,

D. Ghosh (✉) · S. Kar (✉)
Department of Geography, University of Calcutta, Kolkata, India
e-mail: dipig201@gmail.com

S. Kar
e-mail: surajit15213@gmail.com

S. Mukherjee · S. Bandyopadhyay
Institute of Geographical Sciences (Physical Geography), Freie Universität Berlin, Malteserstr. 74-100, 12249 Berlin, Germany

M. Firoz C et al. (eds.), *Urban Planning and Design for Megacities in the Global South*, Advances in 21st Century Human Settlements,
https://doi.org/10.1007/978-981-97-8370-0_11

and Air-conditioning Engineers)," a "technical committee" in the United States has considered a building with more than 91 m/300 ft as high-rise building according to (Ellis and Torcellini 2005; Al-Kodmany 2018a). In Germany, 35–100 m multistory buildings or a building with 12–39 floors are considered as high-rise buildings (Al-Kodmany 2017).

1.1 Literate Review

A strong correlation exists between evaluating human development and its geographical context, a relationship that has remained pivotal through history. The surge of technological advancements and economic influences since the nineteenth century precipitated the rise of high-rise constructions within cities (Zhang et al. 2021). Exploring the factors linked with high-rise constructions, various studies have highlighted the impact of high land costs, which catalyzed vertical expansion, particularly in industrial cities like Chicago—the birthplace of the world's first skyscrapers due to soaring land values (Borck 2016).

High-rise buildings, often synonymous with space allocation, typically anchor within the central business and administrative zones of million-plus cities, sometimes serving as symbols of prestige, inciting a shared pursuit (Karimimoshaver and Winkemann 2018). Examining Shanghai's urban landscape reveals a dual expansion—vertical and horizontal—driven by economic and urbanization forces, respectively (Shi et al. 2009; Yang and Zhao 2022). This evolving urban expansion has triggered multifaceted debates, including one on gentrification. Tokyo's concentration of skyscrapers in densely populated areas has sculpted a distinctive scenery labeled as "gentrification without gentry" (Machimura 2021).

In certain cases, the vertical expansion within a city can significantly reshape its political economy. For instance, Melbourne experienced a surge in high-rise development, a departure from its traditional suburban growth patterns, triggered by economic shifts (Nethercote 2019). Beyond the functional rationale behind high-rise construction, urban planning literature underscores its impact on the thermal balance and wind turbulence within the city region (Nugroho et al. 2023). However, despite these insights, a comprehensive assessment of high-rise constructions across the city regions of the Global South remains elusive. This paper aims to bridge this gap by undertaking a holistic appraisal of high-rise construction within the Kolkata region. It endeavors to analyze trends, draw comparative insights with other major cities, and propose a robust policy framework aimed at fostering a sustainable future for the city.

On the other hand, within the Kolkata city region, the transformation of low-lying areas into towering structures presents a concerning threat to land subsidies (Rumbach 2017). In contrast to horizontal expansion, vertical construction minimizes the disruption of natural habitats and necessitates resource conservation (Generalova and Generalov 2021; Ren et al. 2022). Singapore's incorporation of the "Green Heart" concept with high-rise construction serves as a paradigm, showcasing how public spaces and urban greenery can harmoniously coexist, fostering urban sustainability (Yi 2020; Fujii and Ray 2021). In the Indian context, the structuring of high-rises is approached through various methods aimed at bolstering economic gains while ensuring city sustainability (Mishra and Goel 2019). Among India's major cities, Kolkata bears significant socio-cultural, economic, and colonial connotations. The transformation of traditional buildings into high-rises, while facilitating economic growth and administrative consolidation, simultaneously erodes cultural heritage (Bose 2015). The city's escalating population, coupled with the need to bolster economic activity and provide accommodation, inexorably drives vertical expansion within the region (Majumdar and Chatterjee 2021; Basu and Bhattacharjee 2022b).

As the economic nucleus (Kolkata Urban Agglomeration) of eastern India, Kolkata's economic vitality fuels urban expansion, exemplifying the city's significance as a source of livelihood for its populace. The exponential growth in GDP, from 14.9 billion euros in 2010–2011 to a considerable increase, paralleling urban expansion from 53 to 205 sq./km by 2012, underscores the intrinsic relationship between urban expansion and economic development (Gupta 2011; Census 2011). Hence, the current dilemma revolves around accommodating and managing the burgeoning population (Berardi and Jafarpur 2020; Azian et al. 2020). Providing safe and adequate housing for this burgeoning population necessitates vertical expansion through high-rise constructions (Azian et al. 2020). Thus, this research sheds light on the utility of vertical expansion through high-rise buildings within Kolkata, addressing the critical issues of urban growth, economic development, and sustainable accommodation.

From the preceding discussion on high-rise construction, it's evident that certain explicit components—factors, processes, and sustainable policies—are implemented within specific contexts. However, to date, the study has not achieved a comprehensive evaluation of high-rise construction across the city region of the Global South. With this in consideration, this paper aims to facilitate an encompassing assessment of high-rise construction within the Kolkata region. It endeavors to delve into its trends, comparative distinctions with other major cities, and the policy framework aimed at shaping a sustainable future for the city.

1.2 Aims and Objectives

The study intends to spotlight the trajectory of vertical expansion within the Kolkata city region, specifically comparing it with other million-plus cities in India to foster a holistic understanding of urban development. The study's objectives are delineated as follows:

1. To delineate the inherent nature and trajectory of vertical expansion within Kolkata.

2. To conduct a comparative analysis between Kolkata and other major cities in India regarding vertical expansion.
3. To evaluate diverse policy outcomes geared toward sustainable urban development by scrutinizing the status of vertical expansion across Indian million-plus cities (Fig. 1).

2 Methodology

2.1 Data and Method

Data from secondary and primary sources are collected for the study. Firstly, the required dataset is collected from a questionnaire survey through Google forms to rectify "high-rise building's" utilities based on human perspective. The secondary data sets are collected through published Government documents, published newspaper documents, NATMO, individual reports, and journals, and these data will fulfill the study's aims. The data set from different Published documents, reports, projects, and journals, the frequency of construction rate for "high-rise buildings," as well as their heights were collected chronologically until 2020. All constructed "high-rise building's" total height in a particular year span is divided by the total number of high-rise buildings constructed in that particular year span to find out the "Average height of high-rise buildings of different regions"-

$$= \frac{\text{Constructed high - rise building's total height of a particular year span}}{\text{A total number of high - rise buildings constructed in that particular year span}}$$

The data on human perception of high-rise buildings are collected through the digital questionnaire method. These questions are converted into a questionnaire survey in Google forms. Then this questionnaire is sent to different people. The collected data set of Google forms is converted to Microsoft Excel, and then different facts of vertical expansion are found. Higher to lower concentrations of vertical expansion of Kolkata are analyzed through GIS-based analysis and Google Earth. Using Q GIS of a Kolkata map, the word-wise division is done, and the "high-rise building's" frequencies are implemented to each word of that Kolkata map. After that, the Choropleth method shows the high-rise buildings' consumption area with different shades of color. The boundary of the Kolkata map is geo-referenced and digitized. Then the digitized map (shape file) is opened in Google Earth. Next, the high-rise buildings are marked with the help of a polygon. After that, the KML (Keyhole Mark-up Language) file of Google Earth is opened in Q GIS. The layout is prepared with latitude and longitude, showing the high-rise buildings concentration in the Kolkata region. Different metropolitan cities are located on Google Earth with latitude and longitude, and different high-rise buildings are identified to interpret the high-rise buildings' development status of these cities. The high-rise building's

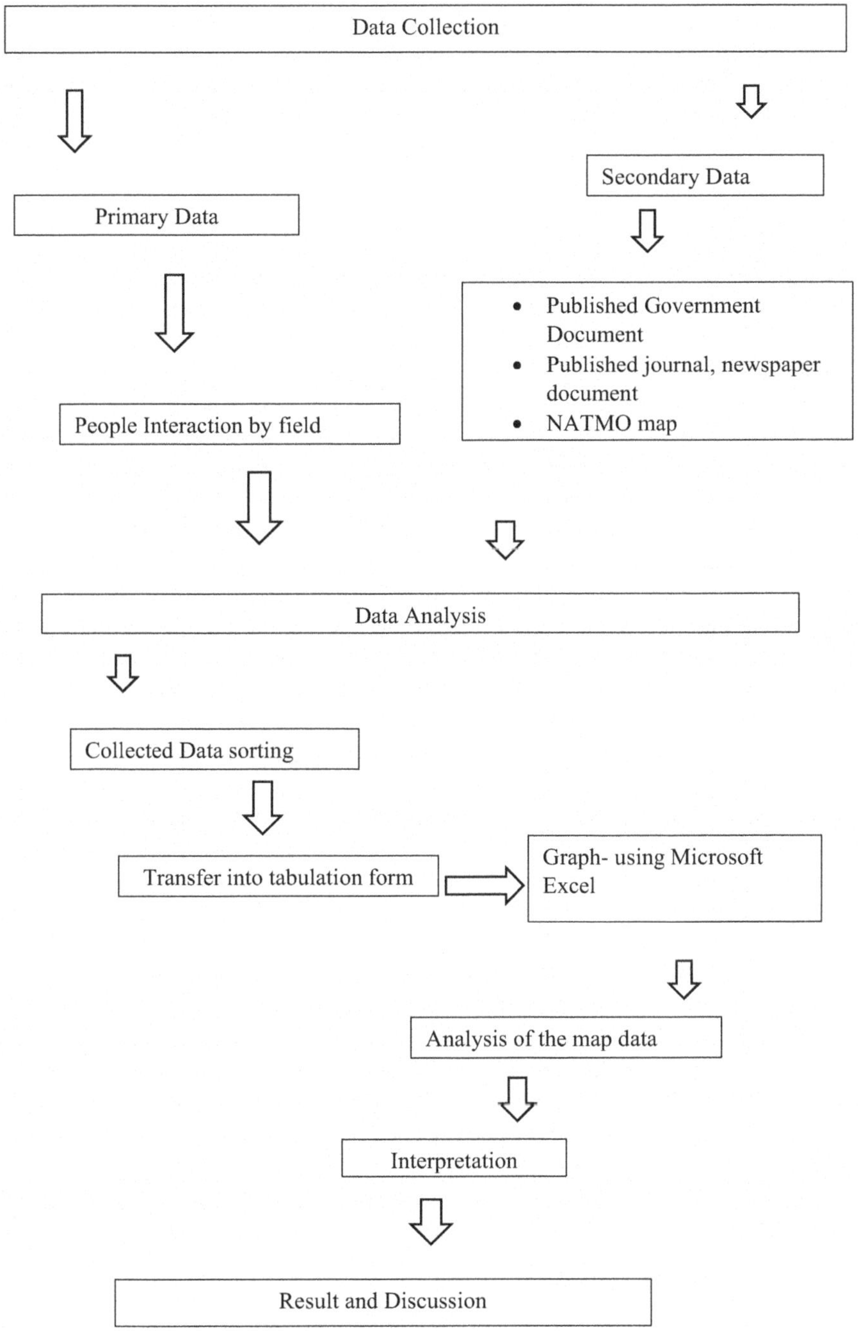

Fig. 1 Flow chart for methodology

related data and information is identified and the data was converted from questionnaire Google forms data into Microsoft Excel for tabulation form. Next, the Excel tabulation is represented with various graphs to represent the different phenomena of high-rise buildings and their effects.

2.2 *Study Area*

Kolkata is in Ganga Deltas lower zones in the eastern parts of India, and the distance of nearest international border of Bangladesh is 75 km, and it takes almost 47 min (Mukherjee et al. 2018). The elevation of Kolkata is 1.5–9 m. The main sources of water in Kolkata are Hooghly River that controls the major part of Kolkata's drainage system and the rest of the part depends on groundwater. Kolkata belongs to "subtropical climatic" zone, that comes under monsoons regimes. Kolkata is a warm temperature zone, and the temperature range is 38 degrees Celsius in April to 27 °C in December. After that in the winter season (November to January) dry seasons take place in Kolkata (Sinha et al. 2019). Kolkata's population was 14,850,066 in 2020, however the population was 4,604,143 in 1950. Besides that, 427,396 population growth was secured in the year 2015 which reflects 0.59% change in annual level that represents rapid urban growth (Smith 2019). The population are estimated by the "UN World Urbanization Prospects revision" projection system. It has preferable weather suitable for living and availability of groundwater becomes a human settlement-prone area which indicates high economic development (Mukherjee and Sundberg 2023) (Fig. 2).

Urbanization in many cities causes encroachment of land area (Barthel et al. 2019); hence vertical expansion is another form of rapid Urbanization of different cities in India (Saxena et al. 2021), as it provides shelter for a more significant proportion of the population within a smaller land (Rondhi et al. 2018). The idea on vertical expansion for urbanization is a strategy of emphasizing environmental conservation (Yu et al. 2020; Ding et al. 2018) through vertical greenery construction's incorporation (Wong and Baldwin 2016; Wang et al. 2020), which brings sustainability in the city regions (Wen et al. 2020). This is the reason why vertical expansion is a much preferable idea in the context of Urbanization in a million-plus city like Kolkata (Majumdar and Sivaramakrishnan 2020), Mumbai (Ramachandra et al. 2014), Delhi (Balala et al. 2020), and Bangalore (Ghawana et al. 2021; Kumara and Gopiprasad 2017; Rathee 2014).

3 Results and Discussion

We have summarized the entire results in few sections where the trend of vertical expansion of Kolkata is rectified and the borough-wise concertation of the high-rise buildings is also identified. The vertical expansion in different metropolitan cities of

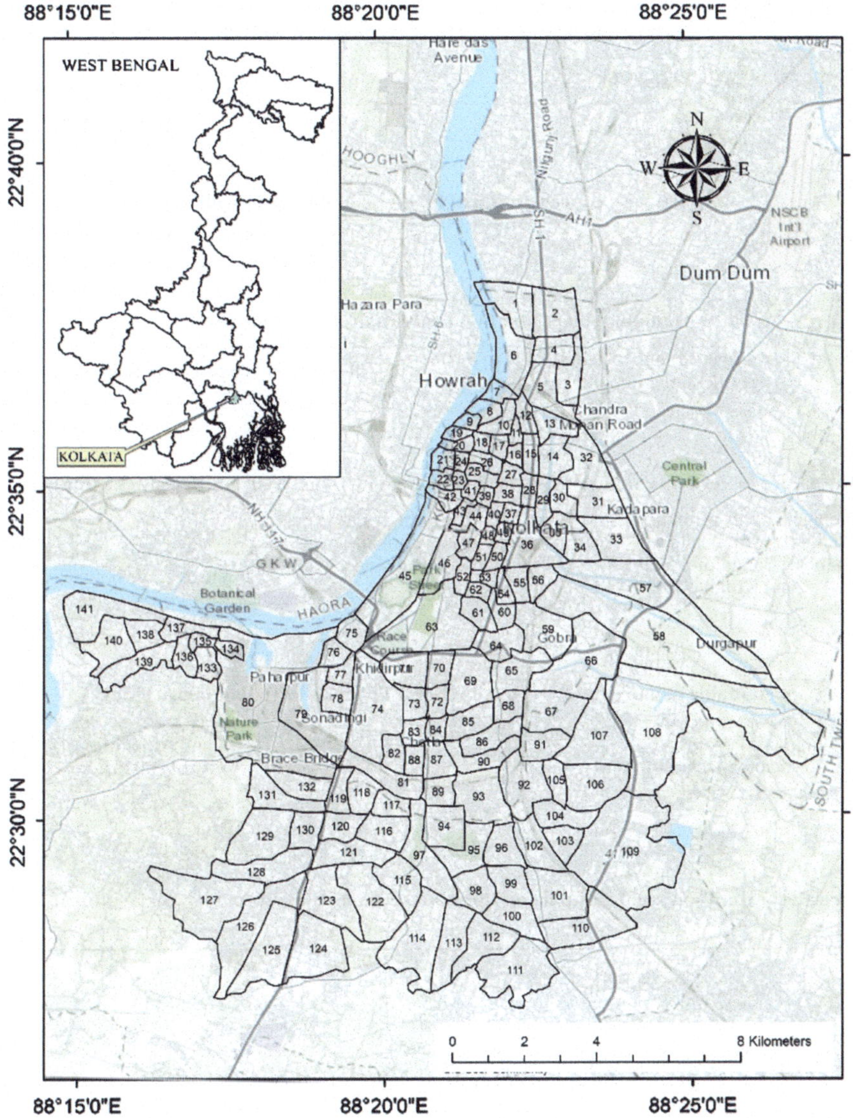

Fig. 2 Location map of Kolkata

India is analyzed in this section and the context of sustainable urban growth by the alignment of the policies is critically illustrated (Fig. 3).

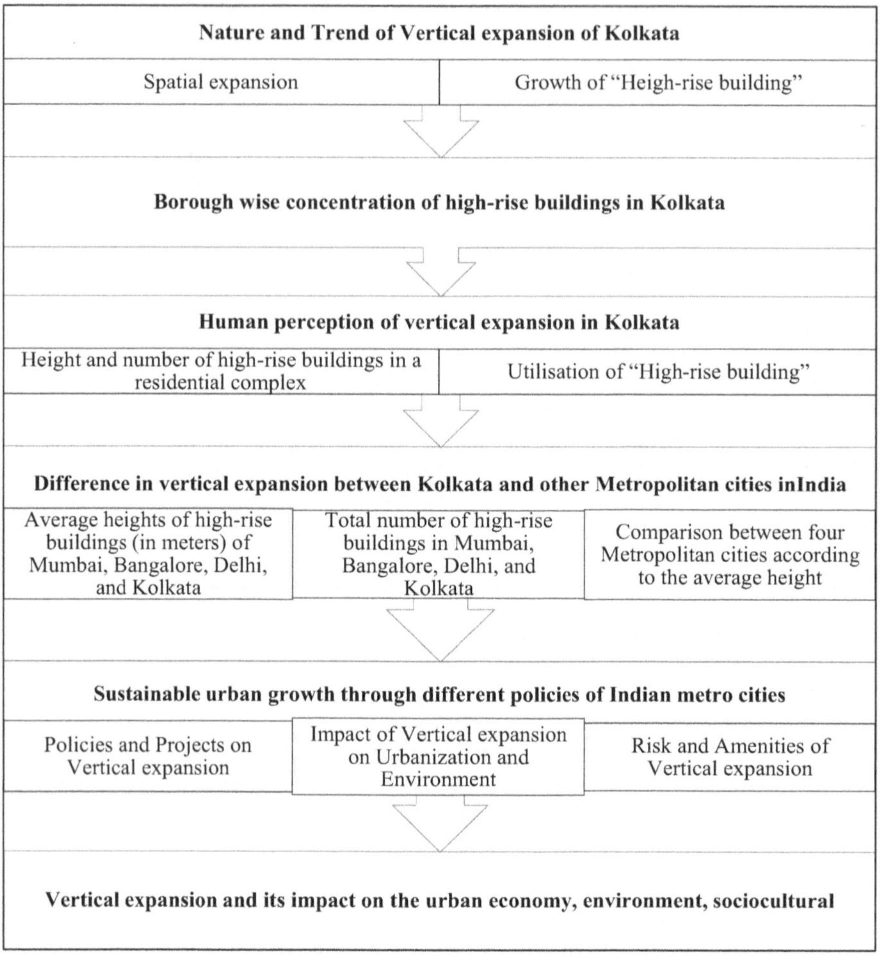

Fig. 3 Steps of findings of the study

3.1 Nature and Trend of Vertical Expansion in Kolkata

Kolkata ranked third in the metropolitan cities of India, and this is reasoning that those high-rise buildings are rapidly following the increasing trend. The urban expansion started in Kolkata in the seventeenth century and it is still expanding, but the expansion mode has changed (Simlandy 2022). As per the previous year, the horizontal expansion was high, but after the year 2012, the tendency of vertical expansion in Kolkata is taking place majorly. After that, in the year 1911, the Indian's capital city was shifted to Delhi and the reunion of West Bengal and the eastern part of India was reunited (India Today, 24th August 2018, 14:14 IST). Kolkata has experienced continuous spatial growth from the very ancient time (mainly from the British ruled time). In a very early period, it covered only, 7.5 km^2 area, and now it is about 187.33 km^2 after increasing. Additionally, Kolkata is able to achieve 10th rank as an effective urban agglomeration in all over the world (UN 2011) (Fig. 4).

Figure 4 represents the temporal-spatial expansion of Kolkata, which started in the year 1690 and then, in the year 1757, expanded its spatial expansion till the year 1996. After 1996, Kolkata continued its spreading till 2015, and after the political boundary is restricted, urban growth brings the concept of Vertical expansion (Zhang et al. 2017). Although in the above diagrams from NATMO, the spatial expansion of Kolkata from the year 1600–1996 is represented but after 1996, the spatial expansion was continued till 2012 as per the Census 2011 (Census 2011). After that, the spatial expansion is restricted by the limitation of Kolkata boundary and gradually the vertical expansion was started.

Frequency of "high-rise buildings" in Kolkata is growing and, the height of tall buildings is increasing simultaneously. Regarding this, Fig. 5 represents that, in Kolkata, the height of high-rise buildings was 60–110 m. Apart from this, the height increased from 110 to 150 in 2013 and after 2013 it increased even more than 150 m and reached above 250 m in 2019. Therefore, where the number of tall buildings is growing equally, the heights is increasing more rapidly. Analysis for the differentiation of height before 2011–2019 represents evidence of the rapid increase in the height of tall buildings of Kolkata.

As per this analysis in Fig. 6, there are high-rise buildings with different heights as represented. 50–100 m in height tall building's existence is highest, and then comes tall buildings with 100–150 m height. On the other hand, the buildings with 150–200 m are much larger than the previous category of 100–150 m, and the buildings above 200 m have a minimum quantity in Kolkata and there is only one building with a height above 200 m.

From the High-rise (more than 7 floors) concentration map (Fig. 7) of Kolkata, the highly High-rise concentrated area is central Kolkata. On the other hand, north and east Kolkata is a semi-high, High-rise (7 or less than 7 floors) concentration area. South Kolkata is a moderate High-rise concentration area as only one High-rise can be found (for example: South city). Lastly, the western part has no such high-rises buildings. Instead of this, some high-rise buildings can be found in the surrounding area of Kolkata, mainly in the Newton and Rajarhat area. Hence, the central and northern part of Kolkata is mainly the High-rise and high buildings (6–7 floor) concentration area.

3.1.1 Borough-Wise Concentration of High-Rise Buildings in Kolkata

Kolkata can be divided into 15 boroughs (Gupta 2011). After identification of the high-rise building's concentration area through the previous choropleth map, the borough-wise concentration of high-rise buildings of Kolkata can be measured. According to the map (Fig. 8), the high-rise buildings concentration is highest in central Kolkata, which is in the darkest shade of color. Mainly borough number 7 has the highest concentration of high-rises buildings. Here the number of high-rise

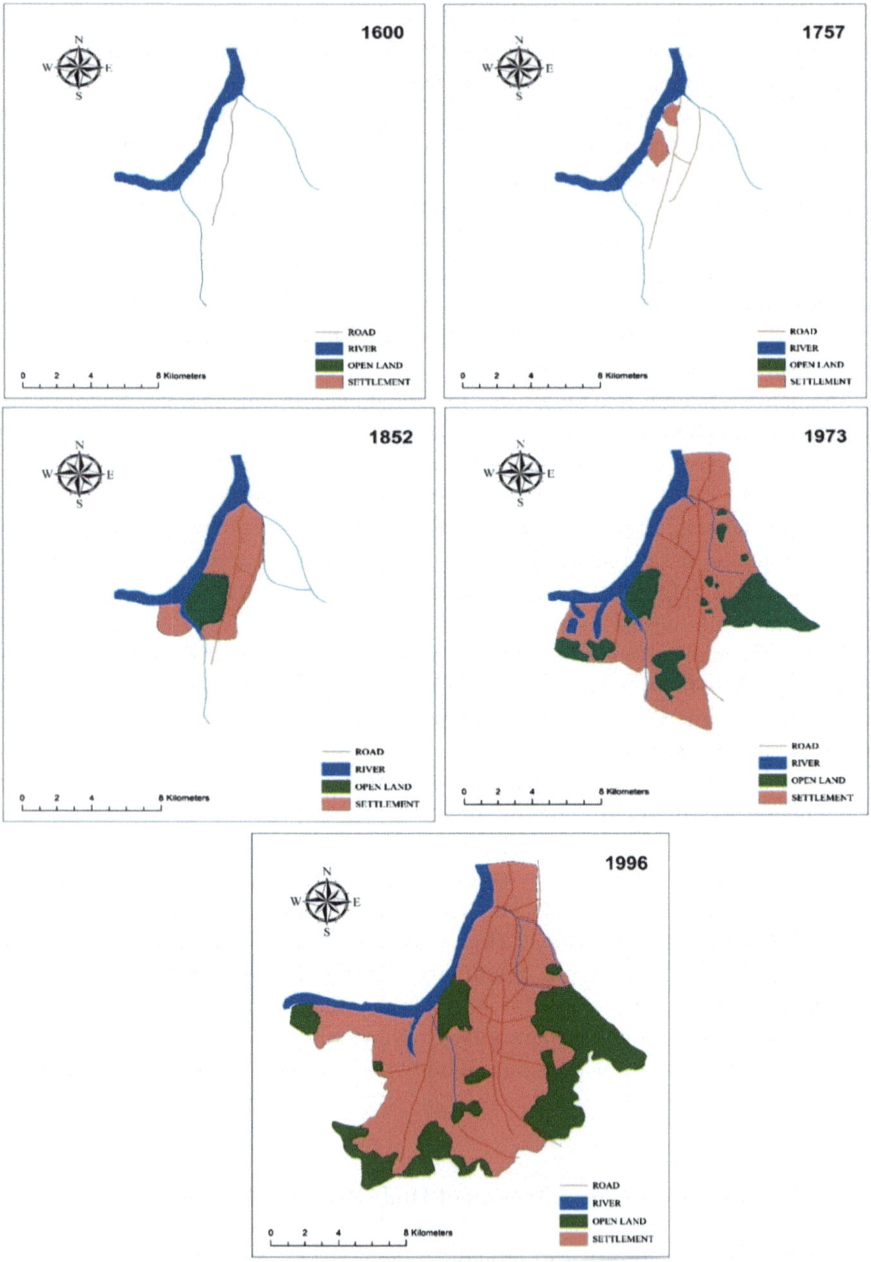

Fig. 4 Temporal expansion of Kolkata from pre-independence to 1996. *Source* NATMO

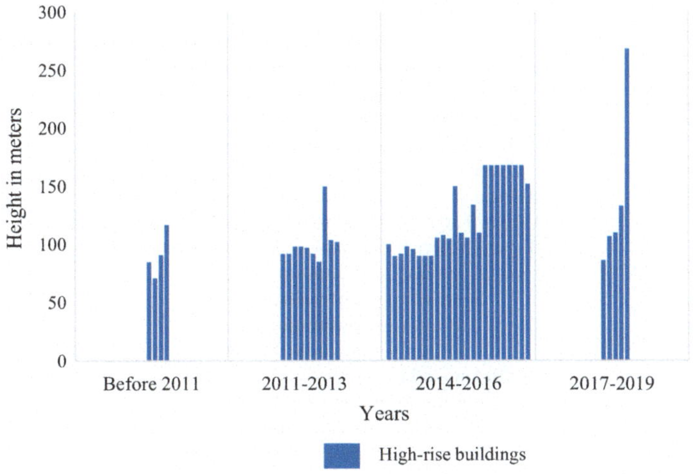

Fig. 5 Height of high-rise buildings in Kolkata from 2011 to 2019

buildings is 6–9. Then the comparatively lighter shade of color shows the moderate concentration of high-rise buildings. This category includes boroughs 8, 10 (the southern part of Kolkata), and 3 (the northern part of Kolkata). Here the numbers of high-rise buildings are 3–6. Last but not the list, the lightest shade of color shows the minor concentration of high-rises. This concentration can be shown in the 1, 2, 4, 5, 6, 9, 11, 12, 13, 14, and 15 boroughs of Kolkata and frequencies of tall buildings is 0–3.

3.1.2 Human Perception of Vertical Expansion of Kolkata in the Present Context

Human perception refers to the practical thought of a specific aspect. The importance of human perception is the primary concern for constructing a factor (Frank 2019), like the execution of vertical expansion in Kolkata. As per the human perception of high-rise buildings, dynamic thoughts are present. The thinking is the reflection of the desired requirements for high-rise buildings. Additionally, human perception signifies the contemporary importance of high-rise buildings. In this regard, the importance of high-rise buildings can be estimated for the execution of sustainability, but the functional interaction with the different people can measure the practical scenario behind the execution. Hence the analysis of the human perception is established for strengthening the urban growth concept through vertical expansion for sustainability.

Figure 9 represents the frequency of “high-rise buildings” at the resident’s complex area. In Kolkata, high-rise construction is taking place and almost 49.09% of residents have high-rise buildings in their complex area. Most of the places have 1–3 high-rise buildings (48.15%), and some of the places have 4–6 buildings (33.33%),

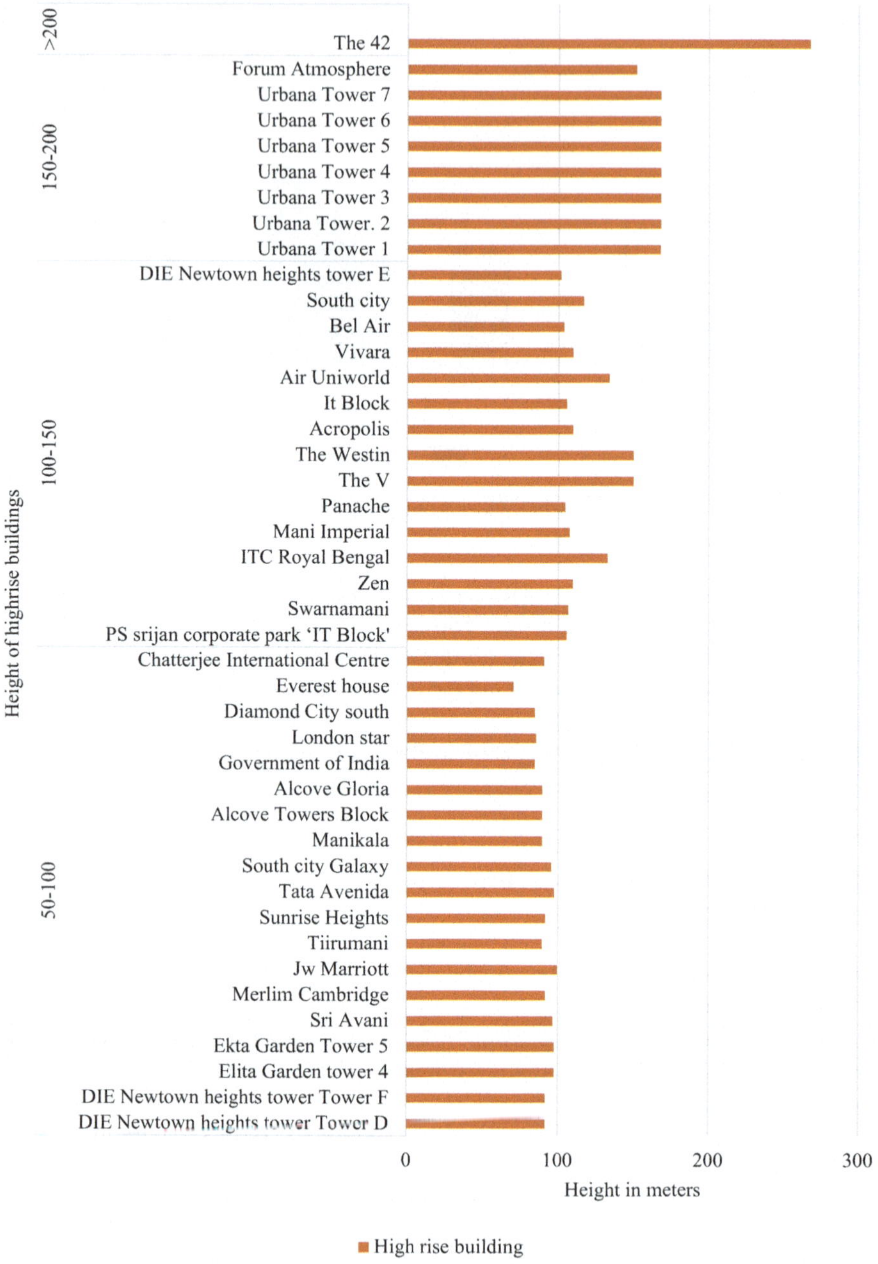

Fig. 6 Name of the high-rise building in Kolkata with height (in meters)

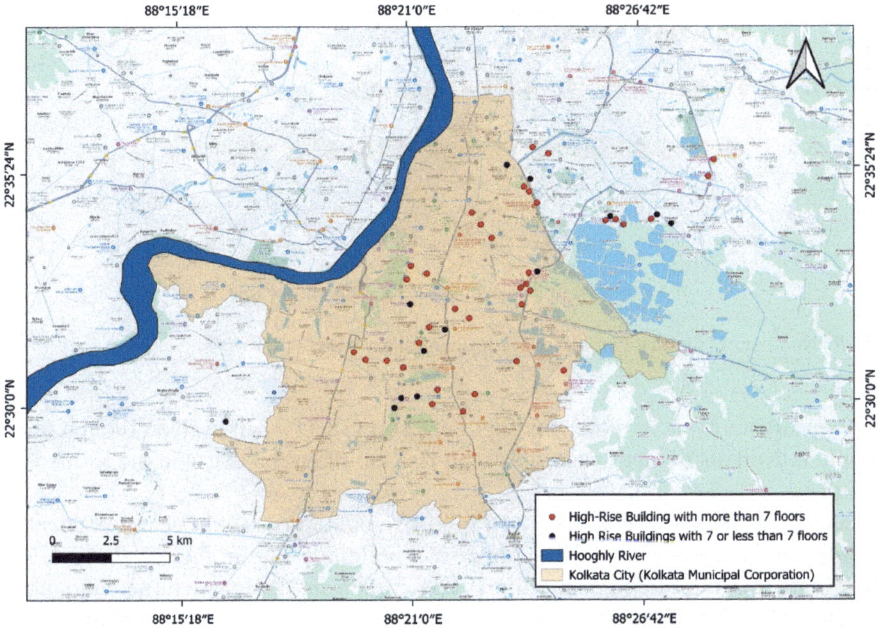

Fig. 7 "High-rise buildings" as well as other "high buildings" of Kolkata and surrounding area

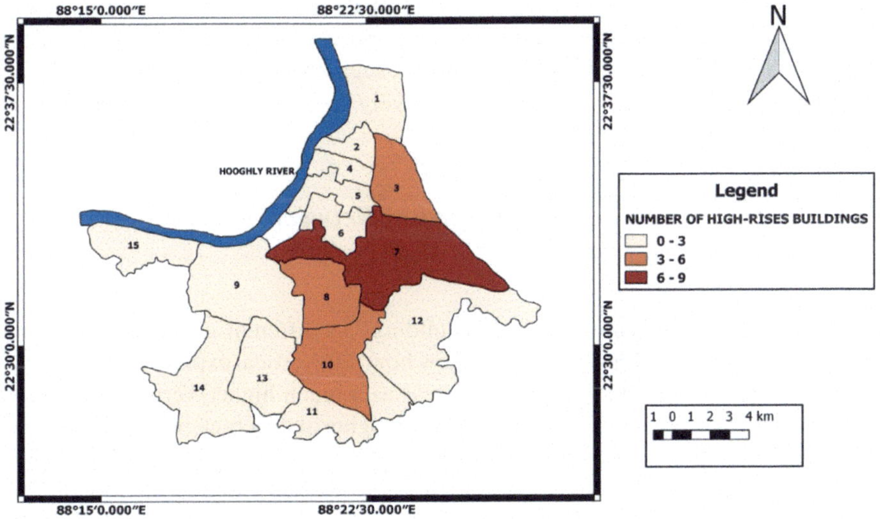

Fig. 8 Borough-wise concentration of high-rise buildings in Kolkata

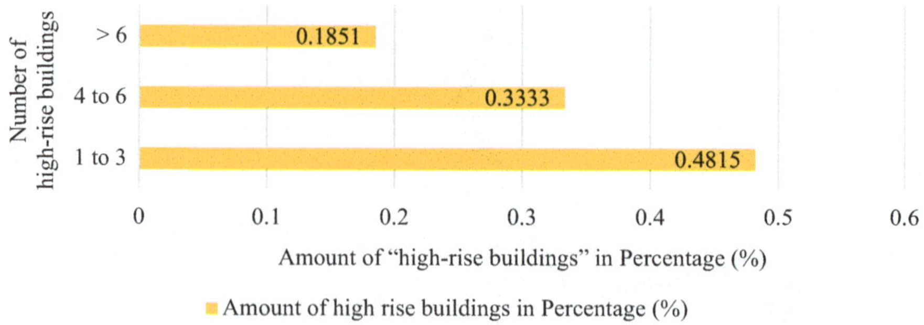

Fig. 9 Number of "high-rise buildings" in residential complexes (%)

and appearance of six or more buildings are lesser (18.51%). Therefore, though some people in Kolkata do not live in the "high-rise buildings," it exists in the residential complex.

As per the analysis (Fig. 10), most high-rise buildings' height near a residential complex area is 61–80 m. 120-m "high-rise buildings" are in the lowest (7.41%). Regarding this, the "old high buildings" (lesser than six floors) are turned into much higher within few years. Based on the analysis from Fig. 11 represents that high-rise development emphasizes quite a positive impact on "urban development," "rapid urban growth," "residential ownership," human "comfort," and "population demand." From the survey, it can be measured that rapid urban growth (67.50%) fulfill housing demand through residential ownership (60%), population demand (57.50%), which is significant for sustainable urban growth, and the segment of human comfort (34.17%) and urban development (31.66%) are respectively lesser significant than others.

3.2 *Difference in Vertical Expansion Between Kolkata and Other Metropolitan Cities of India*

There are several studies on high-rise buildings in different cities with different perspectives, but the studies have yet to show how this Vertical expansion is growing. It can be stated that Kolkata will also appear as a much higher Vertical expansion than how it is at present, and by comparing the heights of different metropolitan cities high-rise buildings with Kolkata, the prospects of vertical expansion of Kolkata and the changing human perspective about high-rise buildings could be identified. That is the reason for comparing different Vertical expansions of other cities to Kolkata, the utility of vertical expansion, and the higher concentration of high-rise buildings in Kolkata that emphasizes sustainable urban growth.

Figure 12 represents that Mumbai is expected to be the world's second-largest city in 2015, containing 22.6 million people (UN Population Division 2006; Pavri 2009). The tallest building in India, the World one building, 442 m high from the surface, is in Mumbai. It could also become the tallest building in the world (The Telegraph

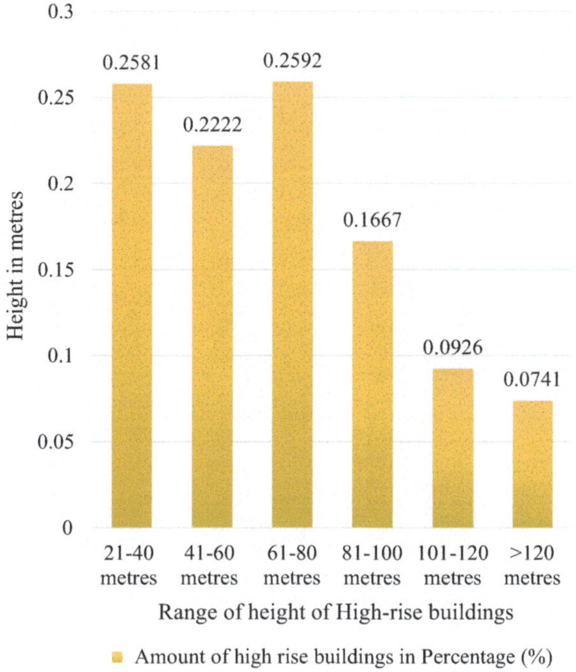

Fig. 10 Height of high-rise buildings in a residential complex in percentage

2010). The height of high-rise buildings and the number of high-rise buildings are growing from 2012 to 2020. At the same time, the vertical expansion of Delhi is lower than in Mumbai but higher than in Bangalore and Kolkata but in 2015–2017, it was highest. On the other hand, in Kolkata, the current year's (2015–2017) vertical growth height and the number of high-rise buildings are lesser than the previous year's (2015–2017). The average height of the vertical expansion of Kolkata is 115.08 m, and in Bangalore, with the increasing rate of the built-up area, the vertical expansion furthermore grows simultaneously.

Figure 13 is evidence of the increase in several high-rise buildings in different Metropolitan cities in different years. From 2012 to 2014, only three high-rise buildings emerged in Delhi, and the number increased to 26 in 2015–2017. After that, in 2018–2019, 13 more buildings were constructed in Delhi. In the context of Kolkata, there was 14 high-rise building in 2012–2014, and 16 buildings were constructed in 2015–2017, which decreased to 12 high-rise building construction in 2020. In Bangalore, four buildings were constructed in 2014, and the number increased to 15 at the end of 2017. After that, within the year 2020, 37 high-rise building was constructed in Bangalore. For Mumbai, there was 15 building in 2014. In 2015–2017, 16 high-rise building was additionally constructed, and at the end of 2020, 33 new high-rise building was constructed. Therefore, it can be stated that the highest number of high-rise buildings were constructed in Mumbai from 2018 to 2020. However, the trend of construction of high-rise buildings from 2017 to 2020 was highest in Bangalore.

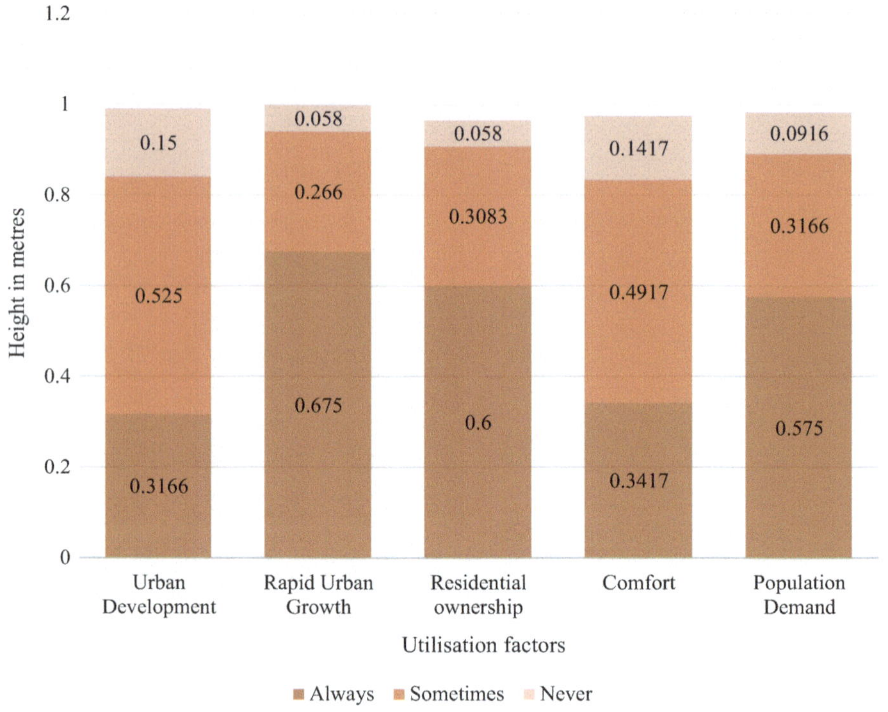

Fig. 11 Emphasis of "high-rise buildings" in diverse sectors (%)

From the analysis of Fig. 14, it can be measured that the average height of high-rise buildings from 2012 to 2018 was the highest in Mumbai. Delhi, the national capital region, is in the second position. Kolkata is in the fourth position after Bangalore, but Kolkata and Bangalore's average height of high-rise buildings are almost similar. On the other hand, in Mumbai, there are the highest number of high-rise buildings being constructed, and then in Bangalore and Delhi, respectively, and then Kolkata takes place. Therefore, the highest vertical expansion is occurring in Mumbai as it is the financial capital of India and Mumbai is the most emigrated region for better career opportunities and better social livelihood (Choithaniet al. 2021). Kolkata is in the primary stage of vertical expansion as compared to Mumbai and Delhi. However, the increasing rate of high-rise buildings is relatively high, and the rapid construction of high-rise buildings is the emergence of different IT industries and economic sectors (Gollin et al. 2016). The IT industries and innovative City projects (Antenucci 2021; Tiwari et al. 2022) focused on human development, and it is emphasizing vertical expansion as well to form these industries a considerable place is needed, which is only possible through high-rise buildings such as TCS, Wipro Ltd., HCL Technologies Ltd., Tech Mahindra Ltd., Accenture, Cognizant, Cap Gemini, and IBM (Sadhukhan 2020), etc., as the horizontal space is limited. On

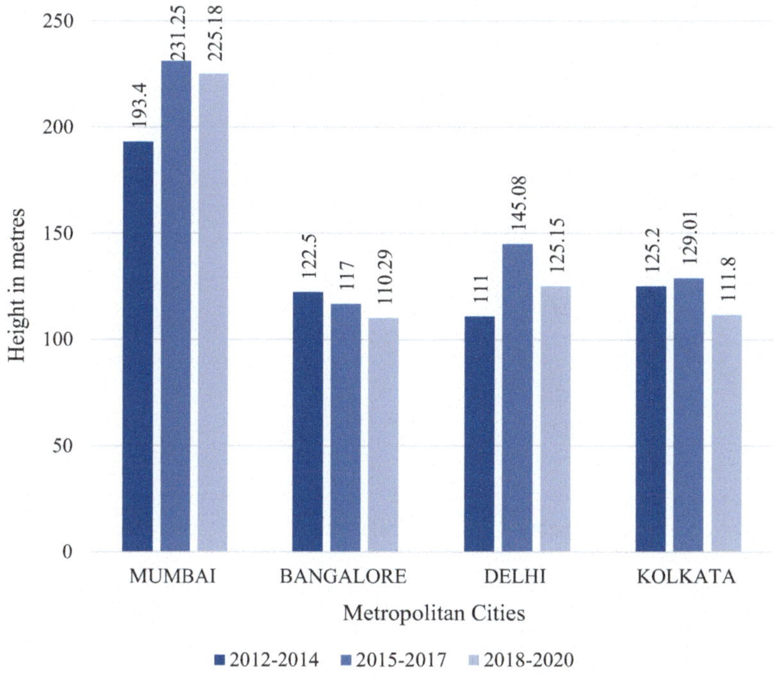

Fig. 12 Average heights of high-rise buildings (in meters) of Mumbai, Bangalore, Delhi, and Kolkata from 2012 to 2020

the other hand, the economic condition of Kolkata is improving, and several businessmen and wealthy people are emerging daily (Chakrabarti and Sinha 2021). They are very much attracted to this high-rise building, which is much more convenient for business and residential purposes. Therefore, from this comparison of different metropolitan cities' Vertical expansion, the prospect of vertical expansion of Kolkata can be implemented that in the future, Kolkata's vertical growth will increase even if it can overtake the present height of Mumbai City, which indicates that there is a modification of human acceptance about high-rise buildings which is changing positively.

3.3 Sustainable Urban Growth Through Different Policies of Indian Metro Cities

The execution of urban growth in a sustainable way is the outcome of adopting appropriate policies with the affiliation of the government, especially for the Indian Metropolitan cities. Additionally, successful implementation is also essential for the ultimate result of the accomplishment of Sustainable urban growth. Regarding this,

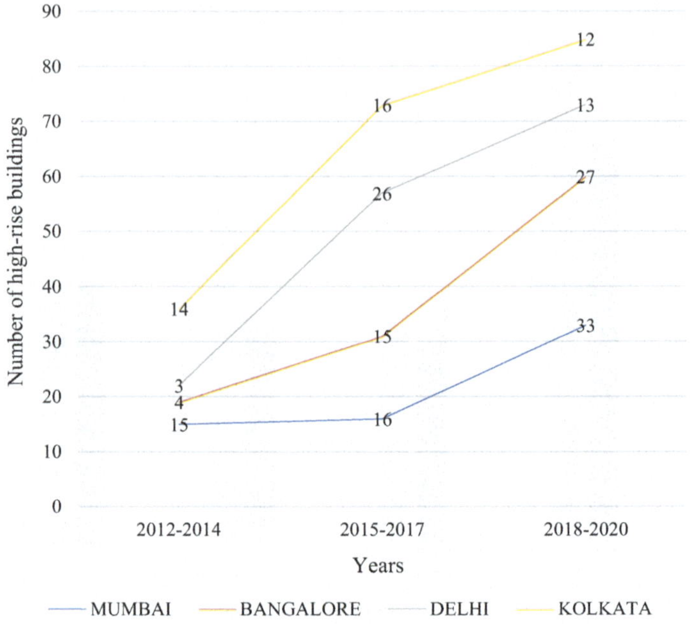

Fig. 13 Total number of high-rise buildings in Mumbai, Bangalore, Delhi, and Kolkata from 2012 to 2020

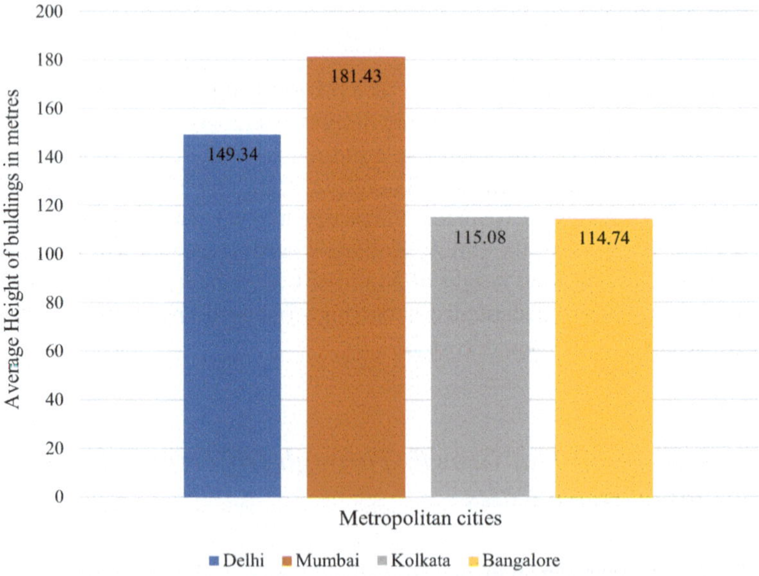

Fig. 14 Comparison between four Metropolitan cities according to the average height of buildings in meters from 2012 to 2018

in constructing policies for Indian Metropolitan cities to establish urban development and Sustainability, the assessment of the risk of the urban area is essential. The assessment of urban amenities and the urban environment is equally essential for the appropriate policy implementation. The policies to secure Sustainability in urban areas reflect structured management for measuring the different activities. In this regard, vertical expansion is correlated with risk reduction as well as the effectiveness of urban amenities aspect. A planned and environment-friendly action for the construction of vertical expansion can successfully establish sustainable urban growth. There are different types of the chance of risk that can occur due to the high construction of high-rise buildings. Earthquakes (Wu and Lin 2022), groundwater problem (Attard et al. 2016), water logging, urban flood (O'Donnell and Thorne 2020), deforestation (Digdowiseiso and Sugiyanto 2020), and even pollution (Liang et al. 2019), can occur. However, the chances and intensity of these kinds of risks can be reduced by adopting several technologies (Al-Kodmany, 2018b) and knowledge (Pan and Hon 2020) because some risks could be occurred by vertical expansion, but some at which present day this is one of the most important aspects to balance the economy and environment (Ferdous et al. 2019; Ghiselliniet al. 2018). Amenities are an essential part of human living (Hartig and Kahn 2016), and when many populations live together, the importance and number of amenities increase (Albouy and Stuart 2020). Amenities means internet access (Pires et al. 2019), parks, pools, workout places (Latham and Layton 2019), greenery (Black and Richards 2020), cultural places, market, parking area, etc. The security of a household is as significant as the amenities (Voskresenskaya and Vorona-Slivinskaya 2018). Henceforth, the caretakers or the house guards (Adu-Gyamfi 2021), and CCTV cameras (Au-Yong et al. 2018), are essential and can be considered as a part of amenities too (Beames et al. 2018), and most of the high-rise buildings provide these kinds of amenities and security which is a better sign (Arican and Bahar 2020). Environment and urbanization aspects sometimes conflict with each other (Seto et al. 2012; Güneralp and Seto 2013) but with the help of modern technology and knowledge (Fu et al. 2020); presently plantation is taking place vertically (Patil et al. 2021) as well. The green roof (Wang et al. 2020; Dwivedi and Mohan 2018), green terrace (Wang et al. 2020; Dwivedi and Mohan 2018), green building (Sharma 2018), Green wall (Patil et al. 2021; Wong and Baldwin 2016) concepts are the result of this. Hence it is a newly appearing thought, especially in cities like Kolkata (Chatterjee 2021). Therefore, some environmental challenges could be mitigated (Conejos et al. 2019); if the mitigation strategies are established with proper knowledge, then this kind of plantation could help to reach the sustainability aspects in the future (Sharma 2018) (Table 1).

From the above-mentioned discussion, it is evident that the policy and practices applied for sustainable urban infrastructure are primarily restricted to green infrastructure, eco-friendly residential areas, and economic and administrative units, which can support high-rise construction in its optimal stage. The metro cities, especially the high-rise construction, symbolize economic, administrative, and residential units, which become obligatory for contemporary urban infrastructure. The sustainability

Table 1 Different policies and projects on vertical expansion

Policies and "high-rise buildings" projects	Country	Objective	Source
Atal Mission for Rejuvenation and Urban Transformation (AMRUT) June 2015–2022/2023	500 cities of India (West Bengal—-60 cities including Kolkata and Delhi, Mumbai, etc. from different states)	It aims to provide better livelihood and amenities (water supply, sewerage, urban transport, green space, parks, storm water drainage) in the urban areas	Atal Mission For Rejuvenation and Urban Transformation-Official website
The Green City Mission by New Town Kolkata Green Smart City Corporation Limited" or NKGSCC 25 June 2015–2025	(Newtown, Rajarhat) Kolkata, India	Modification of the city with modern technology	Newtown Green Smart City officials and Tiwari et al. (2022)
"National smart cities mission" was initiated by "Union Ministry of Urban Development" 2015–2023	100 cities in India (like—Bangalore, New Delhi, etc.)	Modification of cities for optimum technological outputs for well-being of urban citizens	"The Smart City Mission official" website
The Imperial 3, 2016–2025	Mumbai, India	To accommodate residential places with an eco-friendly architecture	Zhang (2018)
Palais Royale 2008–2020/2021	Mumbai, India	Focus on green building concept and achieved platinum rating from Green Building Council	Balasubramaniam (2018) and Nagawkar et al. (2014)
Namaste Tower	Mumbai, India	To construct luxurious hotels and accommodation	Bargh and Mansouri (2023)
Supernova Spira 2011–2020	Noida, India	Focus on green building concept and achieved gold rating from Green Building Council	Behera et al. (2004)
The Summit 2025	Sector V, Kolkata	To build an IT hub in the metropolis and tallest commercial hub of Kolkata	The Economic Times, April 15, 2022, 12:04 p.m. IST
West Bengal Urban Affordable Housing and Habitat Policy 2015	West Bengal	To offer affordable residential support for common poor population	Devalt (2015)
Bengal N R I Township Project (URBANA) 2012–2022	Kolkata	To offer residential support	Arun et al. (2013)

(continued)

Table 1 (continued)

Policies and "high-rise buildings" projects	Country	Objective	Source
South city project 2004–2008	Kolkata	To execute economic center and residential support	Bose (2014)
Uniworld City Residential Township project 2008	Kolkata	To support city accommodation	Botti (2023)

concept is primarily considered through protecting and maintaining the natural environment. Regardless of consider blue-green (water-vegetation) infrastructure development for sustainable urban development (Krishnan and Priya 2022), another group of scholars calls for the human well-being issues associated with urban vertical infrastructure (Marques et al. 2020). A group of literature on China represents their empirical observation of how urbanization through the association of mounting high-rise residential complexes has a positive relation with human depression due to declining neighborhood social capital (Li and Sunikka-Blank 2021; Zhang et al. 2021). The driving factor behind the high-rise construction is the economy, which provides the citizens with the affordability of their residence (Ekici et al. 2021) and serves trade and administration (Al-Kodmany 2023). Hence, from the configuration to the course of development of high-rise construction in cities, there are a total of three explicit but interconnected issues associated with it, i.e., (1) *Economic issues*, (2) *Environmental issues*, and (3) *Sociocultural issues*, which are interlinked with high-rise construction through affirmative and off-putting comportment.

(1) *Economic issues*: Instead of framing the argument of environmental consequences due to urban growth, it is undeniable that the high-rise construction let loose economic potential for the growing population in the cities. Especially in the Indian context, high-rise construction is associated with multiple economic activities that deal with the influx of skilled individuals from the surrounding areas. Considering the Kolkata city region, the high-rise construction not only serves as a residential unit that offers an aesthetic view of the surrounding area, but multiple other types of high-rises are utilized for business and administration. Areas like Tollygunge, E.M. Bypass, high-rise construction are utilized for commercial and office space. In the adjacent to the core city area, in Newtown, the emerging high-rise units separately serve as residential and office units for companies like Tata Consultancy Service. Over the past decade, the New Town and Rajarhat area have grown with the association of high-rises that serve as the business and administrative unit due to the saturation of the core city area through the pre-existing activities service. The prime cluster of high-rise hotel accommodations for tourists with lavish amenities are concentrated in the E.M bypass area (Chakravorty 2022).

(2) *Environmental issues*: The critical urban policy literature considered the downbeat consequences like depletion of groundwater (Nath et al. 2021), exposure to

cyclonic storms (Pal et al. 2022), and land subsidizing (Basu and Bhattacharjee 2022a) associated with high-rise construction, whereas on the other hand, high-rises are also putting some of the favorable aspects like consuming less space by population holding capacity (Ekici et al. 2021). Considering the negative aspects of high-rise construction, the geo-lithology and structural control are disrupted through high-rise construction. For example, groundwater availability is sometimes disrupted due to the interruption in in-ground lithology by the vertical and underground concrete structures (Attard et al. 2016), which provides the high-rise factor of land subsidence. For example, Sahu and Sikdar (2011) show how the over-drifting of groundwater accelerates the rate of land subsidence (mean subsidence is 3.28 cm) in Kolkata and its surrounding regions. To prevent such unwanted issues, some new technologies (micro hydro power) are used to reduce water wastage and water leakage, which can decrease the water pressure to distribute water in high-rise buildings (Du et al. 2017). Another risk factor associated with urban planning is earthquakes and their threat, especially in big north Indian cities, i.e., Delhi. Not because Delhi is already under the seismic 4th zone but because there is little seismic activity in Nepal, which increases the threat of seismic events for north Indian cities shortly (Sandhu et al. 2020). For that reason, the "US-based Taylor Devices Inc." was initiated as an "Indian subsidiary" that helps to measure earthquakes and construct "earthquake-proof" tall buildings. The construction is implemented by "aerospace technology" installed with the wall of the tall building. The technology easily absorbs the destructive shocks of earthquakes (Reddy et al. 2023). In many of the case studies in India and in context to other countries, green cover and high-rise construction have a well-bond relationship, i.e., Vertical terraces, vertical forests, and green roofs are the leading modern ideas in the present urban world, which integrates sustainable vertical urban growth (Wang et al. 2020; Yang and Zhao 2022). More specifically, green wall formation in high-rise buildings emphasizes the insulation system places the air conditioner and saves electricity (Wong and Baldwin 2016). In opposition to high-rise green infrastructure, Conejos et al. (2019) show how the greenery in many of the above-mentioned forms may cause waterlogging, fungus, and dumping in concrete structures.

(3) Sociocultural *issues*: Sociocultural relations are not individual. Rather, it is an integral part of the whole urban environment, as the construction of a high-rise increase's privacy and social contact, which decreases the collectiveness and group-based decision-making ability for the environmental allied aspects. The transformation of physical architecture in the urban set is apparent with individual separation by taking apart residence in the name of privacy. This catalyzes declining social capital, which may be gained through social contact, cultural norms, and practices. However, the transformation of a cooperative to a nuclear family reduces the strength of social ties, such as close friends and family members. The social capital in the urban system is not only meant for emotional and material support (Lin 2017); it may also increase the resilience level to mitigate any natural calamity (Aldrich et al. 2018). Although sustainable urban policy in the contemporary era of climate change only deals with the

conservation of natural resources, there is a gap in academia where human psychological behavior is not being sketched alongside urban growth. Mental health is directly or indirectly connected with high-rise living due to the decline of social well-being, which integrates with variables like the sense of belonging, community, and togetherness (Barros et al. 2019).

The construction of high-rises deals with technologies to provide the best facilities for humans (Baldwin Hess et al. 2018). However, the construction where sustainable land use can go head-to-head with high-rises is limited in a regional setup. Hence, obtaining sustainable land use with the association of high-rises has become a challenge in some of the specific favorable areas, even within an urban set up.

4 Conclusion

In the context of globalization, cities hold pivotal roles as economic engines driving a significant portion of a country's GDP, particularly in the service and industrial sectors. This is crucial for nations in the Global South striving for economic growth, given their historical challenges like overpopulation, colonial legacies, and technological disparities compared to the Global North. However, the pace of technological advancement in these regions hasn't always kept pace with urbanization and economic expansion, leading to potential environmental challenges associated with high-rise structures in urban centers.

The ongoing debate of whether economic development and environmental protection can harmoniously coexist persists, particularly in the realm of vertical urban expansion. Our policy analysis reveals numerous case studies demonstrating that vertical expansion can indeed align with sustainable urban development, especially when integrating concepts like urban green buildings. While sustainability traditionally emphasizes environmental conservation, there's a growing need to address citizen mental health, often overlooked in urban sustainability discourse. Despite this ongoing debate, the concentration of high-rises remains a crucial aspect of urbanization, with urban sustainability intricately tied to artificial green spaces, open areas, and aesthetically pleasing urban design for overall human well-being.

Mature metropolitan cities like Kolkata, Delhi, Mumbai, and Chennai could greatly benefit from monitoring high-rise developments through rigorous Environment Impact Assessments (EIA) and Social Impact Assessments (SIA) to ensure long-term sustainability. Conversely, emerging cities could initiate urban sustainability efforts by conducting thorough site suitability analyses for high-rise constructions from their inception. International collaboration for financial and technological exchanges becomes imperative to assess and mitigate environmental risks associated with high-rise concentration.

Moreover, enhancing legal frameworks, particularly concerning groundwater extraction, carbon emissions, noise pollution, waste management, and sewage treatment in urban systems, is essential for effectively managing the relationship between high-rises and environmental conservation. This joint association requires a more specific and stringent regulatory approach. Collaborative efforts among nations and robust legal mechanisms can ensure that high-rise development aligns with environmental preservation within urban systems.

Looking ahead, this study lays the groundwork for future research and development by highlighting the need for interdisciplinary approaches, continued international collaboration, and evolving policy frameworks. As cities evolve and grow, understanding the interplay between high-rise development, urban sustainability, and environmental conservation will remain critical for fostering resilient and harmonious urban landscapes globally.

Acknowledgement We are gratified to Mr. Rituparna Acharya from Central University of Karnataka for putting his valuable insight in the thematic representation on the manuscript.

References

Abdulghafour SH, Al-Rawe MK (2021) Vertical hierarchy of urban space and place-making. Turkish J Comput Math Educ 12(11): 730–741. https://www.researchgate.net/profile/MohebFleyeh/publication/364675262_Vertical_Hierarchy_of_Urban_Space_and_Place-Making/links/6356dea66e0d367d91c0b4b3/Vertical-Hierarchy-of-Urban-Space-and-Place-Making.pdf. Accessed on 25 July 2023

Adu-Gyamfi A (2021) The role of caretakers in improving housing conditions in peri-urban areas. Cities 110:103049. https://doi.org/10.1016/j.cities.2020.103049. Accessed on 26 July 2023

Albouy D, Stuart BA (2020) Urban population and amenities: the neoclassical model of location. Int Econ Rev 61(1):127–158. https://www.nber.org/system/files/working_papers/w19919/w19919.pdf. Accessed on 28 July 2023

Aldrich DP, Meyer MA, Page-Tan CM (2018) Social capital and natural hazards governance. In: Oxford research encyclopedia of natural hazard science

Ali MM, Armstrong PJ (1995) Architecture of tall buildings. McGraw-Hill, New York

Al-Kodmany K (2017) Understanding tall buildings: a theory of placemaking. Routledge. https://books.google.com/books?hl=en&lr=&id=4i8lDwAAQBAJ&oi=fnd&pg=PP1&dq=AlKodmany+,+K.+(2017).+Understanding+tall+buildings:+a+theory+of+placemaking.+Routledge.&ots=6oQitrCFXc&sig=fASj5K9fizIFdzxTn2SpTQa2vek. Accessed on 14 June 2023

Al-Kodmany K (2018a) The sustainability of tall building developments: a conceptual framework. Buildings 8(1):7. https://www.mdpi.com/2075-5309/8/1/7/htm. Accessed on 20 June 2023

Al-Kodmany K (2018b) The vertical city: a sustainable development model. WIT Press. https://books.google.com/books?hl=en&lr=&id=pTlpDwAAQBAJ&oi=fnd&pg=PP1&dq=AlKodmany,+K.+(2018).+The+vertical+city:+a+sustainable+development+model.+WIT+press.&ots=Mggwgcz6M6&sig=4GwGTr7-EpdfnhqeKPvEftYzKOY. Accessed on 24 Aug 2023

Al-Kodmany K (2023) High-rise developments: a critical review of the nature and extent of their sustainability. In: Pragmatic engineering and lifestyle: responsible engineering for a sustainable future, pp 1–20

Anjan Chatterjee, Nov 14, 2014. https://medium.com/@anjan.chatterjee/an-overview-of-kolkata-city-from-the-drainage-aspect-3e20505c8143. Accessed on 13 July 2023

Antenucci I (2021) Infrastructures of extraction in the smart city zones, finance, and platforms in new town Kolkata. Int J Commun (19328036):15. https://www.researchgate.net/profile/Ilia-Antenucci/publication/355105323_Digital_Infrastructure_Liminality_and_World-Making_Via_Asia_Infrastructures_of_Extraction_in_the_Smart_City_Zones_Finance_and_Platforms_in_

New_Town_Kolkata/links/61ae2272ca2d401f27cdb8e2/Digital-Infrastructure-Liminality-and-World-Making-Via-Asia-Infrastructures-of-Extraction-in-the-Smart-City-Zones-Finance-and-Platforms-in-New-Town-Kolkata.pdf. Accessed on 20 July 2023

Arican M, Bahar HI (2020) Policies concerning urban safety and urbanization in Istanbul. Ann Soc Sci Manag Stud 5(2):37–44.https://doi.org/10.19080/ASM.2020.05.555657Accessed on 10 Nov 2023

Arun PR, Jayapal R, Murugesan M (2013) Impact of construction of high-rise towers on migration of fauna including Avifauna-Bengal NRI township project (Urbana), Anandapur

Attard G, Winiarski T, Rossier Y, Eisenlohr L (2016) Impact of underground structures on the flow of urban groundwater. Hydrogeol J 24(1):5–19. https://doi.org/10.1007/s10040-015-1317-3. Accessed on 10 Nov 2023

Au-Yong CP, Ali AS, Chua SJL (2018) A literature review of routine maintenance in high-rise residential buildings: a theoretical framework and directions for future research. J Facil Manag https://doi.org/10.1108/JFM-10-2017-0051/full/html. Accessed on 11 Nov 2023

Azian FUM, Yusof N, Kamal EM (2020). Problems in high-rise residential building: from management perspective. In: IOP conference series: earth and environmental science, vol 452, no 1, p 012087. IOP Publishing. https://doi.org/10.1088/1755-1315/452/1/012087. Accessed on 10 Nov 2023

Balala ABH, Tanveeruddin K, Uroosa T, Khan AS, Hussain MJ, Nanchari MK (2020) Computer aided design, analyze and drafting of multi Storey RCC building using ETABS. https://www.academia.edu/download/77918586/computer-aided-design-analyze-and-drafting-of-multi-storey-rcc-building-using-etabs-IJERTV9IS030136.pdf

Balasubramaniam H (2018) Makaan. Here's a look at 10 upcoming skyscrapers in India. https://www.makaan.com/iq/news-views/heres-a-look-at-10-upcoming-skyscrapers-in-india. Accessed on 24 July 2023

Baldwin Hess D, Tammaru T, Van Ham M (2018) Housing estates in Europe: poverty, ethnic segregation and policy challenges. Springer Nature, p 429. https://doi.org/10.1007/978-3-319-92813-5. Accessed on 12 Nov 2023

Bargh FH, Mansouri B (2023) Localization of Iranian architectural model in low-cost housing production technology (Tehran). MAJ-Malaysia Arch J 5(1):152–165

Barros P, Fat LN, Garcia LM, Slovic AD, Thomopoulos N, de Sa TH, Mindell JS et al (2019) Social consequences and mental health outcomes of living in high-rise residential buildings and the influence of planning, urban design and architectural decisions: a systematic review. Cities 93:263–272

Barthel S, Isendahl C, Vis BN, Drescher A, Evans DL, van Timmeren A (2019) Global urbanization and food production in direct competition for land: leverage places to mitigate impacts on SDG2 and on the Earth System. Anthropocene Rev 6(1–2):71–97. https://doi.org/10.1177/2053019619856672. Accessed on 9 Nov 2023

Basu A, Bhattacharjee A (2022a) Urban expansion around Kolkata: evaluating urbanogenic interventions in new town, Rajarhat. In: Anthropogeomorphology: a geospatial technology based approach. Springer International Publishing, Cham, pp 571–603

Basu A, Bhattacharjee A (2022b) Urban expansion around Kolkata: evaluating urbanogenic interventions in new town, Rajarhat. In: Anthropogeomorphology. Springer, Cham, pp 571–603. https://doi.org/10.1007/978-3-030-77572-8_28. Accessed on 15 Nov 2023

Beames A, Broekx S, Schneidewind U, Landuyt D, van der Meulen M, Heijungs R, Seuntjens P (2018) Amenity proximity analysis for sustainable brownfield redevelopment planning. Landsc Urban Plann 171:68–79. https://doi.org/10.1016/j.landurbplan.2017.12.003. Accessed on 8 Nov 2023

Behera SK, Mishra SP, Panda S, Siddique M (2004) An autodesk modeling of urbanization by vertical multistoried structures over incompetent soils. City 1:2

Berardi U, Jafarpur P (2020) Assessing the impact of climate change on building heating and cooling energy demand in Canada. Renew Sustain Energy Rev 121:109681. https://doi.org/10.1016/j.rser.2019.109681.Accessed on 10 Nov 2023

Black KJ, Richards M (2020) Eco-gentrification and who benefits from urban green amenities: NYC's high line. Landsc Urban Plann 204:103900. https://doi.org/10.1016/j.landurbplan.2020.103900. Accessed on 10 Nov 2023

Borck R (2016) Will skyscrapers save the planet? Building height limits and urban greenhouse gas emissions. https://doi.org/10.1016/j.regsciurbeco.2016.01.004

Bose PS (2014) Living the way the world does: global Indians in the remaking of Kolkata. Ann Assoc Am Geogr 104(2):391–400

Botti G (2023) Dwelling: new forms of housing, new forms of city. In: Designing emerging markets: a quantitative history of architectural globalisation. Springer Nature Singapore, Singapore, pp 331–382

Census. Data & Resource (2011) https://censusindia.gov.in/census.website/. Accessed on 15th Oct 2023

Census of India (2011) Provisional population totals, paper 1 of 2011 India, series-1, office of the registrar general & census commissioner, New Delhi

Chakrabarti BK, Sinha A (2021) Development of econophysics: a biased account and perspective from Kolkata. Entropy 23(2):254. https://doi.org/10.3390/e23020254. Accessed on 5 Nov 2023

Chakravorty S (2022) From colonial city to globalizing city? The far-from-complete spatial transformation of Calcutta: an update after 25 years. In: Kolkata—the colonial city in transition. Routledge India, pp 3–27

Chatterjee I (2021) From species life to nature's outside: new town "green city", Kolkata. In: The Routledge handbook of development and environment. Routledge, pp 354–371. https://doi.org/10.4324/9780429450310-30/species-life-nature-outside-ipsita-chatterjee. Accessed on 10 Nov 2023

Choithani C, van Duijne RJ, Nijman J (2021) Changing livelihoods at India's rural–urban transition. World Dev 146:105617. https://doi.org/10.1016/j.worlddev.2021.105617. Accessed on 10 Nov 2023

Conejos S, Chew MYL, Azril FHB (2019) Green maintainability assessment of high-rise vertical greenery systems. Facilities. https://doi.org/10.1108/F-09-2018-0107/full/html. Accessed on 10 Nov 2023

Data Standards: high-rise building (ESN 18727), Emporis Standards. Accessed online 16 Oct 2009

Devalt. (2015) Ministry of Housing & Urban Poverty Alleviation (MoHUPA) Government of India. West Bengal Urban Affordable Housing & Habitat Policy. https://www.devalt.org/images/L2_ProjectPdfs/WestBengalUrbanAffordableHousingandHabitatPolicy.pdf?Oid=242. Accessed on 10 Dec 2023

Digdowiseiso K, Sugiyanto E (2020) The effects of multinational companies on deforestation: the building block or stumbling block. J Environ Manag Tour 11(1): 41. https://doi.org/10.14505/jemt. Accessed on 18 Nov 2023

Ding Z, Fan Z, Tam VW, Bian Y, Li S, Illankoon ICS, Moon S (2018) Green building evaluation system implementation. Build Environ 133: 32–40. https://www.sciencedirect.com/science/article/am/pii/S0360132318300787. Accessed on 18 Nov 2023

Du J, Yang H, Shen Z, Chen J (2017) Micro hydro power generation from water supply system in high-rise buildings using pump as turbines. Energy 137: 431–440. https://doi.org/10.1016/j.energy.2017.03.023. Accessed on 18 Nov 2023

Dwivedi A, Mohan BK (2018) Impact of green roof on micro climate to reduce Urban Heat Island. Remote Sens Appl Soc Environ 10:56–69. https://doi.org/10.1016/j.rsase.2018.01.003. Accessed on 18 Nov 2023

Ekici B, Kazanasmaz ZT, Turrin M, Taşgetiren MF, Sariyildiz IS (2021) Multi-zone optimisation of high-rise buildings using artificial intelligence for sustainable metropolises. Part 1: background, methodology, setup, and machine learning results. Solar Energy 224:373–389. https://doi.org/10.1016/j.solener.2021.05.083. Accessed on 18 Nov 2023

Ellis PG, Torcellini PA (2005) Simulating tall buildings using EnergyPlus (No. NREL/CP-550-38133). National Renewable Energy Lab., Golden, CO (US).https://www.osti.gov/servlets/purl/15016731-1cLqj6/native/. Accessed on 18 Nov 2023

EMPORIS, 23 September, 2013. https://www.emporis.com/city/102893/gurgaon-india, https://www.emporis.com/city/102493/ghaziabad-india, https://www.emporis.com/city/106659/noida-india, https://www.emporis.com/city/103730/faridabad-india, https://www.emporis.com/city/162370/greater-noida-india. Accessed on 2 Mar 2023

Ferdous W, Bai Y, Ngo TD, Manalo A, Mendis P (2019) New advancements, challenges and opportunities of multi-storey modular buildings—a state-of-the-art review. Eng Struct 183:883–893. https://doi.org/10.1016/j.engstruct.2019.01.061. Accessed on 10 Nov 2023

Frank T (2019) Determinism and self-organization of human perception and performance. Springer, Berlin. https://doi.org/10.1007/978-3-030-28821-1. Accessed on 9 November 2023

Fu S, Zhuo H, Song H, Wang J, Ren L (2020) Examination of a coupling coordination relationship between urbanization and the eco-environment: a case study in Qingdao, China. Environ Sci Pollut Res 27(19):23981–23993. https://doi.org/10.1007/s11356-020-08683-7. Accessed on 9 Nov 2023

Fujii T, Ray R (2021) Singapore as a sustainable city. https://doi.org/10.4324/9780429266584

Gan VJ, Chan CM, Tse KT, Lo IM, Cheng JC (2017) A comparative analysis of embodied carbon in high-rise buildings regarding different design parameters. J Clean Prod 161:663–675. https://doi.org/10.1016/j.jclepro.2017.05.156. Accessed on 8 Nov 2023

Ganai B, Sarker K (2018) Distribution of high-rise building and its relationship with increase in population of Burdwan municipality West Bengal. Int J Comput Eng Res (IJCER) 08(12):2250–3005

Generalova E, Generalov V (2021) Connecting a skyscraper to urban space: a stylobates typology for tall buildings. https://doi.org/10.1088/1757-899X/1079/2/022034

Ghawana T, Janečka K, Zlatanova S (2021) Development of 3D spatial database enabled IT framework for land agencies: case studies of Delhi, India and Prague, Czech Republic. Land 10(3):248. https://doi.org/10.3390/land10030248. Accessed on 7 Nov 2023

Ghisellini P, Ripa M, Ulgiati S (2018) Exploring environmental and economic costs and benefits of a circular economy approach to the construction and demolition sector. A literature review. J Clean Prod 178:618–643. https://doi.org/10.1016/j.jclepro.2017.11.207. Accessed on 7 Nov 2023

Gollin D, Jedwab R, Vollrath D (2016) Urbanization with and without industrialization. J Econ Growth 21(1): 35–70. https://doi.org/10.1007/s10887-015-9121-4. Accessed on 7 Nov 2023

Güneralp B, Seto KC (2013) Futures of global urban expansion: uncertainties and implications for biodiversity conservation. Environ Res Lett 8(1):014025. https://doi.org/10.1088/1748-9326/8/1/014025. Accessed on 7 Nov 2023

Gupta JD (2011) Impact of high-rise buildings on the social economic and environmental aspects of Kolkata

Hartig T, Kahn PH (2016) Living in cities, naturally. Science 352(6288):938–940. https://parkrxamerica.org/pdf/learn/Hartig-2016-Living-in-cities-naturally.pdf. Accessed on 7 Nov 2023

He Q, He W, Song Y, Wu J, Yin C, Mou Y (2018) The impact of urban growth patterns on urban vitality in newly built-up areas based on an association rules analysis using geographical 'big data'. Land Use Policy 78:726–738. https://doi.org/10.1016/j.landusepol.2018.07.020. Accessed on 7 Nov 2023

He Q, Zeng C, Xie P, Tan S, Wu J (2019) Comparison of urban growth patterns and changes between three urban agglomerations in China and three metropolises in the USA from 1995 to 2015. Sustain Cities Soc 50:101649. https://doi.org/10.1016/j.scs.2019.101649. Accessed on 7 Nov 2023

Ilesanmi AO (2010) Post-occupancy evaluation and residents' satisfaction with public housing in Lagos, Nigeria. J Build Apprais 6(2):153–169. https://doi.org/10.1057/jba.2010.20. Accessed on 7 Nov 2023

Karimimoshaver M, Winkemann P (2018) A framework for assessing tall buildings' impact on the city skyline: aesthetic, visibility, and meaning dimensions. https://doi.org/10.1016/j.eiar.2018.08.007

Kavilkar R, Patil S (2014) Study of high-rise residential buildings in Indian cities (a case study-Pune city). Int J Eng Technol 6(1):86

Krishnan G, Priya RS (2022) Blue Green Infrastructure: a panacea for urban environmental challenges. Case study: Thiruvananthapuram city, Kerala, India. Materi Today Proc 68:2646–2652

Kumara HS, Gopiprasad S (2017) Transfer of development rights as an instrument for spatial planning implementation: a case of Bengaluru Metropolitan Area. Initiatives 18. https://www.researchgate.net/profile/HKumara/publication/344584284_Transfer_of_Development_Rights_as_an_instrument_for_Spatial_Planning_implementation_A_Case_of_Bengaluru_Metropolitan_Area/links/5f81b9e492851c14bcbc23ca/Transfer-of-Development-Rights-as-an-instrument-for-Spatial-Planning-implementation-A-Case-of-Bengaluru-Metropolitan-Area.pdf. Accessed on 6 Oct 2023

Latham A, Layton J (2019) Social infrastructure and the public life of cities: studying urban sociality and public spaces. Geogr Compass 13(7):e12444. https://doi.org/10.1111/gec3.12444. Accessed on 6 Oct 2023

Li X, Sunikka-Blank M (2021) Urban densification and social capital: neighbourhood restructuring in Jinan, China. Build Cities 2(1)

Liang L, Wang Z, Li J (2019) The effect of urbanization on environmental pollution in rapidly developing urban agglomerations. J Clean Prod 237:117649. https://doi.org/10.1016/j.jclepro.2019.117649. Accessed on 6 Oct 2023

Lin N (2017) Building a network theory of social capital. Soc Capital, 3–28

Machimura T (2021) Gentrification without gentry in a declining global city?: vertical expansion of Tokyo and its Urban Meaning. https://doi.org/10.1111/ijjs.12126

Majumdar S, Chatterjee U (2021) Modelling urban growth using urban growth deterministic model in Kolkata Metropolitan Area: a geo-statistical approach. https://doi.org/10.1007/s40808-020-00985-6

Majumdar S, Sivaramakrishnan L (2020) Mapping of urban growth dynamics in Kolkata metropolitan area: a geospatial approach. In: Urbanization and regional sustainability in South Asia. Springer, Cham, pp 9–24. https://www.researchgate.net/profile/Md-Saidur-Sumon/publication/337154475_HIVAIDS_Vulnerability_Among_Street_Children_in_Urban_Bangladesh_A_Reality_of_Old_Dhaka/links/6175acf8a767a03c14a90b2a/HIV-AIDS-Vulnerability-Among-Street-Children-in-Urban-Bangladesh-A-Reality-of-Old-Dhaka.pdf#page=27. Accessed on 6 Oct 2023

Marques B, McINTOSH J, Chanse V (2020) Improving community health and wellbeing through multi-functional green infrastructure in cities undergoing densification. Acta Horticulturae et Regiotecturae 23(2):101–107

Ministry of Housing and Urban Affairs, Government of India (2019) Atal mission for rejuvenation and urban transformation. http://amrut.gov.in/content/innerpage/the-mission.php. Accessed on 22 July 2023

Ministry of Housing and Urban Affairs, Government of India (2022) SmartCity. https://smartcities.gov.in/#:~:text=The%20purpose%20of%20the%20Smart,SmartCities%20Dashboard. Accessed on 22, 2023

Mishra K, Goel A (2019) Energy analysis of high rise building integrated with BIM. https://doi.org/10.17485/ijst%2F2019%2Fv12i6%2F141917

Mukherjee S, Bebermeier W, Schütt B (2018) An overview of the impacts of land use land cover changes (1980–2014) on urban water security of Kolkata. Land 7(3):91

Mukherjee S, Sundberg T (2023) A transdisciplinary and collaborative urban water security framework: developed through an interdisciplinary study in Kolkata, India. World Water Policy 9(3):519–549

N.K. Sinha, Swapna Banerjee-Guha, 2019—https://www.britannica.com/place/Kolkata. Accessed on 6 Oct 2023

Nagawkar J, Ghosh S, Kataria R, Nashit A, Deora A (2014) Effect of sky scrapers on natural ventilation patterns and human comfort index in low-risebuildings-a CFD analysis over central Mumbai. ARPN J Eng Appl Sci 9(3):293–295

Nath B, Ni-Meister W, Choudhury R (2021) Impact of urbanization on land use and land cover change in Guwahati city, India and its implication on declining groundwater level. Groundw Sustain Dev 12:100500

Nethercote M (2019) Melbourne's vertical expansion and the political economies of high-rise residential development. https://doi.org/10.1177/0042098018817225

Newtown Kolkata Green Smart City Corporation Limited (2018) NewtownGreenSmartCity. https://www.newtowngreencity.in/#. Accessed on 20 July 2023

Nugroho F, Sugara A, Priana A, Suci ANN (2023) Monitoring the impact of land cover change on urban heat Island with remote sensing & GIS. https://doi.org/10.0000-0002-9743-6948

O'Donnell EC, Thorne CR (2020) Drivers of future urban flood risk. Philos Trans Roy Soc A 378(2168):20190216. https://doi.org/10.1098/rsta.2019.0216. Accessed on 6 Oct 2023

Pal I, Ghosh S, Tuladhar N (2022) Risk governance perspectives for compounding hazards: a case study in Megacity Kolkata. In: Pandemic risk, response, and resilience. Elsevier, pp 335–349

Pan W, Hon CK (2020) Briefing: modular integrated construction for high-rise buildings. In: Proceedings of the institution of civil engineers-municipal engineer, vol 173, No 2, pp 64–68. Thomas Telford Ltd. https://doi.org/10.1680/jmuen.18.00028

Patil S, Pradhan S, Sangle P, Sharma S (2021) Analysis & development of vertical plantation to the building structures (No. 5093). EasyChair. https://easychair.org/publications/preprint_download/Pk5d. Accessed on 6 Oct 2023

Pavri F (2009) Urban expansion and sea-Level rise related flood vulnerability for Mumbai (Bombay), India using remotely sensed data. In: Geospatial techniques in urban hazard and disaster analysis. Springer, Dordrecht, pp 31–49. https://doi.org/10.1007/978-90-481-2238-7_3. Accessed on 6 Oct 2023

Pires FM, Mendes LDS, Quiñonez LL (2019) Integrated system architecture for decision-making and urban planning in smart cities. Int J Distrib Sen Netw 15(8):1550147719867829. https://doi.org/10.1177/1550147719867829. Accessed on 6 Oct 2023

Ramachandra TV, Bharath HA, Sowmyashree MV (2014) The urban footprint of Mumbai-the commercial capital of India. J Urban Regional Anal 6(1):71. https://www.proquest.com/openview/e82e5b323e02d097699c67613f3cc43f/1?pq-origsite=gscholar&cbl=506347. Accessed on 6 Oct 2023

Rathee G (2014) Trends of land-use change in India. In: Urbanization in Asia. Springer, New Delhi, pp 215–238.https://doi.org/10.1007/978-81-322-1638-4_13. Accessed on 6 Oct 2023

Reddy MM, Rao CH, Reddy KR, Kumar GK (2023) Deaggregation of seismic hazard for Amaravati capital region in Peninsular India. Asian J Civ Eng 24(4):1077–1095

Ren Q, He C, Huang Q, Shi P, Zhang D, Güneralp B (2022) Impacts of urban expansion on natural habitats in global drylands. https://doi.org/10.1038/s41893-022-00930-8

Rondhi M, Pratiwi PA, Handini VT, Sunartomo AF, Budiman SA (2018) Agricultural land conversion, land economic value, and sustainable agriculture: a case study in East Java, Indonesia. Land 7(4):148. https://doi.org/10.3390/land7040148. Accessed on 6 Oct 2023

Rumbach A (2017) At the roots of urban disasters: planning and uneven geographies of risk in Kolkata, India. https://doi.org/10.1080/07352166.2017.1282771

Sadhukhan S (2020) Sustainable human development: a study on work-life balance with special reference to information technology sector in Kolkata. Special Volume 20. http://admin.iaasouthbengalbranch.org/journal/20.pdf#page=26

Sahu P, Sikdar PK (2011) Threat of land subsidence in and around Kolkata city and east Kolkata wetlands, West Bengal, India. J Earth Syst Sci 120:435–446

Sandhu M, Sharma B, Mittal H, Yadav RBS, Kumar D, Teotia SS (2020) Simulation of strong ground motion due to active Sohna fault in Delhi, National Capital Region (NCR) of India: an implication for imminent plausible seismic hazard. Nat Hazards 104(3):2389–2408

Saxena A, Jat MK, Clarke KC (2021) Development of SLEUTH-Density for the simulation of built up land density. Comput Environ Urban Syst 86:101586. https://doi.org/10.1016/j.compenvurbsys.2020.101586. Accessed on 6 Oct 2023

Seto KC, Güneralp B, Hutyra LR (2012) Global forecasts of urban expansion to 2030 and direct impacts on biodiversity and carbon pools. Proc Nat Acad Sci 109(40):16083–16088. https://doi.org/10.1073/pnas.1211658109. Accessed on 6 Oct 2023

Sharma M (2018) Development of a 'Green building sustainability model' for Green buildings in India. J Clean Prod 190:538–551. https://doi.org/10.1016/j.jclepro.2018.04.154. Accessed on 6 Oct 2023

Shi L, Shao G, Cui S, Li X, Lin T, Yin K, Zhao J (2009) Urban three-dimensional expansion and its driving forces—a case study of Shanghai, China. https://link.springer.com/article/10.1007/s11769-009-0291-x

Simlandy S (2022) Colonial origins of modernity in India. BFC Publications. https://books.google.co.in/books?hl=en&lr=&id=3nIEAAAQBAJ&oi=fnd&pg=PA45&dq=Simlandy,+S.+(2022).+Colonial+Origins+Of+Modernity+In+India.+BFC+Publications.&ots=dL9FD7BQB5&sig=wxDDAJVaSBmE9Gbjk4Oz5GOuN9I&redir_esc=y#v=onepage&q=Simlandy%2C%20S.%20(2022).%20Colonial%20Origins%20Of%20Modernity%20In%20India.%20BFC%20Publications.&f=false. Accessed on 6 Oct 2023

Sinha P, Gaughan AE, Stevens FR, Nieves JJ, Sorichetta A, Tatem AJ (2019) Assessing the spatial sensitivity of a random forest model: application in gridded population modeling. https://doi.org/10.1016/j.compenvurbsys.2019.01.006

Smith ME (2019) Energized crowding and the generative role of settlement aggregation and urbanization, vol 8, pp 37–58. State University of New York Press, New York. https://www.academia.edu/download/58172286/Smith_Proof_20180915.pdf. Accessed on 6 Oct 2023

The Economic Times (2022, April 15). https://m.economictimes.com/industry/services/property-/-cstruction/merlin-launches-tallest-skyscraper-in-kolkatas-it-hub/articleshow/90859577.cms. Accessed on 11 June 2023

The Economic Times (2022, April 15).https://m.economictimes.com/industry/services/property-/-cstruction/merlin-launches-tallest-skyscraper-in-kolkatas-it-hub/articleshow/90859577.cms. Accessed on 3 June 2023

The Telegraph (2010) Vertical city. https://www.telegraphindia.com/7-days/vertical-city/cid/496845. Accessed on 28 May 2023

Tiwari V, Sarkar A, Jana A (2022) Integrating master planning with smart city initiative: challenges and opportunities in Newtown, Kolkata. Smart Master Plann Cities 251–267. https://doi.org/10.1007/978-981-19-2386-9_7

United Nation Population Division (2006) https://www.un.org/development/desa/pd/sites/www.un.org.development.desa.pd/files/files/documents/2020/Jan/un_2006_world_population_prospects-2006_revision_volume-i.pdf. Accessed on 4 Aug 2023

United Nation (2011) https://www.un.org/en/development/desa/publications/2011.html. Accessed on 20 July 2023

Voskresenskaya E, Vorona-Slivinskaya L (2018) Development of national standards related to the integrated safety and security of high-rise buildings. In: E3S web of conferences, vol 33, p 03052. EDP Sciences. https://doi.org/10.1051/e3sconf/20183303052. Accessed on 6 Oct 2023

Wang X, Gard W, Borska H, Ursem B, Van De Kuilen JWG (2020) Vertical greenery systems: from plants to trees with self-growing interconnections. Eur. J. Wood Wood Prod 78(5):1031–1043. https://doi.org/10.1007/s00107-020-01583-0. Accessed on 6 Oct 2023

Wen B, Musa N, Onn CC, Ramesh S, Liang L, Wang W (2020) Evolution of sustainability in global green building rating tools. J Clean Prod 259:120912. https://doi.org/10.1016/j.jclepro.2020.120912. Accessed on 6 Oct 2023

Wong I, Baldwin AN (2016) Investigating the potential of applying vertical green walls to high-rise residential buildings for energy-saving in sub-tropical region. Build Environ 97:34–39. https://doi.org/10.1016/j.buildenv.2015.11.028. Accessed on 6 Oct 2023

Wu X, Lin D (2022) Seismic assessment of high-rise buildings through dynamic nonlinear analysis. Ann Geophys 65. https://doi.org/10.4401/ag-8770. Accessed on 6 Oct 2023

Yang C, Zhao S (2022) Urban vertical profiles of three most urbanized Chinese cities and the spatial coupling with horizontal urban expansion. Land Use Policy 113:105919. https://doi.org/10.1016/j.landusepol.2021.105919. Accessed on 6 Oct 2023

Yu S, Chen Z, Yu B, Wang L, Wu B, Wu J, Zhao F (2020) Exploring the relationship between 2D/3D landscape pattern and land surface temperature based on explainable eXtreme Gradient Boosting tree: a case study of Shanghai, China. Sci Tot Environ 725:138229. https://doi.org/10.1016/j.scitotenv.2020.138229. Accessed on 6 Oct 2023

Zhang R, Liu S, Li M, He X, Zhou C (2021) The effect of high-density built environments on elderly individuals' physical health: a cross-sectional study in Guangzhou, China. Int J Environ Res Public Health 18(19):10250

Zhang W, Li W, Zhang C, Ouimet WB (2017) Detecting horizontal and vertical urban growth from medium resolution imagery and its relationships with major socioeconomic factors. Int J Remote Sens 38(12):3704-3734. https://doi.org/10.1080/01431161.2017.1302113. Accessed on 6 Oct 2023

Zhang Y (2018) The credibility of slums: informal housing and urban governance in India. Land Use Policy 79:876–890

Urban Growth Patterns and Spatial Sustainability in Coastal Cities: The Case of Visakhapatnam in India

Nunna Tagore Sai Priya and Ankhi Banerjee

1 Introduction

Human overexploitation has emerged as a significant threat to coastal regions globally. Coastal cities, strategically situated within 100 kms of coastlines, have historically been magnets for human habitation due to their pivotal role in economic activities such as shipping, tourism, and fishing. This allure of coastal urbanization has led to rapid urban growth, setting the stage for a complex interplay between economic development and environmental preservation (UN 2018; Balk et al. 2009). According to data from 2018, approximately 55% of the world's population lived in urban areas, with a significant proportion, almost 60%, residing in low-lying coastal zones (Cummins et al. 2004; UN 2018). Urbanization in coastal areas surpasses that of inland areas, expanding along the coastline and inland due to their unique topographic attributes (Balk et al. 2009; Forman 2010; Yılmaz and Terzi 2021). This phenomenon has led to the growth of cities, making them epicentres of population density and economic activity (Balk et al. 2009; Small and Nicholls 2003). Projections indicate that coastal populations will become more concentrated and interconnected due to the ongoing expansion of human populations and the progressively tighter integration of trade networks in the coming century (Neumann et al. 2015). Asia, notably India along with a few other developing nations, bears the brunt of sea-level rise and flooding (Neumann et al. 2015). Among the 20 most flood-prone cities projected for 2050, 13 are in Asia, underscoring the region's vulnerability (Hallegatte et al. 2018). These cities are grappling with the convergence of rising

N. T. Sai Priya (✉) · A. Banerjee
Ranbir and Chitra Gupta School of Infrastructure Design and Management, IIT Kharagpur, Kharagpur, India
e-mail: tagoresaipriyanunna@iitkgp.ac.in

A. Banerjee
e-mail: ankhi@infra.iitkgp.ac.in

M. Firoz C et al. (eds.), *Urban Planning and Design for Megacities in the Global South*, Advances in 21st Century Human Settlements,
https://doi.org/10.1007/978-981-97-8370-0_12

sea level, land subsidence, and climate change impacts (Cao et al. 2021; Nicholls et al. 2021). The magnitude of this challenge is exemplified by projections indicating that some coastal megacities, including Mumbai and Chennai, could be submerged by the end of the century as per the NASA Sea Level Projection Tool (available at https://sealevel.nasa.gov/). While coastal areas are vital hubs for trade and economic growth, their expansion often comes at the cost of environmental degradation. Rapid urbanization gives rise to slums, inadequate infrastructure, air pollution, and urban sprawl, thereby exerting pressure on the coastal and marine ecosystems. Therefore, balancing the economic benefits of urbanization with environmental conservation is imperative, aligning with the 11th goal of Sustainable Development (i.e. sustainable cities and communities).

2 Literature Review

2.1 Urban Growth in Coastal Cities

Coastal cities serve as intricate intersections where natural landscapes blend with built environments (Forman 2010; Yılmaz and Terzi 2021). These cities face significant growth pressure due to the socio-economic, cultural, and demographic attractiveness of their coastal settings witnessing rapid urbanization and the emergence of extensive urban settlements (Forman 2010). The spatial layout and developmental trajectories of these cities are influenced by an array of factors, encompassing topography, transport networks, natural landscapes, and spatial policies (Yılmaz and Terzi 2021). Ports have historically acted as pivotal catalysts in the transformation of coastal regions into vibrant urban centres (Yeung 2001; Lee et al., 2008). Their role as trade gateways has been instrumental in shaping the evolution of small coastal towns into bustling metropolises, driving economic growth and urban development (Merk 2013). This transition has not only reshaped the economic landscapes but has also triggered substantial expansion in urban areas, particularly notable in developing economies like India. The symbiotic relationship between ports and cities is evident in the proliferation of diverse industries in these areas, ranging from transportation, logistics, and manufacturing to the emergence of sectors like renewable energy (Hoyle 1989). Their presence acts as a nucleus, fostering not just economic activities but also influencing the social and demographic fabric of these regions (ASCELA 2020). These industries, thriving in proximity to ports, contribute significantly to the linear growth pattern along the coastal line. This unique characteristic underscores the critical necessity to devise growth strategies and policies that specifically address the impact of coastal features on the developmental trajectories of these cities (Boulos 2016). Notably, population density also plays a pivotal role in shaping the form and growth of these coastal urban centres, notably evident through higher population growth rates (Balk et al. 2009). Understanding the dynamics of this evolution is crucial in planning and policy formulation to ensure sustainable urban development, leveraging the historical significance of ports while accommodating the evolving needs of burgeoning coastal cities.

2.2 Coastal Regulations in India

India's efforts to sustain coastal areas involve the Coastal Regulation Zone (CRZ) notifications, which have evolved from 1991 to 2019. These notifications aim to balance development with environmental protection along the coast. The CRZ has undergone numerous amendments, with changes in categorization and permissible activities. Notably, the years 2011 and 2018 brought crucial modifications to the CRZ notification. The former opened the door for the construction of tourism infrastructure like hotels and resorts along eco-sensitive areas of CRZ II, along with land reclamation. The latter, in 2018, introduced a draft CRZ notification for public scrutiny, heralding substantial changes in regulations. These changes involved categorizing CRZ areas and reducing NDZ limits in densely populated regions, permitting the establishment of tourism facilities close to shorelines, and allowing land reclamation for specific purposes such as ports and harbours. Strategic projects like Sagarmala, Bharatmala, and Coastal Economic Zones (CEZs) received exemptions from CRZ provisions, potentially accelerating developmental activities (Table 1).

These transformative shifts in the CRZ hold profound implications for coastal regions. On the one hand, they offer leeway for greater coastal development, potentially boosting economic growth and employment opportunities. However, these changes also carry the risk of heightened environmental damage and a decline in biodiversity. For example, a reduced NDZ could exacerbate erosion and flooding in densely populated areas. The redevelopment of buildings within CRZ zones may contribute to pollution and degradation of coastal ecosystems. As these alterations in the CRZ framework continue to be a subject of discussion and critique by environmentalists and coastal communities, the delicate equilibrium between development and ecological preservation comes to the forefront. The evolution of the CRZ from 1991 to 2019 underscores the imperative of vigilant monitoring and adept management to navigate the implications of these changes on the growth patterns of coastal regions.

2.3 Application of Spatial Metrics for Assessing the Urban Growth

The application of spatial metrics derived from remote sensing data has emerged as a fundamental tool for quantifying and comprehending urban growth patterns. Initially, landscape metrics were devised to assess vegetation patterns in natural landscapes but have since transitioned into pivotal tools within landscape ecology

Table 1 Transforming coastal regulations in India

Coastal Regulation Zone (CRZ) in India	
CRZ 1991	CRZ 1—No construction zone within 500 m from HTL except the utility infrastructure for treated waste, oil and gas pipelines and other basic necessities CRZ II—Developed area up to or close to the shoreline CRZ III—Rural areas and other built-up areas that are not under CRZ I or CRZ II. No new constructions are allowed other than repairs of existing structures. Tourism infrastructure, construction of houses with specified measurements along with alteration of existing authorized buildings between 200 and 500 m of high tide line (HTL) CRZ IV—covers the areas excluded from CRZ I, II, and III, which include restrictions on construction within 200 m from HTL and prohibiting the use of corals and sand from beaches and coastal waters
CRZ 2011	CRZ I—No development zone areas that include ecologically sensitive areas CRZ II—Developed areas up to or close to the shoreline CRZ III—Areas that are relatively undisturbed and do not fall under CRZ 1 or II CRZ IV—Area between low tide line (LTL) and 12 nautical miles into the sea/tidal influenced water bodies NDZ—No development zone that extends up to 200 m from HTL towards land in the CRZ III area
CRZ 2019	CRZ I A—Eco-sensitive areas CRZ I B—Inter-tidal areas CRZ II—Developed areas up to or close to the shoreline CRZ III A—CRZ—III areas with population density above 2161 km^2 CRZ III B—Areas with a population density below 2161 per sq. km CRZ IV A—12 nautical miles from LTL towards the sea CRZ IV B—Tidal-influenced water bodies NDZ—50 m from HTL in CRZ III A areas, 200 m from HTL in CRZ—III B areas

Source MoEFC, GoI (Available at: https://environmentclearance.nic.in/report/CRZ_Notifications.aspx)

(Gustafson 1998; McGarigal et al. 2002). These metrics, blending information theory and fractal geometry, offer a patch-based representation of landscapes, measuring spatial heterogeneity across various landscape levels (Gustafson 1998). While rooted in landscape ecology, these concepts have gained prominence in urban studies due to their capacity to quantify spatial patterns, dynamics, and structures within urban environments. Spatial metrics serve as quantitative frameworks to characterize landscape heterogeneity and are increasingly employed to understand urban growth, land-use changes, and their implications for sustainable urban planning and management (McGarigal et al. 2002). Notably, they enable the assessment of fragmentation, connectivity, composition, and other critical factors influencing urban morphology and dynamics (Schneider et al. 2008). These investigations demonstrate how metrics like patch size, shape, edge length, and fractal dimension offer profound insights into the spatial heterogeneity of urban expansion. Furthermore, the utility of these metrics extends to delineating urban expansion, understanding changing landscape structures, and even aiding in urban modelling efforts (Batty et al. 1999; Schneider et al. 2008). Evaluating historical urban growth through these metrics is a precise and quantifiable method facilitating a deeper understanding of the diverse socio-economic levels found within coastal cities (Yılmaz and Terzi 2021). These metrics stand as indispensable tools for planners, policymakers, and researchers, facilitating a deeper understanding and effective management of urbanization processes. However, despite their prevalence in landscape ecology and urban studies, there remains a gap in the potential integration of spatial metrics with socio-economic and environmental data which promises a more comprehensive understanding of the drivers and impacts of urban growth.

2.4 Significance of SDG 11.3.1 for Sustainable Growth in Coastal Cities

While the expansion of urban area brings economic benefits, it also creates numerous problems due to poor planning and uncontrolled expansion which result in inefficient urban land use (Long and Qu 2018). Therefore, this section of literature underscores the significance of the indicator 11.3.1, 'land-use efficiency', which serves as a critical measure in quantifying the relationship between urban population growth and the increase in urban land, pivotal for monitoring sustainable urbanization (UN-Habitat 2018a). Numerous studies have analysed land-use efficiency indicator, identifying gaps in urban land-use policy and aiming for sustainable urban growth (He et al. 2020; Melchiorri et al. 2019). The studies cover diverse facets, from measuring land-use efficiency to examining the consequences of land consumption on sustainability in coastal areas like the Mazandaran region (Dadashpoor and Hasankhani 2022). Assessments of land consumption patterns reveal varied efficiencies among regions, with Europe and North America exhibiting lower land-use efficiency, contrasting regions in Africa experiencing higher population growth rates (Estoque et al. 2021; Mudau et al. 2020). The literature also emphasizes the environmental impact of land consumption, leading to changes in land cover characteristics, spatial imbalances, and reduced environmental quality (Angel et al. 2011; Bencardino and Iovino 2014). Bobylev (2009) underscores that prioritizing efficient urban land use stands as a paramount strategy for fostering enduring sustainable development in cities for the future. Thus, this comprehensive literature highlights the urgent need to understand and address land consumption as part of SDG 11 (i.e. sustainable cities and communities) in coastal cities for sustainable urban development.

In the light of these multifaceted dynamics, this chapter embarks on an exploration of the spatial growth patterns of coastal cities, focusing on the case of Visakhapatnam. By delving into urbanization trends, port-city relationships, and the evolution of coastal regulations, this study seeks to unravel the interplay between economic growth and environmental sustainability. This investigation could potentially contribute to achieving a delicate equilibrium, fostering harmonious and sustainable development in the coastal regions.

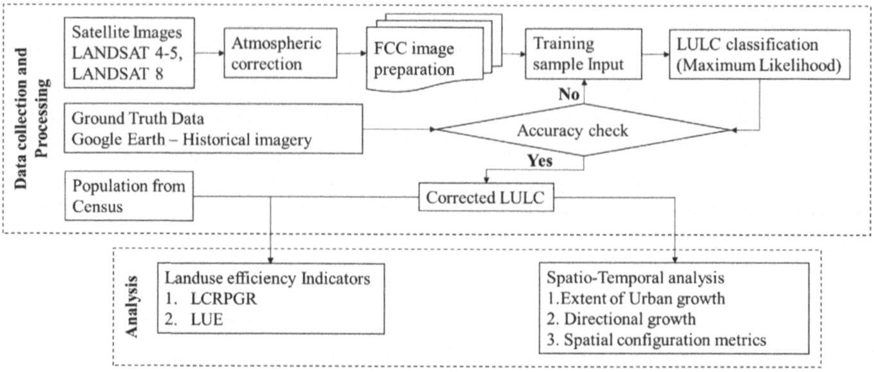

Fig. 1 Methodology adopted for the study

3 Materials and Methods

The research methodology, depicted in Fig. 1, spans three decades (1991–2021) and integrates two quantitative approaches to analyse urban growth in Visakhapatnam.

Initially, it examines overall and direction-wise urban growth, followed by employing six spatial metrics to assess the configuration of urban patches. Additionally, established land-use efficiency indicators from UN-Habitat's SDG 11.3.1 and data from the Global Human Settlement Layer (GHSL) are utilized to evaluate spatial sustainability. This comprehensive approach captures the intricate dynamics of Visakhapatnam's urban growth patterns and their implications for sustainable development.

4 Study Area

Visakhapatnam, or Vizag, has transformed from a small fishing village into a major port city over the years. Strategically positioned on the 'Golden Quadrilateral', Vizag is intersected by NH-16, linking Kolkata in the East to Chennai in the South and spanning an area of 515 km^2. It boasts robust connectivity through various modes of transportation. It has international accessibility through major ports and airports. This advantageous location, coupled with port-driven export–import facilitation, has attracted substantial investments in industry, real estate, and tourism (Fig. 2).

Vizag serves as a central hub for the Visakhapatnam-Kakinada Petroleum, Chemicals and Petrochemicals Investment Region (VK-PCPIR) and the Visakhapatnam Chennai Industrial Corridor (VCIC). The city's strategic geographical location, complemented by rail, road, and air connectivity, has propelled it to become a hotspot for significant commercial and industrial investments. This, in turn, has fuelled a

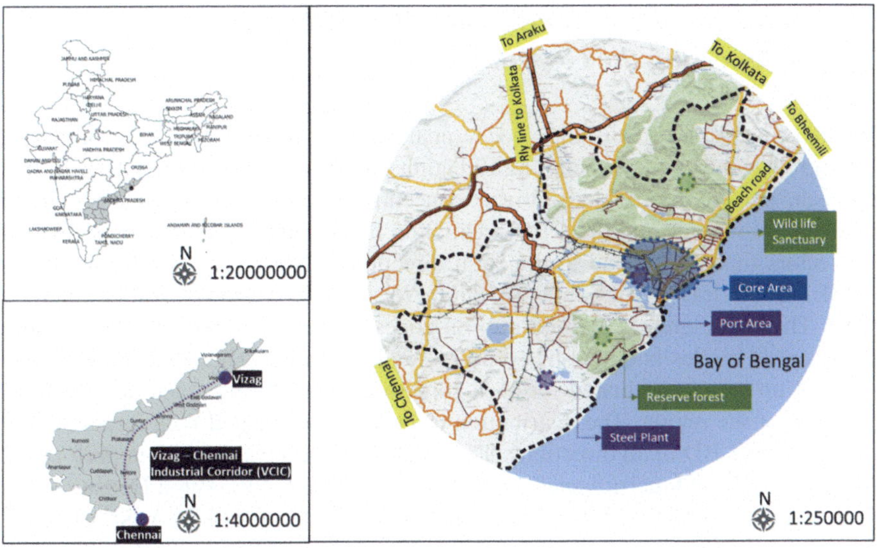

Fig. 2 Study area. *Source* Author

surge of investments within and around the city, propelling its transformation into a future megacity. With a lengthy coastline, Vizag features an extensive beachfront road and is framed by the Eastern Ghats, offering immense potential for tourism development. Following bifurcation, Vizag has been a focal point for development, receiving top priority in the state's investment agenda across IT, industrial, and tourism sectors.

4.1 *Data Collection and Processing*

Demographic data and jurisdictional boundaries of the study area were meticulously sourced from the Census of India and other authorized secondary sources, ensuring the accuracy and reliability of foundational data. A 30-m resolution satellite data from LANDSAT 4–5 for the years 1991, 2001 and 2011, and LANDSAT 8 for 2021, were acquired from the portal of United States Geological Survey's (USGS) Earth Explorer (https://earthexplorer.usgs.gov/). The satellite images underwent necessary pre-processing steps like Dark Object Subtraction (DOS) correction to rectify atmospheric effects. Supervised classification, particularly the maximum likelihood classifier, was employed to categorize the imagery into four key classes: water bodies, vegetation, built-up areas, and other land types and then validated the classification through the ground truth data obtained from Google Earth's historical imagery to ensure the accuracy above 80% and kappa value airport, positions it as a gateway to the global reaching at least 0.7–0.8 The images are then transformed into binary images differentiating between urban and non-urban areas.

4.2 Spatio-Temporal Analysis and Patch Metrics for Assessing the Growth Patterns

The study incorporated distance decay analysis to investigate the distribution of urban growth over time concerning direction and distance from the CBD. In this approach, the area is divided into 'n' concentric circles with increasing radii of about 1 km intervals and the CBD area is the focal point. The circles are divided into sectors that correspond to the eight directions—North (N), North East (NE), East (E), South East (SE), South (S), South West (SW), West (W), and North West (NW). The entire area is divided into 8n zones, where 'n' represents the number of concentric circles bounded by study region boundaries. The percentage change in a built-up area is evaluated for each zone to understand the potential directions of growth of the region. The distribution of growth is evaluated with respect to the distance from CBD in all directions. The patch-based analysis was then performed, utilizing established metrics such as Number of Patches (NP), Patch Density (PD), Mean Patch Area (AREA_MN), Patch Size Standard Deviation (AREA_SD), Largest Patch Index (LPI), and Area-Weighted Mean Fractal Dimension Index (FRAC_AM). These metrics provided quantitative insights into the spatial configuration in terms of urban form, distribution, heterogeneity, size, shape, and complexity of urban growth patterns (Table 2).

Table 2 Selected patch-based metrics for the study

Metric	Description
No. of patches (NP)	Total count of patches in the landscape Higher/Increase—Dispersion Lower/Decrease—Compaction
Patch Density (PD) (patches per 100 hectares)	Total count of patches in an area of 100 hectares Increase—Scattered Decrease—Aggregated
Mean Patch Area (AREA_MN) (in hectares)	The average of the areas of all patches low for small fragmented patches, high for aggregated large patches Increase—growing or aggregation of patches, decrease—shrinking or dispersion into small patches
Patch Size Standard Deviation (AREA_SD)	The standard deviation of the distribution of patch areas
Largest Patch Index (LPI) (per cent)	The proportion of the area occupied by largest patch
Area-Weighted Mean Fractal Dimension Index (FRAC_AM)	The average of the fractal dimension of patches, weighted by area of the patch

Source McGarigal et al. (2002)

4.3 Land-Use Efficiency (LUE) Indicators for Assessing Sustainability

The study evaluated the SDG 11.3.1 indicator recommended by UN-Habitat in 2015 and land-use efficiency indicators proposed by GHSL in 2017 for understanding the spatial sustainability of the city, referring to the benchmarks recommended by UN-Habitat. SDG 11.3.1 indicator measures the ratio of land consumption to population growth rate (LCRPGR) and provides insights into the balance between land consumption and population growth for sustainable urbanization.

Ratio of land consumption rate to population arrival growth rate (LCRPGR) = $\frac{(\mathrm{LCR})}{(\mathrm{PGR})}$
where

LCR = Land Consumption rate = $\frac{\mathrm{LN}(\mathrm{BU}_{(t+n)}/\mathrm{BU}_{(t)})}{(y)}$,

PGR = Population growth rate = $\frac{\mathrm{LN}(\mathrm{Pop}_{(t+n)}/Pop_{(t)})}{(y)}$

$_{(t)}$ = Built-up area in km^2 for past/initial year '*t*'

$_{(t+n)}$ = Built-up area in km^2 for current/final year '(*t*+*n*)'

$\mathrm{Pop}_{(t)}$ = Population in the past/initial year '*t*'

$\mathrm{Pop}_{(t+n)}$ = Population in the past/initial year '(*t*+*n*)'

y = The number of years between the two measurement periods

The ideal value for LCRPGR for achieving sustainable urbanization is recommended as 'one' for all cities. If LCRPGR is greater than one, it indicates increased land consumption while LCRPGR less than one indicates the decreased land consumption. The interpretation of LCRPGR in relation to efficiency of the land utilization is provided with respect to population density in the city as given in Table 3.

The GHSL recommended a built-up per capita indicator to measure land-use efficiency to support SDG 11.3.1 indicator for understanding the status of the city for achieving sustainable urbanization. Built-up per capita is defined as the extent of built-up area available per each resident in the city and the change in built-up per capita indicates the LUE of the region. The percentage measure of land-use efficiency indicates the distance of the city from achieving sustainable urbanization. The value of LUE equal to zero indicates ideal value for achieving efficient land use, while

Table 3 Benchmarks for SDG 11.3.1 recommended by UN-Habitat

City density (persons per hectare)	LCRPGR <1	LCRPGR >1
10–150	Efficient land use	Inefficient land use
151–250	Moving towards efficiency	Moving away from efficiency
>250	Insufficient land per person	Moving towards sufficient land per person

Source UN-Habitat (2018b)

the positive indicates decreasing land consumption per person leading to compact growth and negative values indicate increased land consumption per person leading to sprawl.

Land-use efficiency (LUE) = $(Y_t - Y_{t+n})/Y_t$,
where

Y_t = Built-up per capita in the past/initial year 't' (year) = $\frac{(BU(t))}{(Pop(t))}$
Y_{t+n} = Built-up per capita in the current/final year '$t + n$' (year) = $\frac{(BU(t+n))}{(Pop(t+n))}$
$_{(t)}$ = Built-up area in km^2 for past/initial year 't'
$(t + n)$ = Built-up area in km^2 for current/final year '$(t + n)$'
$Pop_{(t)}$ = Population in the past/initial year 't'
$Pop_{(t+n)}$ = Population in the past/initial year '$(t + n)$'

5 Results and Discussion

5.1 *Extent of Urban Growth*

Over the course of the past three decades, Visakhapatnam has experienced a remarkable surge in urbanization, leading to significant transformations and expansions. This growth is exemplified by the extension of the city's jurisdictional boundaries, which grew from 310.97 to 513.61 km^2 by 2011. This expansion was a direct outcome of the amalgamation of various administrative entities, including the Visakhapatnam Municipal Corporation (VMC), Gajuwaka Municipality, and 32 neighbouring villages. This amalgamation resulted in the establishment of the Greater Visakhapatnam Municipal Corporation (GVMC). In parallel with this territorial enlargement, the city's population escalated from 1 million (10 lakhs) in 1991 to 1.7 million (17 lakhs) in 2011, as reported by the Census of India and further projected to increase around 2.1 million (21 lakhs) by 2021 (city development plan) (Fig. 3).

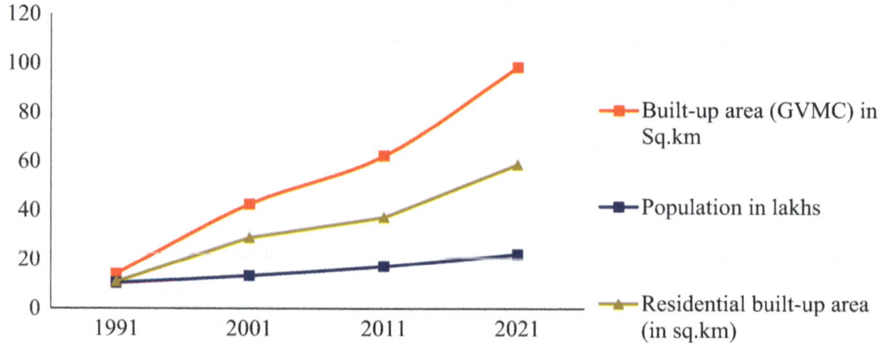

Fig. 3 Urbanization trends in Visakhapatnam. *Source* Author

The allure of Visakhapatnam as an industrial, trade, and tourism hub has attracted a substantial influx of migrants and tourists, amplifying the pace of growth. This multifaceted appeal has fuelled prominent growth trends, as evident in the escalating proportion of developed land within the study area over the past four decades. Specifically, the built-up area witnessed a striking escalation of 197.77% from 1991 to 2001, followed by a significant 119.85% upswing from 2001 to 2011. The momentum continued with a substantial 190.14% increment from 2011 to 2021. Notably, this expansion encompasses not only the overall urban landscape but is also underscored by a noteworthy augmentation in residential built-up areas contributing to 59% of overall built-up area by 2021, signifying the pivotal role of housing development in the city's expansion.

5.2 Directional Analysis of Urban Growth

Until the 1970s, Visakhapatnam (Vizag) was primarily concentrated around its port area, gradually transforming into an industrial hub with the establishment of industries and a shipyard. This evolution also positioned it as a prominent centre for trade, handling both exports and imports. During this period, the port area played a pivotal role in driving growth, with a substantial portion of the city's built-up area centred around it by 1991. However, a significant shift occurred starting from 1991, marked by the emergence of substantial residential developments along the city's coastal stretch. This residential expansion was accompanied by the establishment of institutional and recreational facilities along the NH-5 and the beach road, connecting Vizag to Bheemunipatnam. Notably, this phase of growth saw pronounced development in the East (E) and South–East (SE) directions, which encompassed the coastal region. Concurrently, the North (N), North–East (NE), and North–West (NW) sectors also experienced residential developments between 1991 and 2001. This pattern persisted through 2001–2011, albeit with minimal expansion due to industrial advancements, such as the steel plant and petroleum industries (Fig. 4).

However, the year 2021 marked a significant milestone, witnessing substantial built-up area expansion in all directions due to residential sprawl along the highways, barring certain constraints imposed by the region's topography. Natural features such as Kailasagiri Hills, Simhachalam Hills, and the Kambalakonda Wildlife Sanctuary in the north, the Bay of Bengal in the East and South East, as well as Yarada Hills in the South, have limited expansion potential. In this context, the city's growth prospects are notably favourable in the South-West (SW), West (W), North-West (NW), and North–East (NE) directions. Furthermore, the period following the 2012 state bifurcation saw Visakhapatnam emerge as a significant player in the East Coast Economic Corridor strategy, including the development of the Vizag-Chennai Industrial corridor passing through the NE and SW sectors. Additionally, proposals for an IT Zone further underscore the city's potential for growth and economic development. The direction-wise analysis presented here emphasizes the dynamic urbanization patterns of Vizag. The city's transformation from an industrial and trading hub

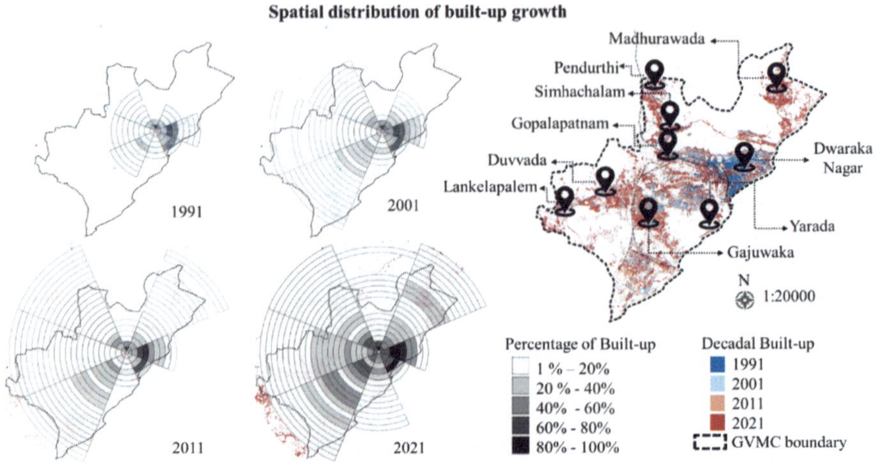

Fig. 4 Spatial expansion trends of Visakhapatnam. *Source* Author

to a multifaceted urban centre has been influenced by factors ranging from industrial developments to strategic economic corridors.

5.3 Spatial Configuration of the Urban Growth Patterns

Six patch-based metrics are examined in a directional context to gain a detailed understanding of growth patterns and the underlying factors driving such patterns. Urban growth in the city has been influenced by a variety of factors, and this is evident in the distribution of these metrics across different directions (Fig. 5).

Notably, the South-West (SW) and West (W) direction has witnessed a considerable extent of outlying growth from 2001 to 2011, which is often associated with industrial expansion. This growth pattern, however, has shown signs of reduction from 2011 to 2021 with development of industrial townships and residential areas, suggesting a shift towards more compact and aggregated growth relating to urban agglomeration. A distinct trend emerged in the North-East (NE) direction, with substantial increases observed in both NP and PD in 2011 and 2021, particularly along the beach road and NH-16. This implies that development expanded and became denser in these areas, potentially due to factors such as increased residential and commercial activities. The growth of the city's built-up area in the East (E) and South–East (SE) directions is closely linked to the intensification of residential land use. These areas have experienced more centralized and compact growth, with the core city located in the East and the port area in the South East, contributing to a dense urban form.

Furthermore, the NW sector witnessed a significant increase in the LPI between 2011 and 2021. This could be a result of the agglomeration of surrounding villages

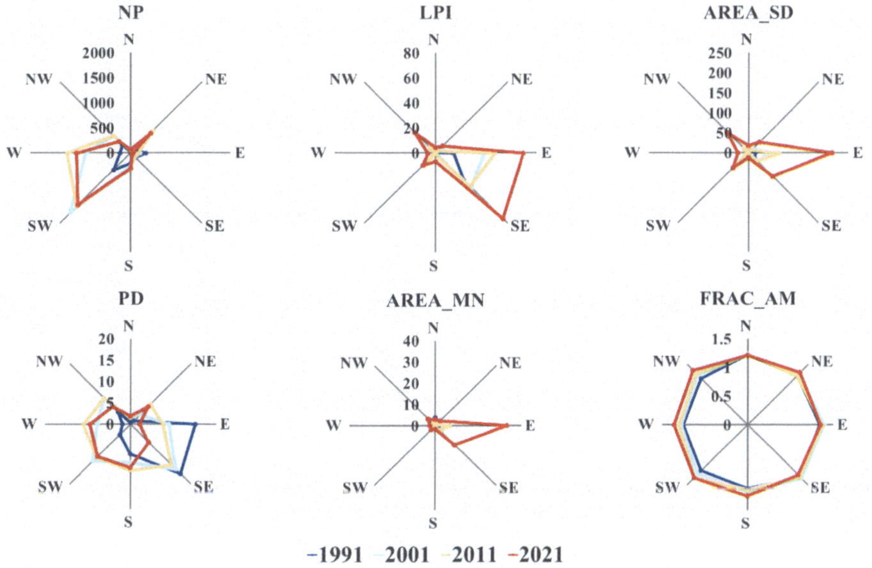

Fig. 5 Direction-wise analysis of patch metrics. *Source* Author

and towns in this direction, leading to larger and more consolidated patches. The AREA_MN and AREA_SD metrics provide insights into the sizes and variations of patches. These metrics are relatively higher towards the E direction, indicating the dominance of larger patches and a higher degree of variability in growth. Conversely, both metrics are lower in the South and North sectors, reflecting the presence of smaller and relatively uniform patches. FRAC_AM indicates the complexity of shape in growth patterns. This metric shows a gradual increase in complexity in the NW, W, and SW sectors. This complexity suggests intricate growth dynamics that might be harder to predict accurately. The analysis of these patch-based metrics reveals that the NW, W, and SW sectors exhibit fragmented growth patterns. On the other hand, the E and SE sectors demonstrate more compact and centralized growth, influenced by factors like the city's core location, industrial activities, and residential development.

In the broader context of the entire study area, the analysis of the NP metric in conjunction with PD metric reveals an initial increasing trend up until 2011, signifying a phase of fragmented and spontaneous growth. However, this trend transitions to a rapid decrease from 2011 to 2021, indicating a notable shift towards the aggregation of patches and a more organized growth pattern. Consequently, the LPI value increased significantly from 2011 to 2021, indicating the expansion of the largest urban patch dominating other patches, which is evident alongside the coast. Examining the Area_MN metric, there is a notable increase over time, particularly from 1991 to 2021. This indicates the expansion or aggregation of smaller patches into larger ones, contributing to a more cohesive urban landscape. Meanwhile, the Area_SD metric, which measures the variability in patch sizes, demonstrates an increase

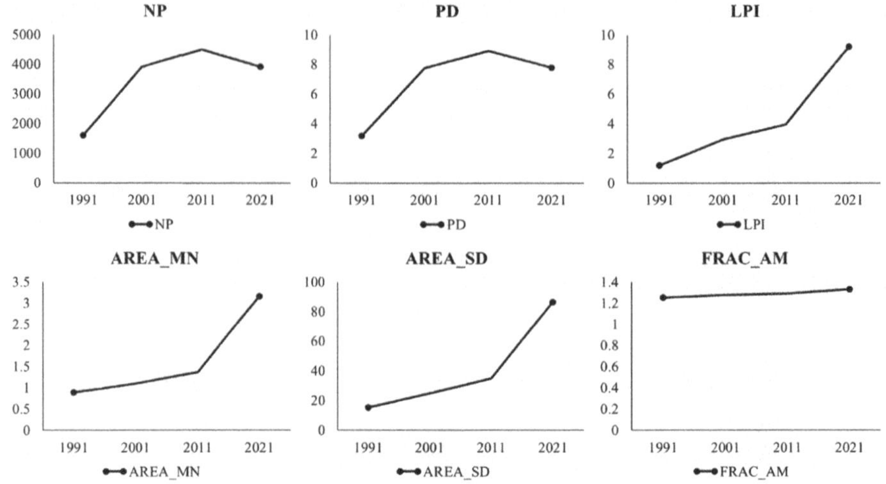

Fig. 6 Temporal change of patch metrics in the study area. *Source* Author

alongside the growth in the Area_MN. This implies that while larger patches are forming, there remains a degree of variability in the sizes of individual patches. The FRAC_AM, which assesses the complexity of the growth patterns' shapes, has remained relatively stable and close to 1 throughout the three decades This suggests that the city's growth has generally followed a more predictable and straightforward trajectory without significant deviations in shape complexity (Fig. 6).

Implications of these findings highlight a significant shift in urban growth dynamics over time. The city's growth trajectory has undergone a noteworthy transformation, shifting from scattered expansion to a more cohesive and concentrated pattern, particularly in the E and SE sectors, encompassing the core urban regions along the coastline. In contrast, the NW, W, and SW sectors exhibit more intricate and fragmented growth, albeit with a trend towards densification due to the aggregation of urban developments from neighbouring areas like Simhachalam, Pendurthi, Lankelapalem, Duvvada, Gopalapatnam, Anakapalle, and Gajuwaka. The NW sector stands out as a hotspot for residential growth, characterized by planned encroachments radiating around the hills. In the periphery of the NW directions, residential expansion has followed linear formations along road networks, branching inwards, contributing to urban sprawl, particularly in the NW direction. The city's expansion is constrained by natural features such as Simhachalam, Kailasagiri hills, and Kambalakonda Wildlife Sanctuary in the NW, N, and NE directions, resulting in growth concentrated around these geographic features. Moreover, the proposed IT Zone in Madhurawada has spurred urban sprawl in the NE sector along the connecting highway to Kolkata. Historical industrial establishments like Hindustan Shipyard, HPCL, Hindustan Zinc, and Hindustan Polymers have influenced growth dynamics, contributing to fragmented development in the SW sector but also fostering residential growth near areas like Gajuwaka, Duvvada, and Gopalapatnam in the SW and W sectors. These findings underscore the complex interplay of factors shaping the city's urban growth and the necessity for thoughtful planning and sustainability considerations.

5.4 *Dynamics of Land Consumption and Spatial Sustainability*

The analysis of the SDG 11.3.1 indicator and land-use efficiency (LUE) for different decadal periods reveals insightful trends in the urban growth dynamics of the city. During the initial phase from 1991 to 2001, the ratio of LCR and PGR is 1.68 indicating a concerning trend of scattered land consumption, primarily driven by industrial expansion. This resulted in sprawl, leading the city away from efficient land-use practices. However, a positive shift in urban growth patterns was observed from 2001 to 2011, with an LCRPGR value of 1.09. This period witnessed a move towards more compact growth, especially in residential areas. This transition suggested a reduction in sprawl and an inclination towards efficient land use, marked by a decrease of 14% in LUE (Table 4).

Nonetheless, the subsequent decade from 2011 to 2021 brought new challenges. The LCRPGR value of 1.14 indicated a high-density urban sprawl of residential areas into suburban and peripheral regions. Furthermore, the infilling growth within the core city contributed to compaction. This trend, reflected in the 23% reduction in LUE, signalled a move away from efficiency due to increased sprawl. The implications of these values are significant. The city's trajectory showcases a shift from scattered and inefficient land consumption towards more organized and compact growth, highlighting the potential for improved land-use efficiency. The positive period of compact growth during 2001–2011 demonstrates the feasibility of managing urban sprawl through strategic planning and targeted development. However, the

Table 4 Temporal change in growth patterns and land-use efficiency

Decadal period	LCRPGR	LUE (%)	Growth pattern	Efficiency
1991–2001	1.68	−131	Scattered land consumption driven by industrial development led to sprawl	Away from efficient land use
2001–2011	1.09	−14	A shift towards compaction in residential areas with some sprawl in suburban and peripheral areas due to industrial developments	Moving towards efficiency and reduction of sprawl
2011–2021	1.14	−23	High-density urban sprawl of residential areas into suburban and peripheral areas and infilling growth in the core city leading to compaction	Moving away from efficiency due to increased sprawl

Source Author

subsequent increase in sprawl and reduction in efficiency during 2011–2021 raises concerns about the city's ability to balance growth and sustainability, despite its potential for competing with other global cities in economic development.

6 Implications

The results of this study have significant implications for sustainable development in Visakhapatnam's urban expansion. The post-bifurcation scenario in Andhra Pradesh has ushered in significant changes in Visakhapatnam's urban landscape. The merging of nearby villages and towns into GVMC has allowed for the growth of suburban and peripheral regions, resulting in the expansion of the urban area. This has brought forth both opportunities and challenges. The advent of a proposed IT Zone in Madhurawada and the establishment of an industrial corridor along the National Highway connecting Kolkata to Chennai have triggered a real estate boom. This surge in economic activity has attracted both residential and industrial townships, leading to the growth of residential areas along major National and State Highways connecting prominent cities. The growing residential built-up areas, particularly in the direction of the city's core and suburban areas, signify the significance of housing development in driving urban growth. The escalating population, driven by the city's attractiveness as an industrial, trade, and tourism hub, places a strain on infrastructure, resources, and environmental quality (Fig. 7).

The directional analysis of urban growth patterns provides a nuanced understanding of the spatial dynamics. The shift from scattered industrial-driven expansion to more compact and centralized growth patterns indicates progress towards efficient land use. However, challenges emerge with the emergence of high-density sprawl

Fig. 7 Image showing the high-rise sprawl around the hills. *Source* Author

in peripheral areas and the core city's infilling growth. This trend demands careful urban planning to balance the need for accommodating growth while preserving green spaces, minimizing sprawl, and fostering a sense of community. The analysis of patch-based metrics further highlights the evolving urban landscape's complexity. The fragmentation observed in certain sectors suggests the need for targeted interventions to promote more cohesive and integrated development. The dominance of larger patches and increased variability in growth size raises questions about equity in land distribution and access to resources. The trend towards higher complexity in growth shape emphasizes the need for adaptive and flexible planning strategies that can accommodate evolving urban dynamics.

Furthermore, the assessment of land-use efficiency shows a fluctuating trend, with a shift towards increased efficiency during 2001–2011 while reduced efficiency with increasing sprawl during 2011–2021. It is observed that the Coastal Regulation Zone (CRZ) amendments permitting tourism infrastructure closer to Visakhapatnam's shoreline have significantly influenced the city's spatial dynamics. This change has accelerated land consumption for tourism and recreational activity, particularly along the beachfront, intensifying the challenge of balancing coastal development with ecological preservation. Also, the encroachments around sensitive ecological zones, notably the undulating terrain and hills, pose significant threats to the city's ecological equilibrium and its ability to withstand environmental changes. Therefore, the study implies to revisit CRZ policies for city-specific measures and implement strict zoning policies aligning with CRZ for sustainability of the region. Also, fostering sustainable land utilization practices becomes imperative to curtail urban sprawl, safeguard vital natural landscapes, and promote the development of neighbourhoods that prioritize pedestrian-friendly and environmentally conscious living.

Moreover, this expansion comes at the cost of green spaces, transforming them into built-up areas, directly impacting the safety and well-being of city residents. This transition from natural features to concrete structures not only compromises Visakhapatnam's climate resilience but also heightens its vulnerability to climate-related hazards like floods, landslides, and extreme temperatures. Hence, incorporating green infrastructure within urban planning strategies emerges as a pivotal approach to fortify Visakhapatnam against environmental risks. Integrating green spaces into city planning not only enhances the city's resilience to climate-related challenges but also serves as a critical buffer against the adverse impacts of rapid urbanization. Collaborative urban planning involving governmental authorities, urban planners, developers, and local communities is essential. This collective effort is instrumental in crafting an inclusive and resilient urban environment that aligns with global sustainability objectives while addressing the multifaceted challenges arising from CRZ policies, climate change risks, and steering Visakhapatnam towards sustainable and resilient development.

7 Conclusion

In summary, the study of Visakhapatnam's urban growth dynamics presents a comprehensive understanding of its evolution from a fishing village to a burgeoning port city. Over three decades, Visakhapatnam witnessed significant urbanization, expanding both territorially and demographically. The post-bifurcation scenario in Andhra Pradesh has positioned Visakhapatnam as a pivotal player in the state's development landscape. The amalgamation of surrounding villages and towns into the Greater Visakhapatnam Municipal Corporation (GVMC) has expanded the city's boundaries, presenting both opportunities and challenges. The extensive spatio-temporal analysis revealed notable trends in the city's expansion, showcasing a shift from scattered industrial-driven sprawl towards more organized, yet occasionally dense growth patterns. The directional analysis revealed growth patterns, with sectors exhibiting diverse expansion trajectories. The East and South East experienced compact growth, while the North-West, West, and South-West sectors exhibited fragmented expansion. Spatial configuration metrics like NP, PD, LPI, and FRAC_AM offered insights into growth complexity and distribution. They highlighted the emergence of larger patches and increased variability, emphasizing the need for equitable land distribution and adaptable planning strategies. Overall, the city evolved from scattered expansion to more cohesive growth patterns, particularly in residential areas. However, challenges emerged with increased high-density sprawl and infilling within the core city, demanding strategic urban planning. Moreover, the assessment of land consumption and sustainability indicators (i.e. LCRPGR and LUE) revealed fluctuating trends, signalling the city's potential for efficient land use but also indicating challenges in managing urban sprawl. While there were periods of more compact growth, subsequent decades witnessed increased sprawl, impacting land-use efficiency negatively. This trend raises concerns about the city's ability to balance growth and sustainability. The findings underscore the necessity for thoughtful urban planning to balance growth while preserving green spaces and minimizing sprawl. Therefore, the study underscores the imperative for collaborative efforts among stakeholders to foster sustainable land-use practices, integrate green infrastructure, and mitigate the city's vulnerability to natural hazards, thereby steering Visakhapatnam towards a resilient and balanced urban future.

The findings of this study carry significant policy implications for the urban development of Visakhapatnam. They underscore the pressing need for a reevaluation of urban planning strategies, emphasizing a balanced approach that harmonizes development initiatives with ecological preservation. Mitigating the challenges posed by high-density sprawl and encroachments on ecologically sensitive areas requires stringent zoning regulations and conservation efforts. Policymakers must implement targeted interventions to ensure equitable and integrated urban development. Additionally, the integration of sustainable land-use practices and green infrastructure within urban planning frameworks is crucial to bolster the city's resilience against natural hazards and safeguard its ecological balance. Collaboration among governmental entities, urban planners, developers, and local communities is pivotal in formulating inclusive policies that align with global sustainability goals while effectively managing the complexities arising from rapid urbanization and environmental conservation.

However, this study holds limitations in terms of its sole reliance on historical data and satellite imagery, which might not capture recent developments accurately. Future studies could integrate real-time data sources and advanced remote sensing techniques for more precise analyses. Additionally, exploring the social and economic implications of urban growth and assessing residents' perspectives would provide a comprehensive understanding. Exploring land-use optimization methods considering climate change effects and fostering community engagement in decision-making processes stand as critical areas for future investigation. This research lays the foundation for assessing the spatial sustainability while addressing the complex challenges posed by rapid urbanization in Visakhapatnam. In conclusion, this study emphasizes the pivotal role of strategic urban planning with zoning policies in fostering sustainable growth, balancing economic development with ecological preservation. By embracing these findings, policymakers and urban planners can pave the way for a resilient, inclusive, and environmentally conscious urban environment, steering Visakhapatnam towards balanced urbanization in an ever-evolving landscape.

References

Angel S, Parent J, Civco DL, Blei A, Potere D (2011) The dimensions of global urban expansion: estimates and projections for all countries, 2000–2050. Prog Plan 75(2):53–107

ASCELA (2020) Port-city as an engine of port development—case of major ports in India. ASCELA Advisors Private Limited. Retrieved from: https://ascelaadvisors.com/

Bobylev N (2009) Mainstreaming sustainable development into a city's Master plan: a case of urban underground space use. Land use policy 26(4):1128–1137

Balk D, Montgomery MR, McGranahan G, Kim D, Mara V, Todd M, Dorélien A et al (2009). Mapping urban settlements and the risks of climate change in Africa, Asia and South America. Popul Dyn Clim Change 80:103

Batty M, Xie Y, Sun Z (1999) Modeling urban dynamics through GIS-based cellular automata. Comput Environ Urban Syst 23(3):205–233

Bencardino M, Iovino G (2014) Analysing and managing urban sprawl and land take (No. 131). CELPE-Centre of Labour Economics and Economic Policy, University of Salerno, Italy

Boulos J (2016) Sustainable development of coastal cities-proposal of a modelling framework to achieve sustainable city-port connectivity. Procedia Soc Behav Sci 216:974–985

Burgi M, Hersperger A, Schneeberger N (2004) Driving forces of landscape change-current and new directions. Landscape Ecol 19:857–868

Cao A et al (2021) Future of Asian Deltaic Megacities under sea level rise and land subsidence: current adaptation pathways for Tokyo, Jakarta, Manila, and Ho Chi Minh City. Curr Opin Environ Sustain 50:87–97

Cummins V, O'Mahony C, Connolly N (2004) Review of integrated coastal zone management & principals of best practice prepared for the heritage council by the coastal and marine resources centre. Environmental Research Institute University College Cork, Ireland, 8 4 p. Website: http://www.heritagecouncil.ie/fileadmin/userupload/Publications/Marine/coastalzonereview.pdf. Last Accessed 15 Jun 2009

Dadashpoor H, Hasankhani Z (2022) Exploring patterns and consequences of land consumption in a coastal city-region. Ecol Process 11(1):49. https://doi.org/10.1186/s13717-022-00391-z
Estoque RC, Ooba M, Togawa T, Hijioka Y, Murayama Y (2021) Monitoring global land-use efficiency in the context of the UN 2030 Agenda for Sustainable Development. Habitat International, 115, 102403.
Forman RTT (2010) Coastal regions: spatial patterns, flows, and a people-nature solution from the lens of landscape ecology. In: Mininni M (eds) La costa obliqua: un atlante per la Puglia (The Oblique Coast: An Atlas for Puglia). Donzelli editore, Roma, Italy, pp 249–265
Greater Visakhapatnam Municipal Corporation (n.d.). Visakhapatnam city development strategy and slum upgrading action plan (Report No. P095174). Cities Alliance Project Output.
Gustafson EJ (1998) Quantifying landscape spatial pattern: what is the state of the art? Ecosystems 1(2):143–156
Hallegatte S, Rentschler J, Walsh B (2018) Building back better: achieving resilience through stronger, faster, and more inclusive post-disaster reconstruction. World Bank, Washington, DC, USA
He S, Yu S, Li G, Zhang J (2020) Exploring the influence of urban form on land-use efficiency from a spatiotemporal heterogeneity perspective: evidence from 336 Chinese cities. Land Use Policy 95:104576
Hoyle BS (1989) The port—city interface: trends, problems and examples. Geoforum 20(4):429–435
IBI Group (2016). Visakhapatnam NMT city specific plan. Prepared for Ministry of Urban development, Government of India
Lee SW, Song DW, Ducruet C (2008) A tale of Asia's world ports: the spatial evolution in global hub port cities. Geoforum 39(1):372–385
Long H, Qu Y (2018) Land use transitions and land management: a mutual feedback perspective. Land Use Policy 74:111–120
McGarigal K, Cushman SA, Neel MC, Ene E (2002) FRAGSTATS v3: spatial pattern analysis program for categorical maps. Computer software program produced by the authors at the University of Massachusetts, Amherst. http://www.umass.edu/landeco/research/fragstats/fragstats.html
Melchiorri M, Pesaresi M, Florczyk AJ, Corbane C, Kemper T (2019) Principles and applications of the global human settlement layer as baseline for the land use efficiency indicator—SDG 11.3. 1. ISPRS Int J of Geo-Inf 8(2):96
Merk O (2013) Port-city governance: results from a global survey. OECD Regional Development Working Papers (9):1–68. Retrieved from https://www.oecd-ilibrary.org/urban-rural-and-regional-development/port-city-governance_5k3txf7z6p6d-en
Mudau N, Mwaniki D, Tsoeleng L, Mashalane M, Beguy D, Ndugwa R (2020) Assessment of SDG indicator 11.3. 1 and urban growth trends of major and small cities in South Africa. Sustainability 12(17):7063
Neumann B, Vafeidis AT, Zimmermann J, Nicholls RJ (2015) Future coastal population growth and exposure to sea-level rise and coastal flooding-a global assessment. PLoS ONE 10(3):e118571
Nicholls RJ et al (2021) A global analysis of subsidence, relative sea-level change and coastal flood exposure. Nat Clim Chang 11(4):338–342
Schneider A, Woodcock CE (2008) Compact, dispersed, fragmented, extensive? A comparison of urban growth in twenty-five global cities using remotely sensed data, pattern metrics and census information. Urban Stud 45(3):659–692
Small C, Nicholls RJ (2003) A global analysis of human settlement in coastal zones. J Coastal Res 19:584–599
UN-HABITAT (2018a) Metadata on SDGs Indicator 11.3.1 Indicator category: Tier II. United Nations Human Settlement Program, Nairobi
UN-HABITAT (2018b) SDG Indicator 11.3.1 training module: land use efficiency. United Nations Human Settlement Programme (UN-Habitat), Nairobi.
United Nations (2018). World urbanization prospects: the 2018 revision. Online Edition.

Visakhapatnam Metropolitan Region Development Authority (2021). Draft master plan for VMR 2041. Government of Andhra Pradesh.
Yeung YM (2001) Coastal mega-cities in Asia: transformation, sustainability and management. Ocean Coast Manag 44(5–6):319–333
Yılmaz M, Terzi F (2021) Measuring the patterns of urban spatial growth of coastal cities in developing countries by geospatial metrics. Land Use Policy 107:105487

Sustainable Megacities and Planning for Social Resilience

Investigating the Housing Preferences of Slum Dwellers in Slum Relocation Projects: A Community-Based Operations Research Framework for Megacities in the Global South

Debanjali Saha, Ruchita Gupta, and Haimanti Banerji

1 Introduction

The rapid urbanization and significant migration from rural areas to megacities have led to a growing demand for housing, which impacts accommodation patterns and living standards (Makinde 2012). This influx of people has significantly strained the housing delivery system, resulting in severe housing shortages (Mahdi and Mazumder 2023). As a result, the formal housing sector has struggled to keep pace with the escalating demand for housing solutions. In response to this inadequacy, informal settlements like slums have emerged as coping mechanisms for individuals seeking accommodation (Butala et al. 2010; Das et al. 2021). These slums, proliferating in megacities throughout the Global South, with India being a prime example, have presented challenges in enhancing infrastructure due to their chaotic and close-knit nature (Verma et al. 2019). This situation underscores the urgent need for comprehensive policies addressing the diverse and multifaceted issues faced by the slum population (Ooi and Phua 2007).

Slums have become the home for approximately 880 million individuals, mainly in urban areas of the Global South, with India hosting 154 million, constituting one-third of its urban population (Killemsetty et al. 2022). The projections suggest a rise to 2 billion slum dwellers by 2030 (Das et al. 2021; Boateng and Adams 2023), emphasizing the critical role of housing in achieving Sustainable Development Goal 11.1, which aims to ensure access to adequate, safe, and affordable housing for all by

D. Saha (✉) · H. Banerji
Department of Architecture and Regional Planning, Indian Institute of Technology, Kharagpur, India
e-mail: ddebanjali18@kgpian.iitkgp.ac.in

R. Gupta
Department of Housing, School of Planning and Architecture, New Delhi, India

M. Firoz C et al. (eds.), *Urban Planning and Design for Megacities in the Global South*, Advances in 21st Century Human Settlements,
https://doi.org/10.1007/978-981-97-8370-0_13

2030. Over the decades, attitudes toward slum dwellers have evolved from outright eviction to upgradation, resettlement, and relocation. Various policy interventions, including slum upgrading programs, affordable housing schemes, and relocation projects, have been implemented at local, state, and federal levels (Bardhan et al. 2015). Extensive studies on resettlement and relocation have highlighted diverse impacts on re-settlers. Critics argue that such initiatives adversely affect communities and the local economy (Patel 2013), while proponents advocate for utilizing prime slum land for more productive purposes, potentially benefiting the broader urban economy (Viratkapan and Perera 2006).

In megacities, where urbanization challenges are intensified due to the immense population and limited resources, the issue of relocation takes on added significance (Patel 2013). The success of these efforts depends on understanding the diverse needs and preferences of slum dwellers. Implementing a top-down approach without adequate consideration for each community's unique circumstances has led to adverse consequences, including social dislocation, economic hardships, and resistance from the affected population (Patel et al. 2014; Killemsetty et al. 2022). Diverse research methodologies have been employed to address these challenges. These include ethnographic studies on community interventions (Doshi 2012; Ghosh et al. 2016; Mwiti and Goulding 2018), estimation studies on demographical data (Patel et al. 2014; Bardhan et al. 2015; Killemsetty et al. 2022), the use of satellite imagery to identify slum locations and housing structures (Kit and Lüdeke 2013; Verma et al. 2019), and simulations to test different scenarios (Roy et al. 2014; Thomson et al. 2022). Despite the wealth of information gathered through these studies, a critical gap persists in research methodology: fully capturing the intricate lives of slum dwellers and their complex relationships with the community and the environment. The comprehensive review of the literature has revealed significant gaps that underline the importance of comprehending and considering the preferences of slum dwellers as an integral aspect of effective and sustainable urban development strategies in megacities.

To address these gaps and contribute to the existing body of knowledge, this study aims to investigate the complex dynamics of slum relocation projects through the lens of community preferences and experiences. Specifically, this research seeks to answer three key research questions: (1) What are the key housing challenges encountered in a slum relocation project? (2) What housing decisions are of utmost importance to slum-relocated dwellers based on their preferences? (3) What policy recommendations or strategies are imperative to augment the success of the relocation project in Indian megacities? To answer these research questions, this study employs a multi-faceted approach with two primary objectives: (1) to employ a community-based operations research (CBOR) framework for analyzing slum relocation challenges and (2) to utilize two analytic methods—Strategic Options Development and Analysis (SODA) and Value-Focused Thinking (VFT)—for exploring slum dwellers' preferences and decision-making processes. These approaches facilitate an in-depth exploration of the intricacies shaping the choices and needs of the slum population undergoing relocation. By bridging the gap between theoretical frameworks and practical implementation of slum relocation initiatives, this research aims to address critical gaps in current research methodologies and contribute to more effective and sustainable urban development strategies in megacities.

This study seeks to provide valuable insights into slum relocation projects by addressing critical gaps in the current literature and applying decision sciences to tackle socially relevant challenges. The research offers a holistic exploration of the challenges and opportunities associated with slum relocation projects in megacities, particularly within the Indian context, by emphasizing the subjective experiences and preferences of slum dwellers. The methodological approach, combining CBOR, SODA, and VFT, enhances the robustness and reliability of the findings. This methodological richness extends the study's contribution beyond academia, offering practical implications for policymakers, urban planners, and development practitioners involved in slum relocation projects. By bridging the gap between theoretical research and practical application, this study aspires to inform evidence-based policy interventions. Ultimately, the research aims to contribute to sustainable and inclusive urban development strategies that can effectively address the complex challenges of rapid urbanization and housing shortages in megacities.

The paper is structured as follows: Sect. 2 provides a brief overview of the relevant literature on CBOR and PSMs, including SODA and VFT, to highlight the gaps in the existing literature. Section 3 outlines the research methodology, commencing with details regarding the conceptual framework, followed by an elaboration on the case study area and the approaches employed for data collection. Section 4 summarizes the study findings, employing the methodologies of SODA and VFT within the CBOR framework. Section 5 presents a comprehensive discussion describing the policy implications derived from this study and concludes the paper by summarizing the key findings and their significance.

2 Literature Review

2.1 Community-Based Operations Research (CBOR) Framework

This section offers an overview of the methodologies employed in operational research (OR), particularly CBOR. Traditionally, public-sector OR applications have relied heavily on mathematical modeling and quantitative methods, which cannot fully capture the complexities of social issues affecting marginalized communities (Midgley et al. 2018). In response to this limitation, CBOR has emerged as an approach integrating quantitative and qualitative methods to address complex and ill-structured problems (Killemsetty et al. 2022). When utilized, these methodologies empower local communities and generate actionable recommendations to enhance economic efficiency and social equity, and reduce administrative burdens for the population in megacities (Midgley et al. 2018).

Furthermore, CBOR advocates for a 'bottom-up' approach involving communities in decision-making, improving their adaptive and coping abilities (Smit and Wandel 2006). The method employs multiple analytical approaches to address problems, such as participatory methods, decision analytics, and mathematical programming (Midgley et al. 2018). The CBOR's engagement with vulnerable communities and its use of diverse methodologies make it crucial for enhancing OR frameworks.

2.2 Problem Structuring Methods (PSMs)

Around four decades ago, researchers developed qualitative modeling approaches known as Problem Structuring Methods (PSMs) to tackle complex and poorly structured problems (Mingers and Rosenhead 2004). Notably, this method includes Strategic Options Development and Analysis (SODA), Value-Focused Thinking (VFT), Soft Systems Methodology (SSM), Robustness Analysis (RA), and Strategic Choice Approach (SCA) (Ghosh et al. 2016; Killemsetty et al. 2022). This approach significantly contributes to clarifying intricate situations, focuses on actionable problems, and achieves consensus on commitments to address them (Smith and Shaw 2019).

Within the spectrum of PSM approaches, SODA and VFT are particularly compatible with analyzing vulnerable communities like slum dwellers and supporting open dialogue (Mingers and White 2010). Given their relevance, this study employs SODA and VFT to clarify challenging situations, identify actionable problems, and foster agreements on commitments to address and resolve them. SODA focuses on identifying diverse stakeholder viewpoints, while VFT elicits stakeholder values, making them highly relevant to the research (Killemsetty et al. 2022). Combining these methods provides a comprehensive view of decision situations, enhancing efficiency in problem identification and generating solutions.

2.3 Strategic Options Development Analysis (SODA)

The SODA approach focuses on social negotiation and cognitive psychology, utilizing a dynamic model that continuously evolves and represents complex problems (Eden and Ackermann 2004, 2006). It is a valuable tool for problem exploration before determining potential solutions (Ackermann and Eden 2010). The SODA methodology employs a structured framework incorporating decision-makers viewpoints, considering their experiences, aspirations, and subjectivity (Guarnieri et al. 2016). The preference for SODA over other PSMs mainly stems from its ability to highlight diverse subjective viewpoints of the problem by constructing a cognitive map. Thus, creating cognitive maps through SODA leads to a comprehensive compilation of issues, challenges, methods, and possibilities, providing insights for holistic decision-making possibilities (Killemsetty et al. 2022).

2.4 Value-Focused Thinking (VFT)

VFT was created by Keeney (1996) as an analytical technique tailored for complex decision-making situations that involve multiple stakeholders and high uncertainty. While VFT has been mainly associated with quantitative modeling, integrating it into PSM addresses stakeholders' concerns by combining quantitative and qualitative analysis. This integration has led to a more concise, manageable, and measurable set of factors (Keeney 1996). Moreover, when following Soft Operational Research principles, VFT gives precedence to value judgments and encourages active involvement from various stakeholders (Eden and Ackermann 2004). It articulates goals in a hierarchical network format, uncovering hidden objectives and producing valuable information. Integrating VFT within a comprehensive decision framework empowers decision-makers to achieve two key purposes: (1) define, categorize, and structure their values and objectives, and (2) assess and choose from various alternatives (Arvai et al. 2001).

The existing literature has extensively employed CBOR to examine vulnerable communities in domains such as urban planning (Brennan 2017), disaster relief (Goulding et al. 2018), and housing issues (Killemsetty et al. 2022), primarily in the Global South. However, there is a scarcity of studies employing CBOR to understand the challenges experienced by vulnerable communities. This chapter contributes to the expanding body of research by utilizing CBOR to investigate vulnerable communities in the megacities of the Global South. In addition to CBOR, several studies have employed PSM in urban planning to assess housing requirements and sustainable urban development (Bender et al. 2000; Mulliner et al. 2013; Fernandes et al. 2017). The existing literature typically involved a top-down approach with housing experts, urban planners, policymakers, and NGOs (Mulliner et al. 2013; Chan and Adabre 2019) but often excluded dwellers from the decision-making process. Exceptions like Bender et al. (2000) and Killemsetty et al. (2022) engaged dwellers through household surveys to comprehend their preferences.

Similarly, this research utilizes PSMs to engage with existing slum communities and those affected by relocation projects. While some studies have employed multiple PSMs in a single study for a holistic perspective on decision-making situations (Marttunen et al. 2017; Small and Wainwright 2018), this approach remains relatively uncommon in the Global South (Françozo et al. 2021). Notably, this research represents the second instance in India of employing multiple PSMs to analyze slum dwellers, following the study conducted by Killemsetty et al. (2022) in the slum areas of Odisha, India. The key difference lies in the scope: while Killemsetty et al. (2022) focused solely on existing slum dwellers, this study considers both the residents of existing slums and those affected by slum-related projects, providing a more comprehensive analysis of the relocation process.

SODA and cognitive mapping have been widely implemented across various domains, including the public sector (Eden and Ackermann 2004), management studies (Guarnieri et al. 2016), and planning (Hjortsø 2004; Killemsetty et al. 2022).

While prior research has utilized SODA to investigate housing preferences and challenges, this study distinguishes itself as the first to specifically employ SODA to gain insight into slum relocation housing projects. Moreover, this research goes beyond the application of SODA alone by incorporating VFT, another powerful problem-structuring method. By applying both VFT and SODA in the novel context of slums in Indian megacities, this research contributes significantly to the existing body of literature. The combination of these methodologies provides a comprehensive framework for analyzing the complex issues surrounding slum relocation, offering new perspectives on urban planning challenges in rapidly growing cities.

3 Research Methodology

This section presents the conceptual framework, outlining a four-step analytical process of CBOR as implemented by Killemsetty et al. (2022). Subsequently, it delves into the discussion of the case study area and provides a comprehensive explanation of the data collection strategies utilized in the research.

3.1 Conceptual Framework

The CBOR process involves four steps: identifying the problem, formulating the problem, developing the solution, and implementing the solution (Johnson 2011; Killemsetty et al. 2022). This framework addresses the complex decision-making process in ambiguous contexts. Two problem-structuring methods, SODA and VFT, were employed effectively at each stage of CBOR. By utilizing this approach, the unique challenges faced by the slum dwellers were identified, and their diverse objectives to improve housing stability were structured. The data collection also provided profound insights into their motivations and values. In the following sections, the authors comprehensively describe each step in the CBOR process (refer to Fig. 1).

3.1.1 Step 1: Problem Identification

In the first step of the research process, the SODA method was employed for problem identification, utilizing semi-structured interviews to understand diverse factors contributing to challenges in the slum community (Johnson 2011; Killemsetty et al. 2022). The collected information was used to create an aggregate cognitive map, which organized the challenges thematically (Ackermann and Eden 2010).

The SODA analysis revealed interconnections between various issues. Action-oriented phrases were identified from interviews, forming a causal hierarchy network.

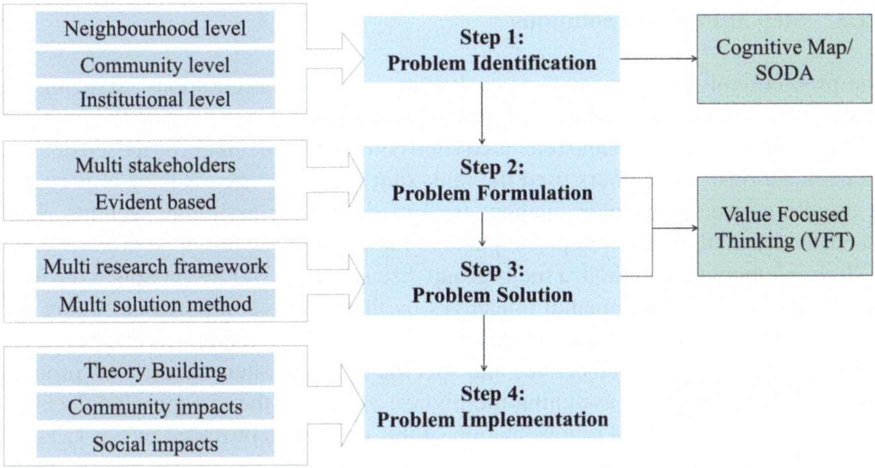

Fig. 1 Methodological framework of CBOR utilizing PSM—reproduced from Killemsetty et al. (2022) and Johnson (2011)

This network represented the collective input of survey participants, including challenges related to livelihoods and social connections in relocated areas (Fregonese et al. 2020).

3.1.2 Step 2: Problem Formulation

In the problem formulation step of CBOR, the key objective was to deepen the understanding of the problem context and establish factors that connect actions to outcomes (Killemsetty et al. 2022). For example, a value identified through VFT, such as maximizing community belonging (action), can reduce disputes and enhance social cohesion (outcome).

A thematic evaluation and pattern-matching technique were employed to analyze interview responses, highlighting challenges in the cognitive maps and revealing slum dwellers' values in VFT networks. Pattern-matching is a widely used method for analyzing qualitative data to identify central themes and group similar responses. Participants' interview responses were examined to identify specific objectives. Subsequently, these objectives were categorized and organized into groups using the matching technique. This process condensed a list of fundamental objectives and means-ends objectives. These means-ends objectives served as a basis for identifying alternative decision actions, providing a framework for exploring various strategies and interventions to address the challenges identified in the research.

3.1.3 Step 3: Problem Solutions

The problem-solving step involves evaluating system configurations, agreeing on process changes, and establishing shared goals (Johnson 2011). During the problem formulation stage, existing and relocated slum community members proposed diverse decision actions. The authors used the rank-sum method to assess their importance as an alternative to complex mathematical modeling for assigning weights. This user-friendly method allowed participants to rank objectives based on preferences.

To implement this method, a focus group discussion (FGD) was conducted where participants ranked fundamental objectives by their relative importance. Using the rank-sum method, the authors assigned weights to these fundamental objectives, followed by means-ends objectives and specific decision alternatives. This process normalized the rankings, assigning the highest weight to the most crucial decision actions. The mathematical representation of this process is provided in Eq. 1.

$$Wi = \frac{2(n + 1 + i)}{n(n + 1)}, i = 1, 2, 3, \ldots, n \tag{1}$$

Here, Wi represents the rank weight, n is the number of objectives, and i is the rank of the objective (with 1 being the highest rank). Using this equation, the authors calculated final scores to determine the relative significance of individual decision actions. These actions were then categorized into relevant themes based on their interconnections, providing a structured framework for analyzing the preferences and priorities of the slum dwellers.

3.1.4 Step 4: Problem Implementation

The final step of CBOR involves implementing solutions, ranging from an increased understanding of the problem to revised policy interventions. However, the implementation of the identified solutions is beyond the scope of this study, primarily due to the requirement for government involvement and relevant resources for the decision actions. Despite this limitation, the value structures and weighted hierarchies derived from the analysis offer valuable insights for designing better policy frameworks, particularly for slum housing programs.

3.2 Case Study Area: Kolkata

According to the 2011 Census of India, cities with populations over one million comprised 43% of India's urban population, with Mumbai and Delhi ranking among the world's top 10 most populous cities and Kolkata securing a place in the top 15. Kolkata, located in West Bengal (88° 30′E—22° 33′N), is India's third largest megacity, with approximately 14.85 million people. The Kolkata Metropolitan Area

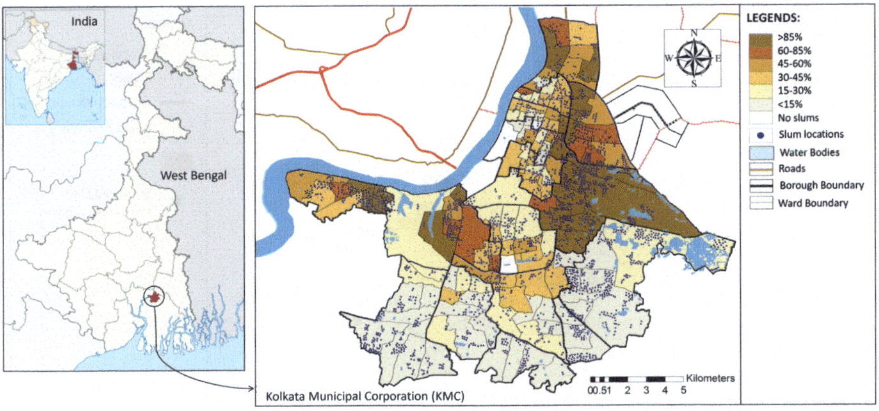

Fig. 2 Map showing slum location and its density in Kolkata Municipal Corporation (KMC)

extends beyond the administrative boundaries of the Kolkata Municipal Corporation (KMC), which includes neighboring districts like North 24 Parganas, South 24 Parganas, Howrah, and Hooghly. The scope of this study is concentrated on the KMC area, which has a population of around 4.4 million. Approximately 31.35% of the urban population of KMC (refer to Fig. 2) reside in slums. This significant slum population necessitates various strategies to address the associated challenges (Census 2011).

Various government interventions in Kolkata, including Valmiki Ambedkar Awas Yojana (VAMBAY) in 2001, the Basic Services for the Urban Poor (BSUP) in 2005, and the Pradhan Mantri Awas Yojana (PMAY) in 2015, have aimed at addressing slum community issues (Bardhan et al. 2015). While PMAY focused on slum redevelopment (Kumar and Shukla 2022), the other two included relocation and slum redevelopment projects (Patel 2013). Each policy has encountered challenges, especially regarding the dwellers' basic needs. Relocating slum dwellers requires careful consideration of factors like proximity to original settlements, livelihood opportunities, basic services, and social networks, as unsuitable sites may lead to resistance.

Given these considerations, extensive research was conducted in the KMC area, which includes forty-eight notified slums[1] and two BSUP relocation projects (refer to Fig. 3). The authors aimed to actively include slum dwellers' voices in policy-making, utilizing SODA and VFT methodologies to explore housing, infrastructure, and policy-related objectives and decisions.

[1] All specified areas in a town or city notified as 'Slum' by State/Local Government and Union Territory (UT) Administration under any Act including a 'Slum Act'.

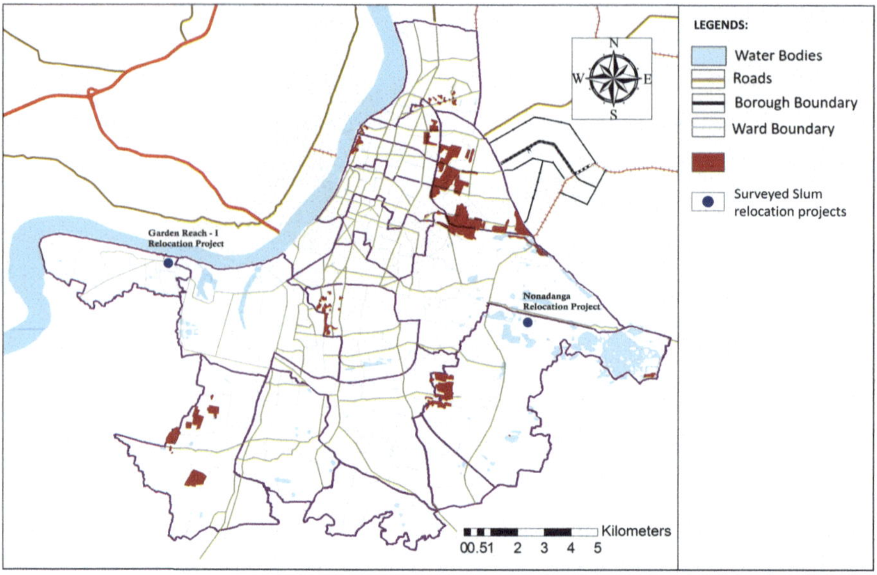

Fig. 3 Location of case study sites in KMC

3.3 Data Collection

The research employed the VFT approach to capture diverse viewpoints in the context of slums. The survey instrument predominantly comprised ordinal scales and open-ended questions, allowing for structured quantification and in-depth exploration of respondents' perspectives. To gain a comprehensive understanding of the complex housing challenges experienced by slum residents, the researchers opted for semi-structured interviews as their primary data collection method. This approach allowed participants to freely express their views and experiences without being unduly influenced by the researchers' preconceived notions or the limitations of a rigid questionnaire. To ensure statistical rigor and representativeness, the study utilized Cochran's formula (Cochran, 1977) to determine an appropriate sample size. Applying the formula with a 95% confidence level (z-value = 1.96) and a 5% margin of error yielded a minimum required sample size of 384 participants. This sample size was sufficient to capture the diverse perspectives within the slum population while maintaining a balance between statistical power and the feasibility of data collection in the challenging urban environment of Kolkata's slums.

The data collection process employed a multi-stage sampling approach. Initially, simple random sampling was used to select participants, followed by snowball sampling to reach data saturation. The sample was diversified across genders, age groups, and occupations to enhance external validity. A researcher-developed questionnaire, comprising 12 closed-ended and 18 open-ended questions, was administered through interviews lasting 10–12 min. The open-ended questions were designed

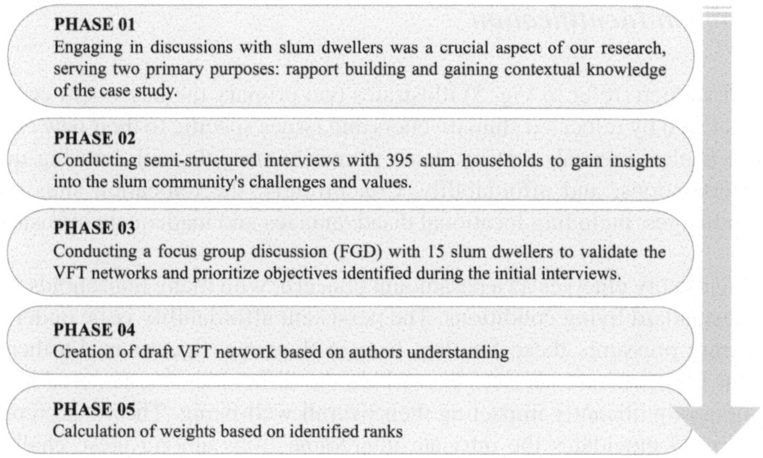

Fig. 4 Phases of research

to identify housing challenges, construct cognitive maps, and elicit housing preferences and values, thereby aiding in the identification of specific objectives related to housing and infrastructure needs. The data collection, conducted from December 5, 2019, to March 8, 2020, yielded 410 samples. After rigorous data cleaning, which involved removing outliers, missing data, and addressing disparities between stated perceptions and revealed data, 395 samples were retained for final analysis (Fig. 4).

A second phase of the survey involved in-depth interviews with a subset of 15 participants selected from the cleaned sample of 395. This phase aimed to validate the preliminary VFT networks established earlier and facilitate a collective understanding of dwellers' priorities. The process involved ranking identified objectives through deliberations and consensus-building, providing a more nuanced insight into the community's housing preferences and needs.

4 Analysis and Results

This section provides a comprehensive overview of the findings from the research on slum dwellers in Kolkata, following the four-step CBOR framework outlined in the methodology.

4.1 Problem Identification

The cognitive map (refer to Fig. 5) illustrates two primary themes: socio-economic challenges faced by relocated slum dwellers and issues specific to their new environment. The analysis reveals that residents primarily struggle with housing quality, policy interventions, and affordability. Concurrently, the relocation sites present unique challenges, including locational disadvantages and inadequate infrastructure services.

Housing quality emerges as a paramount concern, with many households experiencing substandard living conditions. The persistent affordability issue underscores the economic pressures these families face in their new locations. Furthermore, insufficient infrastructure services exacerbate the difficulties encountered by these communities, significantly impacting their overall well-being. The visual representation in Fig. 5 elucidates the intricate interconnections among these challenges, providing a holistic understanding of the multifaceted issues confronting the relocated slum community. This comprehensive depiction aids in grasping the complexity of the situation and the interplay between various factors affecting the residents' quality of life in their new settlements.

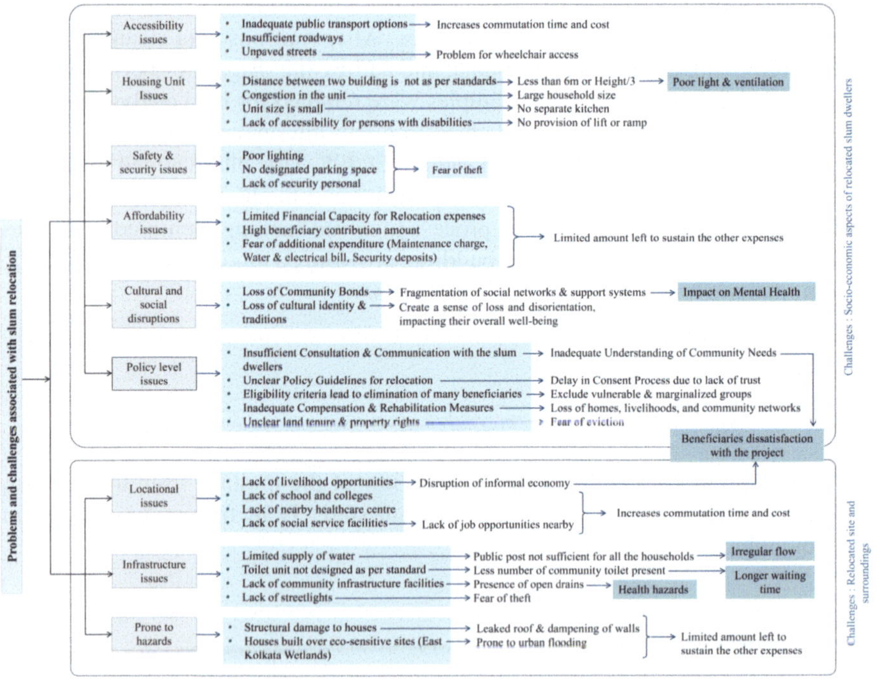

Fig. 5 Cognitive map illustrates the housing challenges identified by relocated slum dwellers

4.2 Problem Formulation

In Fig. 6, the authors present the VFT network developed for relocated slum dwellers to improve their quality of residential life. The analysis identified three fundamental objectives: (i) maximizing the quality and adequacy of the housing unit, (ii) maximizing the quality of infrastructure, and (iii) maximizing coordination among governmental bodies. These fundamental objectives have been derived from more specific objectives, leading to the identification of means-end objectives within the network. These means-end objectives represent potential solutions or decision actions to address the challenges faced by relocated slum dwellers.

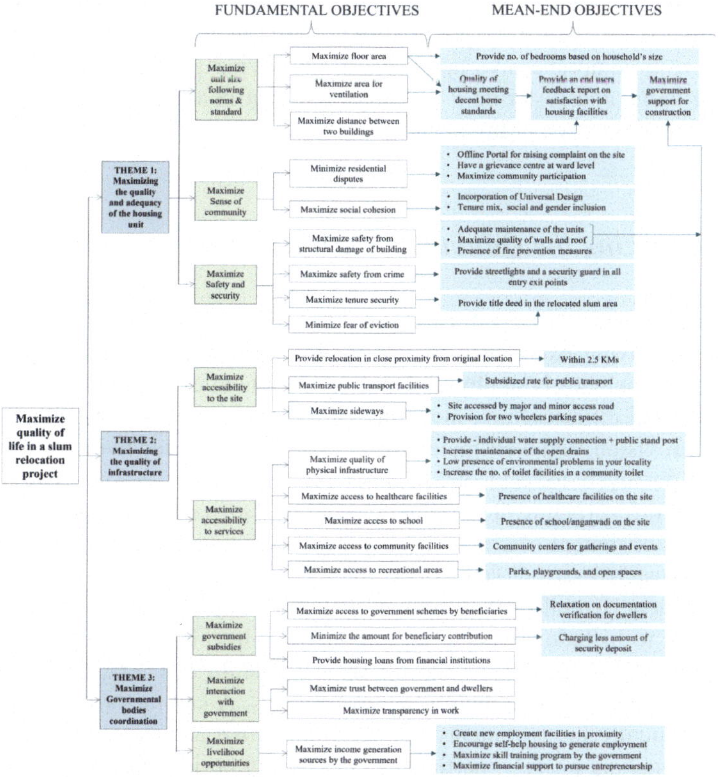

Fig. 6 Value-focused thinking network on housing preferences of relocated slum dwellers

4.3 Theme 1: Maximizing the Quality and Adequacy of the Housing Unit

The primary goal of maximizing housing unit quality focuses on enhancing living conditions and meeting the basic needs of slum dwellers. This objective incorporates several key aspects. Firstly, structural improvements are essential, involving renovations that enhance structural integrity, ensure proper ventilation, and improve overall comfort (Ghosh et al. 2016). Secondly, safety measures are crucial, including the implementation of assessments and protocols for structural stability and fire safety. Thirdly, the provision of essential facilities, particularly access to sanitation and clean water, is vital for maintaining health and hygiene. Lastly, appropriate sizing of housing units is considered, taking into account the adequacy of space based on household size (Chan and Adabre 2019). These interrelated aspects collectively contribute to a comprehensive approach in improving the quality and suitability of housing units, and fostering a profound sense of belonging for relocated slum dwellers in their new site (Ahmed and Salam 2022).

4.4 Maximizing the Quality of Infrastructure

Maximizing the quality of infrastructure has emerged as a crucial objective in enhancing the living experience of relocated slum dwellers (Butala et al. 2010). This approach includes both physical and social amenities. Significant investments in basic physical infrastructure, including clean water systems, sanitation facilities, electricity provision, and well-maintained roads, have elevated living standards and contributed positively to residents' health. Complementing these physical improvements, social infrastructure development has played a vital role in community cohesion. The establishment of community spaces, recreational areas, parks, and playgrounds has fostered a strong sense of belonging and created a more cohesive community environment (Ezebilo 2017; Aalbers et al. 2019). Furthermore, extending social infrastructure to include access to educational and healthcare facilities has significantly enhanced the quality of life for relocated slum dwellers, facilitating their adaptation and prosperity in their new surroundings. A key factor in the success of these infrastructure improvements has been the active engagement of dwellers in the decision-making process, particularly in housing project planning, ensuring the implementation of effective and relevant infrastructure solutions (Narayanan et al. 2017). This participatory approach has been instrumental in aligning infrastructure development with the actual needs and preferences of the community.

4.5 Maximize Governmental Bodies' Coordination

The objective emphasizes the importance of fostering cooperation between various government agencies to create a supportive environment for the relocated communities. Collaboration among housing, urban development, employment, and social welfare agencies has proven crucial in formulating inclusive policies and programs that holistically address the multifaceted challenges faced by relocated slum dwellers (Ghosh et al. 2016; Mahdi and Mazumder 2023). By aligning efforts and resources, these agencies have effectively tackled the diverse issues arising in newly relocated sites.

A key aspect of this coordinated approach involves the implementation of incentive structures to support slum dwellers through their transition. These incentives encompass financial support, tax breaks, and subsidies for essential services such as housing, education, and health care (Stone 2006). Such measures have provided tangible support and encouraged active participation in the relocation process. Furthermore, the focus on creating employment opportunities has been pivotal for economic empowerment. This has involved collaborations with private enterprises and NGOs to establish vocational training, skill development, and job placement initiatives (Viratkapan and Perera 2006). By providing access to training and employment opportunities, the government has been able to empower slum dwellers, enhance their livelihoods, and contribute to the overall upliftment of their socio-economic status. The subsequent section of this study presents an analysis of potential solutions, framed as decision actions, to determine which strategies are most preferred by relocated slum dwellers. This analysis aims to provide insights into the most effective and acceptable approaches to addressing the challenges faced by these communities in their new environments.

4.6 Problem Solution

The VFT network analysis yielded a comprehensive list of 38 decision actions presented in Table 1. These actions are categorized into three key components: housing unit improvements, infrastructure enhancements, and policy-level interventions. To assess the relative importance of each decision action, the study employed the rank-sum method, as previously described in the methodology section (Eq. 1). This analytical approach allowed for the quantification of each action's significance. The resulting weights were then converted into percentage shares, clearly representing their comparative importance in enhancing the quality of residential life for relocated slum dwellers.

The analysis of decision actions reveals distinct priorities among slum dwellers across three key areas: housing, infrastructure, and policy initiatives. Housing-related actions emerged as the second most significant category, accounting for 32% of the

Table 1 Rank-sum weights

Component	Decision actions	% share of weights (Wi X 100)	Aggregate % share of weights
Housing unit level	Decent quality of housing meeting standards	7.65	32.18
	Adequate maintenance of existing houses	6.68	
	Number of bedrooms based on the household's size	4.4	
	Incorporation of universal design in the housing units	3.8	
	Low presence of environmental problems in the locality (litter, open drains, traffic)	3.5	
	Tenure mix, social and gender inclusion	2.66	
	Place-making and sense of community	2.25	
	Provision for two-wheelers parking spaces	1.17	
	Provision of space for livelihood activities	0.05	
	Presence of fire prevention measures	0.02	
Infrastructure provision level	Construct individual water supply connection	8.1	47.18
	Provision for public transportation facilities	7.6	
	Construct close drain	6.6	
	Provide individual toilet facilities with dwelling units	6.1	
	Increase the no. of toilet and bathing facilities in a community toilet	5.7	
	Proximity to healthcare facilities	5.4	

(continued)

Table 1 (continued)

Component	Decision actions	% share of weights (Wi X 100)	Aggregate % share of weights
	Increase the number of streetlights in the community	2.2	
	Provide more public stand posts on the site	1.8	
	Site accessed by the major and minor access road	1.32	
	Increase maintenance of the open drains	1.06	
	Presence of school/ *Anganwadi* on the site	0.8	
	The presence of common community facilities on the site	0.5	
Policy level	Maximize government subsidies for slum relocation	3.47	20.64
	Provide tenure security for the relocated housing unit	3.2	
	Relocate within proximity to the original slum site	2.6	
	Provide housing loans from financial institutions	2.2	
	Government to provide skill-based training for women	1.84	
	Maximize financial support to pursue entrepreneurship	1.76	
	Provide Rental vouchers	1.33	
	Create new employment facilities in proximity	1.03	
	Offline Portal for raising complaints on the site	0.18	
	Have a grievance center at the ward level	0.16	

(continued)

Table 1 (continued)

Component	Decision actions	% share of weights (Wi X 100)	Aggregate % share of weights
	Encourage public participation in the planning process	0.12	
	Encourage self-help housing to generate employment	0.8	
	Charging less amount of security deposit	0.79	
	Provide an end-user satisfaction report on housing facilities	0.66	
	Relax documentation verification for house eligibility	0.5	

total weight. Within this category, residents emphasized the importance of high-quality housing that adheres to established norms and guidelines, with particular stress on regular building maintenance (6.6%). Additionally, they highlighted the need for appropriate living space, advocating for the determination of bedroom numbers based on household size and composition to ensure adequate privacy and comfort for family members (4.4%). Infrastructure provision emerged as the top priority, commanding 47% of the total weight and underscoring the critical importance of social infrastructure. A noteworthy aspect within this category is the emphasis on public transportation (7.6%), reflecting the residents' desire for improved mobility and connectivity to essential services and employment opportunities. This focus on transportation not only enhances accessibility but also has the potential to significantly reduce commuting costs, particularly benefiting those with limited financial resources. Interestingly, policy-level initiatives received the least emphasis, accounting for only 21% of the total weight. This lower prioritization might stem from past experiences, potentially indicating reduced expectations regarding significant government involvement in policymaking and implementation. Despite this, the weights derived from the analysis provide valuable insights for developing targeted strategies and recommendations that align closely with the preferences and needs of slum dwellers.

4.7 Problem Implementation—Future Implications

The identified decision actions in the slum relocation project, based on dwellers' preferences, hold significant potential for transforming the lives of relocated communities. Implementing these community-driven priorities can lead to multifaceted positive outcomes. Improved housing quality, characterized by permanent structures and secure tenure, can provide stability and a sense of ownership. Enhanced infrastructure services, including water, sanitation, and electricity access, can create a more conducive living environment. Furthermore, the focus on socio-economic facilities such as healthcare, education, and public transportation can open avenues for improved livelihoods and better access to essential services.

However, the realization of these benefits faces challenges in implementation and impact assessment. The success of these initiatives depends heavily on factors such as government commitment, adequate resource allocation, and sustained community engagement. The long-term impact on the well-being of relocated dwellers may take considerable time to manifest fully. This underscores the importance of the research in providing evidence-based strategies that align with the aspirations of slum dwellers. To ensure the efficacy of these interventions, continuous monitoring and evaluation are crucial. This ongoing assessment will help understand the true impact of the implemented policies and allow for necessary adjustments. The study's transformative potential lies in its inclusive and collaborative approach, incorporating community input into policy formulation. This method enhances the relevance of the interventions and promotes a sense of ownership among the beneficiaries. Ultimately, this approach aims to secure the future well-being and improvement of slum dwellers in relocation projects, marking a significant step toward more effective and humane urban development strategies.

5 Discussion and Conclusion

The study employed a Community-Based Operations Research (CBOR) framework, integrating Strategic Options Development and Analysis (SODA) and Value-Focused Thinking (VFT), revealing critical insights into slum relocation projects in Kolkata. The key findings reveal the critical importance of housing quality, infrastructure development, and community engagement in successful relocation initiatives. The weighted analysis of slum dwellers' preferences highlighted the paramount importance of infrastructure provision (47%) and housing-related decisions (32%). This quantification of priorities addresses a significant research gap by providing a detailed understanding of relocated communities' needs. The findings confirm that successful relocation initiatives require a holistic approach that considers physical infrastructure and social factors. This comprehensive understanding challenges the traditional top-down approach to slum relocation, emphasizing the need for community-centric solutions.

The findings of this study align with previous research highlighting the importance of housing quality and infrastructure in slum relocation projects (Patel 2013; Ghosh et al. 2016). However, this study further quantifies the relative importance of different factors based on the slum dwellers' priorities. The emphasis on community engagement and participatory decision-making processes in this study echoes the work of Narayanan et al. (2017) and Mwiti and Goulding (2018). Therefore, this research extends these findings by directly integrating community preferences into policy recommendations through the CBOR framework. Furthermore, the novel combination of SODA and VFT within the CBOR framework represents a methodological advancement in slum relocation research, offering a more comprehensive analysis than studies using single methods.

Based on the findings, several evidence-based strategies were recommended for policy implementation and interventions in slum relocation projects. A primary recommendation is the development of customized housing solutions that prioritize high-quality structures adhering to established norms (Maliene and Malys 2009). This approach considers the unique needs of families, determining the number of bedrooms based on household size and composition. Ensuring decent living spaces and respecting privacy addresses specific familial considerations and enhances the overall quality of life (Akinyede et al. 2017). Significant investment in social infrastructure development emerges as another key strategy. This includes developing essential amenities such as clean water, sanitation, electricity, and well-maintained roads (Butala et al. 2010; Narayanan et al. 2017; Galiani et al. 2017). Moreover, establishing community spaces, recreational areas, parks, and playgrounds fosters a sense of belonging and contributes to residents' overall well-being. This approach directly addresses the high priority slum dwellers give to social infrastructure, potentially enhancing the quality of residential life in relocated areas.

The provision of subsidized public transportation represents a critical intervention to improve mobility and connectivity for dwellers (Tsamboulas 2007; Sterzer 2017; Ghosh et al. 2023). By making commuting more accessible and reducing transportation costs, this strategy enhances access to essential services and employment opportunities (Hanni and Rao 2024). This approach addresses economic considerations, promotes inclusivity, and supports social integration within the broader urban fabric. Community engagement in decision-making processes is emphasized as a crucial component of successful relocation initiatives (Samaddar et al. 2021). Establishing ongoing communication and feedback mechanisms between governmental bodies and the community fosters a participatory approach to decision-making (Killemsetty and Patel 2022). This strategy ensures that policies align with evolving community needs, fostering ownership, trust, and collaboration, which are essential for the long-term success of relocation projects. Finally, establishing robust monitoring and evaluation mechanisms is recommended to continually assess the impact of policies and interventions (Kissi et al. 2019). This approach allows for flexibility and adaptability in policy implementation, enabling adjustments based on evolving dynamics and addressing challenges the relocated community faces. Such responsive interventions are crucial for contributing to sustainable and inclusive urban development. These strategies offer a roadmap for more effective and equitable relocation initiatives, potentially transforming the approach to urban development in rapidly growing cities.

Despite its strengths, the study has limitations. The findings are context-specific to the Megacity of Kolkata and may not directly apply to other megacities. The reliance on self-reported data (like surveys, interviews, or questionnaires) from slum dwellers about housing preferences, challenges, or experiences with relocation might be influenced by some bias. This could potentially affect the accuracy or representativeness of the findings. Additionally, the study does not measure the long-term effects of relocation projects that might have been designed based on the preferences identified. For example, it does not assess whether implementing the recommended strategies may lead to improved quality of life, better integration into urban fabric, or sustainable community development. Therefore, future research in the field of slum relocation should expand upon this study's findings in several key directions. Firstly, applying similar methodologies across diverse urban contexts would test the generalizability of the findings, potentially revealing universal principles and context-specific factors in slum relocation. This broader application could lead to a more robust global framework for addressing urban housing challenges. Building on this, longitudinal studies tracking the long-term outcomes of relocation projects implemented based on these recommendations would provide crucial insights into the effectiveness and sustainability of these strategies over time. Such studies would validate the approach and allow for iterative improvements in relocation policies. As these longitudinal studies progress, they likely uncover various implementation challenges, such as resource constraints and political dynamics. Thus, a parallel stream of research should investigate strategies to overcome these obstacles, ensuring that theoretical recommendations can be effectively translated into practical solutions. Finally, to support all these research directions and enhance the overall effectiveness of slum relocation projects, exploring the potential of technology and data-driven approaches in monitoring and evaluation would be invaluable. These technological solutions could provide real-time insights, improve decision-making processes, and facilitate more responsive and adaptive management of relocation initiatives, ultimately contributing to more successful and sustainable urban development outcomes.

References

Aalbers CBEM, Kamphorst DA, Langers F (2019) Fourteen local governance initiatives in greenspace in urban areas in the Netherlands. Discourses, success and failure factors, and the perspectives of local authorities. Urban For Urban Green 42:82–99. https://doi.org/10.1016/j.ufug.2019.04.019

Ackermann F, Eden C (2010) Strategic options development and analysis. In: Systems approaches to managing change: a practical guide, pp 135–190. https://doi.org/10.1007/978-1-84882-809-4_4

Ahmed S, Salam M (2022) Rental housing policies and associated legal covers: case of middle income formal housing in Karachi. J Urban Manag 11(4):488–499. https://doi.org/10.1016/j.jum.2022.09.002

Akinyede IJ, Fapohunda JA, Haldwang R (2017) Evaluation of critical factors for achieving quality housing delivery within budget. Int Conf Sustain Infrastruct 3:153–164. https://doi.org/10.1061/9780784481202.015

Arvai JL, Gregory R, McDaniels TL (2001) Testing a structured decision approach: value-focused thinking for deliberative risk communication. Risk Anal 21(6):1065–1076. Portico. https://doi.org/10.1111/0272-4332.216175

Bardhan R, Sarkar S, Jana A, Velaga NR (2015) Mumbai slums since independence: evaluating the policy outcomes. Habitat Int 50:1–11. https://doi.org/10.1016/j.habitatint.2015.07.009

Bender A, Din A, Hoesli M, Brocher S (2000) Environmental preferences of homeowners. J Prop Invest Fin 18(4):445–455. https://doi.org/10.1108/14635780010345391

Boateng GO, Adams EA (2023) A multilevel, multidimensional scale for measuring housing insecurity in slums and informal settlements. Cities 132:104059. https://doi.org/10.1016/j.cities.2022.104059

Brennan MA (2017) Review: decision science for housing and community development: localized and evidence-based responses to distressed housing and blighted communities By Michael P. Johnson, Jeffrey M. Keisler, Senay Solak, David A. Turcotte, Armagan Bayram, and Rachel Bogardus Drew. J Plann Educ Res 38(2):250–252. https://doi.org/10.1177/0739456x17714026

Butala NM, VanRooyen MJ, Patel RB (2010) Improved health outcomes in urban slums through infrastructure upgrading. Soc Sci Med 71(5):935–940. https://doi.org/10.1016/j.socscimed.2010.05.037

Chan APC, Adabre MA (2019) Bridging the gap between sustainable housing and affordable housing: the required critical success criteria (CSC). Build Environ 151:112–125. https://doi.org/10.1016/j.buildenv.2019.01.029

Census of India (2011) Office of the registrar general & census commissioner, India. Retrieved April 25, 2022, from https://www.loc.gov/item/lcwaN0017959

Cochran W (1977) Sampling techniques, 3rd edn. Wiley.

Das M, Das A, Giri B, Sarkar R, Saha S (2021) Habitat vulnerability in slum areas of India—what we learnt from COVID-19? Int J Disaster Risk Reduct 65:102553. https://doi.org/10.1016/j.ijdrr.2021.102553

Doshi S (2012) The politics of the evicted: redevelopment, subjectivity, and difference in Mumbai's slum frontier. Antipode 45(4):844–865. Portico. https://doi.org/10.1111/j.1467-8330.2012.01023.x

Eden C, Ackermann F (2006) Where next for problem structuring methods. J Oper Res Soc 57(7):766–768. https://doi.org/10.1057/palgrave.jors.2602090

Eden C, Ackermann F (2004) Cognitive mapping expert views for policy analysis in the public sector. Eur J Oper Res 152(3):615–630. https://doi.org/10.1016/s0377-2217(03)00061-4

Ezebilo E (2017) Evaluation of affordable housing program in Papua New Guinea: a case of Port Moresby. Buildings 7(3):73. https://doi.org/10.3390/buildings7030073

Fernandes IDS, Ferreira FAF, Bento P, Jalali MS, António NJS (2017) Assessing sustainable development in urban areas using cognitive mapping and MCDA. Int J Sust Dev World 25(3):216–226. https://doi.org/10.1080/13504509.2017.1358221

Françozo R, Paucar-Caceres A, Belderrain MCN (2021) Combining value-focused thinking and soft systems methodology: a systemic framework to structure the planning process at a special educational needs school in Brazil. J Oper Res Soc 73(5):994–1013. https://doi.org/10.1080/01605682.2021.1880298

Fregonese E, Lami IM, Todella E (2020) Aesthetic perspectives in group decision and negotiation practice. Group Decis Negot 29(6):993–1019. https://doi.org/10.1007/s10726-020-09692-0

Galiani S, Gertler PJ, Undurraga R, Cooper R, Martínez S, Ross A (2017) Shelter from the storm: upgrading housing infrastructure in Latin American slums. J Urban Econ 98:187–213. https://doi.org/10.1016/j.jue.2016.11.001

Ghosh S, Roy S, Sanyal MK (2016) Soft system methodology as a tool to understand issues of governmental affordable housing programme of India: a case study approach. J Inst Eng (India): Series A, 97(3):343–354. https://doi.org/10.1007/s40030-016-0177-8

Ghosh T, Kanitkar T, Srikanth R (2023) Affordable and sustainable transportation: key drivers and policy choices for a megacity in India. Case Stud Transp Policy 13:101061. https://doi.org/10.1016/j.cstp.2023.101061

Goulding C, Kelemen M, Kiyomiya T (2018) Community based response to the Japanese tsunami: a bottom-up approach. Eur J Oper Res 268(3):887–903. https://doi.org/10.1016/j.ejor.2017.11.066

Guarnieri P, e Silva LC, Levino NA (2016) Analysis of electronic waste reverse logistics decisions using strategic options development analysis methodology: a Brazilian case. J Clean Prod 133:1105–1117

Hanni CK, Rao KVK (2024) Factors influencing the residential location choice of individuals working at an integrated township. Transp Policy 147:59–69. https://doi.org/10.1016/j.tranpol.2023.12.016

Hjortsø CN (2004) Enhancing public participation in natural resource management using soft OR––an application of strategic option development and analysis in tactical forest planning. Eur J Oper Res 152(3):667–683. https://doi.org/10.1016/s0377-2217(03)00065-1

Johnson MP (2011) Community-based operations research: introduction, theory, and applications. Community-Based Oper Res, 3–36. https://doi.org/10.1007/978-1-4614-0806-2_1

Keeney RL (1996) Value-focused thinking: Identifying decision opportunities and creating alternatives. Eur J Oper Res 92(3):537–549. https://doi.org/10.1016/0377-2217(96)00004-5

Killemsetty N, Johnson M, Patel A (2022) Understanding housing preferences of slum dwellers in India: a community-based operations research approach. Eur J Oper Res 298(2):699–713. https://doi.org/10.1016/j.ejor.2021.06.055

Killemsetty N, Patel A (2022) Slum-dwellers as experts: a problem structuring approach to understand housing challenges in slum communities of India. J Urban Aff 46(5):1020–1038. https://doi.org/10.1080/07352166.2022.2099283

Kissi E, Agyekum K, Baiden BK, Tannor RA, Asamoah GE, Andam ET (2019) Impact of project monitoring and evaluation practices on construction project success criteria in Ghana. Built Environ Proj Asset Manag 9(3):364–382. https://doi.org/10.1108/bepam-11-2018-0135

Kit O, Lüdeke M (2013) Automated detection of slum area change in Hyderabad, India using multitemporal satellite imagery. ISPRS J Photogramm Remote Sens 83:130–137. https://doi.org/10.1016/j.isprsjprs.2013.06.009

Kumar A, Shukla SK (2022) Affordable housing and the urban poor in India. Soc Change 52(1):58–75. https://doi.org/10.1177/00490857211040249

Mahdi Z, Mazumder TN (2023) Re-examining the informal housing problem in Delhi: a wicked problem perspective. Cities 140:104419. https://doi.org/10.1016/j.cities.2023.104419

Makinde OO (2012) Urbanization, housing and environment: megacities of Africa. Int J Dev Sci 3:976–993

Maliene V, Malys N (2009) High-quality housing—a key issue in delivering sustainable communities. Build Environ 44(2):426–430. https://doi.org/10.1016/j.buildenv.2008.04.004

Marttunen M, Lienert J, Belton V (2017) Structuring problems for multi-criteria decision analysis in practice: a literature review of method combinations. Eur J Oper Res 263(1):1–17. https://doi.org/10.1016/j.ejor.2017.04.041

Midgley G, Johnson MP, Chichirau G (2018) What is community operational research? Eur J Oper Res 268(3):771–783. https://doi.org/10.1016/j.ejor.2017.08.014

Mingers J, Rosenhead J (2004) Problem structuring methods in action. Eur J Oper Res 152(3):530–554. https://doi.org/10.1016/s0377-2217(03)00056-0

Mingers J, White L (2010) A review of the recent contribution of systems thinking to operational research and management science. Eur J Oper Res 207(3):1147–1161. https://doi.org/10.1016/j.ejor.2009.12.019

Mulliner E, Smallbone K, Maliene V (2013) An assessment of sustainable housing affordability using a multiple criteria decision making method. Omega 41(2):270–279. https://doi.org/10.1016/j.omega.2012.05.002

Mwiti F, Goulding C (2018) Strategies for community improvement to tackle poverty and gender issues: An ethnography of community based organizations ('Chamas') and women's interventions in the Nairobi slums. Eur J Oper Res 268(3):875–886. https://doi.org/10.1016/j.ejor.2017.12.009

Narayanan S, Rajan AT, Jebaraj P, Elayaraja MS (2017) Delivering basic infrastructure services to the urban poor: a meta-analysis of the effectiveness of bottom-up approaches. Util Policy 44:50–62. https://doi.org/10.1016/j.jup.2017.01.002

Ooi GL, Phua KH (2007) Urbanization and slum formation. J Urban Health 84(S1):27–34. https://doi.org/10.1007/s11524-007-9167-5
Patel A, Koizumi N, Crooks A (2014) Measuring slum severity in Mumbai and Kolkata: a household-based approach. Habitat Int 41:300–306. https://doi.org/10.1016/j.habitatint.2013.09.002
Patel S (2013) Upgrade, rehouse or resettle? An assessment of the Indian government's Basic Services for the Urban Poor (BSUP) programme. Environ Urban 25(1):177–188. https://doi.org/10.1177/0956247812473731
Roy D, Lees MH, Palavalli B, Pfeffer K, Sloot MAP (2014) The emergence of slums: a contemporary view on simulation models. Environ Model Softw 59:76–90. https://doi.org/10.1016/j.envsoft.2014.05.004
Samaddar S, Oteng-Ababio M, Dayour F, Ayaribila A, Obeng FK, Ziem R, Yokomatsu M (2021) Successful community participation in climate change adaptation programs: on whose terms? Environ Manage 67(4):747–762. https://doi.org/10.1007/s00267-020-01421-2
Small A, Wainwright D (2018) Privacy and security of electronic patient records—tailoring multi-methodology to explore the socio-political problems associated with role based access control systems. Eur J Oper Res 265(1):344–360. https://doi.org/10.1016/j.ejor.2017.07.041
Smit B, Wandel J (2006) Adaptation, adaptive capacity and vulnerability. Glob Environ Chang 16(3):282–292. https://doi.org/10.1016/j.gloenvcha.2006.03.008
Smith CM, Shaw D (2019) The characteristics of problem structuring methods: a literature review. Eur J Oper Res 274(2):403–416. https://doi.org/10.1016/j.ejor.2018.05.003
Sterzer L (2017) Does competition in the housing market cause transport poverty? Interrelations of residential location choice and mobility. Eur Transp Res Rev 9(3). https://doi.org/10.1007/s12544-017-0259-3
Stone ME (2006) What is housing affordability? The case for the residual income approach. Hous Policy Debate 17(1):151–184. https://doi.org/10.1080/10511482.2006.9521564
Thomson DR, Stevens FR, Chen R, Yetman G, Sorichetta A, Gaughan AE (2022) Improving the accuracy of gridded population estimates in cities and slums to monitor SDG 11: evidence from a simulation study in Namibia. Land Use Policy 123:106392. https://doi.org/10.1016/j.landusepol.2022.106392
Tsamboulas DA (2007) A tool for prioritizing multinational transport infrastructure investments. Transp Policy 14(1):11–26. https://doi.org/10.1016/j.tranpol.2006.06.001
Verma D, Jana A, Ramamritham K (2019) Transfer learning approach to map urban slums using high and medium resolution satellite imagery. Habitat Int 88:101981. https://doi.org/10.1016/j.habitatint.2019.04.008
Viratkapan V, Perera R (2006) Slum relocation projects in Bangkok: what has contributed to their success or failure? Habitat Int 30(1):157–174. https://doi.org/10.1016/j.habitatint.2004.09.002

From Sustainable to Thrivable Urban Development: Conceptual Frameworks for Designing a Socio-Cultural, Relational, and Educational Ecosystem of Cities

Marina Demchenko, Alexander Laszlo, Marilyn Hamilton, Shweta Srivastav, Ranjani Balasubramanian, Anne-Marie Voorhoeve, and Pieter Wackers

1 Introduction

As we approach mid-century, urban development has become a matter of pivotal importance not only to the future well-being of humanity, but to the interconnected web of life across the entire planet. Megacities belong as much to the Global South as

M. Demchenko (✉)
Living Cities Earth (LCE), Zurich, Switzerland
e-mail: livingcityresearch@gmail.com

A. Laszlo
The Bertalanffy Center for the Study of Systems Science, Buenos Aires, Argentina
e-mail: alexander.laszlo@bcsss.org

M. Hamilton
Integral City Meshworks Inc, Findhorn, Scotland, UK
e-mail: marilyn@integralcity.com

S. Srivastav
The Hague Center for Global Governance, Innovation and Emergence, Vreeland, Netherlands
e-mail: shweta@thehaguecenter.org

R. Balasubramanian
Glocal Colab, Bangalore, India
e-mail: ranjani.bms@gmail.com

A.-M. Voorhoeve · P. Wackers
The Hague Center for Global Governance, Innovation and Emergence, Vreeland, Netherlands
e-mail: anne-marie@thehaguecenter.org

P. Wackers
e-mail: pieter@thehaguecenter.org

M. Firoz C et al. (eds.), *Urban Planning and Design for Megacities in the Global South*, Advances in 21st Century Human Settlements,
https://doi.org/10.1007/978-981-97-8370-0_14

the Global North, and the experiences and insights common to all cities and specific to each stand to greatly enrich efforts to design living cities that balance the ecologies of persons, people, places, and planet throughout the world.

This chapter presents and explores how one such effort seeks to address the challenge of urban development as a matter of living systems innovation. The effort began in 2000 with the creation of Integral City Meshworks, organising the Elab 2012 with an action-based inquiry into the intelligences that constitute an integral city (Hamilton et al. 2013). More recently, it has found expression in such landmark works as "Systemic Innovation for Thrivable Cities" (Laszlo et al. 2021) and related research on the shift "From Sustainability to Thrivability: Transforming Systems with Purpose" (Laszlo et al. 2022).

While this work has many roots with influences from both Western perspectives and Eastern cultures that combine for views on life, community, and development, it also embraces a transcendent unitive narrative of humanity (Currivan 2022; Unitive Narrative 2022a, 2022b; Living Cities Earth Unitive Narrative 2022). As such, it celebrates ways in which built environments offer opportunities for all peoples to express their local genius through full-spectrum humaning. And while we are expanding our insights and experiences to more fully embrace urban development in the Global South, it already provides perspective on more universal patterns of humaning well together for aliveness in megacities across the world. What this means and how it is curated are the focus of the Living Cities Earth (LCE) initiative as a living socio-technical and ecosystemic framework that invites the immersive, experiential, participatory co-creation of urban environments in a praxis of innovation and action.

As a praxis of innovation and action, LCE offers an integral approach to the shaping of a post-postmodern ideology of city development in ways that are inclusive not just of the Global South but, because its axiology is grounded in a unitive narrative, of true planetary thrivability. According to standard usage, an *innovation* is the concretisation of a practical idea that augments human capability for action with societal impact, existing as an intermediate phase between the conceptual *invention* of an idea and its marketable *diffusion* in society. Clearly, advances in science and technology have created unprecedented opportunities for human development and well-being, many of which are showcased in the built landscapes of urban environments. Moreover, the finitude of resources on our planet myopically focuses on extraction, production, distribution, and consumption as solutions to stress and crisis.

The Global South has been a nursery of innovations in urban development as well as a focus of much colonisation. Mahler (2017) suggests that "the Global South captures a deterritorialised geography of capitalism's externalities and serves to account for subjugated peoples within the borders of wealthier countries, such that there exist 'economic Souths' in the geographic north and 'Norths' in the geographic south". It is now incumbent upon all efforts in urban development to consciously, purposefully, and intentionally foster frameworks and practices that ensure pathways to decolonisation as well as to greater integration with and inclusion of nature. Ever since the postcolonial age of modernity, there has been a growing appreciation of the reality of diversity and the need and opportunity to build innovative and resilient systems that are born out of, and cater to, diverse living contexts.

The conceptual framework described here does not pretend to be a panacea for all efforts of urban development. However, as is explored in the proceeding pages, it suggests systemic patterns for human/nature/technology interactions that foster not only dynamics of sustainability and regeneration in urban environments, but extends further to promote ways of being and interbeing that express the capacity for life to celebrate life, for people to meaningfully interact with each other and their surroundings, and for megacities to be greenhouses generating new patterns of aliveness in the Global South, the Global North, and from East to West.

This chapter adopts a qualitative and conceptual research approach rather than a traditional empirical or experimental research methodology. The key characteristics of this qualitative and conceptual research approach are as follows:

- **Conceptual Framework:** This chapter primarily presents a conceptual framework for understanding urban development, systemic innovation, and the Living Cities Earth initiative. It discusses ideas, principles, and perspectives related to these concepts.
- **Case Studies and Examples:** The inclusion of case studies and examples from around the world (e.g. Piplantri Village, Imagine initiative, Ekurhuleni Metropolitan Municipality) exemplifies our qualitative approach. These examples serve to illustrate and contextualise the conceptual framework.
- **Philosophical and Theoretical Reflection:** The chapter engages with philosophical and theoretical reflections on urban development, such as the stages of development (extract, sustain, regenerate, evolve) and the need for decolonisation. It explores the interconnectedness of various dimensions of development, including personal, social, ecological, and cosmic.
- **Narrative and Descriptive Style:** The style of engagement adopted in this chapter is generally narrative and descriptive, conveying ideas through a story-like structure. It emphasises the evolution of cities and the principles behind the Living Cities Earth initiative.
- **Interdisciplinary, Transdisciplinary, and Post-disciplinary Perspectives:** The authors seek to integrate perspectives from various disciplines, including ecology, sociology, and economics, supporting a holistic and systemic approach to urban development. And through the consideration of indigenous and non-Western practices, it suggests the adoption of post-disciplinary perspectives, as well.
- **Absence of Quantitative Data or Experimentation:** This contribution purposefully makes no mention of quantitative data, statistical analysis, or experimentation. Instead, the focus is on qualitative insights, narratives, and the exploration of conceptual frameworks.

In short, the research approach adopted in this chapter is qualitative and conceptual, emphasising the exploration and understanding of complex ideas related to urban development rather than empirical investigation or hypothesis testing.

2 The Evolution of City Impact in Global Sustainable Development

According to the World Urbanization Prospects 2018, published by the United Nations Department of Economic and Social Affairs (2019), the world's urban population is expected to nearly double by 2050. Urbanisation will thus be one of the most transformative trends of the twenty-first century. Populations, economic activities, social and cultural interactions, as well as environmental and humanitarian impacts are increasingly concentrated in cities, and this poses enormous challenges for sustainability in terms of housing, infrastructure, basic services, food security, health, education, decent jobs, safety, and natural resources, among others.

By the end of 2021, over 4.2 billion urban residents, representing 56.6% of the world's population, were living in cities (Total and Urban Population 2023). Today, humanity as a whole and cities as the most concentrated habitat of people are facing a challenge: it is time to reassess the extent of human impact on the Earth and find answers to crucial questions of our time—*how can people become active participants in the regeneration and healing of their habitats, and learn how to co-create conditions that are conducive to life, rather than detrimental to it?*

The urban theme and its role in sustainable development are attracting a growing number of authors. Some talk about eco-cities, eco-settlements, and creative cities; others about aspects of city functioning: renewable energy, transport, and green building; still others write about community development and co-creative leadership. All of these perspectives and viewpoints are undoubtedly important: each reflects the discovery, wisdom, and science of vital niches of urban life. But none of these perspectives offers us a sufficiently capacious, holistic frame of reference that can encompass all models of human systems at the level of comprehensive urban development.

According to Marilyn Hamilton (2008, 2019), in the cities we thought we had tamed, fragmentation and division created disconnected spaces in social sectors that should have naturally functioned as value-adding subsystems and thereby serve as a coherent city system. People have lost contact with the deeply intertwined web of parts of the city system, the development of which now often seems out of control. Indeed, we have created megalopolises with populations of over 20 million that are not only unmanageable or unsustainable, but they have become massive heat emitting furnaces that are changing the global climate and sucking up resources at such a rate that they are destroying the ecology in which they are located (United Nations Human Settlements 2005).

The situation is also exacerbated by the fact that, worldwide, alongside the growth of megacities, small towns are experiencing massive shrinkage in population due to declining birth rates and general population ageing, labour migration to big cities, loss of local identity and social connections, de-industrialisation and economic decline, disparities in health, education and social welfare development, changes in the structure of technological production and agriculture, and other reasons. Given these global demographic changes and the local effects of the many global crises on the global economy, it is likely that the shrinkage of small towns will become even more prevalent in the near future (Großmann et al. 2013).

Moreover, it is necessary to take into account the widespread worldview crisis that impedes the sustainable development of cities and associated ecological systems. Our culturally dominant narrative of 'separation from', 'power over', and 'ownership of' nature creates the mistaken belief that we can manage, technologically innovate and (carbon) trade ourselves out of the evolutionary dead-end we have been heading into for a couple of centuries (Wahl 2021).

As early as the beginning of the twentieth century, Georg Simmel, a prominent German philosopher and sociologist who observed the accelerated industrialisation and urbanisation of European capitals, noted the problems of city dwellers (Simmel 1950). His statements characterise the state of cities at the beginning of the twenty-first century, so they clearly have been somewhat ahead of their time.

According to Simmel, the residents of megacities are affected by negative influences on their worldview, the nature of communication and the quality of their spiritual and emotional life: with a high pace of life, emotional turmoil and atrophy caused by frequent changes, supremacy of material benefit and commercial advantage, objectification of relationships, and devaluation of high ideals. This brings city dwellers to the state of nervousness and mental overload, rational egoism and moral insensitivity, as well as a schematic and formulaic way of life, exacerbating wariness, mistrust and alienation, caused by increasing disconnection with nature, community, neighbours, and finally with their self-identity.

Although we are separated from Simmel by more than a century, contemporary research (Demchenko 2022) only confirms the exacerbation of these trends in the accelerated urbanisation of the world. Nature Deficit Disorder, also known as an 'urban disease', has now been established as a clinical category by the American Psychological Association (Louv 2005). Quite a few of us have turned into overindulged, insensitive bio-robots, often unable to think critically or negotiate, despising each other and desperately suffering from loneliness. Paradoxically, even though we are living closer together than ever before, the population density of modern cities appears to be inversely proportional to the ability to interact and empathise with our neighbours.

In such a situation, how can cities develop enough resilience to thrive in the face of the converging tectonic stresses highlighted by Thomas Homer-Dixon: overpopulation, energy shortages, environmental degradation, climate change, and economic destabilisation (Homer-Dixon 2006)? Today every member of our species must ask themself: *how can we reconnect people and nature; how can we regain our capacity to feel and care; how can we become the embodiment of happiness and healing, rather than manifesting objectification and possessiveness toward the places that we inhabit?*

This is the quest of Living Cities Earth (LCE): *to foster conditions that empower individual and collective humaning well for greater aliveness.* A notion that is central to that quest (as well as an indicator for our success) is the notion of happiness. As discussed below, the highest objective of the integral city model is full-spectrum thrivability. To engage in truly generative dynamics for this to emerge means distinguishing between "happiness of consumption" and "happiness of creation and caring".

"Happiness of consumption", also known as "hedonistic happiness", which can be traced to the ancient Greek philosopher Aristippus (Lampe 2015), reflects the level of citizens' satisfaction with their needs in various areas of urban development; it demonstrates the degree of availability of resources and benefits that give people a subjective sense of happiness and well-being. It is about the pursuit of pleasure, the enjoyment of opportunities and created conditions, so its purpose is to attract and use external resources for one's own goals and interests.

"Happiness of creation and caring", known as "eudaimonic happiness", about which Aristotle wrote in his Nicomachean Ethics (Ross 1925), reflects key value beliefs and the extent to which people realise their creative potential, strive for self-actualisation, and show creative activity for the benefit of themselves, others, their place of living, and the planet as a whole. It is the desire to realise one's true calling and to live according to a higher purpose. Thus, "happiness of creation and care" represents a person's willingness to transform the world around them for the better, to benefit others and leave something meaningful behind for posterity.

3 How Perception of the City Impacts the Role of Residents in Its Development

To reconnect human settlements with nature and increase their happiness and aliveness, it is important first to understand what the city really is, and perhaps, rethink the very notion of this phenomenon.

The dominant modernist perception perpetuated through colonialism is that of the city as *a machine*—a mechanical device that has levers and switches, simplifying manual labour and controlled by humans. This perception of the city detaches people from the structures they have created. This metaphor clearly demonstrates the acute crisis of separateness mentioned earlier, the loss of spirituality, and in general, the limitations of the contemporary perception of the city by its inhabitants (Simmel 1950).

Another similar metaphor for the city is that of the *clockwork*, where the city is described as running like a clock, ticking and humming, being in constant motion, continuously moving forward in time. More intransigent than a machine, once the clock is switched on, it functions independently without outside intervention. This image reflects the competence of its creators and the intelligence of its managers, but it leaves no room for inconsistencies, surprises, and disruptions; it implies that "citizens all march to the same regimented clockwork, which the vicissitudes of daily life in any individual household clearly contradict" (Hamilton 2008, 2019).

Post-colonial thinkers and urbanists, especially from the newly formed independent nation states, attempted to reimagine the city by drawing on local and indigenous worldview and native approaches to the city. As many new nation states built new capitals and cities, the modernist approach began to be debated and alternative imaginations articulated and proposed. Patrick Geddes, a contemporary of Simmel, provided a more applied alternative to the city than Simmel warned against. His work in regions like India and the Middle East further prompted him to consider the regional, cultural, and geographic influences of place on planning cities and he proposed an *ecological approach to city building* (Eisenman and Murray 2017). Geddes, influenced by Herbert Spencer, imagined the city, not just as a biological entity with the family at the centre, but his core approach involved creating inclusive, vibrant, culturally diverse, and ecologically sustainable cities that cater to all individuals, regardless of gender, wealth, and age, while also minimising the disruption of existing infrastructure and avoiding unnecessary expenses (Khan 2011; Welter 2002). Other practitioners like Hasan Fathy in Egypt, Geoffery Bawa in Sri Lanka, and Laurie Baker in India shifted the focus of the city from being a regimented machine to having other qualities—including local traditional knowledge and building techniques, the integration of ecology, and the idea of people as active designers and users of the city.

Another shift in the perspective on the city, its inhabitants, and the very notion of "dwelling" was offered by Martin Heidegger in his essay "Building Dwelling Thinking" (Heidegger 1971). In the meaning of "dweller" and "dwelling" by Heidegger, an important aspect is not simply to exist in a particular location, passively consuming the created goods and using the resources for one's own well-being, but also to participate with care in the development of the habitat and its "earth"—in the "cultivation of the vines". In one of the translations of Heidegger's works, there is a quote that has become a famous aphorism: "to dwell is to garden". In other words, the *city is perceived as a garden and we as its gardeners.*

The Norwegian architect, writer, educator, and architectural theorist Christian Norberg-Schultz went even further in his understanding of urban development and construction with a focus on saving and caring for the land, arguing that the creation of places of residence (locations) should be done through the discovery of the meanings that the environment potentially holds (Norberg-Schulz, 1980). Further to Heidegger's concept, C. Norberg-Schulz refers to the ancient Roman concept of the 'Spirit of Place' (*Genius Loci*)—the intangible essence of a place which determines its uniqueness and position in the general order of things. In order to develop a harmonious relationship with the environment, Genius Loci must be protected and preserved, which according to C. Norberg-Schulz means to concretise its essence in ever new historical contexts (Bazina and Repina 2020). Thus, the city starts to emerge as *a living system, a kind of animated organism*, in which the objective (explicit) infrastructural and architectural development is inseparable from the ideological-subjective (implicit) cultural heritage and historical perspective. Just as consideration of the natural landscape and the characteristics of the local ecosystem is a prerequisite for creating places that are economically and ergonomically comfortable to live in, the 'Spirit of Place' is important for the socio-cultural and psycho-emotional perception of a city by its inhabitants.

An important advantage of perceiving the city as a living organism is that residents in this case are not passive consumers of urban benefits; not soulless "cogs" who ensure the smooth running of the "social apparatus"; not impartial engineers who opportunely maintain the complex urban mechanism. Standing in the shoes of gardeners we become integral life systems in the garden co-existing with the vegetation, the non-human life forms, and the geo-climatic reality.

From this city-as-garden reframe, we can begin to appreciate ourselves as integral organs of the 'living body' of a city. James Lovelock (2009) goes even further by proposing that humans are *Gaia's 'reflective organs'*. Hamilton (2018) takes this proposition to a fractal/holographic refinement, proposing that the *city is the organ, the organisations are organelles and individual humans are cells of these vital organs.* The healthy functioning of each and all such city organs ultimately determines the health of the entire global urban organ system.

Zooming out to a Gaian perspective, the diversity of life and social relationships between people, their collectives, organisations and communities, and the biodiversity of the many other species of living beings that coexist with humans in today's urban areas, we can view *the city as an ecosystem.* By definition, this ecosystem is a self-organising, self-maintaining, self-regenerating, and self-transcending system (Capra 1982).

In this context, humans no longer perceive themselves as something separate from nature and their habitat, but as part of *a biocenosis* (Möbius 1877): a historically evolved dynamic, self-regulating system that is an association of different living organisms (animals, plants, fungi, and microorganisms) which form a closely integrated community and interact with each other, affecting each other's abundance, distribution, adaptation, and existence.

4 The Principles of Regenerativity and Thrivability as the Basis for Ecosystem-Based Urban Development

The transition from a mechanistic to an organic and then to an all-encompassing, ecosystem-based urban development view of the city redefines the place of humans in the urban system. This shift lets us fundamentally reconsider the nature of relationships in society, where the emerging field of regenerative development and design marks a significant evolution in the concept and application of sustainability (Mang and Reed 2012). Such a shift is important not only to reverse the degradation of the earth's natural systems, but also to design human systems that can co-evolve with natural systems: to evolve in a way that ensures mutual benefits and a greater expression of viability, thrivability, and resilience.

To better understand the notion of thrivability and distinguish it from the related concept of flourishing, it is useful to build upon the aforementioned metaphor of the garden. A well-tended and healthy garden is a flourishing domain, and the plants and interdependent subsystems within it are the thriving beings within that domain. A flourishing garden can bring forth thriving plants, but it is also true that thriving plants—when structurally coupled through both auto-catalytic and cross-catalytic feedback cycles—can bring forth a flourishing garden. As such, there is a simultaneity of "bubble up" dynamics and "greenhouse" dynamics: the former express as the upward causation of systemic leverage points among the component parts of the system that can lead to higher levels of synergy, and the latter express as the 'downward causation' of systemic nurturance spaces which elicit higher levels of synergy among the component parts. Flourishing relates to the characteristics of the domain, and thrivability (as with sustainability) relates to the potential realised/ actualised expression of the beings within the domain. As such, it is worth bearing in mind that the term flourish comes from the Latin *florere* meaning to flower or blossom. By contrast, thrive comes from the Norse term *thrífask* meaning to grasp or get hold of. The agency involved in thriving connotes taking an active stand in creating conditions of well-being for oneself and for others in order that everyone, and everything, may flourish. A city is a systemic milieu in which both thrivability and flourishing may take place.

All of humanity's successes and failures in co-creating regenerative futures will depend on this ability to 'take an active stand' and realign with the syntropic[1] evolutionary patterns of life itself—*moving from competitive scarcity toward collaborative abundance*. This requires a comprehensive study of the biosphere and the cultural specificity of each particular area: each community, each ecosystem, and each bioregion (Vilhena and Antonelli 2015).

Paradoxically, it seems that the path to a global regenerative human impact on the Earth lies precisely through local and regional regeneration. It is not a global problem-solving task, but rather an ongoing collective learning and capacity-building process of people and communities in place. Regenerative development does not solve global problems by scaling up universally applicable solutions: rather, the focus is on increasing our capacity for local co-creation and constant transformation in response to changing circumstances (Wahl 2021).

This framing, this understanding, and this perspective on the need for healthy, life-affirming, and future-creating relationships in the constructed environments where human populations live in greatest density are precisely why we have established the Living Cities Earth (LCE) initiative. We invited 100 co-founders from around the world '*to help 10,000 cities to human well… together… for aliveness*'. We are mapping and sharing the ways that LCE can "think, link and act" to build regeneration into cities in individuals, collectives, and collaborations so that we strategically regenerate, evolve, and transcend cities across multiple generations and perspectives in alignment with the pulse of life as expressed in their local ecoregions.

Participation and inclusion are fundamental prerequisites for the emergence of diverse regenerative cultures everywhere. Co-creating a thrivable, regenerative future is about supporting persons, people, places, and cultures to express their unique contribution to the health and vitality of the nested complexity in which we are embedded. To do so simultaneously serves ourselves, our communities, and life on the planet as a whole. After all, what we do at the local level has both regional and

---

[1] Syntropy describes the process of negentropy-importation. In thermodynamics, a syntropic system is a dissipative structure.

global implications. Just as the health of our bodies (individually and collectively) depends on healthy cells and organs, the health of the planet depends on healthy communities, places, ecosystems, and bioregions.

Thus, if we are to evolve thrivable, living cities, we must inquire: *how can we holistically consider the local cultural-historical, natural-climatic, socio-psychological and politico-economic development of each individual territory? How can we bring the opinions, personal energy and creative potential of its many inhabitants into the process? How can we involve them in the active preservation, rehabilitation, regeneration, and creative evolution of the entire bioregion and its ecosystems?*

5 Methods of Decolonisation and Ecosystemic Approaches to Urban Development—Examples from the Global South

As mentioned previously, the Global South has been a nursery of innovations in urban development. Many of the systemic innovations are rooted in a practice of *decolonisation*. Seen from this lens, the practice of urban sustainability innovations can also be understood as one of decolonising. A regenerative approach to urban development as such, like decolonisation, is a pursuit of changing relationships—changing the relationships we have with ourselves, our communities, our cities, ecosystems, and the world at large. In many ways, these shifts in relationships required for realising the ideals and essence of a regenerative approach are the very shifts that come from a decolonising practice (Srivastav and Balasubramanian 2022).

Colonisation practices depend on the objectification of the colony: they depend on a worldview where others and Nature are seen to be outside the coloniser. This perception is what enables the "othering" stance of objectification.

As a bridge to transcend this colonial pattern of othering, the frame of thrivability evokes exploration of, and engagement with, the way in which learning, playing, talking, working, and all aspects of life-as-art connect us to ourselves, to each other, to our more-than-human world, to past and future generations of all beings, and to the deeper dimension of the cosmos that patterns the evolutionary dynamics of emergence and coherence. These are the five intertwingled dimensions of systemic thrivability. Systemic innovation engages evolutionary systems designers across all five dimensions through dynamics that encourage consciously connecting, intertwingling, and cultivating the bigger story of our individual and collective being and becoming. The quality and character of this story depend on the way in which leaders of systemic innovation seek to author their life interactively along these five dimensions. Since these dimensions have been articulated and explored in detail elsewhere in the literature (e.g. Laszlo 2018, 2020), suffice it to list them here:

- The intra-personal dimension of coherence; thrivability within oneself;
- The inter-personal dimension of coherence; thrivability with one's communities and social systems;

- The trans-species dimension of coherence; thrivability with the more-than-human world;
- The trans-generational dimension of coherence; thrivability with past and future generations of all beings;
- The pan-cosmic dimension of coherence; thrivability with the deep dimension of immanent consciousness in the cosmos.

This phenomenon is also addressed by Marilyn Hamilton (2008, 2019) in the '4Ps' concept (Person, People, Place, Planet): like naturally expanding circles of care, persons first begin to make efforts for their own and their families' well-being, then they come together in organisations and communities to provide support for each other (as people), then these (fractal/holographic) structures begin to transform their territories (places), which ultimately ensure prosperity for the whole planet (Figs. 1 and 2). It is important to underline that (paradoxically) *caring, as the highest creative expression of humanity, is spreading out emergently,* just as uncontrolled consumption is now depleting resources—first personal, then public, regional, and eventually planetary.

The perception of the world as living, systemic, and interdependent, in accordance with a holistic paradigm of thought, can change the course of human evolutionary development, making our lives and activities more conscious, ecological, humane, and undoubtedly happier (Krippner 1991). *It is intended to acknowledge, transform and replace brutal industrialisation and its competitive mindset, with a considerate, caring, subtle attitude toward the rational and harmonious development of individuals and society as a whole.*

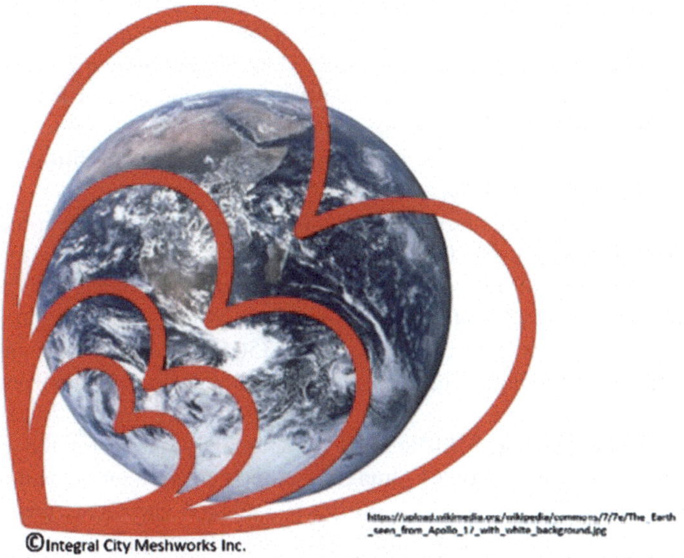

Fig. 1 Four circles of care. Reproduced from Hamilton (2023)

Fig. 2 "4Ps" concept—the four circles of care in social holarchical systems. Reproduced from Demchenko (2022)

The '4Ps' are an articulation of a diverse ecosystem of approaches and ideas that are mushrooming around the world ranging from small, localised initiatives to larger city level plans. These examples from around the world illustrate some of these ideas that could inspire a holistic and multi-dimensional approach to city making.

We have categorised these approaches in the 4P-framework to develop a glossary of initiatives that illustrate the principles of Living Cities Earth. Of course, not all projects will fit the categories neatly, yet the ones offered below help portray the kind of alignment that is possible with the principles of integral city development. Please note that it is not an attempt to appropriate any of these initiatives.

Piplantri Village: Rajasthan, India (Regeneration—Person + People + Place + Planet). Piplantri, a village situated in the Rajsamand district of Rajasthan, was once afflicted by deforested land due to marble mining, depleted vegetation, and gender inequality. Like many traditional Indian villages, daughters were often considered less valuable than sons. However, Piplantri has now gained worldwide acclaim for its extraordinary initiative, where 111 trees are planted each time a girl child is born.

This remarkable transformation was initiated by Shyam Sundar Paliwal, a recipient of the Padma Shri award, who was devastated by the loss of his 17-year-old daughter due to dehydration. To honour her memory, he planted a tree and inspired fellow villagers to do the same for their daughters. Thus, the tradition of planting trees for newborn girls took root.

On the occasion of a girl's birth, the parents and the entire village come together to raise a Fixed Deposit of ₹31,000 ($450) dedicated to supporting the girl's future endeavours. As a result of this collective effort, Piplantri has managed to rejuvenate its land and even generate additional income through the forest produce. Today, the village boasts an impressive count of over 350,000 trees, comprising various indigenous species, making a significant positive impact on the environment and climate.

Imagine: India, Africa, Middle East (A Global Initiative for the Empowerment of Women: Person + People). 'Imagine' empowers women in challenging life circumstances in developing world countries to envision and create new possibilities for their lives through participating in a four-day training programme called the "Empowerment Workshop". This training programme, developed by David Gershon and Gail Straub—pioneers of the field of empowerment and founders and co-directors of Empowerment Institute's School for Transformative Social Change—has successfully been applied over the past forty years in over forty countries worldwide.

This programme addresses the part in women's empowerment that is often missing—agency or self-efficacy—which enables the traditional investments in outer resources such as education, health or income generation to be more effective.

This agency is achieved through the application of the empowerment methodology, which helps individuals increase their self-knowledge so they can discover what's important to them; translate this knowledge into a compelling vision; identify and transform the limiting beliefs that inevitably arise when creating something new; and adopt an actionable growth strategy to attain their goals. 'Imagine' is being delivered by 39 partner organisations certified by the Empowerment Institute across ten countries: Kenya, Nigeria, South Africa, Rwanda, Ghana, Egypt, Morocco, India, and Afghanistan. Over 30,000 women have participated in these workshops who have impacted over 7.5 million people throughout their communities.

Ekurhuleni 2055: Growth & Development Strategy, South Africa (Vision, Purpose and Strategy: Person + People + Place). Ekurhuleni Metropolitan Municipality (EMM) adopted its Growth and Development Strategy (GDS) 2055 in 2005. This was the first of its kind after the amalgamation of the former nine East Rand towns into Ekurhuleni as a Metropolitan Municipality. The main aims of the GDS 2055 were:

- To provide a framework and a point of reference for all the EMM's plans, policies, and strategies—Integral City contributed analysis of input in terms of Person, People, Place;
- To build a common vision and purpose between government, private sector and civil society; Integral City contributed analysis using 4 Voices of the Integral City (citizens, civil society, civic managers, business/innovators).

GDS 2055 was intended to be a road map that guides the Metro's actions for the next 20+ years. It not only reflected the Metro's strategic thinking, but also informed policy and delivery programmes. The adopted strategy was defined in the context of three landscapes (physical development, economic development, and social development), and these three informed the strategic thrust for the ensuing years.

Sustainability Institute: Lynedoch Eco Village, South Africa (Education for Regenerative Communities: Person + Place + Planet). The Sustainability Institute is a renowned educational and research centre based in South Africa, dedicated to promoting sustainable living and fostering positive change in society. Established in 1999 and located in Stellenbosch, the institute operates on the principles of ecological, social, and economic sustainability. It offers innovative programmes and courses, focusing on sustainable development, permaculture, and community engagement. The institute's curriculum emphasises a holistic approach, integrating environmental stewardship, social justice, and ethical practices. Through collaborative efforts with local communities, the Sustainability Institute conducts research and implements practical solutions to address pressing global challenges, including climate change and inequality. By nurturing future leaders and encouraging responsible action, the institute aims to create a more equitable and sustainable world for current and future generations.

Regenerate Costa Rica Initiative: Costa Rica (Collaborative Network: Place + Planet). The Regenerate Costa Rica Initiative is taking shape as regenerative agriculture gains momentum in Costa Rica. This national initiative aims to establish a comprehensive network that facilitates the exchange of skills, knowledge, and solutions. By collaborating with existing projects, the initiative strives to unite and enhance regeneration capabilities throughout the country, granting communities access to the ecological, economic, and cultural benefits of regenerative development. Key focus areas include bolstering educational infrastructure and creating a network of learning sites to foster the sharing of vital knowledge and skills. Spearheaded by Dr. Eduard Müller from the University for International Cooperation, the initiative joins forces with esteemed partners like Capital Institute, Savory Institute, Natural Capitalism Solutions, Economy for the Common Good, and Gaia Education to advance the cause of sustainability and regeneration in Costa Rica.

Encouraging the Heart of Leon: Leon, Mexico (Civil Society Integrates 4 Voices: People + Place). Marilyn Hamilton (2012) had the opportunity to explore with the Leon Mexico Citizens' Observatory, how their Four Voices (see Fig. 3) could work together to improve economic, environmental, and social conditions. A full case study is available on Integral City Meshworks (Hamilton 2012).

Like the gardeners interacting with their garden, the Four Voices discovered they could create a "heart wave" of intelligences by bringing the separate voices together.

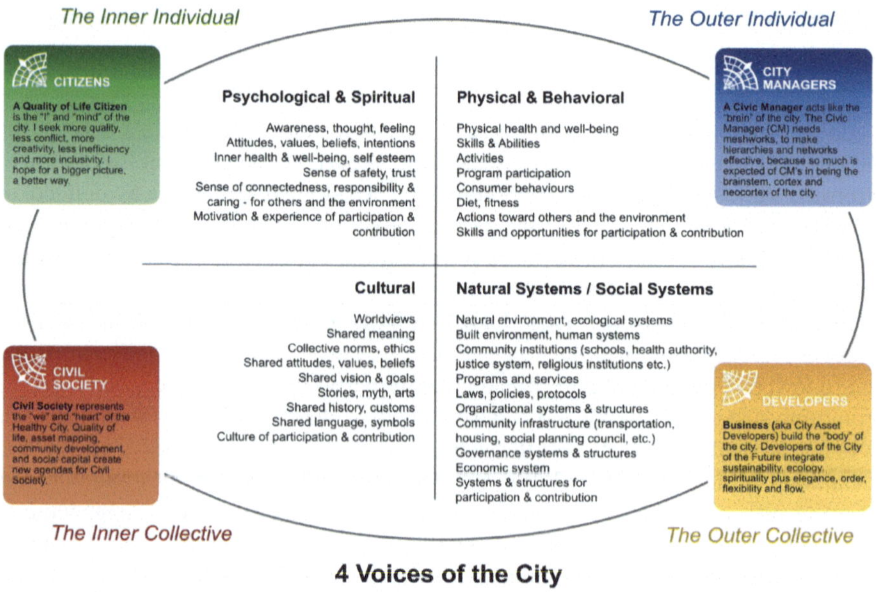

Fig. 3 Four voices of the city

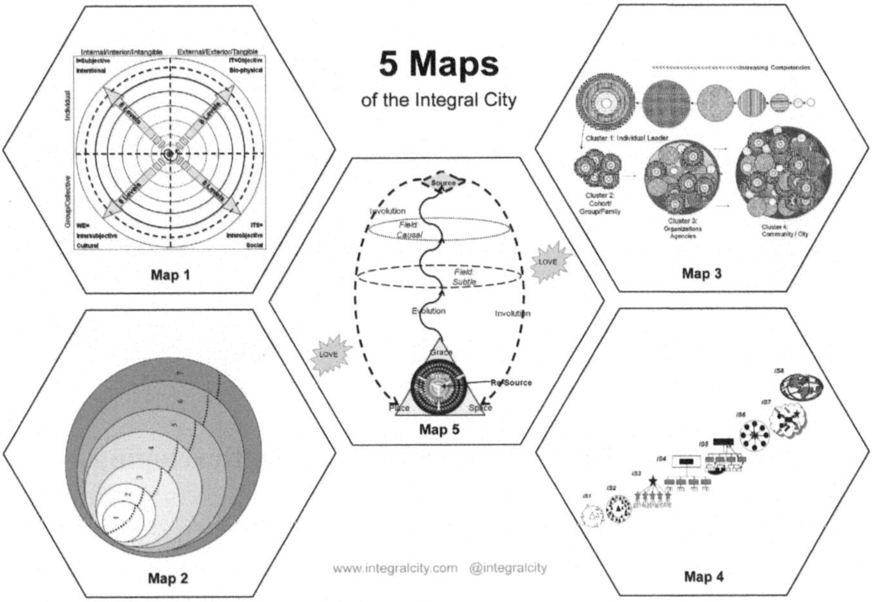

Fig. 4 Five maps of the integral city

They worked with the universities, student interns, and many of Leon's Civil Society organisations, to build bridges amongst the civic manager education, health, business, and social services sectors.

Using Integral City's 5 maps (see Fig. 4) in a discovery process, the Citizen's Observatory was able to map worldviews (Map 1), the holographic nest of city groups that individuals belonged to (Map 2) and working groups who resonated strong caring capacities (Map 3). They realised that they had representations of every cultural-historical form of organisation (Map 4), and they discovered the power of Civil Society to enable caring connections (Map 5).

In working with the Four Voices of Leon, they discovered the power of the collective 'We' to strengthen bridges between people like the rhythms and pulses of a living system. They learned that "in order to improve the health of a system, connect it to more of itself" (Wheatley 1999). In retrospect, this interaction revealed how Leon naturally applied the 12 Intelligences of the city that are embedded in the Integral City Global Positioning System (GPS) Tool (see Fig. 5).

Fig. 5 Global Positioning System (GPS)

6 Implications for Urban Development in the Developing World

The approaches considered in this chapter have both positive and negative aspects with regard to urban development. The positive aspects are those which clearly augment and empower urban development in the developing world without much risk of undesired secondary or tertiary consequences. The negative aspects are those which entail possible risk or undesired impacts if not engaged with from a holistic/systemic perspective. As such, it is important for any consideration of their application that a full assessment of both aspects is considered.

Consideration of positive potentials:

- **Thrivable and Regenerative Cities:** This chapter advocates for a shift towards "thrivability" and regeneration in urban development. The implication explored is that it is possible to engage in a practical focus on creating cities that go beyond sustainability to actively contribute to the well-being of both inhabitants and the environment.
- **Decolonisation and Inclusion:** Emphasising decolonisation in urban development suggests a commitment to addressing historical inequalities and ensuring inclusive practices. This could lead to more equitable development strategies.
- **Systemic and Interconnected Approach:** The conceptual framework emphasises a holistic and systemic view of cities, considering personal, social, ecological, and cosmic dimensions. By following this approach, it is possible to engage in integrated urban planning that accounts for the complex interdependencies within cities.
- **Cultural and Contextual Sensitivity:** The inclusion of case studies from the Global South suggests a recognition of the importance of cultural and contextual sensitivity in urban development. This could lead to solutions that are more attuned to the specific needs of diverse communities.
- **Empowerment of Women:** Initiatives like "Imagine", which empower women in challenging circumstances, can contribute to social and economic development.

Empowering women often have ripple effects on communities and can lead to more resilient societies.
- **Regenerative Agriculture:** The focus on regenerative agriculture in initiatives like the Regenerate Costa Rica Initiative can have positive implications for food security, environmental sustainability, and economic development in developing world contexts.

Consideration of negative potentials:

- **Complexity and Implementation Challenges:** The holistic and interconnected approach outlined in the conceptual framework may not always prove easy to implement. The complexity of considering multiple dimensions makes the creation of actionable plans challenging. However, the LCE insistence on leveraging the power of collective intelligence mitigates these challenges.
- **Resource Intensity:** Implementing regenerative practices and holistic development may require significant resources. Megacities in the Global South often face resource constraints, and therefore caution needs to be exercised so as to avoid engaging in the proposed approach in ways that may be overly resource-intensive. Again, by drawing on a globally interconnected community of urban development initiatives, the LCE approach mitigates against this challenge, as well.
- **Cultural Adaptation:** While cultural sensitivity is emphasised, the universal applicability of certain concepts will require adaptation to fit the specific cultural, social, and economic contexts of different cities in the developing world.
- **Balancing Tradition and Innovation:** In the Global South, there is often a delicate balance between preserving traditional practices and embracing innovation. Striking the right balance is crucial to avoid the unintended consequences of rapid urbanisation.
- **Capacity Building and Education:** The success of these initiatives relies on the capacity of communities to understand and implement them. This highlights the need for education and capacity-building efforts, which might face challenges in resource-limited settings. Fortunately, the emergence of such global action ecosystems, as represented by LCE, goes a long way towards generating and distributing the needed lifewide learning approaches for addressing this challenge.
- **Inequality and Social Fragmentation:** Urban development initiatives need to be carefully designed to avoid exacerbating existing inequalities. As this chapter highlights, without careful planning, certain groups may be marginalised, leading to social fragmentation.

Overall, the implications of addressing urban development challenges through systemic innovation considered in this chapter suggest a potentially transformative approach for the developing world. However, the successful application of these concepts requires careful consideration of local contexts, resource constraints, and the potential for unintended consequences.

7 Conclusions

This chapter has showcased the approach of Living Cities Earth revealing the generation of cities as systemic nurturance spaces that bring aliveness to all their denizens. Indeed, this approach includes and transcends regeneration since it creates conditions for evolutionary emergence, identifying distinct phases in the culture of urban development as it has evolved over the years. These can be charted in terms of the worldviews related to nature and the living systems which comprise cities. These positions have emphasised the following objectives at different times in urban development:

- Extract: Deplete I Deforest I Degrade (1760–1987)
- Sustain: Reduce I Re-use I Recycle (1987–2010)
- Regenerate: Rethink I Restore I Replenish (2010–2020)
- Evolve: Transform I Transmute I Transcend (2020+)

It is the last of these stages that captures the efforts of Living Cities Earth to create conditions that invite the emergence of new ways of being and interbeing, understanding the city as a living whole, akin to an organism. When curated and cared for in this way, the city emerges as an ecosystem of thrivability that engenders qualities of care, life, empathy, relationships, interdependence, interbeing, vitality, and aliveness. These characteristics allow the city to evolve in ways that express ever greater coherence, synergy, and happiness both within itself and in relation to its living bioregion.

Finally, with a focus on urban development patterns in the Global South, this chapter has emphasised how happiness and thrivability are implicit in the dynamics of decolonisation. Decolonisation as a process and praxis is important for humanity to address in and of itself, and is essential to all engagements with urban development, no matter where it occurs because it represents a key need in the framing of planetary thrivability. When deterritorialised, the Global South expresses itself in areas of the Global North with aspects of the Global South appearing in all cosmopolitan cities.

This chapter has pointed to the need to understand the city as an integral evolutionary holon in the interdependent ecosystem of human civilisation and nature. To best address this need, Living Cities Earth uses an approach that offers a coherent and pragmatic conceptual framework for designing a socio-cultural, relational, and educational ecosystem of cities based on an axiology of panhumanism grounded in a unitive narrative of life.

References

Bazina A, Repina E (2020) Ponimanie arhitecturi cherez phenomenologiyu M.Heideggera (Understanding architecture through M. Heidegger's phenomenology). Urban Plann Arch. Section: Theory and History of Architecture, Restoration and Reconstruction of Historical and Architectural Heritage 10(4):85–90. https://journals.eco-vector.com/2542-0151/article/view/62693/45674

Capra F (1982) The turning point: science, society and the rising culture. Simon & Schuster, New York

Currivan J (2022) The story of Gaia: the big breath and the evolutionary journey of our conscious planet. Inner Traditions, Rochester, Vermont

Demchenko M (2022) Philosophia ustoychevogo razvitiya gorodov (The philosophy of sustainable urban development). Paper presented at the 9th Living Cities Forum, Moscow, Russia, 16–17 December 2022

Eisenman TS, Murray T (2017) An integral lens on Patrick Geddes. Landsc Urban Plan 166:43–54. https://doi.org/10.1016/j.landurbplan.2017.05.011

Großmann K et al (2013) Shrinking cities: notes for the further research agenda. Cities 35:221–225. https://doi.org/10.1016/j.cities.2013.07.007

Hamilton M (2008, 2019) Integral city: evolutionary intelligences for the human hive. Amaranth Press, LLC, Minneapolis, MN

Hamilton M (2012) Creating intelligent alive city: creating courageous futures for Leon Mexico. https://drive.google.com/file/d/1cvDCV_VL9hjtZ-ga4r_y54MaGjj59haS/view. Accessed 1 July 2023

Hamilton M, Sanders B (2013) Integral city 2.0 online conference 2012 proceedings: a radically optimistic inquiry into operating system 2.0. In: M. Hamilton (ed). http://www.scribd.com/doc/120713339/Integral-City-2-0-Online-Conference-2012-A-Radically-Optimistic-Inquiry-into-Operating-System-2-0

Hamilton M et al (2013) Integral City 2.0 online conference 2012 appendices: a radically optimistic inquiry into operating system 2.0—36 interviews. In: M. Hamilton (ed). http://www.scribd.com/doc/123005653/Integral-City-2-0-Online-Conference-2012-Appendices-A-Radically-Optimistic-Inquiry-into-Operating-System-2-0-36-Interviews

Hamilton M (2018) Integral city 3.7: reframing complex challenges for Gaia's human hives. Amaranth Press, LLC, Minneapolis, MN

Hamilton M (2023) Waking up the human hive: recalibrating with happiness, published in equinotes: roots of city peace in joy, happiness and care (March 2023). https://integralcity.com/2023/03/19/march-2023-equi-notes-roots-of-city-peace-in-joy-happiness-and-care/

Heidegger M (1971) Building dwelling thinking (from "poetry, language, thought"), (trans: Hofstadter A), Harper Colophon Books, New York

Homer-Dixon T (2006) The upside of down: catastrophe, creativity, and the renewal of civilization. Alfred A. Knopf, Toronto

Khan N (2011) Geddes in India: Town Planning, Plant Sentience, and Cooperative Evolution. Environ Plan D: Soc Space, 29(5):840–856.

Krippner S (1991) The holistic paradigm. World Futur J Gen Evol 30(3):133–140. https://doi.org/10.1080/02604027.1991.9972203

Lampe K (2015) The birth of hedonism. The Cyrenaic philosophers and pleasure as a way of life. Princeton University Press

Laszlo A (2018) Leadership and systemic innovation: socio-technical systems, ecological systems, and evolutionary systems design. In: Ruzzeddu M and Ferone E (ed) International review of sociology, Special Issue on "Governance of Innovation Processes: A Sociological Approach", vol 28, no 3, pp 1–12

Laszlo A (2020) Practices that Ensoul the cosmos: expressions of connectedness, on Medium.com, 27 May. https://medium.com/@Alex8er/practices-that-ensoul-the-cosmos-f36f4217f00b. Accessed 15 July 2023

Laszlo A, Huber-Heim K, Blachfellner S, Luksha P (2021) Systemic innovation for thrivable cities. In: Brears RC (ed) The Palgrave encyclopedia of urban and regional futures. Springer Nature, Switzerland, pp 1–20. https://doi.org/10.1007/978-3-030-51812-7_212-1

Laszlo A, Huber-Heim K, Blachfellner S (2022) From sustainability to thrivability: transforming systems with purpose. In: Ladeja Godina Košir (ed) The Kyoto Post-COVID manifesto for global economics: confronting our shattered society. Springer, Singapore, pp 305–316. https://doi.org/10.1007/978-981-16-8566-8_18

Living Cities Earth Unitive Narrative (2022) https://www.livingcities.earth/unitivenarrative. Accessed 1 July 2023
Louv R (2005) The last child in the woods: saving our children from nature-deficit disorder. Algonquin Books, Chapel Hill
Lovelock J (2009) The vanishing face of Gaia. Harmony Books, New York
Mahler AG (2017) Global south. In: O'Brien E (ed) Oxford bibliographies in literary and critical theory. Oxford Press, New York. https://doi.org/10.1093/obo/9780190221911-0055
Mang P, Reed B (2012) Regenerative development and design. In: Meyers RA (ed) Encyclopedia of sustainability science and technology. Springer, New York, NY. https://doi.org/10.1007/978-1-4419-0851-3_303
Möbius K (1877) Die Auster und die Austernwirtschaft. Verlag von Wiegandt, Hemple & Parey, Berlin. English translation: The Oyster and Oyster Farming (1880) U.S. Commission Fish and Fisheries Report, pp 683–751
Norberg-Schulz C (1980) Genius loci: towards a phenomenology of architecture. Rizzoli, NY
Ross D (1925) The Nicomachean ethics of Aristotle. Oxford University Press, Oxford
Simmel G (1950) The metropolis and mental life. In: Wolff K (ed) The sociology of Georg Simmel. Free Press, New York, pp 409–424
Srivastav S, Balasubramanian R (2022) Sustainability—a decolonizing practice. Seminar Magazine 750:70–74
Total and Urban Population (2023) UNCTAD handbook of statistics 2022 https://hbs.unctad.org/total-and-urban-population/#:~:text=The%20share%20of%20urban%20population,minority%20(35.3%20per%20cent). Accessed 20 June 2023
United Nations Human Settlements Programme 2005. Annual Report. (2006). https://reliefweb.int/report/world/united-nations-human-settlements-programme-2005-annual-report. Accessed 14 Apr 2023
Unitive Narrative (2022a) https://www.evolutionaryleaders.net/unitivenarrative. Accessed 13 May 2023
Unitive Narrative NGO UN Thematic Cluster (2022b) https://sdgthoughtleaderscircle.org/ngo-unitive-thematic-cluster/. Accessed 14 May 2023
Vilhena D, Antonelli A (2015) A network approach for identifying and delimiting biogeographical regions. Nat Commun 6(7848). https://www.nature.com/articles/ncomms7848. Accessed 15 May 2023
Wahl DC (2021) Regenerative action: constructing a regenerative future. https://www.thersa.org/comment/2021/11/constructing-a-regenerative-future. Accessed 20 Apr 2023
Welter V M (2002) *Biopolis: Patrick Geddes and the City of Life.* MIT Press, Cambridge, MA
Wheatley M (1999) Bringing schools back to life: schools as living systems in creating successful school systems: voices from the university, the field, and the community. Christopher-Gordon Publishers, September 1999. https://www.margaretwheatley.com/articles/lifetoschools.html
World Urbanization Prospects 2018: Highlights (2019) United Nations Department of Economic and Social Affairs. United Nations publication. Sales No. E19.XIII.6. New York. https://population.un.org/wup/publications/Files/WUP2018-Highlights.pdf. Accessed 15 July 2023

Exploring Community Vitality: Insights of Ethnic Migrants in the Indian Context

Purbita Samanta, Subrata Chattopadhyay, and Haimanti Banerji

1 Introduction

Urban vitality is an essential tool to identify the relationship between urban space, society, and economy (Wu et al. 2018b; Xia et al. 2020; Li et al. 2022). Jane Jacobs (1961) introduced the idea of urban vitality to explain why certain communities have high levels of well-being and continue to develop while others decline or even disintegrate. The recent researches on urban vitality evolved due to increasing urban sprawl, decline, and city residential vacancy (Jin et al. 2017; Yue et al. 2021). According to Wu et al. (2018a), urban vitality arises from the activities of residents and their interactions within the spatial units of a city. Accordingly, it can be argued that while the urban environment supports urban life and works as an intermediary for urban vitality, individuals are ultimately responsible for creating and sustaining urban vitality (Ou et al. 2021). Urban vibrancy is characterised by its dynamic and energetic quality, which is important for fostering sustainable development and has significant effects on people's well-being (Liu et al. 2022a). The United Nations has emphasised the significance of integrating resilience, security, vitality, and inclusion in the creation of sustainable cities through the New Urban Agenda and Sustainable Development Goal (SDG 11). The New Urban Agenda recognises that cities should be resilient to withstand challenges and shocks, ensure safety and well-being, foster liveliness and dynamism, and inclusive to promote equal opportunities for all residents. Similarly,

P. Samanta (✉) · S. Chattopadhyay · H. Banerji
Indian Institute of Technology Kharagpur, Kharagpur, West Bengal 721302, India
e-mail: purbita95@kgpian.iitkgp.ac.in

S. Chattopadhyay
e-mail: schat@arp.iitkgp.ac.in

H. Banerji
e-mail: haimanti@arp.iitkgp.ac.in

M. Firoz C et al. (eds.), *Urban Planning and Design for Megacities in the Global South*, Advances in 21st Century Human Settlements,
https://doi.org/10.1007/978-981-97-8370-0_15

SDG 11, titled "Sustainable Cities and Communities", highlights the need for cities to be resilient, inclusive, safe, and sustainable.

The literature on urban vitality highlights two distinct principles of research (Yue et al. 2021). The first group examines the various planning principles promoting vitality based on Jacobs' interpretive framework (Delclòs-Alió and Miralles-Guasch 2018; Delclòs Alió et al. 2019; Fuentes et al. 2020; Huang et al. 2022a; Paköz et al. 2022). The second stream looks at urban vitality from the perspective of urban form (Zumelzu and Barrientos-Trinanes 2019; Tu et al. 2020; Lang et al. 2022; Liu and Shi 2022; Xia et al. 2022; Yang et al. 2023). Studies on urban vitality have focused mostly on the influence of built environment factors such as FAR, building form, building density, block size, and intersection density, on urban vitality at various spatial context like cities, districts, neighbourhoods, and streets (Ozbil et al. 2011; Ye et al. 2018; Lu et al. 2019; Li et al. 2020; Yang et al. 2021; Chen et al. 2022; Lang et al. 2022; Yu et al. 2022; Zou et al. 2023; Lin et al. 2023; Wang et al. 2023). Multiple studies have assessed vitality in specific contexts such as vitality of traditional villages (Liu et al. 2022b), vitality of neighbourhood parks and squares (Banchiero et al. 2020; Wang et al. 2022b), vitality of complex ecosystems (Gong et al. 2023), vitality of underground space in metro areas (Dong et al. 2021; Xu and Chen 2021), and vitality of coastal zones (Huang et al. 2022b; Yin et al. 2022). However, the broad literature on urban vitality often falls short in adequately addressing the issue of ethnic migrant communities' vitality within urban contexts. While concepts of urban vitality may touch on aspects such as social diversity and community engagement, the specific challenges and dynamics faced by ethnic migrant communities are not given sufficient attention.

India is a highly diverse country with a multitude of ethnic groups coexisting across its vast expanse. The diasporic character of Indian cities is greatly influenced by both internal migration and external migration. Ethnic migration in India occurs for various reasons, and individuals from different ethnic backgrounds migrate for opportunities, education, and other factors. Previous studies have explored the community resilience based on psychological and socio-cultural factors (Kuttikat et al. 2018), ethnic social structure in urban spaces (Subba 2016, 2018; Goswami and Malik 2019; Roy 2023), and ethnic food safety knowledge (Ghatak and Chatterjee 2018). However, there is a lack of scholarly literature on the quantitative analysis of different factors that contributes to the vitality of ethnic communities using the communities' perception. Therefore, this chapter aims to examine the vitality factors of ethnic migrant communities in the neighbourhoods of Indian cities. The concept of urban vitality in this chapter is used to understand why some ethnic migrant communities continue to thrive and have high levels of well-being while others experience decline. The study is grounded in the fieldwork undertaken among the Nepali and Telugu communities in Kharagpur during August 2022 and January 2023. Kharagpur is a city located in the Paschim Medinipur district of the state of West Bengal, India. While Kharagpur is predominantly inhabited by Bengali-speaking people (Registrar General-India 2011), there is also a significant presence of Telugu-speaking community in the city. The Telugu community in Kharagpur consists of people who have migrated from the Telugu-speaking states of India, Andhra Pradesh, and Telangana. On the other hand, Nepalis in India have a long-standing historical connection, primarily due to the close geographical proximity between Nepal and the Indian states bordering it. The study explores their vitality factors while residing with the ethnic majority and host population, the Bengalis, through primary surveys. This research employs factor analysis to comprehend the variables that contribute to the vitality of the Nepali and Telugu community in Kharagpur.

The rest of the chapter is organised as follows: The literature review is included in Sect. 2. The methods used are explained in Sect. 3. Section 4 presents the findings and a discussion. In Sect. 5, the conclusion is reported.

2 Literature Review

Ethnic migrants face several challenges that affect their integration and well-being. These challenges encompass limited access to the job market, job insecurity, and the potential for exploitation (Zhuang and Lok 2023). Furthermore, ethnic migrants often encounter barriers in accessing culturally-appropriate health and settlement services. Additionally, there is a lack of sufficient ethnic support networks, further complicating their settlement and integration experiences (El-Bialy and Mulay 2015). These challenges collectively underscore the importance of understanding and addressing the specific needs of ethnic migrant communities to foster a more inclusive and supportive urban environment.

Ethnic minorities and migrant communities tend to concentrate in places where specific cultural amenities, such as religious and commercial services, are preserved (Jin et al. 2021). These places are termed as ethnic enclaves. The ethnic enclaves are considered as an initial stage in the settlement process for migrants (Massey and Denton 1985). In cities, different ethnic groups have different spatial clustering patterns. Residential segregation is a form of spatially expressed socio-economic inequality (Křížková and Šimon 2022). Migrants are spatially concentrated in ways that may differ from patterns that formed before the earlier settlers as neighbourhoods become more densely inhabited (Asibey et al. 2021). These enclaves can serve as a starting point for migrants to navigate the challenges of a new environment, including language barriers and cultural differences, and initiate the process of assimilation. Park and Burgess (2019) described assimilation as a process characterised by the intermingling and blending of individuals and groups, wherein they adopt the memories, emotions, and attitudes of others. Through the sharing of experiences and histories, these individuals and groups integrate these elements into their everyday cultural existence. Shafray and Kim (2017) differentiate the concept of multiculturalism from assimilation. Assimilation views multiple cultures as a barrier to harmonious society, whereas multiculturalism recognises the dynamism of different cultures and its value in a society. Multiculturalism can allow people to test the bounds of belonging, giving a technique for negotiating diversity (Stanfield and van Riemsdijk 2019). In other words, multiculturalism refers to situations when individuals who practise "diverse" customs, traditions, languages, and/or religious beliefs coexist in the same social milieu while remaining open to maintaining significant aspects of their distinctions and having them recognised in public (Colombo 2015). However, in the context of multiculturalism, community religious buildings encourage cultural separation (Dwyer et al. 2016).

Multiple researches have highlighted the influence of built environment on the vitality. The physical layout of urban environment can improve their capacity to support vibrant social and community life (Farahani et al. 2022). However, the measurement of urban vibrancy has been conducted in multiple ways. Several researches followed Jane Jacob's measure of pedestrian activities to quantify urban vitality at the urban scale (Sung et al. 2013; Sung and Lee 2015; Delclòs-Alió and Miralles-Guasch 2018; Delclòs Alió et al. 2019; Huang et al. 2022a). Zeng et al. (2018) used the factors of density, accessibility, liveability, and diversity to determine the principles of vitality. According to a study, characteristics used to evaluate urban vibrancy include density, diversified land use, and accessibility (Meng and Xing 2019). Multiple studies have suggested that urban vitality can be cultivated through mixed land use, small blocks, and intensive development (Zumelzu and Barrientos-Trinanes 2019; Li et al. 2022). However, Tu et al. (2020) contradicted to this finding and revealed that the mix of land uses, the size of the buildings, and the distance to the district centre have an impact depending on the location and vitality index. J. Wu et al. (2018b) measured community vibrancy by taking into account the proportion of outside-of-home non-work activities that each person engages in within their immediate area. The study concluded that communities that are more welcoming to immigrants will be more vibrant.

In this chapter, both the indicators from literature on ethnic migrants and urban vitality are considered to examine the significant factors that help in attaining vitality at the community level. Therefore, a pool of indicators is selected from multiple researches and the importance is assessed with the help of perception of community members (refer Appendix A). Here, "community" is referred as a group of human beings engaged in specific social activities within a defined area, sustaining its culture through social interactions among its members (Wang et al. 2020).

3 Materials and Methods

3.1 Study Area

Kharagpur is a city located in the Paschim Medinipur district of the state of West Bengal, India. Kharagpur boasts a diverse population, shaped by various cultural influences and ethnic backgrounds. According to the 2011 Census, the population had the following distribution of first languages: 55.39% spoke Bengali, 12.78% Telugu, 10.65% Hindi, 7.16% Santali, 4.63% Odia, 2.44% Nepali, 2.24% Urdu, and 1.79% Koda (Registrar General-India Registrar General-India 2011). This diversity is primarily attributed to the presence of three major institutions, namely the education institution, Indian Institute of Technology Kharagpur, the extensive Indian Railways network, and the significant Eastern Frontier Rifles police base. Among these institutions, the Indian Railways is notably served by the Telugu community, while the Eastern Frontier Rifles has a substantial representation of the Nepali community. As a result of their historical ties to these key institutions, the Nepali and Telugu communities have established a significant presence in Kharagpur (refer Fig. 1). Their contributions to the local workforce, culture, and social fabric further enrich the city's vibrant tapestry. By selecting the Nepali and Telugu communities as case studies, this chapter aims to delve into the unique dynamics and vitality factors specific to these ethnic migrant groups. Their distinct experiences, challenges, and strengths in the context of Kharagpur's diverse urban landscape make them an ideal focus for understanding the broader concept of community vitality among ethnic migrants.

3.2 Target Group

The objective of this chapter is to identify and explore the underlying structure or dimensions among the 38 community vitality indicators based on the responses from the Nepali and Telugu ethnic migrants with the help of factor analysis (refer Fig. 2). To achieve this, first, the relevant indicators are identified through a literature review. Next, data is collected using questionnaire with specific statements based on different dimensions. The survey was conducted in Hindi among 167 Nepali and 183 Telugu individuals. Finally, a total of 320 responses, 156 Nepali and 164 Telugu responses, were analysed. The surveys were conducted during August 2022 and January 2023. Both surveys employed comparable questionnaire designs and utilised a probability sampling method. The data obtained from the surveys were representative of the populations in their respective communities, rendering them suitable for conducting the study.

3.3 Measurement

To identify and explore the underlying dimensions of vitality of communities of ethnic migrants, a total of 38 statements (indicators) were asked using a 7-point Likert's scale (refer Appendix A). Next, the extracted indicators are analysed using Maximum Likelihood (ML) method with Promax rotation using IBM SPSS 22. Factor analysis aims to uncover the latent constructs or factors that explain the patterns of correlations among the indicators (Palermo et al. 2022). By extracting these factors, the analysis seeks to simplify and represent the data in a more interpretable way, revealing the key dimensions that contribute to the overall concept of community vitality as perceived by the ethnic migrant participants. An initial analysis

Fig. 1 Study area: location of Nepali and Telugu communities in Kharagpur

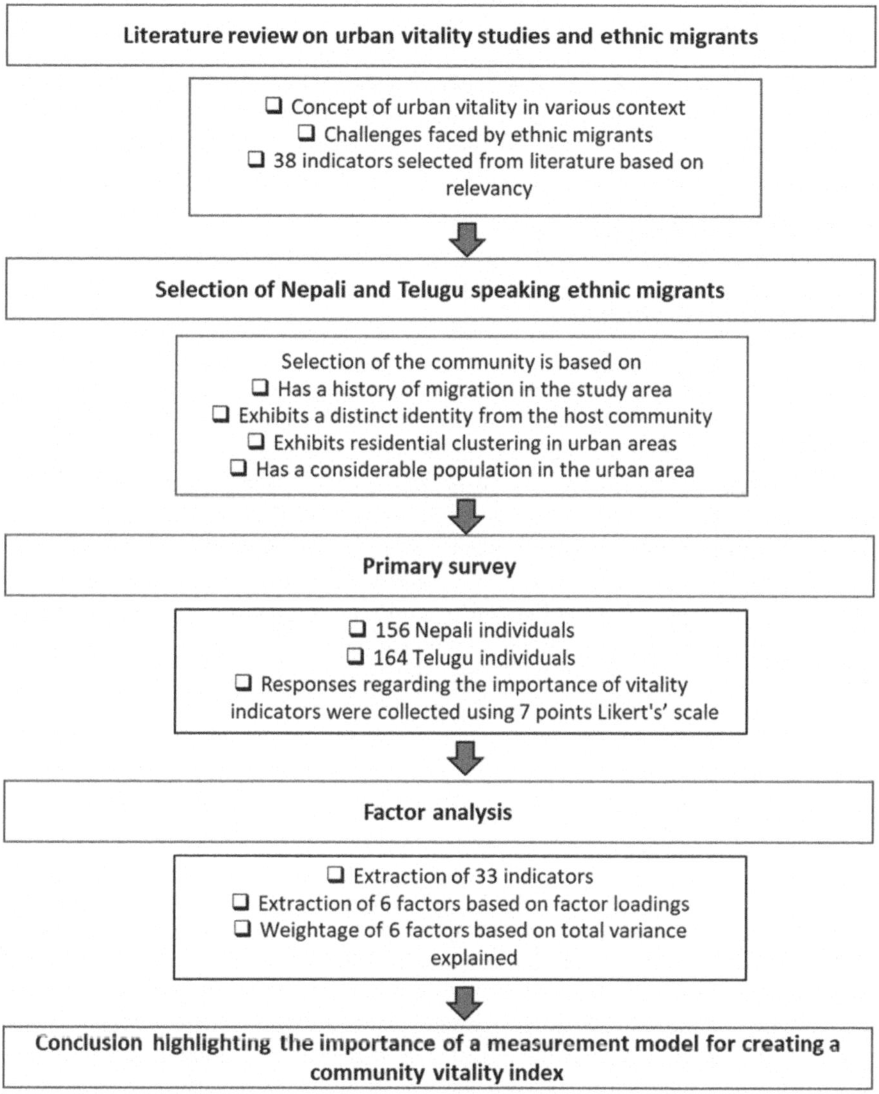

Fig. 2 Flowchart of methodology to identify dimensions of vitality of communities of ethnic migrants

yields eigenvalues, and statements with factor loadings less than 0.5 are removed, resulting in 6 factors. The suitability of the data for factor analysis is assessed by reviewing the commonalities generalised from exploratory factor analysis (EFA). The factor analysis helps in understanding the interrelationships between the indicators and provides valuable insights into the underlying factors that drive the perceptions and importance of community vitality among the ethnic migrant population.

The final questionnaire consists of 38 statements. The study's sample size of 320 respondents is deemed appropriate for conducting EFA, meeting the minimum recommended sample size for such analyses (Boomsma 1985; MacCallum et al. 1999).

4 Results and Discussion

4.1 Descriptive Statistics

Data regarding individual characteristics includes age, gender, community, years of residence, place of birth, marital status, educational qualification, employment, household size, and household per month income. Gender was coded as 1 = "Male" and 2 = "Female". Community was categorised as: 1 = "Nepali" and 2 = "Telugu". Place of birth was divided into three categories: 1 = "Within Kharagpur", 2 = "Outside Kharagpur but not in native place", and 3 = "In native place". Marital status was coded as 1 = "Married" and 2 = "Unmarried". Educational qualification was coded into 5 categories from not attended school to graduation. In total, the survey data obtained 10 variables of individual characteristics. The descriptive statistics of the individual characteristics of the Nepali and Telugu community are mentioned in Table 1, Figs. 3 and 4.

4.2 Data Analysis

Two tests are used to determine whether the data were appropriate for exploratory factor analysis before proceeding with factor extraction. The Kaiser-Meyer-Olkin (KMO) test is first used to determine whether the sample was adequate. The obtained KMO value is 0.922, which is above the minimum threshold of 0.60 and indicated the suitability of the data. Second, the statistical probability of significant correlations among the variables in the correlation matrix is calculated using Bartlett's test of sphericity which also showed the acceptability of the data for framework creation. Since the commonality values are over 0.5 and the correlation matrix indicated significance ($p < 0.05$), the dataset is suitable for Maximum Likelihood (ML). After performing ML on the 38 statements using Promax rotation, a form of oblique rotation, six factors are extracted from 33 statements. The eigenvalues of all six factors exceeded Kaiser's criterion of 1, and collectively, these factors explained 71.881% of the total variance (refer Table 2). The six factors are sequentially labelled AI (accessibility to infrastructure), MC (multiculturalism), SI (social inclusiveness), SA (spatial assimilation), EC (economic opportunities), and RD (residential distinction). The reliability test is performed and Cronbach's alpha value is greater than 0.7

Table 1 Descriptive statistics of responses from Nepali community and Telugu community

Descriptive statistics						
Community	N	Minimum	Maximum	Mean	Std. deviation	
Nepali	Age	156	15	75	42.5385	18.61854
	Years_of_ residence	156	15	58	32.9936	11.61284
	Household_ size	156	2	7	4.4423	1.19267
	Household_ income	156	15,000.00	97,000.00	45,570.51	17,723.36
	Valid N (listwise)	156				
Telugu	Age	164	15	75	44.5488	18.29221
	Years_of_ residence	164	15	61	34.9146	12.12228
	Household_ size	164	2	8	4.4207	1.18262
	Household_ income	164	12,000.00	105000.00	42,384.15	19,604.81
	Valid N (listwise)	164				

for all the factors (refer Appendix B). The factors are named by the authors based on the grouping of indicators.

The pattern matrix displays the linear combinations associated with each statement (refer Appendix B). For instance, the first factor (AI) is the linear combination of seven statements reflecting accessibility to infrastructure in terms of public services, commercial facilities, walkability, role of government, and block size. The second factor (MC) is the linear combination of seven statements pertaining to the multiculturalism aspects of ethnic migrants such as maintenance of ties with people in their native place and efforts to preserve their own culture. The third factor (SI) comprises six statements related to the social inclusiveness such as neighbourhood safety, identity making, psychological inclusivity, and attitude of host towards ethnic migrants. The fourth factor (SA) is represented by five statements concerning the social cohesion, self-rated health, experiences of racism, networking, and interaction with locals to affect the assimilation with host community. The fifth factor (EC) is represented by five statements concerning access to employment opportunities. Lastly, the sixth factor (RD) consists of three statements related to distinct residences in the neighbourhood.

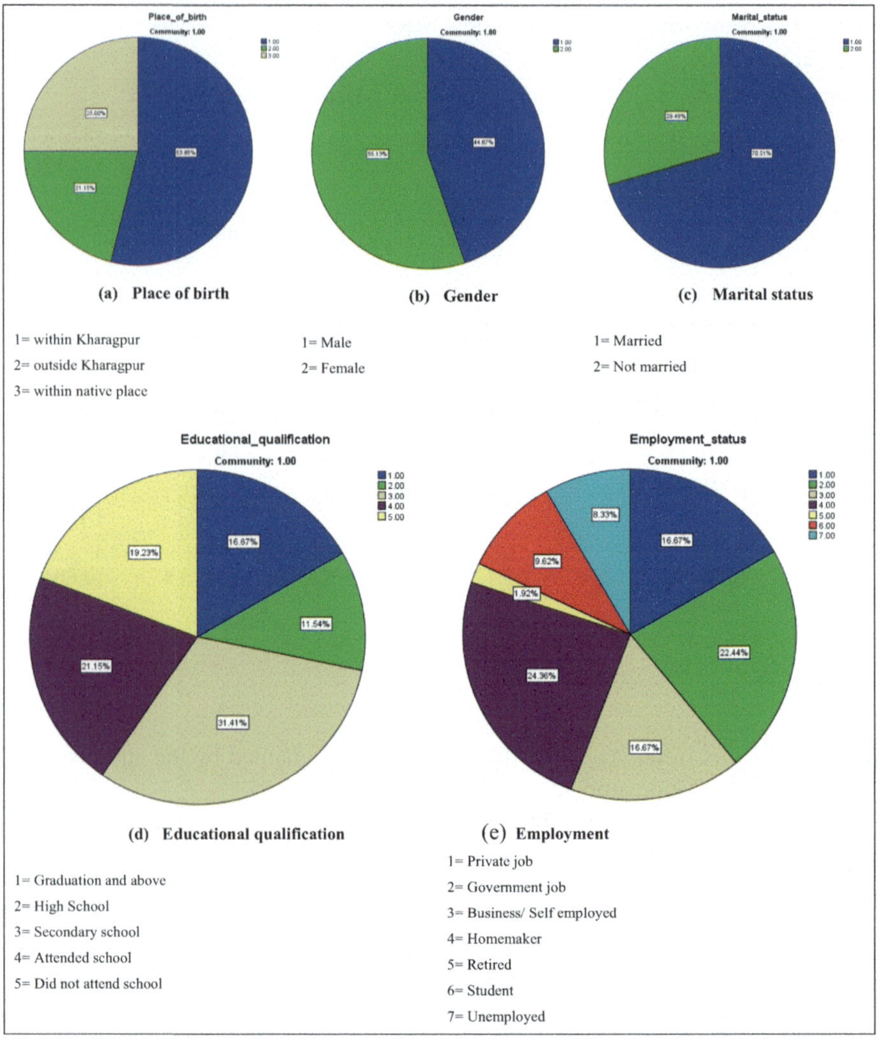

Fig. 3 Descriptive statistics of individual characteristics of Nepali community

4.3 Weightage of the Factors

Six factors are found using the factor analysis, and their weights are assigned based on how much they contributed to the overall variance. The squared loadings from the factor analysis are shown in Table 2, which shows that these factors together accounted for 71.881% of the total variance. Each factor is given a weight that is proportional to the explained variance (Tripathi and Singal 2019); the first factor (AI) is given a weight of 0.31 (0.31 = 22.454/71.881), the second factor (MC) is given a weight of 0.23, the third factor (SI) is given a weight of 0.20, the fourth (SA) is

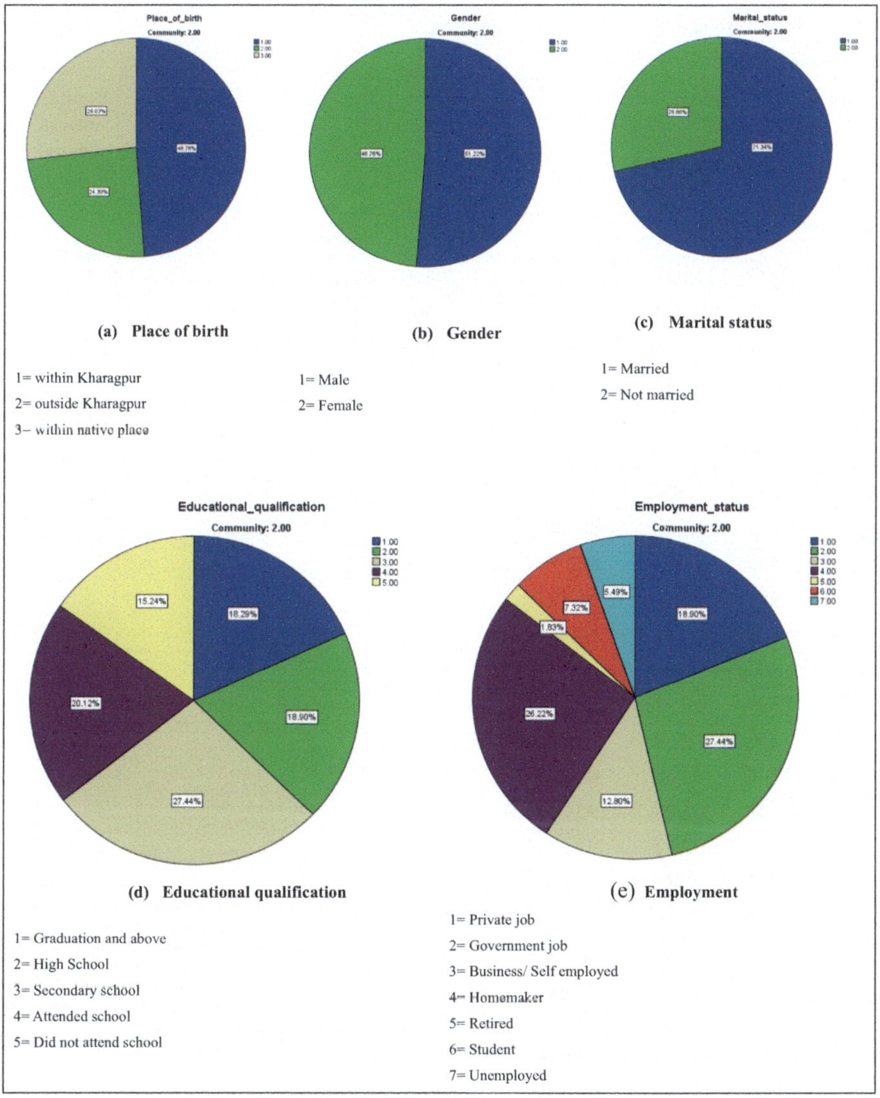

Fig. 4 Descriptive statistics of individual characteristics of Telugu community

given a weight of 0.11, the fifth (EC) is given a weight of 0.10, and the sixth (RD) is given a weight of 0.05 (refer Table 3).

Table 2 Total variance explained of the indicators

Total variance explained							
Factor	Initial eigenvalues			Extraction sums of squared loadings			Rotation sums of squared loadings[a]
	Total	% of variance	Cumulative %	Total	% of variance	Cumulative %	Total
1	7.812	23.674	23.674	7.410	22.454	22.454	7.118
2	6.132	18.582	42.256	5.501	16.671	39.125	5.461
3	5.014	15.194	57.450	4.772	14.462	53.587	4.648
4	2.618	7.934	65.384	2.493	7.554	61.141	5.010
5	2.354	7.134	72.518	2.363	7.160	68.300	4.066
6	1.508	4.570	77.088	1.181	3.580	71.881	2.763
7	0.596	1.805	78.893				
8	0.557	1.687	80.580				
9	0.499	1.511	82.092				
10	0.462	1.400	83.492				
11	0.409	1.239	84.731				
12	0.403	1.220	85.951				
13	0.394	1.193	87.144				
14	0.360	1.089	88.233				
15	0.346	1.048	89.281				
16	0.329	0.998	90.279				
17	0.316	0.957	91.236				
18	0.291	0.881	92.118				
19	0.269	0.815	92.933				
20	0.262	0.793	93.726				
21	0.255	0.772	94.497				
22	0.237	0.718	95.215				
23	0.219	0.664	95.878				
24	0.212	0.643	96.521				
25	0.197	0.598	97.120				
26	0.176	0.533	97.652				
27	0.154	0.467	98.120				
28	0.120	0.365	98.485				
29	0.113	0.343	98.827				
30	0.103	0.312	99.140				
31	0.099	0.300	99.439				
32	0.097	0.294	99.733				

(continued)

Table 2 (continued)

Total variance explained							
Factor	Initial eigenvalues			Extraction sums of squared loadings			Rotation sums of squared loadings[a]
	Total	% of variance	Cumulative %	Total	% of variance	Cumulative %	Total
33	0.088	0.267	100.000				
Extraction Method: Maximum Likelihood							

[a]When factors are correlated, sums of squared loadings cannot be added to obtain a total variance

Table 3 Weightage of 6 factors

	Factor 1 (AI)	Factor 2 (MC)	Factor 3 (SI)	Factor 4 (SA)	Factor 5 (EC)	Factor 6 (RD)
% of variance	22.454	16.671	14.462	7.554	7.16	3.58
Exp. var/ total Exp. var	0.31	0.23	0.20	0.11	0.10	0.05
Weights of factors	0.31	0.23	0.20	0.11	0.10	0.05

4.4 Vitality Factors of Communities of Ethnic Migrants

The key objective of this study is to consider the vitality factors of ethnic migrant communities in Indian context. This study is the first to present a systematic approach to gain a comprehensive understanding in the vitality of a specific community through a measurement model. A multidimensional measurement construct comprising six dimensions (AI, MC, SI, SA, EC, and RD) is developed to assess the level of vitality in these communities. The measurement model proposed in this study can be utilised to evaluate well-being, adaptability, and resilience of the communities formed by ethnic migrants with the help of index. The factors "building age", "house rent", "distance to highway", "distance to water body", and "floor area ratio" had loading of less than 0.5 and were removed from the model. This indicates that although the factors are significant at the city level, they have lesser importance at the community level in our study area.

The first dimension, access to infrastructure, underscores the critical role of providing essential services and amenities to these communities. It encompasses indicators such as access to quality healthcare, education, transportation, and other vital infrastructure. The first factor explains 22.454% of the total variance explained and has a weightage of 0.31. A high factor loading in this dimension indicates the importance of ensuring that ethnic migrant communities have equitable access to resources and services that promote their overall well-being and integration into the host society (Allen et al. 2021). The lack of proper infrastructure in Kharagpur has been reflected through stronger opinion for accessibility to infrastructure in the city. The second dimension, multiculturalism, highlights the significance of embracing and celebrating diversity within the host society. It measures the extent to which the local community values and respects the unique cultural heritage and practices of the ethnic migrant communities. A high factor loading in this dimension with an weightage of 0.23 reflects the importance of fostering an inclusive and tolerant environment that appreciates the richness brought by cultural diversity (McClinchey 2021). During the survey, it was observed that age of the individuals plays an important role. The older generation has a stronger opinion for building community value and preserving their own culture. The third dimension, social inclusiveness, emphasises the significance of social cohesion and community bonding for ethnic migrant communities. It gauges the extent to which these communities are welcomed and accepted by the broader society, as well as their participation in social activities and networks. The weightage of 0.20 for this factor indicates the importance of nurturing a sense of belonging and acceptance, which can contribute to the overall well-being and empowerment of ethnic migrant communities (Mai and Wang 2022). The survey revealed that the community bonding improves with longer years of residence in the city. The fourth dimension, spatial assimilation, underscores the integration and engagement of ethnic migrant communities within the physical urban landscape. It examines indicators such as interaction with the local population and patterns of networking. The factor loading in this dimension signifies the importance of promoting spatial assimilation as a means to facilitate cross-cultural interactions and foster a cohesive and harmonious urban environment. The fifth dimension, economic opportunities, highlights the significance of providing avenues for economic growth and advancement to ethnic migrant communities. It encompasses indicators such as employment prospects, income levels, and access to economic resources. The factor loading in this dimension reflects the importance of creating opportunities for economic mobility and financial stability, which can contribute to the overall prosperity and well-being of these communities. The sixth dimension, residential segregation, underscores the potential challenges posed by segregated residential patterns among ethnic migrant communities. It evaluates the extent to which these communities reside in distinct enclaves or neighbourhoods. A lower factor loading in this dimension calls attention to the importance of promoting mixed residential patterns to avoid social isolation and foster a sense of community and interaction with the broader society.

The identified dimensions of vitality offer valuable insights for urban policymakers, social planners, and community development practitioners. By understanding the multifaceted factors that influence the well-being and vitality of ethnic migrant communities, tailored interventions and policies can be designed to address their specific needs and challenges. Furthermore, the dimensions derived from this factor analysis provide a robust framework for future research and exploration in the realm of urban studies and migration. They offer a comprehensive and systematic approach to understanding the dynamics of ethnic migrant communities' experiences in urban environments, guiding researchers towards a deeper comprehension of the factors that contribute to their resilience and thriving.

5 Conclusion

In conclusion, this chapter aimed to identify and categorise the importance of multiple variables in shaping the vitality of ethnic migrant communities in Kharagpur. As cities continue to become increasingly diverse and multicultural, the insights gained from this study can aid in developing more inclusive and vibrant urban spaces that celebrate and support the vitality of ethnic migrant communities. To achieve this objective, we utilised factor analysis, which allowed us to extract relevant variables from the literature review and uncover the underlying factors influencing community vitality. By analysing the data both quantitatively and qualitatively, the authors present a comprehensive understanding of the intricate dynamics affecting the well-being of these communities in their adopted environment. The results of the factor analysis revealed six key factors of vitality of the communities of ethnic migrants: access to infrastructure (AI), multiculturalism (MC), social inclusiveness (SI), spatial assimilation (SA), economic opportunities (EC), and residential distinction (RD). Each factor plays a pivotal role in shaping the overall vitality and resilience of ethnic migrant communities in Kharagpur. The weightage of the factors obtained from the total variance explained showed that AI obtained the highest value and indicated the importance in community vitality. The weightage of the individual factors obtained has culminated in the development of a robust measurement model to assess community vitality in future research. The weightage can be used to calculate the community vitality index in future and measure the relevant factors of a particular community of ethnic migrants. Further, future studies can conduct longitudinal studies to track the changes in vitality factors over time. This would provide a more comprehensive understanding of the dynamics influencing ethnic migrant communities in Kharagpur.

It is necessary to take into account several limitations while evaluating the findings of our study. Firstly, the study is specific to the context of Kharagpur and may not be fully applicable to other regions with different socio-economic and cultural dynamics. Further, the study captures a snapshot of the vitality factors at a particular point in time. The dynamic nature of communities and cities suggests that these factors might evolve over time, necessitating ongoing research to capture changes. Secondly, the identification of factors influencing community vitality involves some subjectivity, and different researchers may identify alternative factors. Additionally, the data from two communities, namely, Nepali and Telugu, are aggregated to yield the factors. Therefore, the variance explained by the factors may vary at a disaggregated level. To conclude, this model provides a structured approach for evaluating and monitoring the well-being of ethnic migrant communities, offering a valuable tool for researchers and policymakers.

Acknowledgements This work has received funding from the Prime Minister's Research Fellow (PMRF) programme run by the Ministry of Education, Government of India. The authors acknowledge the work facilities and support of Indian Institute of Technology Kharagpur.

Appendix A: List of Indicators

Sl. No.	Indicator	Question asked	Reference
1	Maintenance of ties	Staying connected with our diaspora across India and globally is of great importance	Burnley (2016)
2	Belongingness	Sense of belongingness plays an important role in the adaptation and survivability	Mai and Wang (2022), Liu et al. (2022c), Maruyama et al. (2016) and Yin et al. (2022)
3	Language	It is important to actively speak and preserve the language within our community	Wang et al. (2020)
4	Networking	It is important to actively network with the host population by speaking in the host language to ensure community vitality	Wang et al. (2020) and Liu et al. (2022c)
5	Racism	Addressing issues of caste-based discrimination and promoting social equality within our community is necessary	Devadason (2010), Wessendorf (2020) and Rastogi and Curtis (2020)
6	Neighbourhood composition	It is important to have people of same ethnicity as neighbours for ensuring community vitality	Dhingra et al. (2016), Winkler and Johnson (2016), Zheng et al. (2020) and Li et al. (2009)
7	Interaction with locals	Being inclusive and accommodating towards individuals from different socio-economic backgrounds is necessary	Wang et al. (2020) and Dhingra et al. (2016)
8	Social cohesion	Maintaining social cohesion and a sense of unity within our community is essential	Dhingra et al. (2016) and Mouratidis and Poortinga (2020)
9	Location of relatives	It is important that relatives and friends stay in my neighbourhood	Wang et al. (2017, 2020), Paköz and Işık (2022) and Kathiravelu and Bunnell (2018)

(continued)

(continued)

Sl. No.	Indicator	Question asked	Reference
10	Psychological inclusivity	Collaborating and cooperating with other communities for social and cultural integration is beneficial	Wang et al. (2017) and Kaplan and Moigne (2019)
11	Participation in social programmes	Participation in social programmes (like sports activities, recreational, cultural, etc.) is important	Hui et al. (2021) and Cornelius and Lucio (2012)
12	Place making	Fostering a sense of pride and identity among community members is important	Yin et al. (2022) and McClinchey (2021)
13	Identity recognition	Integrating with the larger Indian society while maintaining our distinct identity is important for our community	Devadason (2010) and Wang et al. (2020)
14	Role of government	It is important to recognise the significant role of the government in providing affordable and accessible housing options for community vitality	Eskelä (2018) and Doyle (2018)
15	Self-rated health	The good health and well-being represent our adaptability in the neighbourhood	Paköz and Işık (2022) and Kearns et al. (2017)
16	Selection of neighbourhood	It is important to select the neighbourhood where all our community members stay	Dhingra et al. (2016), Asibey et al. (2021) and Hume and Hardwick (2005)
17	Neighbourhood safety	Ensuring the safety and security of the community in public spaces and streets is important	Blazek (2014) and Li et al. (2022)
18	Religious buildings	It is important to have religious buildings within the neighbourhood	Rastogi and Curtis (2020), Degli Esposti (2019)
19	Attitude of host	It is important that the host community respects our cultural practice	Li et al. (2009)
20	Practices own culture	Active engagement in cultural and artistic activities, such as music, dance, and theatre, is significant	Zhu et al. (2014)
21	Celebration of festivals	Celebrating and passing down our cultural traditions and festivals to younger generations is crucial	McClinchey (2021) and Bonnerjee (2013)
22	Traditional customs	Preserving and promoting our traditional arts, crafts, and culinary heritage is vital	Wang et al. (2020)
23	Ethnic entrepreneurship	It is important to have business which are specific to our community	Jin et al. (2021) and Yamamura and Lassalle (2020)

(continued)

(continued)

Sl. No.	Indicator	Question asked	Reference
24	Transnational entrepreneurship	It is important to engage in business outside my workplace	Li et al. (2009) and Sirkeci and Zeren (2018)
25	Involvement in business	Supporting entrepreneurship and economic development within our community is important	Yamamura and Lassalle (2020)
26	Local geography of employment	Providing platforms for young talent to showcase their skills and abilities is a valuable endeavour	Zhu et al. (2014) and Joassart-Marcelli (2014)
27	Employment opportunities	Employment opportunities in different sectors help in important for our survival	Joassart-Marcelli (2014)
28	Accessibility to parks & green spaces	Having easily accessible and utilised public spaces, such as parks, and recreational facilities, is important for the community	Wang et al. (2020) and Mouratidis and Poortinga (2020)
29	Pedestrian friendly footpaths	It is important to have pedestrian-friendly footpaths within the neighbourhood	Li et al. (2022)
30	Accessibility to commercial facilities	Having markets or shopping areas that cater to the needs and preferences of the community is important	Wang and Liu (2022) and Wang et al. (2022a)
31	Quality of public services	Ensuring the availability of basic amenities, such as clean water, electricity, and sanitation, to the community is important	Wang et al. (2020) and Dhingra et al. (2016)
32	House rent	It is important to get rented houses to ensure stability and vitality	Parutis (2011)
33	Walkability	Having pedestrian-friendly infrastructure and safe walking paths is essential for promoting social interaction, physical activity, and a sense of belonging among the community	Li et al. (2022)
34	Building age	The mix of older and newer buildings creates a dynamic and vibrant atmosphere that enhances community vitality	Yue et al. (2021) and Wu and Niu (2019)
35	Block size	It is important to have small block sizes to enhance walkability, connectivity, and social interaction	Yue et al. (2021) and Wu et al. (2022)
36	Distance to highway	It is important to stay near highway and have easy accessibility to it	Wu and Niu (2019) and Fuentes et al. (2020)

(continued)

(continued)

Sl. No.	Indicator	Question asked	Reference
37	Distance to water body	It is important to prioritise the preservation and maintenance of waterbodies for the vitality of the community	Wu and Niu (2019) and Fuentes et al. (2020)
38	Floor area ratio	Floor area ratio is an important aspect for our decision to stay in this neighbourhood	Li et al. (2022), Ye et al. (2018) and Lu et al. (2019)

Appendix B: Factor Loadings

	Pattern matrix[a]						
Factors	Cronbach's alpha	Factor					
		0.982	0.951	0.917	0.917	0.836	0.919
Accessibility to infrastructure (AI)	Quality_of_public_ services	0.951					
	Accessibility_to_ commercial_facilities	0.943					
	Walkability	0.942					
	Accessibility_to_ parks__green_spaces	0.936					
	Role_of_government	0.936					
	Pedestrian_friendly_ footpaths	0.935					
	Block_size	0.928					
Multiculturalism (MC)	Religious_buildings		0.902				
	Maintenance_of_ties		0.865				
	Language		0.857				
	Traditional_customs		0.857				
	Practices_own_culture		0.844				
	Celebration_of_ festivals		0.836				
	Belongingness		0.832				
Social inclusiveness (SI)	Neighbourhood_ safety			0.824			
	Participation_in_ social_programmes			0.816			

(continued)

(continued)

	Pattern matrix[a]						
	Place_making			0.809			
	Attitude_of_host			0.807			
	Psychological_ inclusivity			0.784			
	Identity_recognition			0.782			
Spatial assimilation (SA)	Social_cohesion				0.861		
	Self_rated_health				0.840		
	Networking				0.826		
	Racism_RC				0.825		
	Interaction_with_ locals				0.770		
Economic opportunities (EC)	Local_geography_of_ employment					0.727	
	Employment_ opportunities					0.714	
	Transnational_ entrepreneurship					0.690	
	Ethnic_ entrepreneurship					0.687	
	Involvement_in_ business					0.687	
Residential distinction (RD)	Selection_of_ neighbourhood						0.897
	Location_of_relatives						0.892
	Neighbourhood_ composition						0.875
	Extraction Method: Maximum Likelihood Rotation Method: Promax with Kaiser Normalisation						

[a]Rotation converged in 6 iterations

Appendix 3: Scree Plot of Factors

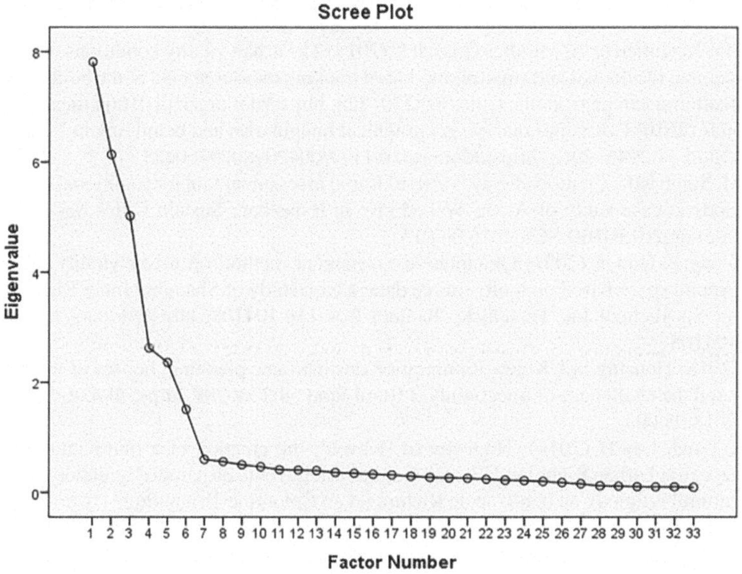

References

Allen J, Farber S, Greaves S et al (2021) Immigrant settlement patterns, transit accessibility, and transit use. J Transp Geogr 96. https://doi.org/10.1016/j.jtrangeo.2021.103187

Asibey MO, Poku-Boansi M, Adutwum IO (2021) Residential segregation of ethnic minorities and sustainable city development. Case of Kumasi, Ghana. Cities 116:103297. https://doi.org/10.1016/j.cities.2021.103297

Banchiero F, Blečić I, Saiu V, Trunfio GA (2020) Neighbourhood park vitality potential: from Jane Jacobs's theory to evaluation model. Sustain 12:1–20. https://doi.org/10.3390/su12155881

Blazek M (2014) Migration, vulnerability and the complexity of violence: experiences of documented non-EU migrants in Slovakia. Geoforum 56:101–112. https://doi.org/10.1016/j.geoforum.2014.07.002

Bonnerjee J (2013) Invisible belonging: Anglo-Indian identity in multicultural Toronto. J Intercult Stud 34:431–442. https://doi.org/10.1080/07256868.2013.821215

Boomsma A (1985) Nonconvergence, improper solutions, and starting values in LISREL maximum likelihood estimation. Psychometrika 50:229–242. https://doi.org/10.1007/BF02294248

Burnley IH (2016) Developments and complementarities in international migration paradigms. J Int Migr Integr 17:77–94. https://doi.org/10.1007/s12134-014-0395-8

Chen Y, Yin G, Hou Y (2022) Street centrality and vitality of a healthy catering industry: a case study of Jinan, China. Front Public Heal 10. https://doi.org/10.3389/fpubh.2022.1032668

Colombo E (2015) Multiculturalisms: an overview of multicultural debates in western societies. Curr Sociol 63:800–824. https://doi.org/10.1177/0011392115586802

Cornelius N, Lucio MM (2012) Introduction: representation, labour markets and immigrant and minority ethnic workers-networking, new forms of representation and politics in the multi-ethnic city. Urban Stud 49:587–594. https://doi.org/10.1177/0042098011431621

Degli Esposti E (2019) Fragmented realities: the 'sectarianisation' of space among Iraqi Shias in London. Contemp Islam 13:259–285. https://doi.org/10.1007/s11562-018-0425-y

Delclòs-Alió X, Miralles-Guasch C (2018) Looking at Barcelona through Jane Jacobs's eyes: mapping the basic conditions for urban vitality in a Mediterranean conurbation. Land Use Policy 75:505–517. https://doi.org/10.1016/j.landusepol.2018.04.026

Delclòs Alió X, Gutiérrez A, Miralles Guasch C (2019) The urban vitality conditions of Jane Jacobs in Barcelona: residential and smartphone-based tracking measurements of the built environment in a Mediterranean metropolis. Cities 86:220–228. https://doi.org/10.1016/j.cities.2018.09.021

Devadason R (2010) Cosmopolitanism, geographical imaginaries and belonging in North London. Urban Stud 47:2945–2963. https://doi.org/10.1177/0042098009360228

Dhingra M, Singh MK, Chattopadhyay S (2016) Rapid assessment tool for traditional Indian neighbourhoods: a case study of Alwar Walled city in Rajasthan. Sustain Cities Soc 26:364–382. https://doi.org/10.1016/J.SCS.2016.06.015

Dong Y, Peng F, Guo T (2021) Quantitative assessment method on urban vitality of metro-led underground space based on multi-source data: a case study of Shanghai Inner Ring area. Tunn Undergr Sp Technol Inc Trenchless Technol Res 116:104108. https://doi.org/10.1016/j.tust.2021.104108

Doyle C (2018) Housing in UK new immigration destinations: planning, houses of multiple occupation and the challenges of uncertainty. J Rural Stud 64:276–288. https://doi.org/10.1016/j.jrurstud.2018.05.003

Dwyer C, Tse J, Ley D (2016) 'Highway to Heaven': the creation of a multicultural, religious landscape in suburban Richmond, British Columbia | 'La route du paradis': création d'un paysage multiculturel religieux en banlieue de Richmond en Colombie-Britannique | 'Carretera hacia el c. Soc Cult Geogr 17:667–693. https://doi.org/10.1080/14649365.2015.1130848

El-Bialy R, Mulay S (2015) Two sides of the same coin: factors that support and challenge the wellbeing of refugees resettled in a small urban center. Heal Place 35:52–59. https://doi.org/10.1016/j.healthplace.2015.06.009

Eskelä E (2018) Housing pathways of skilled migrants: Indian professionals in Helsinki, Finland. Housing, Theory Soc 35:474–494. https://doi.org/10.1080/14036096.2017.1387172

Farahani LM, Izadpanahi P, Tucker R (2022) The death and life of Australian suburbs: relationships between social activity and the physical qualities of Australian suburban neighbourhood centres. City, Cult Soc 28. https://doi.org/10.1016/j.ccs.2021.100426

Fuentes L, Miralles-Guasch C, Truffello R et al (2020) Santiago de Chile through the eyes of Jane Jacobs. Analysis of the conditions for urban vitality in a Latin American metropolis. Land 9:1–17. https://doi.org/10.3390/land9120498

Ghatak I, Chatterjee S (2018) Urban street vending practices: an investigation of ethnic food safety knowledge, attitudes, and risks among untrained Chinese vendors in Chinatown, Kolkata. J Ethn Foods 5:272–285. https://doi.org/10.1016/j.jef.2018.11.003

Gong Y, Ji X, Zhang Y, Cheng S (2023) Spatial vitality evaluation and coupling regulation mechanism of a complex ecosystem in Lixiahe plain based on multi-source data. Sustainability 15:2141. https://doi.org/10.3390/su15032141

Goswami SK, Malik US (2019) Urban ethnic space: a discourse on Chinese Community in Kolkata, West Bengal. Indian J Spat Sci 10:88–95

Huang J, Cui Y, Li L et al (2022a) Re-examining Jane Jacobs' doctrine using new urban data in Hong Kong. Environ Plan B Urban Anal City Sci. https://doi.org/10.1177/23998083221106186

Huang LS, Han Y, Ye Y (2022b) Coastal waterfront vibrancy: an exploration from the perspective of quantitative urban morphology. Buildings 12. https://doi.org/10.3390/buildings12101585

Hui ECM, Chen T, Lang W, Ou Y (2021) Urban community regeneration and community vitality revitalization through participatory planning in China. Cities 110:103072. https://doi.org/10.1016/j.cities.2020.103072

Hume SE, Hardwick SW (2005) African, Russian, and Ukrainian refugee resettlement in Portland, Oregon. Geogr Rev 95:189–209. https://doi.org/10.1111/j.1931-0846.2005.tb00362.x

Jacobs J (1961) The death and life of Great American cities. Vintage

Jin X, Bolt G, Hooimeijer P (2021) Africans in Guangzhou: is the ethnic enclave model applicable in the Chinese context? Cities 117. https://doi.org/10.1016/j.cities.2021.103320

Jin X, Long Y, Sun W et al (2017) Evaluating cities' vitality and identifying ghost cities in China with emerging geographical data. Cities 63:98–109. https://doi.org/10.1016/j.cities.2017.01.002

Joassart-Marcelli P (2014) Gender, social network geographies, and low-wage employment among recent Mexican immigrants in Los Angeles. Urban Geogr 35:822–851. https://doi.org/10.1080/02723638.2014.926634

Kaplan DH, Moigne YL (2019) Multicultural engagements in lived spaces: how cultural communities intersect in Belleville, Paris. City Community 18:392–413. https://doi.org/10.1111/cico.12358

Kathiravelu L, Bunnell T (2018) Introduction: urban friendship networks: affective negotiations and potentialities of care. Urban Stud 55:491–504. https://doi.org/10.1177/0042098017737281

Kearns A, Whitley E, Egan M et al (2017) Healthy migrants in an unhealthy city? The effects of time on the health of migrants living in deprived areas of Glasgow. J Int Migr Integr 18:675–698. https://doi.org/10.1007/s12134-016-0497-6

Křížková I, Šimon M (2022) Measuring residential segregation of non-European migrants using the individualised neighbourhood method: how does Czechia fit to the European landscape? Appl Geogr 144:102730. https://doi.org/10.1016/j.apgeog.2022.102730

Kuttikat M, Vaillancourt A, Massey M (2018) Battered but bold: Sri Lankan Tamil refugee war experiences, camp challenges and resilience. Int J Migr Heal Soc Care 14:245–259. https://doi.org/10.1108/IJMHSC-04-2017-0013

Lang W, Lang H, Hui ECM et al (2022) Measuring urban vibrancy of neighborhood performance using social media data in Oslo, Norway. Cities 131:103908. https://doi.org/10.1016/j.cities.2022.103908

Li S, Wu C, Lin Y, et al (2020) Urban morphology promotes urban vibrancy from the spatiotemporal and synergetic perspectives: a case study using multisource data in Shenzhen, China. Sustainability 12. https://doi.org/10.3390/SU12124829

Li X, Li Y, Jia T et al (2022) The six dimensions of built environment on urban vitality: fusion evidence from multi-source data. Cities 121. https://doi.org/10.1016/j.cities.2021.103482

Li Z, Ma LJC, Xue D (2009) An African enclave in China: the making of a new transnational urban space. Eurasian Geogr Econ 50:699–719. https://doi.org/10.2747/1539-7216.50.6.699

Lin J, Zhuang Y, Zhao Y et al (2023) Measuring the non-linear relationship between three-dimensional built environment and urban vitality based on a random forest model. Int J Environ Res Public Health 20. https://doi.org/10.3390/ijerph20010734

Liu D, Shi Y (2022) The influence mechanism of urban spatial structure on urban vitality based on geographic big data: a case study in downtown Shanghai. Buildings 12. https://doi.org/10.3390/buildings12050569

Liu H, Gou P, Xiong J (2022a) Vital triangle: a new concept to evaluate urban vitality. Comput Environ Urban Syst 98:101886. https://doi.org/10.1016/J.COMPENVURBSYS.2022.101886

Liu S, Ge J, Bai M et al (2022b) Toward classification-based sustainable revitalization: assessing the vitality of traditional villages. Land Use Policy 116:106060. https://doi.org/10.1016/j.landusepol.2022.106060

Liu S, Zhang F, Wu F (2022c) Contrasting migrants' sense of belonging to the city in selected Peri-urban neighbourhoods in Beijing. Cities 120:103499. https://doi.org/10.1016/j.cities.2021.103499

Lu S, Shi C, Yang X (2019) Impacts of built environment on urban vitality: regression analyses of Beijing and Chengdu, China. Int J Environ Res Public Health 16:4592. https://doi.org/10.3390/ijerph16234592

MacCallum RC, Widaman KF, Zhang S, Hong S (1999) Sample size in factor analysis. Psychol Methods 4:84–99. https://doi.org/10.1037/1082-989X.4.1.84

Mai X, Wang J (2022) Situational differences, migratory duration, and social integration of internal migrants in urban China. Cities 125. https://doi.org/10.1016/j.cities.2022.103596

Maruyama NU, Woosnam KM, Boley BB (2016) Comparing levels of resident empowerment among two culturally diverse resident populations in Oizumi, Gunma, Japan. J Sustain Tour 24:1442–1460. https://doi.org/10.1080/09669582.2015.1122015

Massey DS, Denton NA (1985) Spatial assimilation as a socioeconomic outcome. Am Sociol Rev 50:94–106. https://doi.org/10.2307/2095343

McClinchey KA (2021) Contributions to social sustainability through the sensuous multiculturalism and everyday place-making of multi-ethnic festivals. J Sustain Tour 29:2025–2043. https://doi.org/10.1080/09669582.2020.1853760

Meng Y, Xing H (2019) Exploring the relationship between landscape characteristics and urban vibrancy: a case study using morphology and review data. Cities 95:102389. https://doi.org/10.1016/j.cities.2019.102389

Mouratidis K, Poortinga W (2020) Built environment, urban vitality and social cohesion: do vibrant neighborhoods foster strong communities? Landsc Urban Plan 204:103951. https://doi.org/10.1016/j.landurbplan.2020.103951

Ou G, Zhou M, Zeng Z et al (2021) Is there an equality in the spatial distribution of urban vitality: a case study of Wuhan in China. Open Geosci 13:469–481. https://doi.org/10.1515/geo-2020-0249

Ozbil A, Peponis J, Stone B (2011) Understanding the link between street connectivity, land use and pedestrian flows. Urban Des Int 16:125–141. https://doi.org/10.1057/udi.2011.2

Paköz MZ, Işık M (2022) Rethinking urban density, vitality and healthy environment in the post-pandemic city: the case of Istanbul. Cities 124:103598. https://doi.org/10.1016/j.cities.2022.103598

Paköz MZ, Yaratgan D, Şahin A (2022) Re-mapping urban vitality through Jane Jacobs' criteria: the case of Kayseri, Turkey. Land Use Policy 114:105985. https://doi.org/10.1016/J.LANDUSEPOL.2022.105985

Palermo F, Sergi BS, Sironi E (2022) Does urbanization matter? Diverging attitudes toward migrants and Europe's decision-making. Socioecon Plann Sci 83:101278. https://doi.org/10.1016/j.seps.2022.101278

Park RE, Burgess EW (2019) Introduction to the science of sociology. Good Press

Parutis V (2011) "Home" for now or "home" for good? East European migrants' experiences of accommodation in London. Home Cult 8:265–296. https://doi.org/10.2752/175174211X13099693358799

Rastogi A, Curtis K (2020) Beyond the city: exploring the suburban and rural landscapes of racial residential integration across the United States. Popul Res Policy Rev 39:861–888. https://doi.org/10.1007/s11113-020-09610-x

Registrar General-India (RGI) (2011) Census of India 2011: language. Census of India 1–52

Roy K (2023) Sensing and (re) imaging an urban enclave through application of visual streetscape ethnography: a study of twin Chinatowns in Kolkata. In: Ruepert G, Ilić T (eds) Young geographers showcasing research contributions in geography, pp 113–138

Shafray E, Kim S (2017) Divergent migration patterns and foreign community in urban neighbourhood: multiculturalism directions observed from case study in Seoul, South Korea. Environ Urban ASIA 8:22–39. https://doi.org/10.1177/0975425316683860

Sirkeci I, Zeren F (2018) Diaspora marketing revisited: the nexus of entrepreneurs and consumers. Transnatl Mark J 6:139–157. https://doi.org/10.33182/tmj.v6i2.588

Stanfield E, van Riemsdijk M (2019) Creating public space, creating 'the public': immigration politics and representation in two Copenhagen parks. Urban Geogr 40:1356–1374. https://doi.org/10.1080/02723638.2019.1591807

Subba T (2016) Living the Nepali Diaspora in India: an autobiographical essay authors (s): Tanka B. Subba Published by: Dietrich Reimer Verlag GmbH Stable. http://www.jstor.org/stable/25843148. Accessed 27 Mar 2016 08: 57 UTC Living the Nepali Diaspora in Ind. 2:213–232

Subba TB (2018) Race, identity and nationality: relocating Nepali Nationalism in India. Millenn Asia 9:6–18. https://doi.org/10.1177/0976399617753750

Sung H, Lee S (2015) Residential built environment and walking activity: empirical evidence of Jane Jacobs' urban vitality. Transp Res Part D Transp Environ 41:318–329. https://doi.org/10.1016/j.trd.2015.09.009

Sung HG, Go DH, Choi CG (2013) Evidence of Jacobs's street life in the great Seoul city: identifying the association of physical environment with walking activity on streets. Cities 35:164–173. https://doi.org/10.1016/j.cities.2013.07.010

Tripathi M, Singal SK (2019) Allocation of weights using factor analysis for development of a novel water quality index. Ecotoxicol Environ Saf 183:109510. https://doi.org/10.1016/j.ecoenv.2019.109510

Tu W, Zhu T, Xia J et al (2020) Portraying the spatial dynamics of urban vibrancy using multisource urban big data. Comput Environ Urban Syst 80:101428. https://doi.org/10.1016/j.compenvurbsys.2019.101428

Wang L, Zheng W, Wei S (2022a) Assessing urban vitality and its determinants in high-speed rail station areas in the Yangtze River Delta, China. J Transp Land Use 15:333–354. https://doi.org/10.5198/jtlu.2022.2010

Wang T, Li Y, Li H et al (2022b) Research on the vitality evaluation of parks and squares in medium-sized Chinese cities from the perspective of urban functional areas. Int J Environ Res Public Health 19:15238. https://doi.org/10.3390/ijerph192215238

Wang X, Liu Z (2022) Neighborhood environments and inclusive cities: an empirical study of local residents' attitudes toward migrant social integration in Beijing, China. Landsc Urban Plan 226:104495. https://doi.org/10.1016/J.LANDURBPLAN.2022.104495

Wang Y, Feng Y, Han Q et al (2020) Perceived discrimination of displaced people in development-induced displacement and resettlement: the role of integration. Cities 101:102692. https://doi.org/10.1016/j.cities.2020.102692

Wang Z, Lu F, Liu Z et al (2023) Measuring spatial nonstationary effects of POI-based mixed use on urban vibrancy using Bayesian spatially varying coefficients model. Int J Geogr Inf Sci 37:339–359. https://doi.org/10.1080/13658816.2022.2117363

Wang Z, Zhang F, Wu F (2017) Social trust between rural migrants and urban locals in China—exploring the effects of residential diversity and neighbourhood deprivation. Popul Space Place 23. https://doi.org/10.1002/psp.2008

Wessendorf S (2020) Ethnic minorities' reactions to newcomers in East London: symbolic boundaries and convivial labor. Br J Sociol 71:208–220. https://doi.org/10.1111/1468-4446.12729

Winkler RL, Johnson KM (2016) Moving toward integration? Effects of migration on ethnoracial segregation across the rural-urban continuum. Demography 53:1027–1049. https://doi.org/10.1007/s13524-016-0479-5

Wu C, Ye X, Ren F, Du Q (2018a) Check-in behaviour and spatio-temporal vibrancy: an exploratory analysis in Shenzhen, China. Cities 77:104–116. https://doi.org/10.1016/j.cities.2018.01.017

Wu J, Lu Y, Gao H, Wang M (2022) Cultivating historical heritage area vitality using urban morphology approach based on big data and machine learning. Comput Environ Urban Syst 91:101716. https://doi.org/10.1016/j.compenvurbsys.2021.101716

Wu J, Ta N, Song Y et al (2018b) Urban form breeds neighborhood vibrancy: a case study using a GPS-based activity survey in suburban Beijing. Cities 74:100–108. https://doi.org/10.1016/j.cities.2017.11.008

Wu W, Niu X (2019) Influence of built environment on urban vitality: case study of Shanghai using mobile phone location data. J Urban Plan Dev 145. https://doi.org/10.1061/(ASCE)UP.1943-5444.0000513

Xia C, Yeh AGO, Zhang A (2020) Analyzing spatial relationships between urban land use intensity and urban vitality at street block level: a case study of five Chinese megacities. Landsc Urban Plan 193:103669. https://doi.org/10.1016/j.landurbplan.2019.103669

Xia C, Zhang A, Yeh AGO (2022) The varying relationships between multidimensional urban form and urban vitality in Chinese megacities: insights from a comparative analysis. Ann Am Assoc Geogr 112:141–166. https://doi.org/10.1080/24694452.2021.1919502

Xu Y, Chen X (2021) Quantitative analysis of spatial vitality and spatial characteristics of urban underground space (UUS) in metro area. Tunn Undergr Sp Technol 111:103875. https://doi.org/10.1016/j.tust.2021.103875

Yamamura S, Lassalle P (2020) Approximating entrepreneurial superdiversity: reconceptualizing the superdiversity debate in ethnic minority entrepreneurship. J Ethn Migr Stud 46:2218–2239. https://doi.org/10.1080/1369183X.2019.1656058

Yang J, Li X, Du J, Cheng C (2023) Exploring the relationship between urban street spatial patterns and street vitality: a case study of Guiyang, China

Yang Y, Ma Y, Jiao H (2021) Exploring the correlation between block vitality and block environment based on multisource big data: taking Wuhan city as an example. Land 10. https://doi.org/10.3390/land10090984

Ye Y, Li D, Liu X (2018) How block density and typology affect urban vitality: an exploratory analysis in Shenzhen, China. Urban Geogr 39:631–652. https://doi.org/10.1080/02723638.2017.1381536

Yin Y, Shao Y, Lu H, Han Y (2022) Environmental drivers of the vital urban coastal zones: an explorative case study based on the data-driven multi-method approach. Front Ecol Evol 10:962299. https://doi.org/10.3389/fevo.2022.962299

Yu B, Cui X, Li H et al (2022) TOD and vibrancy: the spatio-temporal impacts of the built environment on vibrancy. Front Environ Sci 10:1–17. https://doi.org/10.3389/fenvs.2022.1009094

Yue W, Chen Y, Thy PTM et al (2021) Identifying urban vitality in metropolitan areas of developing countries from a comparative perspective: Ho Chi Minh City versus Shanghai. Sustain Cities Soc 65:102609. https://doi.org/10.1016/j.scs.2020.102609

Zeng C, Song Y, He Q, Shen F (2018) Spatially explicit assessment on urban vitality: case studies in Chicago and Wuhan. Sustain Cities Soc 40:296–306. https://doi.org/10.1016/j.scs.2018.04.021

Zheng S, Song Z, Sun W (2020) Do affordable housing programs facilitate migrants' social integration in Chinese cities? Cities 96:102449. https://doi.org/10.1016/j.cities.2019.102449

Zhu H, Chen X, Qian J (2014) Charting the development of social and cultural geography in Mainland China: voices from the inside | Tracer le développement de la géographie sociale et culturelle en la Chine métropolitaine: Des voix de l'intérieur. Soc Cult Geogr 15:255–283. https://doi.org/10.1080/14649365.2014.883423

Zhuang DZC, Lok RT (2023) Exploring the wellbeing of migrants in third places: an empirical study of smaller Canadian cities. Wellbeing, Sp Soc 4:100146. https://doi.org/10.1016/j.wss.2023.100146

Zou H, Liu R, Cheng W et al (2023) The association between street built environment and street vitality based on quantitative analysis in historic areas: a case study of Wuhan, China. Sustainability 15:1732. https://doi.org/10.3390/su15021732

Zumelzu A, Barrientos-Trinanes M (2019) Analysis of the effects of urban form on neighborhood vitality: five cases in Valdivia, Southern Chile. J Hous Built Environ 34:897–925. https://doi.org/10.1007/s10901-019-09694-8

Sustainable Megacities with Energy Planning and Services

An Institutional Approach for Sustainable Service Provision in Peri-Urban Areas: A Case of Chennai Metropolitan Area

Chetna Singh

1 Introduction

Let me start by quoting Aguilar (2008), to set the context from the beginning itself, 'Contrary to the model of the past decades, when the periphery was a space subordinated to the city, with poor accessibility, deficits in basic services, and lack of amenities, nowadays this space represents another type of city; it has another centrality, it has more provisions, and it represents totally a new form of city expansion' (Aguilar 2008, p. 134). This is true for many peri-urban areas as far as them becoming economically central spaces is concerned. However, when it comes to provision of services there do seem to be variations based on proximity to the core city (Corporation of Chennai in this case), hierarchy of settlements (municipal corporation, municipality, town panchayat or village) and the institutional jurisdiction. This would be further established as we progress with the chapter.

Cities have long been seen as having far reaching developmental impact not only for itself but also for its surroundings, both immediate and distant. Coupled with globalization this specific form of rapid urbanization is resulting in the complex process of 'Peri-Urbanization' on the peripheries of the large developing metropolizes in developing countries (Singh 2020).

Peri-urban has been defined differently by different scholars but there are certain aspects which come out very clearly while defining a peri-urban area. As a place, peri-urban area lies outside the municipal boundary but can be considered as an extension of the cities they surround. As a process, it undergoes continuous transformations with respect to physical, socio-demographic, economic and environmental aspects. As a concept, it is a dynamic mixed space with both urban and rural settlement and activities being carried (Quoted in Singh 2020; Based on study of Mc Gee 1991;

C. Singh (✉)
School of Planning and Architecture, New Delhi 110002, India
e-mail: chetnasingh@spa.ac.in

M. Firoz C et al. (eds.), *Urban Planning and Design for Megacities in the Global South*, Advances in 21st Century Human Settlements,
https://doi.org/10.1007/978-981-97-8370-0_16

Narain 2010; Dupont 2005; Kundu et al. 2002; Moench and Gyawali 2008; Webster 2002; Marshall et al. 2009; Shaw 2005; Arabindoo 2005; Rohilla 2005; Cadene 2005; Oliveau 2005; Aguilar 2008; Narain et al. 2014).

A metropolitan region has several urban and rural constituents, which by virtue of being in the vicinity of the core city are experiencing demographic growth, spatial expansion and economic development. The core city has been the primary reason for these transformations. However, these transformations are both positive as well as negative for the peri-urban areas. The positive ones being, greater economic opportunities, better connectivity with the core and access to many higher-order services located in the core city. The negative ones are, greater exploitation of environmental resources like water and land in peri-urban areas, rapid unplanned development in peri-urban areas, stress on already existing service provision and infrastructure and declining political representation. These negative transformations result in various conflicts and contestations between the core city and the peri-urban areas (Singh 2020). This calls for taking a look at the peri-urban areas from a spatial equity perspective or balanced spatial development perspective. State or the government has an important role to play as far as addressing these negative outcomes of peri-urbanization is concerned, in particular the provision of services such as water and sanitation and bringing about planned development of these areas. Peri-urban areas thus entail a greater role for the state to manage, facilitate and regulate the increasing physical, demographic and economic developments.

2 Methodology

The study focusses more on the peri-urban areas as it is here that the interventions are more so required. Other than secondary data analysis, the study has been carried out through primary data based on household surveys for the four selected peri-urban settlements (micro-case study areas), namely Thiruvottiyur municipality (which lies adjacent to Corporation of Chennai and was annexed to it in 2012), Anakaputhur municipality, Thirumazhisai town panchayat and Irungattukottai village (adjacent to Chennai Metropolitan Area boundary). The concept of a peri-urban area is very dynamic and what is rural today might be peri-urban in times to come. Taking cue from this Irungattukottai was selected from beyond the existing limits of Chennai Metropolitan Area. The Census of India, 2011, identifies it as part of the Chennai Urban Agglomeration. Moreover, land from this village was acquired for developing the SIPCOT (State Industries Promotion Corporation of Tamil Nadu Ltd.) Industrial Park. A sample size of 1% was taken for all the case studies and a total of 350 households were surveyed. A pilot survey was conducted to finalize the questionnaire for the final household surveys in the selected settlements.

Further, a debate is undertaken on the governance of metropolitan regions for efficient, cost effective and equitable provision of services and the way planning needs to be perceived for these areas. This chapter draws upon the gaps and challenges identified by the detailed work drawn from secondary as well as primary sources of data. These are then utilized to build a strategy which can be efficiently applied for better provision of services and planning in peri-urban Chennai in particular and other metropolitan peri-urban areas in general. The discussion on polycentric metropolitan governance versus monocentric metropolitan governance and the concept of 'Home Rule' as given by Bahl (2011) and Bahl and Linn (1992) will also be drawn upon to contribute to strategy building for better service delivery and planning of metropolitan areas. Also, it has been seen how a multistakeholder approach can benefit service delivery with greater role for the peri-urban stakeholders.

The chapter is divided into six sections. It starts with introducing the study area and looking at some of the physical transformations taking place in peri-urban areas of Chennai city. This has been carried out with the help of maps sourced from Bhuvan portal and two existing master plans of Chennai Metropolitan Area (CMA). Next, there is a discussion on the institutional set-up in CMA and on the differential access to services and resources in the peri-urban area. Finally, the last section deals with strategy building and way forward for a more balanced distribution of services in the peri-urban areas using a multistakeholder approach.

3 Chennai Metropolitan Area and Physical Transformations in Its Peri-Urban Areas

The Chennai Metropolitan Area (CMA) comprises of the city of Chennai and contiguous area around it. It has a population of 4.6 million (Census, 2011) (Fig. 1). It was notified in 1974. It extends over 1189 sq.km. CMA comprises of Chennai City and areas to an extent of 376.58 sq.km in Kancheepuram District and 639.39 sq.km in Thiruvallur District. According to 2001 Census, 38.6% of the population of Kancheepuram District and 57.5% of the population in Thiruvallur District live within CMA. Chennai city which was earlier comprised of only the District of Chennai, after the expansion of city limits in 2012 also includes parts of Thiruvallur and Kancheepuram districts. The outlying area consists of one cantonment, 16 municipalities, 20 town panchayats and 213 village panchayats in 10 panchayat unions. In December 2011 the limits of Chennai city were increased to annex certain municipalities, town panchayats and villages from the adjoining districts of Thiruvallur and Kancheepuram after which the outlying or peri-urban area of Chennai city consists of 1 cantonment, 11 municipalities, 8 town panchayats and 181 villages arranged in 10 Panchayat Unions (Fig. 2) (Singh 2020). This is also one form of spatial transformation taking place around metropolitan cities.

Some other types of physical transformations in peri-urban areas include change in built-up and land use. Built-up area has been continuously increasing over the periods from 1973 to 1980 to 1991 and 2006 (Fig. 3). Taking the built-up together it accounts for 83.17% in 2006 as against 71.62% in 1973 for Chennai city. As far as the rest of CMA is concerned, the built-up comprised 13.69% in 1973, which increased to 29.17% in 2006. Thus, there was almost a 100% increase in built-up in

Fig. 1 Settlements in CMA and old Chennai city boundary. *Source* Author based on Census

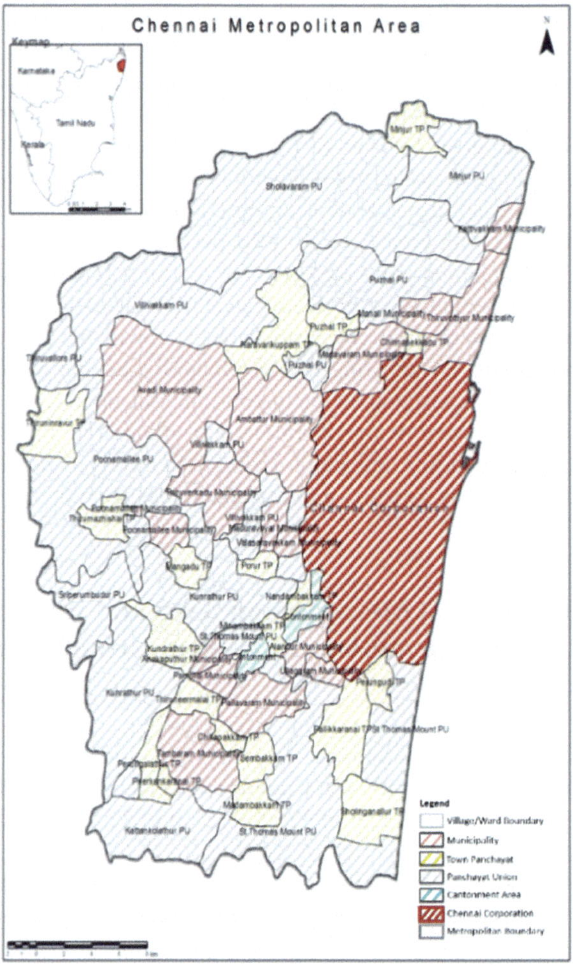

peri-urban Chennai over the 33 year period. Water bodies on the other hand, have registered a decline during the same period. From 1991 onwards more and more areas towards the south and west of Chennai city have experienced rapid increase in built-up. This coincides with the post-liberalization period, when the metropolitan cities underwent rapid economic and spatial transformation.

From 1973 to 2006 there has been a drastic fall in land under agriculture from 80,000 hectares in 1973 to just about 12,000 hectares in 2006, whereas the land under industrial use has increased from around 5,000 hectares in 1973 to around 8,000 hectares in 2006 (Fig. 4). The reason for the drastic fall in agriculture is the fact that farmers have discontinued farming due to easy access to other employment opportunities in the core city as well as the peri-urban areas. Also, farmers have been selling water from their agricultural fields which earns them more than what they

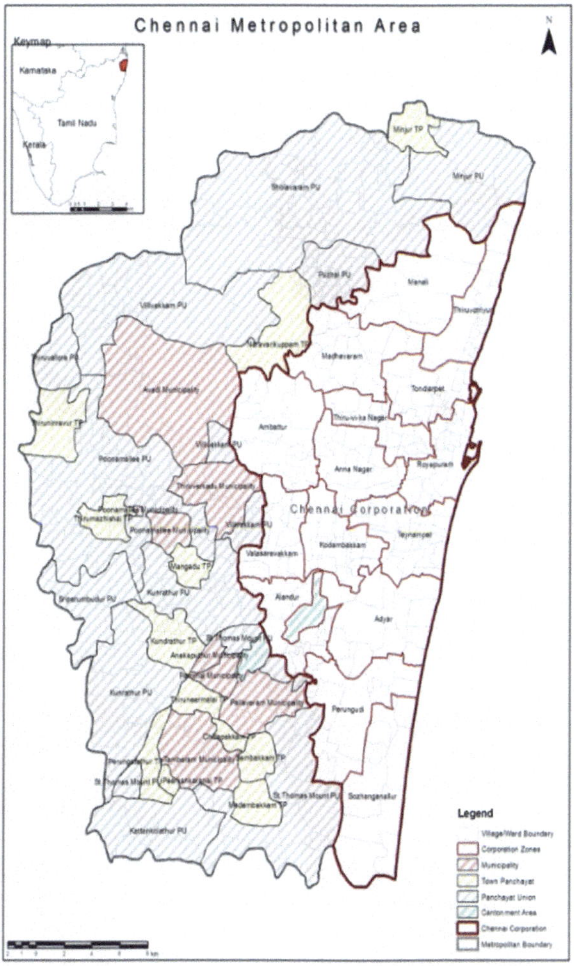

Fig. 2 New Chennai city boundary and CMA. *Source* Author based on Census

used to get from cultivating their fields (Ruet et al. 2007; Primary survey, June 2014). Panchayat Unions have experienced the most radical fall in agricultural land and the increase in industrial land is also the highest for them. It can thus be said that the peri-urban areas have become part of 'city dynamics' over this 33 year period and more so in post-1991 scenario.

Another transformation in the peri-urban areas of metropolitan cities is the rampant growth of census towns. As per Census 2001, Chennai Urban Agglomeration had 50 towns/cities. This number has increased to around 90 in Census 2011. The increase has been mostly by addition of census towns (30) and outgrowths (10). These census towns are administratively still rural and thus not well equipped to deal with the rapid spatial and economic transformations taking place within them. They don't have the resources or capacity to cater to the increase in population they are

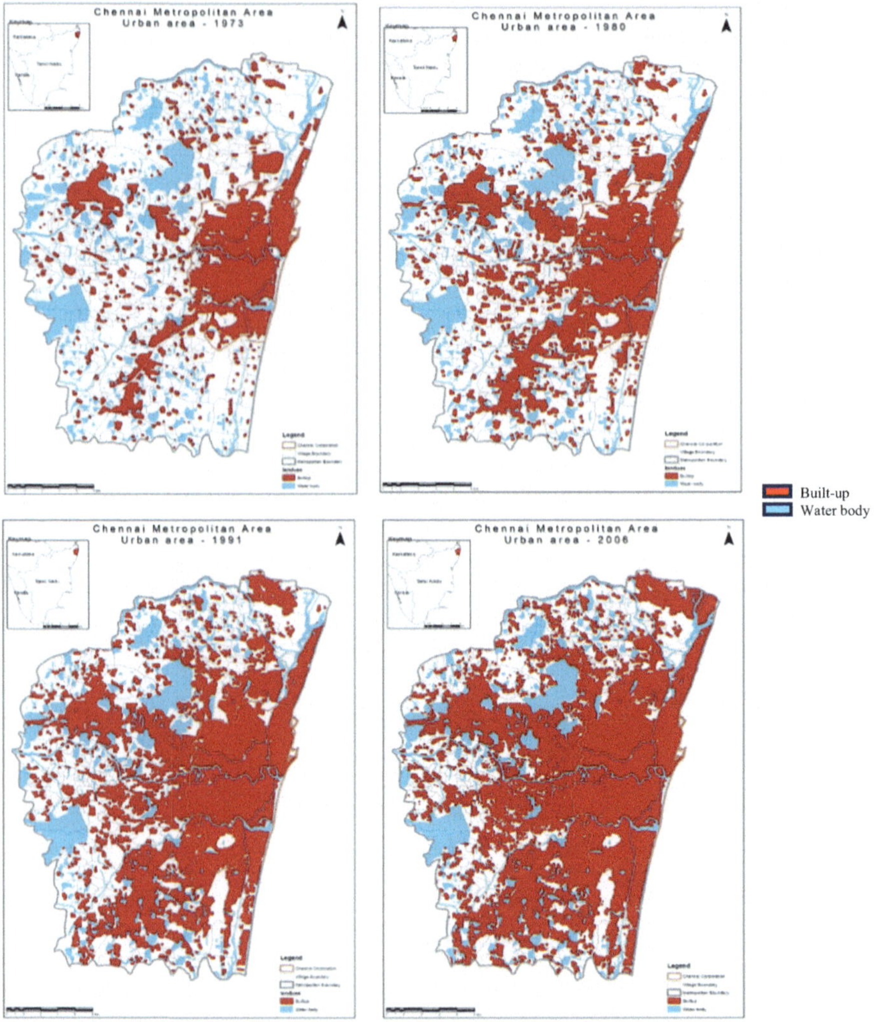

Fig. 3 Changes in built-up area and water bodies for CMA 1973, 1980, 1991, 2006. *Source* Bhuvan and author's own analysis

experiencing. The maximum number of census towns have come up in the Sriperumbudur Taluk to the west of Chennai city, which is becoming the industrial hub of CMA.

Chennai Metropolitan Area has experienced major transformation in the land use for its peri-urban areas. The land under agriculture reduced from around 75% to just about 12%, indicating the pressure exerted on land as a resource. This also highlights the conflicts in use of land when it comes to built up versus agriculture uses (Figs. 5 and 6). Over the 33 year period of 1973–2006 the CMA area has undergone massive

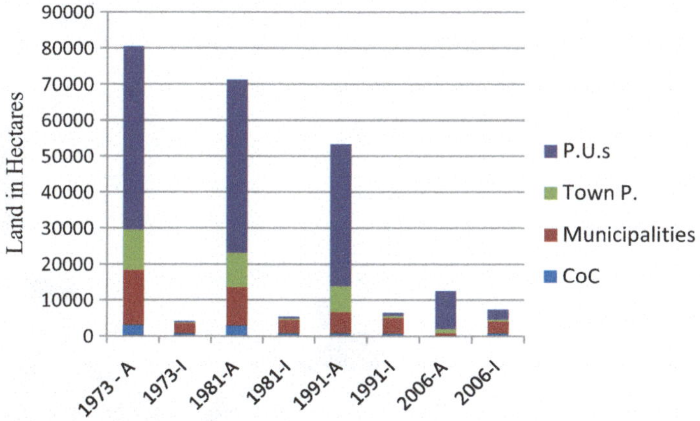

Fig. 4 Comparing area under agricultural (A) and industrial (I) land uses for panchayat unions (P.U), town panchayat, municipalities and Chennai Corporation, 1973–2006. *Source* Unpublished data from CMDA and own analysis

change in land use. It is clearly discernible by comparing maps 4 and 5 that the area of CMA outside the Corporation of Chennai has experienced extensive infilling. The land use for Chennai city has seen the highest change in the residential category, which has increased over a period of time. However, in peri-urban CMA significant change is observed in several categories of land uses such as agriculture, water body, residential, industrial, green.

If we take a look at the change in land use over the two existing plan periods for the peri-urban part of the Chennai Metropolitan Area, the maximum fall has been in case of agriculture from around 70% to 11.72%. The highest increase has been under the others category of land use from 10.3% to 57.31%. Significant increase has also been experienced under the residential category, from 8.67% to 19.92% and industrial category, from 2.82% to 6.14% (refer Figs. 7 and 8). The increase in others category of land use can be attributed to the fall in agriculture as an economic activity in CMA and the farmers becoming seller of water from their now unused agricultural farms. Moreover, the transportation component of land use, which falls in the others category, has increased drastically over the two periods especially in the peri-urban areas.

Other than these two plans no other existing land use plans are available for CMA. Keeping the horizon year for the Master Plan of Chennai as 2026, future demands were worked out and the Second Master Plan was prepared by CMDA. The revised draft Second Master Plan (proposed Chennai Master Plan 2026) was submitted to government by CMDA in December 2005 for approval. Chennai Master Plan 2026 (Second Master Plan) was finally approved on 2 September 2008 and was notified in the Tamil Nadu Govt. Gazette. This proposed land use plan for CMA for 2026 has included even more areas under industrial land use and the green areas have been further reduced. This will add to the land and water stress in the peri-urban areas

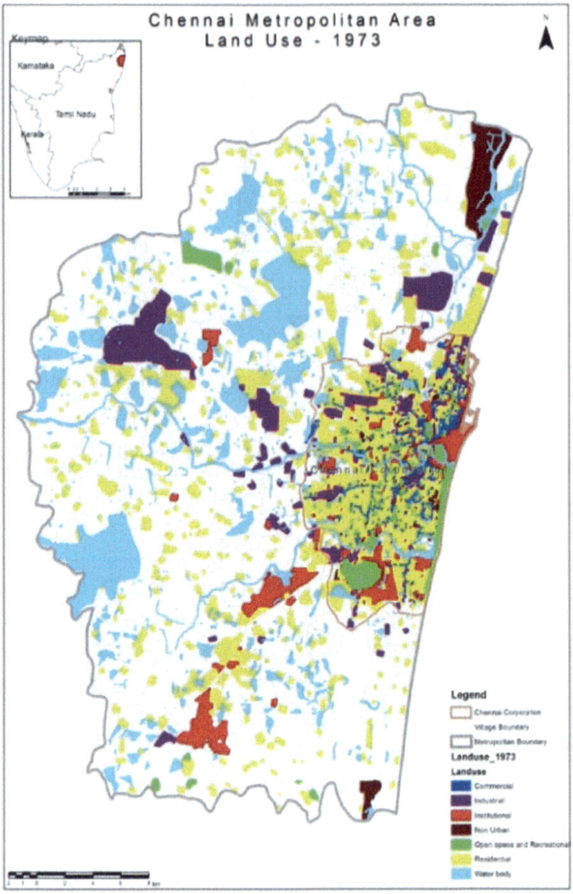

Fig. 5 Existing land use for Chennai Metropolitan Area, 1973

and further impact the access to services with larger inflow of population (Second Master Plan for Chennai Metropolitan Area, 2008).

4 Water Resource and Its Spatial Distribution in CMA

Chennai city has always been water scarce area due to lack of any perennial river in its vicinity and a limited period of rainfall during the north-east monsoon period. Thus, Chennai Corporation has always depended on multiple sources of water for supplying water to the city of Chennai, which is the sixth most populated city in India (Census of India, 2011). It still depends on at least four large reservoirs (Poondi, Puzhal-Red Hills, Chembarambakkam and Cholavaram), at least three river systems (Kaveri, Krishna and Palar), two desalination plants at Minjur (100 MLD) and Nemelli (100 MLD) (with two more being at various stages of planning and execution) and several

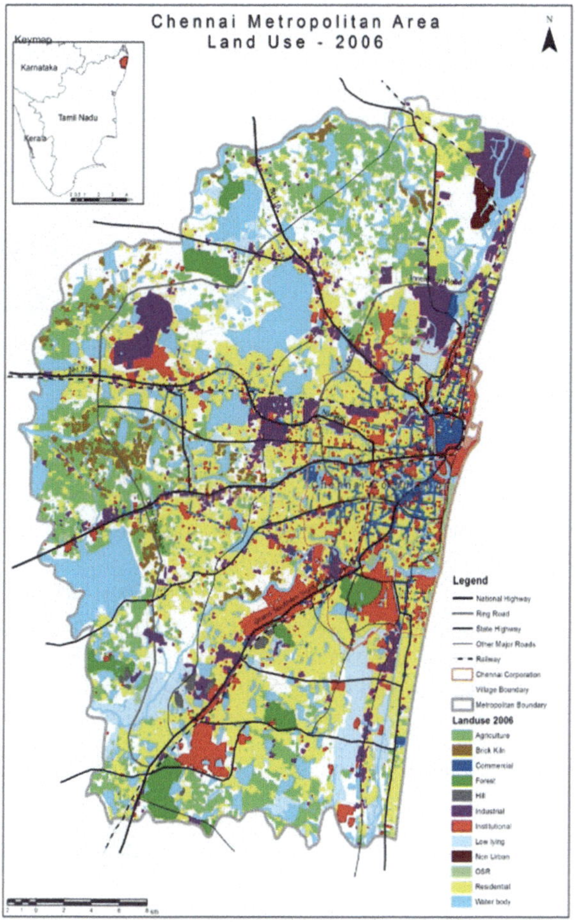

Fig. 6 Existing land use for Chennai Metropolitan Area, 2006

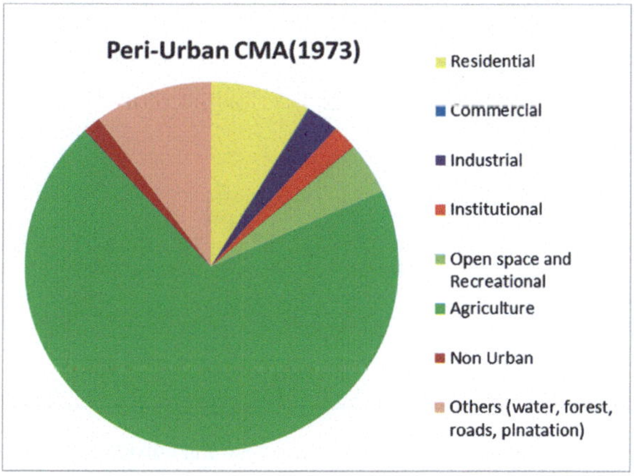

Fig. 7 Land use for peri-urban areas of Chennai Metropolitan Area, 1973. *Source* First Master Plan for Chennai Metropolitan Area, 1973

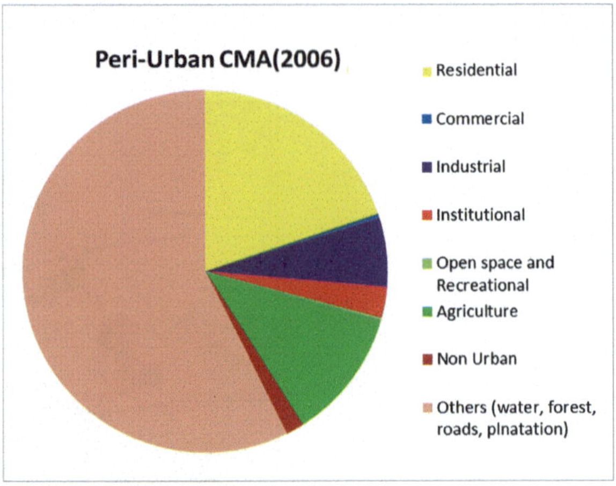

Fig. 8 Land use for peri-urban areas of Chennai Metropolitan Area, 2006. *Source* Second Master Plan for Chennai Metropolitan Area, 2008

hundreds of groundwater aquifers across the city and in its hinterlands (Interview with officials at CMWSSB) (refer Fig. 9).

These various sources of water in the peri-urban areas of CMA and beyond are serving the city of Chennai and its ever-growing population and even catering to the demand of industrial establishments in CMA. Only around 6.8% of the total water being sourced by these multiple sources is being supplied to the peri-urban

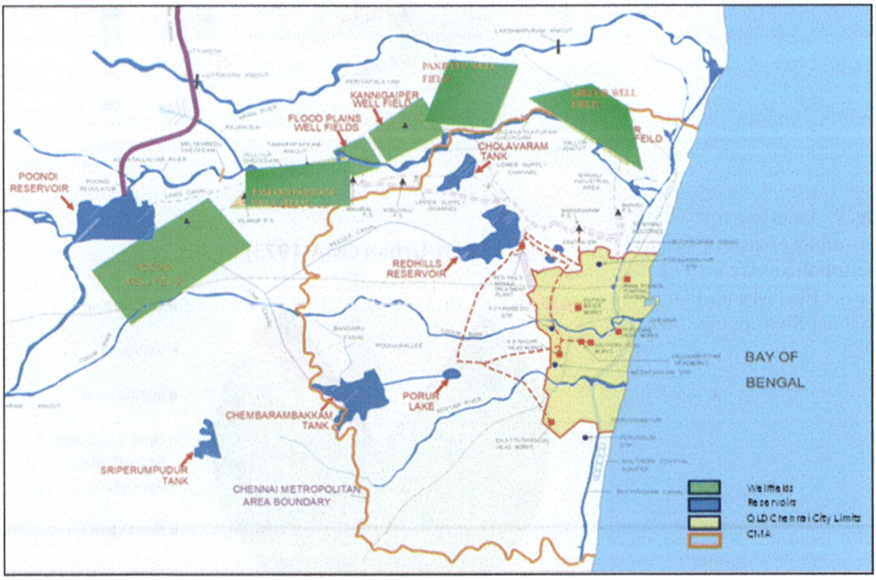

Fig. 9 Various sources of water in CMA and outside its boundary. *Source* Unpublished map from CMWSSB

Water Sourcing	Quantity (in mld)
Redhills to Kilpauk Water Treatment Plant (WTP)	150.75
Redhills to Puzhal WTP	100
Redhills to Surapet WTP	8
Chembarambakkam WTP	120
Veeranam Project	180
Wellfield	15
Porur to KK Nagar	2
South coastal aquifer	4
Desalinated Water from Minjur	80
RO plants	0.25
Total	660

Water supply distribution	Supply (in mld)
City supply through pipeline	565
City supply through tanker lorries	15
Industrial supply	35
Bulk consumers & local bodies	45
Total	660

Fig. 10 Sources of water and water supply in CMA, 2016. *Source* Unpublished data from CMWSSB

settlements in CMA. This is despite the fact that CMWSSB (Chennai Metropolitan Water Supply and Sewerage Board) is supposed to supply water to the entire CMA as per its statute (refer Fig. 9). If we look at the water that is being supplied for industrial use, it comes to almost 5.3% (refer Fig. 10). Further it has been found that more than 60% of the water used in CMA comes from sources other than the public system (Primary Survey 2014; Ruet et al. 2002). Thus there are multiple conflicts and contestations with respect to accessing water in the peri-urban areas of CMA.

5 Multiplicity of Institutions in Peri-Urban Chennai for Service Provision and Planning

The institutional landscape in CMA is as per constitutionally prescribed ULB structure with the municipal bodies at the top. Besides the hierarchy of municipal bodies including municipal corporations (Chennai Municipal Corporation (CMC)) and other municipalities, there are corresponding urban and rural local bodies in town panchayats and village panchayats. The rest of the institutions, especially parastatals such as Chennai Metropolitan Development Authority (CMDA), the planning authority in CMA and Chennai Metropolitan Water Supply and Sewerage Board (CMWSSB) appear to be the institutional responses to the metropolitan scale of challenges in CMA. In addition to these, there are other agencies and line departments conventionally entrusted with provision of various services. A metropolitan region thus holds many different territorial and functional entities (Singh 2020).

6 Micro-Case Studies in Peri-Urban Chennai and Delivery of Services

The status of delivery of services in peri-urban areas has been studied through primary data collected for the four selected peri-urban settlements, namely Thiruvottiyur municipality, Anakaputhur municipality, Thirumazhisai town panchayat and Irungattukottai village (Fig. 11).

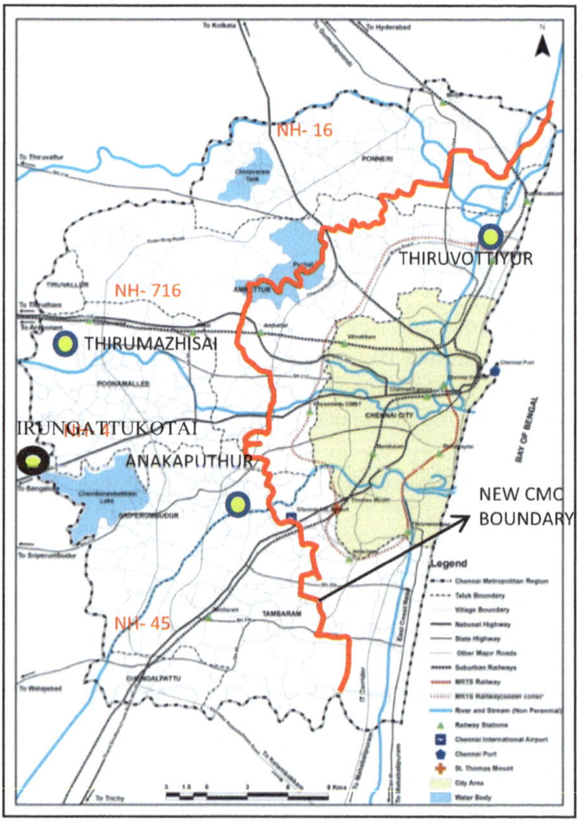

Fig. 11 Location of micro-case studies in CMA

6.1 Water

The situation of availability of water in all the micro-case study areas is very dismal. The supply of tap water is being carried out by the municipal authority in Anakaputhur, by the town panchayat in Thirumazhisai, municipality earlier and now by CMWSSB in Thiruvottiyur and by TWAD Board (Tamil Nadu Water Supply and Drainage Board) in Irungattukottai (Table 1). All the households in these areas are dependent upon multiple sources of water such as tap water, borewell, tanker water and packaged water for their day-to-day needs (Table 1). Anakaputhur has borewells, well and piped water supply sources. In Thiruvottiyur all households have borewells and common taps provided by the municipality are an important additional source. In Thirumazhisai tanker water supply by local government is an important source of water supply (Table 1).

The provision of piped water supply is also the highest in Anakaputhur municipality at 68% followed by Thiruvottiyur and Thirumazhisai (Table 1). Irungattukottai village has no provision for piped water supply. When asked whether water being supplied by government is sufficient for daily needs, more than 69% households in all the selected local bodies felt that it was not sufficient or it just met the necessary needs such as cooking and drinking (Table 2). Another issue related to this is that all the surveyed households stored water for use. This results in wastage of water when compared to a situation where water is supplied regularly. This needs to be avoided in such water-stressed regions.

It was also seen that there was overarching dependence on packaged drinking water in Anakaputhur (75%) and Thiruvottiyur (79%) (Table 2). This was primarily because the purchasing power was higher in these two settlements. The average income of households in Anakaputhur and Thiruvottiyur ranged between Rs. 25,000 and Rs. 40,000 per month as compared to Thirumazhisai and Irungattukottai where

Table 1 Sources of water being used in selected case study areas, 2014

Municipality/town panchayat/ village	Sources of water being used	Percentage of houses having piped water supply and agency supplying it
Anakaputhur	Borewell, piped water supply, tankers (98% have all)	68 (Municipal water supply)
Thiruvottiyur	Borewell (100%), Common taps/handpumps (30%), Piped Water Supply (16%)	30 (now CMWSSB, earlier municipality)
Thirumazhisai	Tanker (59%), Borewell (40%), Piped water supply (28%) [only 16% have more than one source]	27 (Town Panchayat in coordination with TWAD)
Irungattukottai	Borewell, tanker	TWAD—no piped water supply

Source Primary Survey, June 2014

Table 2 Sufficiency of water provided by government sources for daily needs, 2014

Municipality/town panchayat/ village	Percentage of household saying no or that only necessary demands being met	Percentage of households purchasing packaged water for drinking
Anakaputhur	80	75
Thiruvottiyur	69	79 (brackish water being a major issue)
Thirumazhisai	70	45
Irungattukottai	65	22

it was between Rs. 5,000 and Rs. 15,000 per month. Thus, low-income and distant settlements (from Chennai city) of Thirumazhisai and Irugattukottai had relatively less dependence on packaged water for drinking. In addition, the issue of brackish underground water was a major problem in Thiruvottiyur as it is located adjacent to the coast. The problem in Thiruvottiyur thus gets further complicated because the other areas still have some dependence on open wells and borings.

Chennai Metropolitan Area has a major issue when it comes to frequency and duration of water supply. Anakaputhur and Thiruvottiyur, both have institutional tap water supply only once a week for a duration of 2 h. These areas have heavy dependence water supply by private tankers. The situation for Thiruvottiyur was largely the same in the pre-annexation scenario. Thirumazhisai has everyday supply of water by municipal pipelines for one or two hours. Irungattukottai also has everyday supply of water through community taps located in the village (Table 3).

Quality of water being supplied by government agencies is better in the nearby municipalities as compared to the far-off ones. In Anakaputhur and Thiruvottiyur more than 82% households found the quality to be good (Table 3).

Almost 25–50% of the total household income is spent on accessing water from sources other than government (Table 4). Fifty percent households in Anakaputhur and 61% in Thiruvottiyur spent 25–50% of their total household expenditure on these other sources of water. There is lesser expenditure on these other sources in

Table 3 Frequency, duration and quality of piped water supply by government, 2014

Municipality/town panchayat/ village	Frequency and duration of piped water supply	Percentage of households saying quality is good or average
Anakaputhur	Once a week and two hours a day	82
Thiruvottiyur	Once a week and two hours a day	83
Thirumazhisai	Everyday	60
Irungattukottai	Every day through community taps	58

Source Primary Survey, 2014

Thirumazhisai and Irungattukottai. Again the reason being low purchasing power in these settlements.

As already discussed, there are multiple state institutions supplying water in Chennai Metropolitan Area. Nevertheless, it is found that more than 81% of households in all the selected local bodies wanted that a government institution should only be made responsible for supplying water to them (Table 5).

Various reasons have been cited by the respondents in the four micro-case study areas for choosing the government agencies over private ones. The main reasons are regularity of water supply, greater access to water, better quality of water and reduced expenditure on water (Table 5).

Regarding the preference of people with respect to the various state agencies supplying water in Chennai Metropolitan Area, namely CMWSSB, TWAD or local body, it came out that the town panchayat of Thirumazhisai and Irungattukottai village have much greater support for CMWSSB. More than 90% of households surveyed in these areas wanted CMWSSB as the water supplying agency (Table 5). Anakaputhur and Thiruvottiyur, on the other hand, did not have any specific preference between the municipal body supplying water or the CMWSSB. After annexation of Thiruvottiyur

Table 4 Household expenditure on other than institutional sources of water, 2014

Municipality/Town Panchayat/village	Percentage of total household expenditure on other than institutional sources of water
Anakaputhur	50% households with 25–50% expenditure
Thiruvottiyur	61% households with 25–50% expenditure
Thirumazhisai	26% households with 25–50% expenditure
Irungattukottai	12% households with 25–50% expenditure

Source Primary Survey, 2014

Table 5 Percentage of households wanting government agency/CMWSSB to supply water to them and reasons for it, 2014

Municipality/town panchayat/village	Percentage	Major reasons for supporting government supply	Percentage of households preferred CMWSSB
Anakaputhur	95	Regularity/greater access to water, good quality	40
Thiruvottiyur	91	Regularity/greater access to water, less expenditure	30
Thirumazhisai	81	Reduced bill	90
Irungattukottai	91	Greater access to water, good quality, less expenditure	93

Source Primary Survey, June 2014

Table 6 Percent of households having toilet facility within their premises and source of water for toilets, 2014

Municipality/town panchayat/ village	Households having toilet facility (in percent)	Source of water for households having toilet facility
Anakaputhur	100	Groundwater and wells (100%)
Thiruvottiyur	99	Groundwater and wells (88%), tankers (12%)
Thirumazhisai	84	Groundwater and wells (60%), tankers (27%), tap (17%)
Irungattukottai	15 (rest have community toilets)	Community taps (100%)

in Corporation of Chennai, new pipeline for water supply is being laid for the entire area by CMWSSB.

6.2 Sanitation

Sanitation is closely linked with household water supply. Anakaputhur had all the households with toilet facility within their premises. Thiruvottiyur had 99% households with toilet facility (Table 6). Almost 29% of these household have sewer connection as well, which have been laid after CMWSSB took over. The other local bodies just have septic tanks for the toilets. However, Thirumazhisai has only 84% households with toilets. The worst situation is in Irungattukottai where only 15% households have toilet facility within their premises. When the multinational company of Hyundai acquired land in this village they constructed several community toilets. Yet, these toilets do not have tap water connections due to dismal water supply condition in the village. These toilets are thus not being used by the villagers.

Source Primary Survey, June 2014

Goundwater and wells are the main sources of water supply for the toilets in Anakaputhur and Thiruvottiyur (Table 6). This is so because the frequency of tap water supply is very less in these areas. For Thirumazhisai, tap water supply also features as one of the sources because as given in Table 3, frequency of tap water supply is every day.

6.3 Planning and Governance

Chennai Metropolitan Development Authority (CMDA) is responsible for preparing spatial development plans (Land Use Plans) for the CMA. The town planning departments in the local bodies have no role to play in the plans being prepared by CMDA.

Neither do they have the resources nor the capacity to prepare a plan for their respective settlements. More than 81% households in the selected case study areas have no ideas about any spatial plan being prepared for their area (Table 7).

In Thiruvottiyur and Anakaputhur, which are adjacent to old Chennai city limits, more than 90% households want to be part of any such spatial planning process being undertaken for their area. However, in Thirumazhisai town panchayat only around 38% households want to be part of such plan making process. The reason cited is that they feel that there is no use of being part of any such spatial plan making process (Table 7).

In all the micro-case study areas almost 84% or more population have seen/met their councillors/sarpanch (Table 8).

When asked about the frequency of the visits of the councillor, 14–33% of the households say that the councillors come only once in five years (Table 8). In case of Thiruvottiyur 50% of households said that the councillors visited more often in the pre-annexation scenario. This brings out the fact that representation was better in the period when it was not annexed to Chennai city as one of its wards. This becomes significant given the fact that the percentage of households which say that they want to be part of the spatial plan making process is also high.

Table 7 Awareness of households about spatial plan being prepared, 2014

Municipality/town panchayat	Percentage of households not aware	Percentage of households saying they would want to be part of such plan making process
Anakaputhur	90	100
Thiruvottiyur	81	90
Thirumazhisai	98	38
Irungattukottai	99	35

Source Primary Survey, June 2014

Table 8 Percentage of households that had seen/met their councillor/sarpanch and the frequency of their visit, 2014

Municipality/town panchayat	Percentage of households saying yes	Percentage of households saying the councillor comes only once in five years
Anakaputhur	97	24
Thiruvottiyur	92	33 (in pre-annexation scenario 50% say the councillors visited more often)
Thirumazhisai	84	14
Irungattukottai	100	Sarpanch husband is available

Source Primary Survey, June 2014

When asked whether the household's suggestions get incorporated by the councillors/sarpanch it comes out that only Thiruvottiyur is performing well with 89% households saying yes (Table 9). In case of Irungattukottai village it is the lowest and people in the village say that the previous Sarpanch's wife is the current sarpanch and it is actually her husband who manages all the affairs.

The respondents were also asked about the improvements required in their town, village or ward. The issues which came out very predominantly in all the case studies are that water, sewerage and road need to be improved/upgraded. Water is the most pressing demand in all the settlements followed by sewerage and roads (Table 10).

Water, pollution and overcrowding due to migration have emerged as the most important issues leading to further transformations in peri-urban areas. These put more pressure on land and water resources which are being sourced from peri-urban areas of metropolitan cities. Thus, resulting in conflicts amongst the various local bodies. It is in this context that role of the state assumes great importance. Distance from the core city is also a consideration when one looks at the access to services. The heavy dependence on private and informal sources of water also got established from the primary survey.

Table 9 Percentage of households which Say that their suggestions get incorporated by the councillors, 2014

Municipality/town panchayat	Percentage of households saying yes
Anakaputhur	54
Thiruvottiyur	89
Thirumazhisai	49
Irungattukottai	39

Source Primary Survey, June 2014

Table 10 Improvements required in your ward/town

Municipality/town panchayat	Improvements required (in order of preference)
Anakaputhur	Water, sewerage, road
Thiruvottiyur	Water, sewerage, road
Thirumazhisai	Water, sewerage, road
Irungattukottai	Water, sanitation

Source Primary Survey, June 2014

7 Institutional Set-Up in Chennai Metropolitan Area and Discussion on Polycentric Versus Monocentric Governance System

As discussed in the preceding sections of the chapter, metropolitan regions are very complex and are undergoing multiple transformations. Moreover they are comprised of hierarchy of settlements which are seldom ready to deal with the scale of these changes; spatially, environmentally as well as financially. It thus becomes imperative to discuss how these complexities and transformations can be managed by better institutional set-up so that the access to various services is improved especially for the peri-urban areas.

The main aim of metropolitan cities of large size with respect to service provision is based on three factors: efficiency, economy and equity (Howell-Moroney 2008). It also needs to be mentioned here that metropolitan region is largely a legal non-entity (Ostrom et al. 1961). In India, 74th Constitutional Amendment Act provided for the creation of Metropolitan Planning Committees, thus giving some level of legal status to metropolitan areas. However, most of the state governments have shied away from creation of these committees due to political and economic reasons. This holds true for CMA as well. This act gave constitutional recognition to decentralized service provision as well (Singh 2020).

In CMA the urban local bodies, Tamil Nadu Water Supply and Drainage Board (TWAD) and CMWSSB are all involved in provision of water and sanitation facilities. District Planning Committee and Chennai Metropolitan Development Authority are present for integrated planning; however, there is no provision for a Metropolitan Planning Committee. Also, the District Planning Committee is not making any plan in the CMA region. The CMDA prepares plans for the entire CMA without any representation of the urban and rural settlements in the peri-urban Chennai.

A political system with a single dominant centre for making decisions is termed as a ‘Gargantua’ for a metropolitan region (Howell-Moroney 2008). This is a form of consolidated governance system for metropolitan regions, whereby only the metropolitan level authority has the power to take decisions with respect to service provision and development. In context of Chennai the plan making process currently underway and the annexation carried out in 2012 is similar to this kind of a political system.

On the other hand, a pattern of governance in a metropolitan area with multiplicity of political jurisdictions is termed as a ‘polycentric political system’. It is taking into account these factors that polycentric or consolidated or a mix of both kinds of governance structures can be considered for metropolitan areas for development of a strategy for planning and service delivery (Singh 2020).

Over a period of time with greater urbanization and globalization there has been an emergence of small towns in the periphery of large metropolitan core as new economic hubs which has resulted in the formation of polycentric functional structures within urban regions (Growe 2012). This is true for Chennai as well. Various towns and villages which act as economic hubs, like those on IT Corridor along East

Coast Road, Electronics and Engineering Industries in Tambaram, Poonnamalle, Sriperumbudur, etc., have developed all around the city of Chennai.

A polycentric system of governance which brings about separation of powers creating a system of checks and balances (Laffont 2009 quoted in Pethe et al. 2011, p. 189) would be more suitable for a metropolitan region like the CMA. As far as planning in CMA is concerned, CMDA has a kind of unfettered control over the city which is rather unrepresentative of the peri-urban local bodies. Only civic services are being looked after by the local bodies in CMA and not planning function. The set-up of CMDA Board clearly depicts the low representation of the local bodies (Primary survey, 2014). Only four representatives out of 22 are from the local authorities. Moreover, final say with respect to acceptance, changes and rejection of master plan lies with the State Government due to its predominant presence in the board. Also, though the State Government passed an Act regarding the constitution of the 'Metropolitan Planning Committee' in 2009, it has not yet been notified (Singh 2020).

Further, drawing from Bahl's (2013) concept of 'home rule' it can be said that annexation and a monocentric government hampers citizen's control over their own resources. Bahl and Linn (1992) suggest that under a jurisdictional fragmentation approach, many local governments operate in the same metropolitan area with some degree of independence in choosing their financing arrangements. In some cases, there also is an overlying metropolitan government or region-wide special district, but the emphasis in service delivery is on the role of the lower-tier governments. The advantage of the jurisdictional fragmentation model is that it keeps government close to the people. However, the welfare gains from this home rule model will come at the cost of failure to capture economies of scale, and operating within a set of boundaries that are arguably too small to allow coordinated service delivery. The traditions of home rule are particularly strong in the United States where urban services are delivered by municipalities or counties (OECD 2009a quoted in Bahl 2013).

8 Conclusion

It comes out clearly from the discussion above that water supply and sanitation are the two most important improvements sought by households in the peri-urban areas of Chennai city. The condition of water supply is particularly dismal and because sanitation is closely linked, it gets impacted as well. The local authorities are supplying water but, both, the quantity as well as the quality are not up to the mark. CMWSSB emerges as the most preferred agency to supply water. The quality of water and pricing emerges as two clear reasons for preferring CMWSSB. It has more resources as well as the capacity to cater to the entire CMA. However, even CMWSSB is not supplying the entire area it is mandated to supply. Thus, the private water tankers and other private actors fill the space left void by the institutional service providers. However, if we consider the case of Thirumazhisai town panchayat, the picture looks different. The town panchayat has been efficiently supplying water to its residents and has a system of local government tankers as well as piped water supply which cater to the area. This proves the point that equity aspect can be addressed in a decentralized governance system.

Moreover, the current plan making exercise which is being carried out by CMDA is more of a top-down approach, rather than a bottom-up approach, which does not consider these stresses on available resources in the peri-urban areas. As discussed in Sect. 16.3 on transformations, the proposed land use plan for 2026 has added more industrial areas in peri-urban Chennai which would further add to in-migration and result in even more demand for water and land resources. The green areas around the well-fields in the northern part of CMA have also been reduced in the proposed 2026 plan. It is thus required that some sort of arrangement is developed so that these peri-urban areas can address the challenges that they face due to the rapid transformations being experienced by virtue of their location near a metropolitan city.

9 Strategy for Metropolitan Governance and Planning

To start with, it is required that the metropolitan areas have some autonomy from the state control so that their development can be guided by forces of economy and decision by the citizens of the peri-urban areas. For better coordination amongst the local bodies, given the fact that the metropolitan areas are large entities and have multiple jurisdictions, some metropolitan level governing authority like the CMDA can be continued but with a more democratic and decentralized mandate. Further, the local bodies in the peri-urban areas should have enough say in the matters concerning their fate as far as service delivery, environment, use of land/water resources and economic development are concerned. Complete functional fragmentation might thus not work well with jurisdictional fragmentation and some level of coordination is desirable for metropolitan regions.

However, this does not mean that one has to go back on the issue of functional decentralization as enshrined in the 73rd and the 74th Constitutional Amendment Acts. The local bodies should have functional as well as financial autonomy. However, there is a lot of disparity when it comes to financial status of these local bodies because of which the level and quality of service provision varies from settlement to settlement. The metropolitan level body, in this case the CMWSSB, should be made responsible for easing out these inequalities by way of facilitating equitable levels of service delivery in the peri-urban areas. Thus, the sourcing, treatment, laying of network and distribution at local body level should be given to CMWSSB. TWAD should be limited to areas outside CMA. The local bodies need to be made responsible for local distribution of water within the local body jurisdiction and also try to reduce the dependence of households on other informal sources, such as groundwater and tankers. Continuous monitoring by the citizens through e-participation by developing a system of e-feedbacks should also be encouraged by the local body and CMWSSB. A robust system of grievance redressal also needs to be developed for the service provision in local bodies.

As far as planning is concerned, MPC should be set-up in these areas with enough power to enforce plan making in the various constituent local bodies of the peri-urban areas as well as the core city. This again requires political will and autonomy of the Metropolitan Planning Committees (MPC). The planning authorities in local bodies need to be created with ample capacity to prepare local plans in peri-urban areas. They can also function as a support to the MPC as far as overseeing, compiling and implementation of plans is concerned. If a metropolitan level planning body, like the CMDA has to remain true to its central tasks, it cannot be overwhelmed by this 'default planner' responsibility. Hence, the capacity and responsibility for spatial planning should be developed at the local body level in the metropolitan. The metropolitan development authority, in this case the CMDA, can also provide financial support and capacity building at the local body level. The responsibility for plan making will lie with the local bodies and they will in turn make use of e-community as a platform to get inputs from the citizens on efficient planning (Singh 2020).

References

Aguilar AG (2008) Peri-urbanization, illegal settlements and environmental impact in Mexico City. Cities 25(3):133–145

Arabindoo P (2005) Examining the peri-urban interface as a Constructed primordialism. In: Peri-urban dynamics: population, habitat and environment on the peripheries of large Indian metropolises—a review of concepts and general issues. CHS Occasional Paper No.14. Centre de Sciences, New Delhi

Bahl RW (2011) Financing metropolitan cities. In Local government finance: The challenges of the 21st century, ed. United Cities and Local Governments. Barcelona: United Cities and Local Governments.

Bahl R, Linn J (1992) Urban public finance in developing countries. Oxford University Press, New York

Cadene P (2005) Dynamics of peri-urban areas: from the French case to the developing countries. In: Peri-urban dynamics: population, habitat and environment on the peripheries of large Indian metropolises—a review of concepts and general issues. CHS Occasional Paper No.14. Centre de Sciences Humaines, New Delhi

Chennai Metropolitan Development Authority (1976) First Master Plan for Chennai Metropolitan Area, 2001

Chennai Metropolitan Development Authority (2008) Second Master Plan for Chennai Metropolitan Area, 2026

Dupont V (ed) (2005) Peri-urban dynamics: population, habitat and environment on the peripheries of large Indian metropolises—a review of concepts and general issues. CHS Occasional Paper No.14. Centre de Sciences Humaines, New Delhi

Growe A (2012) Emerging polycentric city-regions in Germany: Regionalisation of economic activities in metropolitan regions. Erdkunde 66(4):295–311

Howell-Moroney M (2008) The Tiebout hypothesis 50 years later: lessons and lingering challenges for metropolitan governance in the 21st century. Public Adm Rev 68(1):97–109

Kundu A, Pradhan BK, Subramaniam A (2002, December 14) Dichotomy or continuum: analysis of impact of urban centres on their periphery. Econ Polit Wkly: 5039–5046

Marshall F et al (2009) STEPS working chapter 35. STEPS Centre, Brighton

Mc Gee TG (1991) The emergence of Desakota regions in Asia. Expanding a hypothesis. In: Ginsburg NJ, Koppel B, Mc Gee TG (eds) The extended metropolis. Settlement transition in Asia. University of Hawaii Press, Honolulu, pp 3–26

Moench M, Gyawali D (2008) Desakota: reinterpreting the urban-rural continuum. Accessed on 24 January 2012 at: https://www.espa.ac.uk/files/espa/Final%20Report%20Desakota%20Part%20II%20A%20Reinterpreting%20Urban%20Rural%20continuum_0.pdf

Narain V (2010) Periurban water security in a context of urbanization and climate change: a review of concepts and relationships. Peri Urban Water Security Discussion Chapter Series, Chapter No. 1. SaciWATERS.

Narain V, Banerjee P, Anand P (2014) The shadow of urbanization: the peri-urban interface of five Indian cities in transition. Environmental Change, Vulnerability and Governance, No. 68. East-West Center, Hawaii

Oliveau S (2005) Peri-urbanization in Tamil Nadu: a quantitative approach. CHS Occasional Chapter No.15. Centre de Sciences Humaines, New Delhi

Ostrom V, Tiebout CM, Warren R (1961) The organization of government in metropolitan areas: a theoretical inquiry. Am Polit Sci Rev 55(4):831–842

Pethe A, Gandhi S, Tandel V (2011, June 25–July 8) Assessing the Mumbai metropolitan region: a governance perspective. Econ Polit Wkly 46(26/27):187–195

Rohilla S (2005) Defining 'Peri-urban'—a review. In: Peri-urban dynamics: population, habitat and environment on the peripheries of large Indian metropolises—a review of concepts and general issues. CHS Occasional Chapter No. 14. Centre de Sciences Humaines, New Delhi

Ruet J, Saravanan VS, Zérah M-H (2002) The water and sanitation scenario in Indian metropolitan cities. Occasional Chapter. Centre de Sciences Humaines, New Delhi

Ruet J, Gambiez M, Lacour E (2007) Private appropriation of resource: impact of peri-urban farmers selling water to Chennai Metropolitan Water Board. Cities 24:110–121

Shaw A (2005, August) Peri-urban interface of Indian cities growth, governance and local initiatives. Econ Polit Wkly: 129–136

Singh C (2020) Metropolitan and peri urban governance in India: case study of Chennai Metropolitan Area. Indian J Public Adm 66(1):60–76

Webster D (2002) On the edge: shaping the future of peri-urban East Asia. The Asia Pacific Research Center

Augmenting City Planning Approaches Through Energy Retrofit: A Roadmap to Sustainable Megacities

Priyanka Dey and Shalini Keshri

1 Introduction

The energy crises of the 1970s were a major accelerator for research on energy consumption and demand patterns across the world (AEEE 2021; Kumar 2017). The increasing need to transition from using fossil fuels as a primary energy source to using renewables has been extensively discussed (Bayulgen and Benegal 2019). At present, the built environment accounts for a major share of the global energy demand (International Energy Agency 2022). The energy demand of the building sector displayed a growth from 194 exajoule (EJ)[1] in 1973 to 418EJ in 2019 (International Energy Agency 2021), accounting for approximately 40% of global energy consumption and more than 50% of global electricity demand (Ala-Juusela and ur Rehman 2020; International Energy Agency 2022; Pyloudi et al. 2015; TERI 2013).

It has become a widely accepted fact that changes in design and planning modus operandi can yield substantial savings in energy. However, even though the urban centers are considered to be the main consumers of energy and resources, the area available for the production of primary energy is limited (Stoeglehner et al. 2011). As the cities of the future prepare to accommodate a large population, the questions about the shape, size, density of cities, and distribution of energy become more complex (Polesello and Johnson 2016). The energy-spatial structure relationship and land-use planning were initially discussed in the 1980s to identify potential energy crisis solutions (Owens 1992). According to The Intergovernmental Panel on

[1] 1 exajoule (EJ) = 10^{18} J, SI unit for energy.

P. Dey · S. Keshri (✉)
Department of Architecture and Regional Planning, Indian Institute of Technology Kharagpur, Kharagpur, West Bengal, India
e-mail: shalini.keshri.96047@kgpian.iitkgp.ac.in

P. Dey
e-mail: priyanka@arp.iitkgp.ac.in

M. Firoz C et al. (eds.), *Urban Planning and Design for Megacities in the Global South*, Advances in 21st Century Human Settlements,
https://doi.org/10.1007/978-981-97-8370-0_17

Climate Change (IPCC), currently urban built environment is responsible for 71–76% of energy-induced global warming and 67–76% of global energy consumption (Seto et al. 2014).

Different studies have discussed urban energy demands, and the issues, and challenges related to the exercise of making cities energy efficient (Conticelli et al. 2017; Espinoza et al. 2018). However, research patterns on energy-conscious cities reveal segmented development. The energy-conscious practices have mostly concentrated on building-level applications, whereas the concept of examining the implications of energy policies at the urban scale through 'energy master planning' is slowly gathering momentum around the globe (Polesello and Johnson 2016; Santillan et al. 2022). The inclusion of renewable energy sources along with concepts of energy efficiency through buildings and infrastructure management systems to achieve '*net zero*' and '*carbon zero*' status serve as a solution to the urban energy crises. With settlements becoming more dynamic, the inclusion of net-zero energy and carbon–neutral ideas in energy and economic policies observe a rise across the world (Cole and Fedoruk 2015; Uspenskaia et al. 2021). In addition, organizations like IBPSA (International Building Performance Simulation Association), AEEE (Alliance for an Energy-Efficient Economy), etc. are initiating research and capacity building for practitioners to spread awareness about the cause (AEEE 2023; IBPSA 2023). As efficiency in urban areas is of the highest premise, the need for material management, exploration of the relation of energy demand to land-use, and analyzing the effectiveness of existing planning strategies, is imperative. Major city and building rating standards worldwide consider curbing the energy demand of buildings, water, waste, and transport as key contributors to the increasing energy gap (LEED 2023).

The study endeavours to emphasize that, to effectively modify energy policy and planning, it is necessary to evaluate cities' impending demands. To achieve this objective, the energy consumption by the built environment and water demand was evaluated, as two variables, to depict a picture of the future energy demand. The cases of two Indian cities, Surat and Puri, were analyzed in two phases. First, the two cities were analyzed to quantify the potential energy demand based on their land-use, both existing and proposed, as per their respective development plans. The second section of the study assessed the associated energy demand generated by the water supply and distribution, based on the per capita need to care for the urban population. Section 2 of the article provides a concise overview of the need to consider the energy demand associated with land-use planning and water, as well as the potential of development plans as a tool for energy-efficient city planning. Section 3 outlines the methodology used for the analysis of Surat and Puri as two cases for understanding the potential energy demand at an urban scale. Section 4 provides a discussion of the findings from the study. Section 5 discusses the recommendations based on the findings to strengthen the implementation of renewable energy strategies and energy-efficient net-zero policies at the urban scale through the use of development plans. The investigation is summarized in Sect. 6.

2 Background

The urban sector is changing at a steady pace, transforming from organic to planned development, to controlled growth patterns, while incorporating sustainable development and resource utilization. The megacities are major economic centers, where population growth has been exponential in the previous few decades. The bourgeoning growth in these megacities has exerted environmental pressures leading to imbalances between the demand for resources and their availability. This may also be attributed to the prevailing concept of '*discard and dispose*' in previous practices. The current life cycle of natural resources and the duration needed to renew these energy sources have presented a formidable problem of energy crises (Liao et al. 2020). Given that the nations observe population growth and increased energy demands, the pursuit of sustainable socio-economic development becomes imperative. However, the adaptation of energy-efficient policies and practices at the urban scale is a multi-faceted exercise, with a high number of variables involved to achieve the desired outcome.

The idea of urban energy management has been discussed since the early 1980s. Different nations have undertaken different approaches to implement policies and strategies to achieve their energy-related goals (Table 1). These measures indicate a clear prioritization of carbon neutrality, low emission developments, and clean, affordable energy. Despite several efforts, the energy consequences of human spatial design remain uncertain (Cheung et al. 2019; Owens 1992). It has also been noted that urban energy planning is usually developed at a very late stage in comparison with urban land-use planning (Chen and Chen 2015). The development of governance and planning systems at a city scale has always been a challenge in the context of energy (Dixon et al. 2014; Dobbelsteen et al. 2020). Energy inefficiency, lack of coordination between different aspects of planning, inequitable energy distributions, and missed climate targets are some of the most common issues caused by the imbalance in the environmental and energy-related developments at the city and regional scales (Cajot et al. 2015).

Recent research has brought to light that energy master planning is becoming imperative for cities. The studies especially emphasize large-scale metropolitans, where cities are already at the peak of their development phase and improper planning would result in urban decay (Dobbelsteen et al. 2014). However, preparing and implementing an energy master plan cannot be performed without a prior stock check of the available energy resources and energy potential mapping for the cities. These factors, being affected by multiple variables, are not easy to quantify, thereby creating major barriers in the pathway of energy planning. Studies suggest that the process of urban energy and land use planning also should be at par with the climate goals, with long-term goals being supported by near-term action plans on social, environmental, and economic targets (Fankhauser et al. 2022; Hepburn et al. 2021). These challenges in energy policy implementations and their effectiveness raise important questions, which have been investigated in this research article:

Table 1 Actions undertaken by different nations for energy policy implementations

Country	History	Plan/policy currently active (2023)	Source
China	• Five-year plans starting from 1953 outlined the development strategies, • Energy became a focus under the 7th Five-year plan (1978), • Current policy focuses on the generation of clean energy.	14th Five-year plan on Renewable energy development (2021–2025)	Zhou et al. (2022)
The United States of America	• The first initiative undertaken through '*Project Independence Blueprint*' (1974) was implemented to target energy-related policy measures, • The formation of the U.S. Department of Energy (U.S. DOE) in 1977 led to formulation of the National Action Plans for energy-related concerns, • At present, the focus is to achieve cost-effective energy efficiency by 2025.	National Action Plan for Energy Efficiency (2025)	U.S. Department of Energy (1974, n.d.)
Germany	• Nuclear Energy policy, introduced in the 1970s was the first initiative toward renewable energy sources, • Under the '*Energiewende*' policy, the use of nuclear power has been phased out, and coal power plants are expected to be retired by 2038.	Energiewende (2038)	International Energy Agency (2023), Mez (2004), World Nuclear Association (2020)
The United Kingdom	• Initiated Energy Policy frameworks through the '*Statement on Energy Policy*' in 1974, • At present, '*Energy White Paper*' has multiple editions, targeting energy concerns in the UK, with the latest focused on the Net-Zero future.	Energy White Paper-2050	Government of the United Kingdom (2020)

(continued)

Table 1 (continued)

Country	History	Plan/policy currently active (2023)	Source
India	• Initiated energy action plans through 'The Integrated Energy Policy' of 2006, • At present, The National Action Plan on Climate Change (NAPCC), launched in 2008 focuses on energy-related challenges and reduction of emissions intensity.	NAPCC (2008–continued), Energy Conservation Building Codes (Commercial and Residential)—2017, 2018	NAPCC—National Action Plan on Climate Change (2021) and International Energy Agency (n.d.)
Japan	• Developed '*Basic Energy Plans*' in 2003, which were formulated in response to Japan's energy security concerns and need for strategic energy policies, • The Sixth Strategic Energy Plan currently focuses on achieving carbon neutrality by 2050.	Sixth Strategic Energy Plan—2050	METI (2021) and Duffield and Woodall (2011)
South Korea	• The First Energy master plan was in 2008, to target renewable energy-related concerns, • Currently, the Third Energy Master Plan focuses on small-scale energy distribution, energy access, and consumer concerns related to energy.	Korea Energy Master Plan-2035	MOTIE (2019), Nachmany et al. (2015), and Republic of South Korea (2014)

1. *Is there any relation between land-use, population growth, and energy demand of the cities? If so, what, how, and why?*
2. *Which urban planning tool can be employed to improve energy-related policy implementations at the city scale?*

To fill this gap, the primary step in pursuing energy planning for cities is to determine the present and future energy demand to ascertain a basic idea of the magnitude of energy requirements. A review of existing literature suggests, that there is a positive relationship between energy patterns, population growth, and urban land-use pattern, with significant increases being observed in energy consumption with changes in land-use and population (Zhao et al. 2017; Ke et al. 2022; Wang et al. 2023). Some studies suggest that urban planning tools can be used to control land-use and increased energy demand at the urban scale to a certain extent (Cheung et al. 2019). Given that urban planning is a collective effort, energy policies may be developed at the municipal or regional levels (Brandoni and Polonara 2012; Neves et al.

2015). To focus on the implementation of energy-efficient policies, it is preferable to utilize master plans, city development plans, and town planning schemes as an implementation tool (Leal and Azevedo 2016). Studies also suggest that planning of energy master plans at neighborhood scales can be considered as an optimum-sized area to implement policies for net-zero communities and zero-carbon emission targets (Isaac et al. 2020). The concept of master plans or development plans is commonly understood as a systematic and regulated approach to the development of a certain geographic region. The holistic approach encompasses several dimensions of development, including physical, economic, social, and political factors (Marudachalam and Jothilakshmy 2010). Similarly, the town planning scheme is a process of 'land pooling or readjustment' and 'plot reconstitution' that aids in controlling and anticipating urban development in an area and making allowances for the planned development of neighborhood-scaled spaces in cities (Keshri and Dey 2022).

The purpose of this empirical examination through the case study of two Indian cities is to portray a picture of the potential energy savings and determine whether planning instruments such as master plans and development plans can significantly curb the city's energy demands. Also, how can they be realized by embracing net-zero policies at the micro level of development plans, if applicable? The economic effects of energy production and water extraction were outside the scope of this study.

3 Methodology

The literature suggests that the assessment and comparison of predicted energy demand and present energy demand at the initial planning phase are a better approach for energy master planning of cities (Fig. 1). As the first step of planning for energy-efficient net-zero communities, it is important to understand the current energy consumption and to project demands for the future. To get a better understanding of how energy demand changes at the city scale, this study analyzed the land-use planning and water demand. A hypothetical projection-based analysis was then performed using secondary data sources (Fig. 2). The study was conducted for two cities located along different coastlines of India. The first city, Surat, located in Gujarat, displays a radial growth, whereas, the second city, Puri, located in Odisha, displays a linear growth. The details of both cities have been discussed in Table 2. The existing land-use patterns of both cities have been illustrated in Fig. 2.

To identify the existing and predicted energy demand, the land-use information from respective development plans and population data from the Census of India (1971–2011) were collected. The land-use data from the development plans was validated through a physical examination, and categorized into two classes, namely residential and non-residential. Considering this was a hypothetical investigation, the typical energy demands for residential and non-residential buildings, in addition to the energy intensity required for water delivery, were calculated using national averages. The average energy consumption for the residential and non-residential built environment was considered to be 80 kWh/sq. m/year and 160 kWh/sq. m/year

Table 2 Study area details

Details	Case 1	Case 2
City and state	**Surat**, Gujarat	**Puri**, Odisha
City coordinates	21.1702° N, 72.8311° E	19.8135° N, 85.8312° E
City area[a]	1059.71 sq. km	233.85 sq. km
Population[b]	4,805,101	369,935
Growth pattern	Radial	Linear
Period of establishment[c]	1500–1520 A.D	1229–1230 A.D.
Average population density[a]	49,971 persons per sq. km	9,395 persons per sq. km
Development plan proposed for the year	2035	2031
Land-use percentage	19.25% residential 80.75% non-residential	4.48% residential 95.52% non-residential
Development Authority	Surat Urban Development Authority (SUDA)	Puri-Konark Development Authority (PKDA)

Note
[a]As per respective development plans (PKDA 2013; SUDA 2017)
[b]As per Census 2011 (Office of the Registrar General and Census Commissioner, India 2011)
[c]As per city history details by respective development authorities (Government of Odisha 2023; SUDA 2023)

while conducting the analysis (Kaja 2015). Similarly, the average energy intensity of 0.14 kWh/cu. m associated with water supply through a pumped water distribution system was considered (UNFCC 2020).

To calculate the energy demand, first, the land-use pattern of the two cities was assessed and projections from the year 2021–2071 were made for decadal increments. Figure 3 illustrates the methodology used for the projection of energy demand for the two categories of land-use.

The existing energy demand for the base year of 2011 was analyzed by multiplication of land-use area (sq. m.) and average national energy consumption for residential and non-residential land-uses respectively. Following this, to make projections for energy demands as shown in Fig. 3, the growth rate of connected energy load in India was calculated for 2009–2019. Now based on this growth rate and base year energy demand, decadal projections were made till 2071 for both land-use categories. This is considering that India as a nation targets to become net zero with zero-carbon emissions by 2070 (The United Nations Climate Change Conference 2021). The Eqs. (1)–(5) describe the formulae used to predict the connected load energy demand for the urban built environments in the study areas (Keshri and Dey 2022).

For existing energy demand in the base year 2011:

$$E_{\text{Existing}} = E_{eR} + E_{eNR} \tag{1}$$

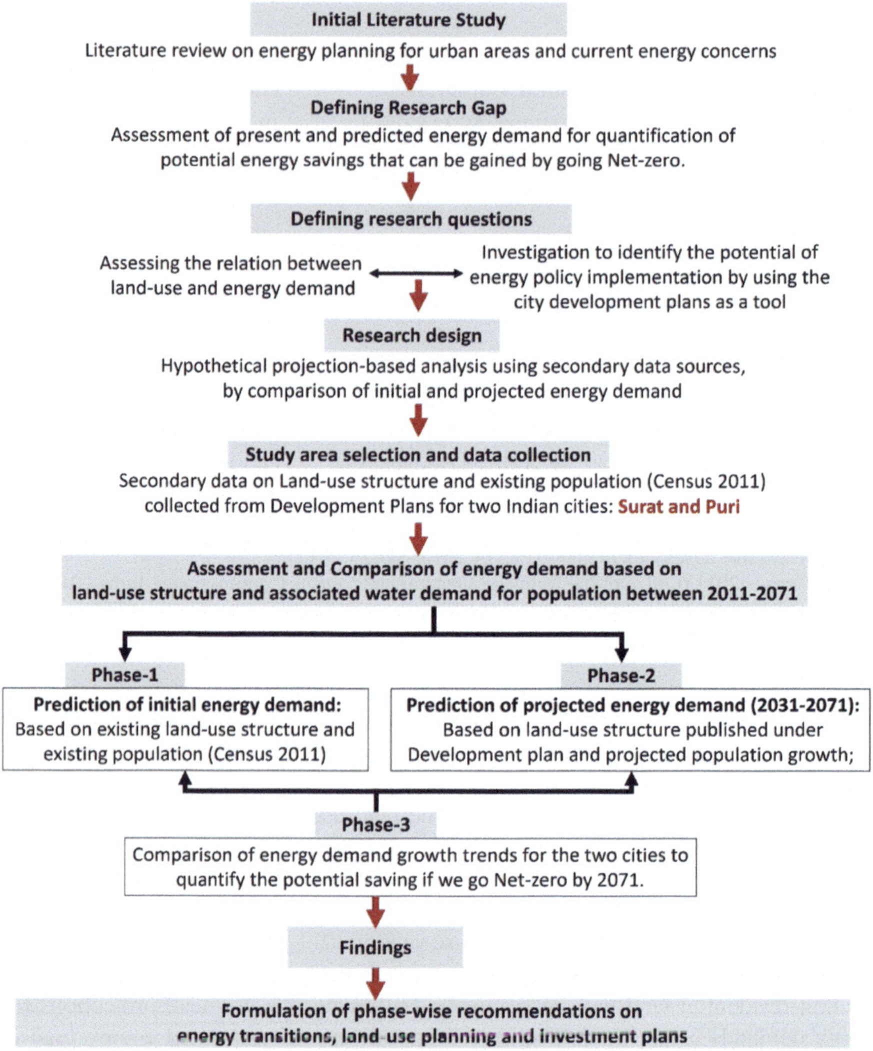

Fig. 1 Framework adopted for this study. *Source* Authors

where E_{Existing} (kWh/year) is the existing total energy demand for the base year 2011, which was subdivided as $E_{(\text{Existing, Residential})}$ (E_{eR}) and $E_{(\text{Existing, Non-residential})}$ (E_{eNR}), where:

$$E_{eR} = \text{Area(sq.m.)} * \text{Average energy consumption}_{\text{Residential}} \text{(kWh/sq.m /year)} \quad (2)$$

$$E_{eNR} = \text{Area(sq.m.)} * \text{Average energy consumption}_{\text{Non-Residential}} \text{(kWh/sq.m /year)} \quad (3)$$

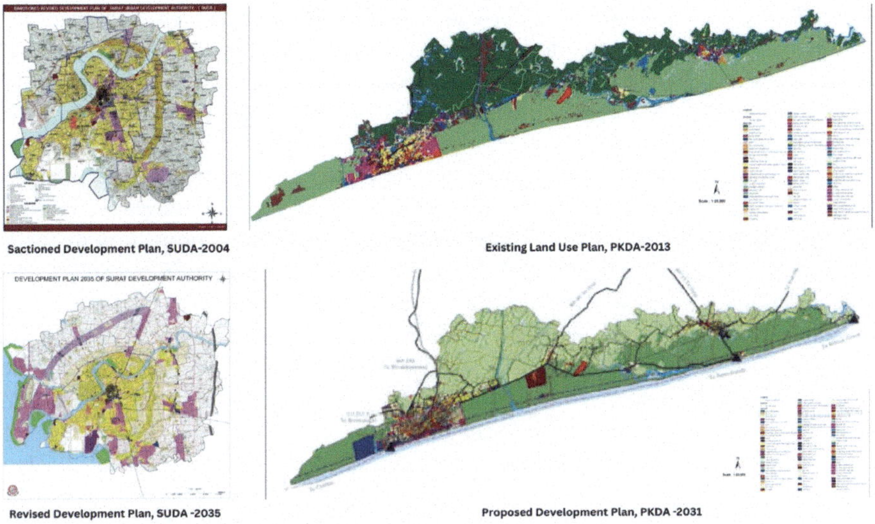

Fig. 2 Existing vs. proposed land-use plan as per development plans of SUDA (2004–2035) and PKDA (2013–2031). *Source* PKDA (2013) and SUDA (2017)

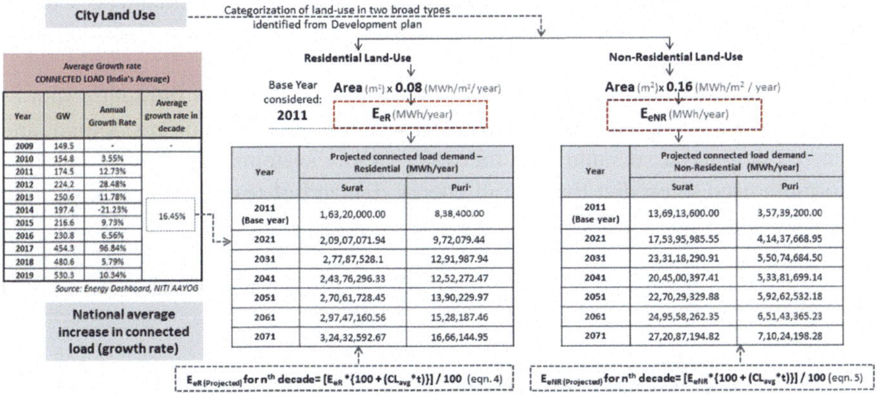

Fig. 3 Methodology for projection of energy demand based on land-use plan for the study areas. *Source* Authors

For projection of energy demand:

$$E_{eR(\text{Projected})}\ \text{for n}^{\text{th}}\ \text{decade} = [E_{eR} * \{100 + (CL_{\text{avg}} * t)\}]/100 \tag{4}$$

$$E_{eNR(\text{Projected})}\text{for n}^{\text{th}}\ \text{decade} = [E_{eNR} * \{100 + (CL_{\text{avg}} * t)\}]/100 \tag{5}$$

where 'CL_{avg}' is considered as a constant value of '**16.45%**' which is the national average increase in connected load (growth rate) calculated for 2009–2019, and '*t*' denominates the change in average connected load for each nth decade, where '*t*' varies from **1, 2,..., 6**; for example, the value of $t = 3$ for the year 2041, w.r.t. the base year 2011, where $t = 0$.

Similarly, to project the energy demand associated with water supply through a pumped distribution system, the population data for respective cities were considered, and projections were made based on the average of three projection methods (arithmetic, geometric increase, and incremental increase projection methods). The projected population was considered to receive 135 L per capita per day (lpcd) of water supply, which predicted the water demand (Ministry of Jal Shakti 2020). This was further multiplied by the average energy intensity as defined by the UNFCC (2020). The energy intensity of water distribution was predicted through the methodology discussed in Eq. (6) and Fig. 4.

Energy Intensity of water demand for each projected census year (2021–2071):

$$\mathrm{EI}_{\mathrm{Water}} \text{ for } \mathrm{n}^{\mathrm{th}} \text{ year} = [(P_t * D_{pr} * 365)/1000] * 0.14 \quad (6)$$

where $\mathrm{EI}_{\mathrm{Water}}$ is Energy Intensity for each projected year in kWh/year, *Pt* is the projected population for the *n*th year; D_{pc} is the per capita water demand, where *D*pc = 135Lpcd is constant and multiplied by 365 days to get Annual Demand per Capita; 0.14 kWh/cu.m is the average energy intensity considered for pump distribution of water (Tables 3 and 4).

The total energy demand for each census year was calculated through a sum of the projected energy demand for urban land-use (residential and non-residential) and the energy demand associated with the water supply systems. Equation (7) defines the formula used to predict the overall energy demand of respective cases of Surat and Puri for the years 2021–2071.

$$E_{\mathrm{Total}} = E_{eR(\mathrm{Projected})} + E_{eNR(\mathrm{Projected})} + \mathrm{EI}_{\mathrm{Water}} \, (\text{for } n^{\mathrm{th}} \text{ year}) \quad (7)$$

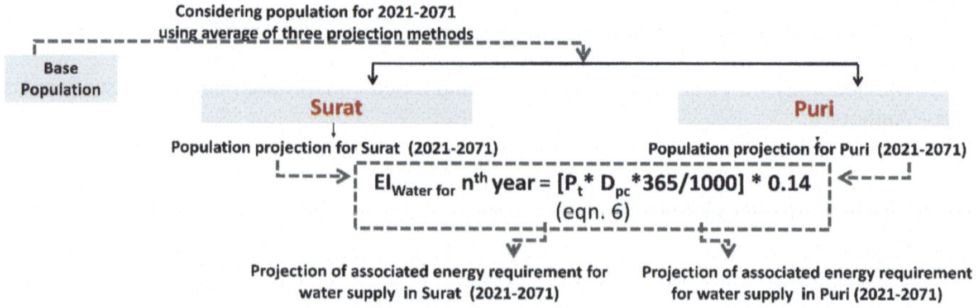

Fig. 4 Methodology for projection of energy demand associated with pumped-water supply. *Source* Authors

Table 3 Calculation of associated energy demand for water supply in Surat

Census year	Existing population	Population projection			Projected population considered	Water demand/year (cu.m.) (w.r.t. 135 Lpcd)	Energy Intensity (GWh/year)
		Arithmetic increase	Geometric increase	Incremental increase			
(1)	(2)	(3)	(4)	(5)	(6) = Average [(3) + (4) + (5)]	(7) = {(6) *135*365]/1000}	(8) = (7) * 0.14 * 0.001
1981	1,120,366	–	–	–	1,120,366	55,206,034.65	7.72
1991	1,779,636	–	–	–	1,779,636	87,691,563.90	12.27
2001	3,105,124	–	–	–	3,105,124	153,004,985.10	21.42
2011 (Base year)	4,805,101	–	–	–	4,805,101	236,771,351.78	33.14
2021	–	2,928,222	7,158,841	6,553,700	5,546,921	273,324,531.81	38.26
2031	–	5,384,712	10,665,543	7,781,945	7,944,066	391,443,876.37	54.80
2041	–	9,069,447	15,889,974	9,010,190	11,323,204	557,950,857.27	78.11
2051	–	13,982,427	23,673,552	10,238,435	15,964,804	786,665,736.81	110.13
2061	–	20,123,652	35,269,853	11,466,680	22,286,728	1,098,178,522.78	153.74
2071	–	27,493,122	52,546,509	12,694,925	30,911,519	1,523,165,074.36	213.24

Source Authors

Table 4 Calculation of associated energy demand for water supply in Puri

Census year	Existing population	Population projection			Projected population considered	Water demand/ year (cu.m.) (w.r.t. 135 pcd)	Energy Intensity (GWh/ year)
		Arithmetic increase	Geometric increase	Incremental increase			
(1)	(2)	(3)	(4)	(5)	(6) = Average [(3) + (4) + (5)]	(7) = {(6) *135*365]/ 1000}	(8) = (7) * 0.14 * 0.001
1971	149,824	–	–	–	149,824	7,382,577.60	1.03
1981	193,434	–	–	–	193,434	9,531,460.35	1.33
1991	240,197	–	–	–	240,197	11,835,707.18	1.65
2001	289,050	–	–	–	289,050	14,242,938.75	1.99
2011 (Base year)	369,935	–	–	–	369,935	18,228,547.13	2.55
2021	–	1,309,130	393,346	6,553,700	2,752,059	135,607,684.29	18.98
2031	–	3,765,620	418,239	7,781,945	3,988,601	196,538,318.61	27.51
2041	–	7,450,355	444,707	9,010,190	5,635,084	277,668,751.63	38.87
2051	–	12,363,335	472,850	10,238,435	7,691,540	379,000,620.80	53.06
2061	–	18,504,560	502,774	11,466,680	10,158,004	500,535,667.21	70.07
2071	–	25,874,030	534,591	12,694,925	13,034,515	642,275,742.11	89.92

Source Authors

The compound annual growth rate method (Khan et al. 2020) was used to compare and understand the effect of land-use and water demand on the energy consumption pattern of the two cities. The compound annual growth rate (CAGR) for the energy demands was calculated using Eq. (8).

$$\text{CAGR} - (E_{2071}/E_{2021})^{(1/\text{number of years})} - 1 \tag{8}$$

where E_{2021} refers to the energy demand or energy intensity projected for the year 2021, and E_{2071} refers to the energy demand for energy demand in the year 2071. The number of years was considered to be '2071–2021 = 60 years', for this study.

4 Findings

The study attempted to predict the energy demands of two major coastal cities in India to understand their respective land-use and water-related energy demand. The results of the study suggest that significant changes in energy demand can be observed with changes in land-use and water demand over time. The results of the study are discussed in three sections, where the first section discusses the comparison of energy demand for the two types of land-uses in the city. The second section discusses the energy intensity associated with water supply systems followed by the comparison of the overall energy demand of the cities as projected between 2011 and 2071.

4.1 Comparison of Energy Demand for Residential vs Non-residential Land-Uses

The first part of the study compared the energy demand of the two cities based on the land-use distribution, where two main categories, residential and non-residential, were considered. The residential land-use consisted of the existing residential built environment and that proposed for future development. All other land-use types were considered together under the non-residential category for simplification of analysis.

The results from the projections displayed that a considerable increase in energy consumption for both land-use types is expected over the study period. A slight reduction in the energy demand was found during 2031–2041 in both cities, corresponding to changes in land-use planning through development plans during the years 2031 and 2035 in the case of Puri and Surat, respectively (Fig. 5).

The comparison of energy demand for both cities displays that Surat, being a larger city in comparison with Puri, has a higher energy demand throughout the projection phase. However, the comparison of results from the CAGR analysis shows that the residential energy demand for Puri is higher in comparison with that of Surat, whereas the energy demand associated with non-residential land-use is almost the same for the two cases (Table 5). The higher growth rate in energy demand noted in the case of Puri suggests that the linear development pattern in the city has led to unstructured development, whereas in the case of Surat, the development pattern was controlled through the use of development plans and town planning schemes (SUDA 2017), which has enabled some control over the increase in energy demand.

However, in the case of non-residential energy demand, a very slight difference is observed between the two cities. This suggests that the non-residential sectors have similar growth patterns in both cases, which may be a result of both cities being major urban centers with rapid growth in terms of industrial and economic activities (PKDA 2013; SUDA 2017).

4.2 Comparison of Energy Demand Associated with Water Demand Resulting from Population Growth

In the case of prediction for change in water demand through population projections, the result suggested that a significant amount of water supply would be required for both cities in the future. The city of Surat, having a higher land area and larger

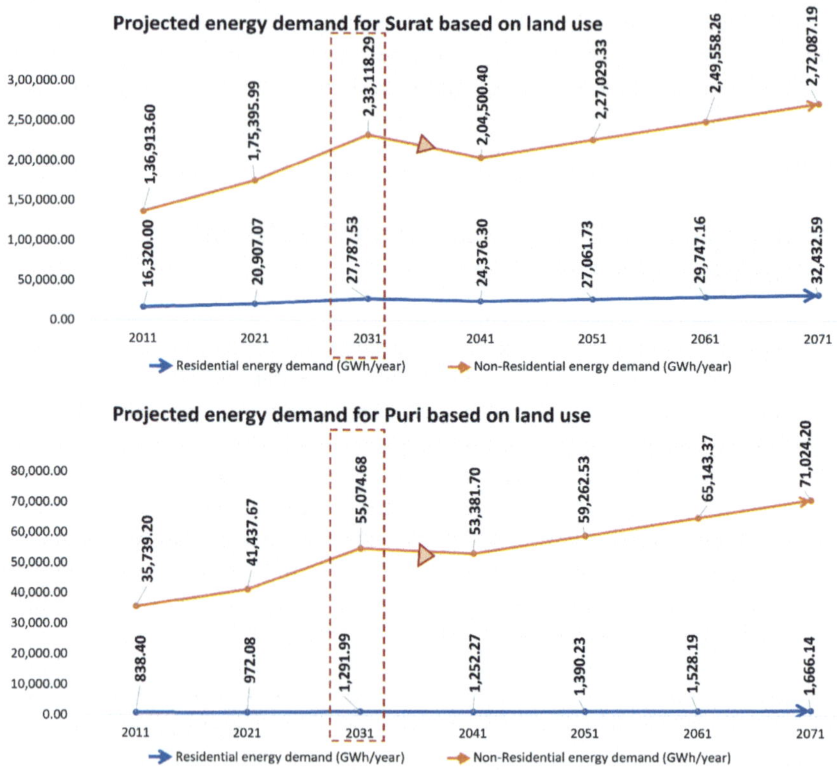

Fig. 5 Graphical comparison of projected energy demand for the two study cases was performed based on their land-use information. *Source* Authors

Table 5 Comparison of rate of increase for the predicted energy demand through a change of land-use in the two cases through the CAGR method

City	Land-use type	Beginning value (GWh/Year) (2011)	Ending value (GWh/Year) (2071)	Number of years	CAGR (%)
Surat	Residential	16,320.00	32,432.59	60 years	1.56%
	Non-residential	136,913.60	272,087.19		1.95%
Puri	Residential	838.40	1,666.14		1.80%
	Non-residential	35,739.20	71,024.20		1.93%

Source Authors

population, resulted in a higher per capita per day water requirement in comparison with Puri. However, both cities display a linear growth in the demand for energy required for water supply considering pumped distribution systems (Fig. 6).

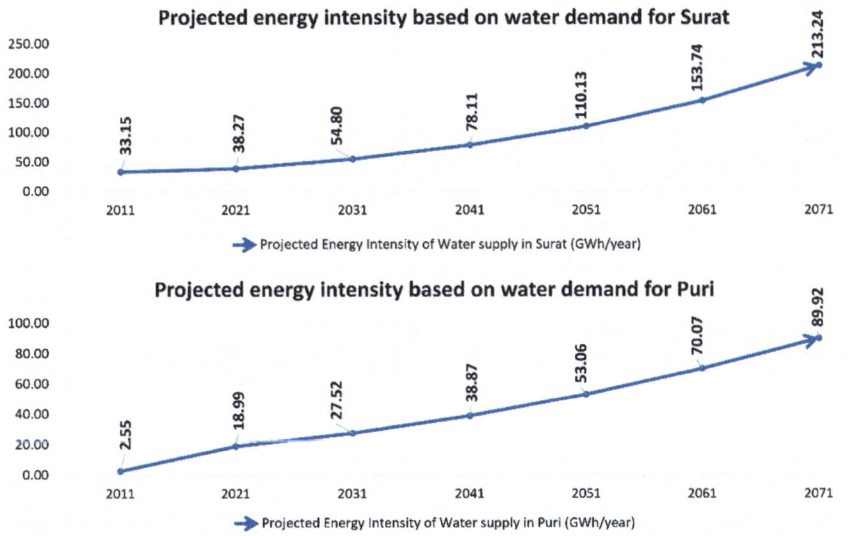

Fig. 6 Graphical comparison of projected energy demand for the two study cases were performed based on their water supply requirement. *Source* Authors

The comparative CAGR analysis for energy demand associated with water supply and distribution suggests a nearly 2.91% higher growth rate for Puri in comparison with that of Surat (Table 6). Considering that the previous analysis of land-use displayed a higher growth rate in residential energy demand, the water demand for Puri displays higher associated energy intensity for Puri. This is also supported by the fact that in a linear planned city, the plumbing network has a greater length in comparison with a radial planned city, leading to higher energy requirements for the transport of water to build structures. The same plumbing system in the case of Surat has been divided into various pump houses located across different town planning zones of the city, leading to a reduction in the length of the network and thereby reducing the energy demand for water supply.

Table 6 Comparison of the rate of increase for the predicted energy demand through the change in water demand for the two cases through the CAGR method

City	Beginning value (GWh/Year) (2011)	Ending value (GWh/Year) (2071)	Number of years	CAGR (%)
Surat	33.15	213.24	60 years	12.43%
Puri	2.55	89.92		9.52%

Source Authors

4.3 Comparison of Overall Energy Demand for the Projected Period of 2011–2071

The projections for the overall energy demand suggested that a significant quantity of energy would be required to support our cities in the future, and as these cities develop over time, it becomes imperative to resort to alternate solutions for energy demands. The comparison of the two cities shows that even if the land-use patterns and growth patterns of the cities are different, there is a common pattern in the increase of their energy demand. As previously identified in Sect. 4.1, the effect of change of land-use as per respective development plans in the period of 2031–2041 is notable, showing sharp change in the case of Surat, whereas a slight variation in the case of Puri (Fig. 7). The gap in effect is a result of the difference in land area, growth pattern, and population sizes of the two cities.

The comparison of CAGR analysis of the two cases displays that both cities have a similar growth in energy demand (Table 7). A slightly higher rate in the case of Puri is observed, resulting from the higher energy growth rate for the residential built environment as well as higher energy intensity associated with water supply and the unstructured land-use pattern from the linear development. This suggests that even though Puri is a comparatively smaller town based on both land area and population, it is more energy-intensive than Surat.

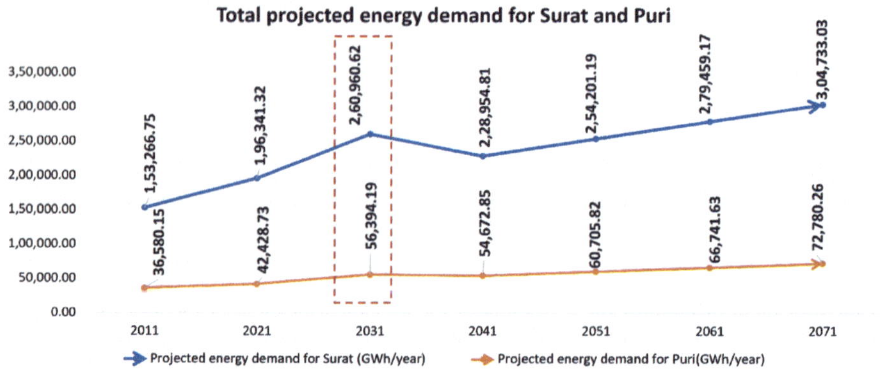

Fig. 7 Graphical comparison of overall projected energy demand for the two study cases. *Source* Authors

Table 7 Comparison of the rate of increase for the predicted overall energy demand for the two cases through the CAGR method

City	Beginning value (GWh/Year) (2011)	Ending value (GWh/Year) (2071)	Number of years	CAGR (%)
Surat	1,53,266.75	3,04,733.03	60 years	1.51%
Puri	36,580.15	72,780.26		1.53%

Source Authors

5 Recommendations

The three sections of the study suggest that Puri has a slightly higher energy demand growth rate in comparison with Surat. This suggests that the development plans and town planning schemes for control of land-use pattern helps in a controlled development of the cities and also has a positive effect on the energy consumption patterns of the built environment, as in the case of Surat. This further justifies that the use of development plans as an implementation tool for energy policies and energy efficiency programs at a city scale would be more beneficial to achieve the set targets of sustainability and net-zero communities. Additionally, town planning schemes allow the implementation of said energy policies at a community level in a controlled manner.

The idea of development plans would provide a boost to the idea of energy master planning, which is highly imperative for both existing cities as well as upcoming urban spaces. The study also helps to identify certain general goals, visions, and key action points that can be adapted at the city scale through integrated master planning for control of both land-use development and energy consumption patterns. Based on the findings of this study, certain recommendations have been listed in two parts, first highlighting certain recommendations for the two cities considered for this study (Table 8), and the second, considering general recommendations that can be replicated for any urban area (Table 9).

6 Conclusion

The study attempted to develop a method for predicting the energy demand in urban areas and investigated the savings potential of energy-efficient policies within the context of urbanization. A comparison of two Indian cities was conducted to determine whether energy consumption is related to and if it can be predicted through assessment of the urban land-use pattern and population growth. The cases of Surat and Puri were investigated in three phases, during which the energy demand of the cities was projected using an analytical method based on hypothetical projections.

The purpose of this study was to determine the potential energy savings that may be realized through the application of development and master plans as an instrument for implementing energy policies, such as net-zero and zero-carbon emission. The results of the study indicated that cities like Surat, with a master plan and town planning initiatives, have better chances of curbing their future energy demands. It was also noted that the ability to predict future energy demand is facilitated by the existence of a fixed land-use plan for the study location. To achieve a holistic approach to sustainability, the study also emphasized the need for a hybrid integrated planning system that integrates the energy, land-use, and economic aspects of city development.

Table 8 Recommendations for energy-efficient development of Surat and Puri based on findings from results

City		Energy transitions	Land-use and Building bye-laws	Investment plans
Surat	Goals—short term (3–5 years)	Investments in renewable energy sources, Encouraging **community participation** toward improvement of energy-efficient behavior in the residents.	Refining development plans and town planning schemes, **Systematically control urban development** to manage energy and water demand efficiently.	Providing **incentives on energy and water efficient behaviors** inculcated by the users, especially for the residential developments, Metering of water intake and wastewater discharge and management systems for residential and non-residential building types.
	Vision—long term (10–15 years)	Inclusion of energy-efficient measures and preparation of **energy mapping plans,** Collaboration with technical and institutional knowledge centers to enable better planning opportunities through research.	Inclusion of energy targets into development plans, Planning for energy efficiency tracking of the city through data logging and real-time data collection networks for identification of energy usage trends.	Collaborations with multi-national companies and industries located in the city area to provide technical and financial support to energy-efficient development of infrastructure networks, Formulating **incentive and subsidy-based models** for operations of the energy and water infrastructures.

(continued)

Table 8 (continued)

City		Energy transitions	Land-use and Building bye-laws	Investment plans
Puri	Goals—short term (3–5 years)	Encourage **energy-efficient measures for the residential sector** in the city to curb the high growth rate in energy demand, Establishment of community-driven initiatives for energy-efficient development.	Encourage the installation of renewable energy and wastewater management systems for existing and future developments, Promoting energy retrofits for existing residential and tourism-related infrastructure to reduce energy demand Formulate real-time data logging for energy and water usage to track energy usage patterns in the city.	Provide incentives for the construction of environment-friendly and energy-efficient structures for upcoming tourist facilities.
	Vision—long term (10–15 years)	Given the significant reliance of the economy on tourism, planning for floating population, and the development of **energy strategies for tourist-related infrastructure** such as lodgings and transit stations are crucial, Inclusion of renewable energy sources.	Land-use redevelopment can be considered to provide more structured development, **Introducing zoning models** and guides to provide efficient residential and non-residential land-use developments.	Formulating energy and water management policies for the city's residential and tourism infrastructure.

Source Authors

The study's findings provide policymakers, urban planners, and built environment stakeholders with vital insights. However, the study assumed hypothetical scenarios, regarding future trends in population, and energy demand based on previous growth trends noted through secondary data. Additionally, the scope of the present study does not include criteria like the dynamic nature of urban environments, economic, and social barriers, and the challenges associated with implementing hybrid policy frameworks for sustainable urban development. Therefore, by considering this study as an initial exploration, primary data collection for the cities should be undertaken to formulate location-specific energy potential mapping and policies for urban areas.

The integrated master planning recommendations provide a road map for attaining energy efficiency and net-zero communities. Successful energy transitions require

Table 9 General recommendations for integrated master planning for land-use and energy-controlled development

Category of recommendation	Energy transitions	Land-use and building bye-laws	Investment plans
Goals—short term (3–5 years)	Achieving energy efficiency through small and easy-to-implement approaches.	Small-scale repairs and changes in everyday practices of city residents.	Phasing of investment for refurbishments and energy retrofitting for buildings in the next 3–5 years through short-term action plans
Vision—long term (10–15 years)	Integration of renewable energy sources at an urban scale along with the development of resilience and reliability.	Inclusion of in-situ and community-scale renewable energy sources, infrastructure planning considering life cycle costs, and 3R's concepts.	Development of integrated master plans inclusive of efficient land-use, energy plans, and economy-boosting action plans for the next 20–30 years.
Key action points	**Community engagement** to spread awareness of energy crises and the need for efficient practices.	Public participation in planning and implementation stages of planning strategies.	Incentives and benefits provided to communities for adaption of energy-efficient practices.
	Regulatory compliance for building owners and renters across the city jurisdictions.	Regular updation and monitoring of programs for checking and verifying the efficiency of the existing regulatory frameworks.	Introduction of energy incentives, regularization of building tax revenues, etc., to support local governing authorities.

(continued)

Table 9 (continued)

Category of recommendation	Energy transitions	Land-use and building bye-laws	Investment plans
	Job creation and economic growth through capacity building of trainers and skilled practitioners in the energy domain.	Promotion and inclusion of technical support from energy auditors and energy managers in the planning phases for the policies.	Improvements in the investment in training and employment sector targeting increase in interest and numbers of energy practitioners.
	Environmental benefits due to a reduction in the need for non-renewable resources and control of land-use supporting a better-built environment.	Control of green spaces and open spaces in the city to maintain the environmental and climatic requirements, promotion of local purchases, and transit-oriented development patterns for the cities.	Promotion of local area planning to support transit-oriented development and local economy development to reduce travel requirements.

Source Authors

public participation, regulatory compliance, investment plans, and capacity building in the energy domain. As global energy demands continue to rise, assessment of potential in our spatial planning of incorporating net-zero and low-carbon technologies is not just an option but a necessity for a sustainable future. The study concludes by emphasizing the importance of a coordinated and comprehensive approach to urban energy master planning. Cities can work toward a sustainable future with reduced environmental impact by implementing energy-efficient policies and incorporating renewable energy sources. The findings of this study contribute to the expanding corpus of knowledge on sustainable urban development, and offers valuable guidelines for designing energy-efficient and resilient urban communities.

Acknowledgements We acknowledge the information provided by the Surat Urban Development Authority and the Puri-Konark Development Authority for the information collected from their respective development plans published by them on their respective websites, which have been used as secondary source information for the analysis in this study and have been duly acknowledged in references.

We confirm that this work is original and has not been published elsewhere, nor is it currently under consideration for publication elsewhere. We have no conflict of interests to disclose. No funding was received for this project.

References

AEEE (2021) Energy conservation and efficiency: a time capsule. Alliance for an Energy Efficient Economy. https://aeee.in/energy-conservation-and-efficiency-a-time-capsule/

AEEE (2023) About AEEE. Alliance for an Energy Efficient Economy (AEEE). https://aeee.in/who-we-are/

Ala-Juusela M, ur Rehman H (2020) PEB as enabler for consumer centred clean energy transition- shared definition and concept; EXCESS Deliverable 1.1. Excess, Proposal No. 870157. https://positive-energy-buildings.eu/fileadmin/user_upload/Materials/EXCESS_D1.1_PEB_definition_and_concept.pdf

Bayulgen O, Benegal S (2019) Green priorities: how economic frames affect perceptions of renewable energy in the United States. Energy Res Soc Sci 47(August 2018):28–36. https://doi.org/10.1016/j.erss.2018.08.017

Brandoni C, Polonara F (2012) The role of municipal energy planning in the regional energy-planning process. Energy 48(1):323–338. https://doi.org/10.1016/j.energy.2012.06.061

Cajot S, Peter M, Bahu JM, Koch A, Maréchal F (2015) Energy planning in the urban context: challenges and perspectives. Energy Procedia 78:3366–3371. https://doi.org/10.1016/j.egypro.2015.11.752

Chen S, Chen B (2015) Urban energy consumption: different insights from energy flow analysis, input-output analysis and ecological network analysis. Appl Energy 138:99–107. https://doi.org/10.1016/j.apenergy.2014.10.055

Cheung G, Davies PJ, Trück S (2019) Transforming urban energy systems: the role of local governments' regional energy master plan. J Clean Prod 220:655–667. https://doi.org/10.1016/j.jclepro.2019.01.179

Cole RJ, Fedoruk L (2015) Shifting from net-zero to net-positive energy buildings. Build Res Inf 43(1):111–120. https://doi.org/10.1080/09613218.2014.950452

Conticelli E, Proli S, Tondelli S (2017) Integrating energy efficiency and urban densification policies: two Italian case studies. Energy Build 155:308–323. https://doi.org/10.1016/j.enbuild.2017.09.036

Dixon T, Eames M, Hunt M, Lannon S (2014) Urban retrofitting for sustainability, 1st edn. Routledge. https://doi.org/10.4324/9781315850184

Duffield JS, Woodall B (2011) Japan's new basic energy plan. https://scholarworks.gsu.edu/cgi/viewcontent.cgi?article=1093&context=political_science_facpub

Espinoza VS, Guayanlema V, Martínez-Gómez J (2018) Energy efficiency plan benefits in Ecuador: long-range energy alternative planning model. Int J Energy Econ Policy 8(4):42–54

Fankhauser S et al (2022) The meaning of net zero and how to get it right. Nat Clim Chang 12(1):15–21. https://doi.org/10.1038/s41558-021-01245-w

Government of Odisha (2023) History of Puri. Puri Konark Development Authority (PKDA). https://puri.nic.in/history/

Government of the United Kingdom (2020) Energy white paper: powering our net zero future. Department for Energy Security and Net Zero and Department for Business, Energy & Industrial Strategy. https://www.gov.uk/government/publications/energy-white-paper-powering-our-net-zero-future

Hepburn C, Qi Y, Stern N, Ward B, Xie C, Zenghelis D (2021) Towards carbon neutrality and China's 14th Five-Year Plan: clean energy transition, sustainable urban development, and investment priorities. Environ Sci Ecotechnology 8. https://doi.org/10.1016/j.ese.2021.100130

IBPSA (2023) Key functions. The International Building Performance Simulation Association (IBPSA). http://www.ibpsa.org/about/

International Energy Agency (2021) Key World Energy Statistics—2021. https://iea.blob.core.windows.net/assets/52f66a88-0b63-4ad2-94a5-29d36e864b82/KeyWorldEnergyStatistics2021.pdf

International Energy Agency (2022) Buildings. https://www.iea.org/reports/buildings

International Energy Agency (2023) Germany's Renewables Energy Act. https://www.iea.org/policies/12392-germanys-renewables-energy-act
International Energy Agency (n.d.) Integrated Energy Policy. Retrieved December 6, 2023, from https://www.iea.org/policies/4518-integrated-energy-policy
Isaac S, Shubin S, Rabinowitz G (2020) Cost-optimal net zero energy communities. Sustainability 12(6):1–15. https://doi.org/10.3390/su12062432
Kaja N (2015) An overview of energy sector in India. Int J Sci Res 6(3):2319–7064. https://www.ijsr.net/archive/v6i3/ART20171783.pdf
Ke H, Yang B, Dai S (2022) Does intensive land use contribute to energy efficiency?—Evidence based on a spatial Durbin model. Int J Environ Res Public Health 19(9). https://doi.org/10.3390/ijerph19095130
Keshri, S., Dey, P. (2024). Net Zero Potential at Neighborhood Level: A Case Study of Town Planning Scheme-45, Surat. In: Hodge, BM., Prajapati, S.K. (eds) Proceedings from the International Conference on Hydro and Renewable Energy. ICHRE 2022. Lecture Notes in Civil Engineering, vol 391. Springer, Singapore. https://doi.org/10.1007/978-981-99-6616-5_23
Khan ZA, Ullah A, Ullah W, Rho S, Lee M, Baik SW (2020) Electrical energy prediction in residential buildings for short-term horizons using hybrid deep learning strategy. Appl Sci 10(23):1–12. https://doi.org/10.3390/app10238634
Kumar GS (2017) Anatomy of Indian energy policy: a critical review. Energy Sources Part B: Econ Plan Policy 12(11):976–985. https://doi.org/10.1080/15567249.2017.1336814
Leal VMS, Azevedo I (2016) Setting targets for local energy planning: critical assessment and a new approach. Sustain Cities Soc 26:421–428. https://doi.org/10.1016/j.scs.2016.04.010
LEED (2023) LEED for cities and communities. https://www.usgbc.org/leed/rating-systems/leed-for-cities-communities
Liao X et al (2020) Comparing water footprint and water scarcity footprint of energy demand in China's six megacities. Appl Energy 269(January):115137. https://doi.org/10.1016/j.apenergy.2020.115137
Marudachalam V, Jothilakshmy N (2010) Development plan and urban design policies. J Inst Town Plan, India 7–1(January–March):65–69. https://www.itpi.org.in/uploads/journalfiles/jan5_10.pdf
METI (2021) Outline of the 6th Strategic Energy Plan. Ministry of Economy, Trade and Industry. https://www.enecho.meti.go.jp/en/category/others/basic_plan/pdf/6th_outline.pdf
Mez L (2004) Renewable energy policy in Germany—institutions and measures promoting a sustainable energy system. 15(4). https://foes.de/pdf/EN_Mez.pdf
Ministry of Jal Shakti (2020) Per capita availability of water. https://pib.gov.in/PressReleasePage.aspx?PRID=1604871
MOTIE (2019) A new energy paradigm for the future: Third Energy Master Plan. Ministry of Trade, Industry and Energy; South Korea. https://www.etrans.or.kr/ebook/05/files/assets/common/downloads/Third%20Energy%20Master%20Plan.pdf
Nachmany M et al (2015) Climate change legislation in South Korea. https://www.lse.ac.uk/GranthamInstitute/wp-content/uploads/2015/05/SOUTH_KOREA.pdf
NAPCC—National Action Plan on Climate Change (2021) https://doi.org/10.1007/978-3-030-87745-3_300442
Neves AR, Leal V, Lourenço JC (2015) A methodology for sustainable and inclusive local energy planning. Sustain Cities Soc 17:110–121. https://doi.org/10.1016/j.scs.2015.04.005
Office of the Registrar General & Census Commissioner, India (2011) Census of India. https://censusindia.gov.in/census.website/
Owens SE (1992) Land-use planning for energy efficiency. Appl Energy 43(1–3):81–114. https://doi.org/10.1016/0306-2619(92)90075-M
PKDA (2013) Draft Comprehensive Development Plan-2031, Puri Konark Development Authority (PKDA) Area: Vol. I. https://pkdapuri.nic.in/CDP/CDP_vol1.pdf

Polesello V, Johnson K (2016) Energy-efficient buildings for low-carbon cities. International Center for Climate Governance, vol 47/March. http://www.iccgov.org/wp-content/uploads/2016/03/47_Energy-efficient-buildings-for-low-carbon-cities_Viola-Katie.pdf

Pyloudi E, Papantoniou S, Kolokotsa D (2015) Retrofitting an office building towards a net zero energy building. Adv Build Energy Res 9(1):20–33. https://doi.org/10.1080/17512549.2014.917985

Republic of South Korea (2014) Republic of Korea: Korea Energy Master Plan: Outlook and Policies to 2035 (Second Energy Master Plan). Ministry of Trade, Industry and Energy (MOTIE). https://policy.asiapacificenergy.org/node/1358

Santillan MR, Syn JW, Charani Shandiz S, Huang Y, Pires de Lacerda M, Rismanchi B (2022) A technology assessment approach for achieving sustainable communities: an energy master plan for a new urban development. Appl Sci 12(8):1–28. https://doi.org/10.3390/app12083860

Seto KC et al (2014). Human settlements, infrastructure, and spatial planning. In: Climate change 2014: Mitigation of climate change. Contribution of Working Group III to the Fifth Assessment Report of the Intergovernmental Panel on Climate Change. https://www.ipcc.ch/site/assets/uploads/2018/02/ipcc_wg3_ar5_chapter12.pdf

Stoeglehner G, Niemetz N, Kettl KH (2011) Spatial dimensions of sustainable energy systems: new visions for integrated spatial and energy planning. Energy Sustain Soc 1(1):1–9. https://doi.org/10.1186/2192-0567-1-2

SUDA (2017) City Development Plan, Surat. https://www.sudaonline.org/wp-content/uploads/2017/03/SUDA-DP-2035-report-final-22.02.2017.pdf

SUDA (2023) History of Surat. Surat Urban Development Authority. https://www.suratmunicipal.gov.in/TheCity/History

TERI (2013) Roadmap for incorporating energy efficiency retrofits in existing buildings. The Energy and Resources Institute Press, New Delhi. http://cbs.teriin.org/pdf/Energy_Efficiency_Retrofits_in_Existing_Buildings.pdf

The United Nations Climate Change Conference (2021) Conference of Parties 26. https://www.un.org/en/climatechange/cop26

UNFCC (2020) Default Energy Intensity Factors for Water Supply Systems (Issue October). https://unfccc.int/sites/default/files/resource/AHSA-004_Default%20Energy%20Intensity%20Factors%20for%20Water%20Supply%20Systems_v1.pdf

U.S. Department of Energy (1974) Project Independence Blueprint: final task force report. An historical perspective. https://www.osti.gov/biblio/5168259

U.S. Department of Energy (n.d.) National Action Plan for Energy Efficiency. Office of State and Community Energy Programs. https://www.energy.gov/scep/national-action-plan-energy-efficiency

Uspenskaia D, Specht K, Kondziella H, Bruckner T (2021) Challenges and barriers for net-zero/positive energy buildings and districts—empirical evidence from the smart city, Project SPARCS. Buildings 11(2):1–25. https://doi.org/10.3390/buildings11020078

Van Den Dobbelsteen A, Tillie N, Broersma S, Fremouw M (2014) The energy master plan: transition to self-sufficient city regions by means of an approach to local energy potentials. In: 30th International PLEA Conference: Sustainable Habitat for Developing Societies: Choosing the Way Forward - Proceedings, 2(January 2019), pp 252–259

Van Den Dobbelsteen A, Roggema R, Tillie N, Broersma S, Fremouw M, Martin CL (2020) Urban energy masterplanning—approaches, strategies, and methods for the energy transition in cities. In: Droege P (ed) Urban energy transition renewable strategies for cities and regions, 2nd edn. Elsevier, pp 635–660. https://doi.org/10.1016/B978-0-08-102074-6.00045-0

Wang M, Yu H, Jing R, Lin X (2023) Prediction and comparison of urban electricity consumption based on grey system theory: a case study of 30 southern China cities. Cities 137(March):104299. https://doi.org/10.1016/j.cities.2023.104299

World Nuclear Association (2020) Germany's Energiewende. https://world-nuclear.org/information-library/energy-and-the-environment/energiewende.aspx

Zhao J, Thinh NX, Li C (2017) Investigation of the impacts of urban land use patterns on energy consumption in China: a case study of 20 provincial capital cities. Sustainability 9(8). https://doi.org/10.3390/su9081383

Zhou F, Peng L, Li J (2022) China's 14th Five-Year Plans on Renewable Energy Development and Modern Energy System. Energy Foundation China. https://www.efchina.org/Blog-en/blog-20220905-en

Embracing the Complexity of Cities: Agent-Based Modelling for Energy Planning and Policy

Manas Vijayan, Akshay Patil, and Vijay Kapse

1 Introduction

Energy planning in each and every city today calls for a non-identical approach that can comprehend the multiple facets of an urban energy system: sources of supply, socio-economic characteristics of the residing population, climatic conditions, environmental deterioration, social justice, etc., and their complex interplay. This complexity is further exacerbated in the cities of developing nations, in their pursuit of boosting social and economic development at extortionate environmental costs. Planning for such nexus of problems is extremely challenging, if not impossible to accomplish using the traditional top-down deterministic notion of urban planning (Jacobs 1961; Weaver 1991).

Urban energy planning today is in dire need of a revival; a new approach where the complex order of energy consumption behaviour in cities can be understood, acknowledged and comprehended through an integrated planning process. Numerous attempts have been made in the past few decades to explain urban complexity using the different concepts of complexity science; to breakdown and understand the naturally arisen complexity in cities, and the ways in which a functional complexity can be generated using urban planning and design (Batty and Marshall 2012; Marshall 2012; Portugali 2012).

This study explores the concepts of complexity science for deconstructing an urban energy system into its constituent parts and processes-household electricity consumption (HEC) as a macro-level urban phenomenon emerging from the micro-level behaviours and processes. Further, the study explores the application of agent-based modelling as a tool for developing computational models that can be used for

M. Vijayan (✉) · A. Patil · V. Kapse
Department of Architecture and Planning, Visvesvaraya National Institute of Technology, Nagpur, Maharashtra 440010, India
e-mail: manas.vijayan@ceew.in

M. Firoz C et al. (eds.), *Urban Planning and Design for Megacities in the Global South*, Advances in 21st Century Human Settlements,
https://doi.org/10.1007/978-981-97-8370-0_18

conducting detailed simulations of HEC in cities; to gain insight into its complexity, which is otherwise impossible to perform in the real world. Four Indian cities, Delhi, Mumbai, Bangalore and Bhopal, are selected for demonstrating the model and its applications.

2 Conceptual Framework

Amongst the different types of energy systems in cities of developing nations, one of the most critical one is electricity consumed in households. For instance, Centre for Policy Research (CPR) mentions that electricity consumption in Indian households has increased fifty times between 1971 and 2018, and the consumption rates are outpacing the growth observed in commercial, industrial and agricultural sectors. With the nation currently undergoing one of the largest ever urban transitions witnessed in history, these striking statistics are expected to skyrocket (Khosla and Chunekar 2017). The sheer scale of growth that is expected in the urban residential sector, and to meet the resultant electricity demand in a sustainable manner will be one of the greatest challenges of energy planning in the nation.

2.1 Literature Review

2.1.1 Criticality of HEC in Developing Nations

As a general statement, it can be stated that the HEC in developing nations accounts for a great share of their total electricity consumption. As for the case of India, the current share is 24%, and an exponential growth is expected in the immediate future (IEA 2015; Singh et al. 2018). At the present time, the most electricity-consuming appliances in households of Indian cities—air-conditioners, air-coolers, geysers and ceiling fans (Chunekar and Sreenivas 2018; Tewathia 2014; Verma et al. 2012).

The fact that a great cut of the electricity share is already consumed by residential sector, even when most of the nation's states haven't achieved 100% electricity accessibility in households, indicates on the criticality of addressing the situation for the future. Further, improving access and affordability of electricity in developing nations can offer a lessened reliance on the traditional sources of energy. This can significantly contribute towards reducing emissions and providing environmental benefits, enhance social development and improved health, and broaden the reach of education (Olaniyan and Mclellan 2018). It also holds great potential in generating employment opportunities to the residents (Khraief et al. 2018).

Even so, more than 95% of research and development (R&D) on world's energy sector were performed by OECD nations, as of the year 2000 (Dooley and Runci 2000). Conducting research in developing nations usually entails numerous research constraints in the form of insufficient funds, lack of readily available data and shortage of resources (Khanna and Rao 2009). This general lack of research studies also mean that spotting relevant peer-reviewed research from literature, to take reference from may not be feasible. Taking reference from research conducted in developed nations is also not advisable, taking into consideration the significant differences in their characteristics across all spheres of society. A general lack of consensus on the research findings on household energy transitions in developing nations also exists because it is observed by various researchers that the research conclusions generally arise from extrapolating the results of a single, or a few cities (Barnes et al. 2010).

2.1.2 Current Policies

Possibly as a consequence of these research constraints, bulks of the energy efficiency practices in cities of developing nations were found to be severely underperforming. These practices were unable to achieve the desired positive impacts in reducing energy consumption and associated GHG emissions, nor were they able to bring forth any substantial social and economic development to the residents (Vijayan et al. 2021). This is an extremely alarming situation, especially because energy efficiency in cities of developing nations is considered to be the need of the hour for a sustainable and healthy future (Bose 2010; UNSDG 2015).

Another serious issue regarding the current practices was that a major share of the policies was mere adaption of success stories from elsewhere, without having a detailed understanding of the local constraints and opportunities. It was observed that cities those are located in distinct geographic locations, experiencing dissimilar environmental conditions, and constituting of diverse socio-economic strata, were all adopting similar strategies to address energy efficiency (Vijayan et al. 2021).

The situation implies that there is a need for comprehensive tools that can encompass the complex dynamics of HEC behaviour in cities, and make one identify and understand its underlying issues and potentials. At the same time, such tools should be flexible enough to be modified w.r.t the individual characteristics and specific set of conditions faced by each city.

2.1.3 Agent-Based Models

The complexity of HEC as an emergent property in the cities of developing nations and its various planning challenges were discussed above. Agent-based Models (ABM) are useful for problems like these that concern 'emergence', i.e. how system dynamics arise from the manner in which, a system's individual elements interact with and respond to each other and their environment; and how the system's behaviour arises from, and is linked to, the characteristics and behaviours of individual elements (Ramalingam et al. 2008).

Use of ABMs for studying HEC process in developing nations has already been proven effective (Mahmood et al. 2020). Computational models like ABMs provide opportunities to incorporate the complex nature of HEC, through readily available theoretical, conceptual, and real-world data and observations, for conducting computer-based simulation experiments; thereby overcoming the prevalent issues of conducting research in cities of developing nations.

Simulation-based experiments conducted using ABMs such as *Sensitivity Analysis* and *What-if* scenario generation experiments allow users to programme individual agents with qualitative choices and adaptation in their behaviour, and programme individual variables with experimental values, which are otherwise impossible to perform in the real world. The major challenge while simulating an urban-level phenomenon is to deal with the sheer scale of a city and to deal with an ocean of numbers that might emerge from these experiments.

2.2 Research Objectives

The objectives of this study are:

a. To develop a computational model for simulating HEC behaviour of Indian cities.
b. To validate the model findings and explore its applications in addressing HEC for different cities.

3 Materials and Methods

The overall methodology of this study is demonstrated as a flowchart in Fig. 1.

The computational model is developed using NetLogo 6.2.0, a programming language and integrated development environment for agent-based modelling. Model validation is carried out using 'predictive-validation' method.

3.1 ODD Protocol

The model design is explained using the standardized method of describing ABMs; called the ODD protocol, which stands for "Overview, Design concepts, and Details" (Steven and Grimm 2019). In the ODD protocol, the model design is explained using the different types of agents that comprise a NetLogo model world—*turtles, patches, links* and *observers.*

3.1.1 ODD: Purpose

The purpose of the model is to simulate HEC of an urban precinct in an Indian city, based on the city's socio-economic and environmental variables, and occupants'

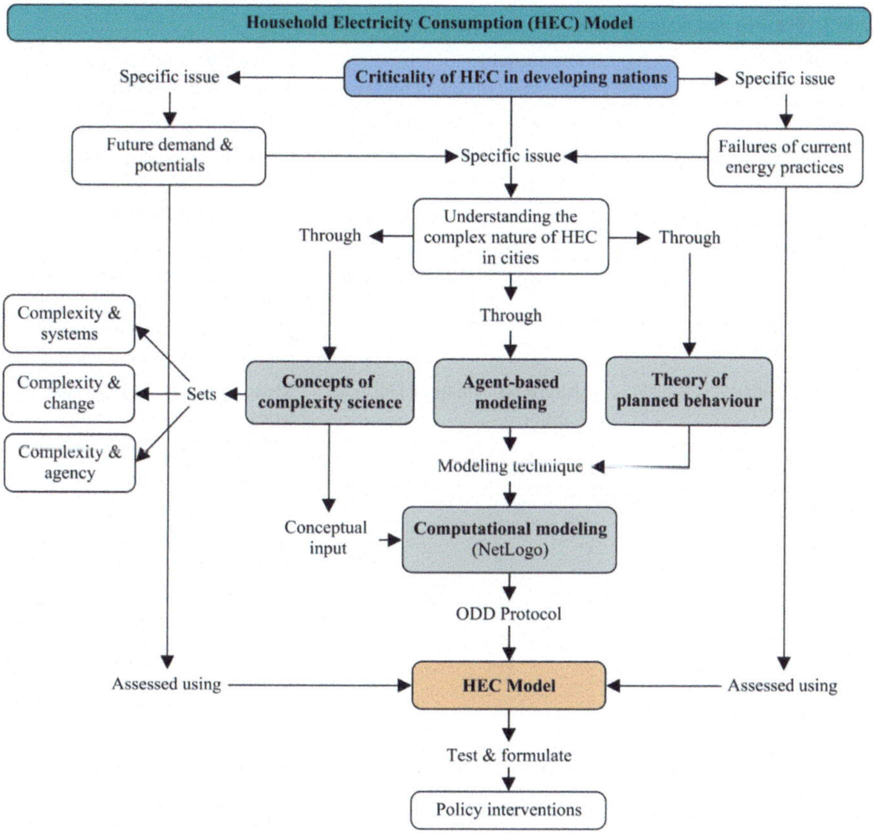

Fig. 1 Methodology flowchart

behavioural variables. In the model, HEC is assumed to be an emergent property of the individual characteristics of elements that comprises the precinct; their interrelations and interactions, micro-level behaviours, and recurring changes to their state, happening over time.

3.1.2 ODD: Entities, State Variables and Scales

The model has two types of entities: households and occupants, which are represented in the model using *patches* and *turtles* respectively. *Links* are used to represent the relationship between individuals, i.e. family members, close confidants, etc. In the model world each patch represents a household, but the size of patches are not indicating the area of buildings. Links are represented using arrowed lines between occupants, which represent a connection between them. The state variables of different entities in the model design are listed down in Table 1.

Table 1 Entities and state variables in the model

Sl	Variables	Data source	Distribution and change process in the model
Globals			
1	Households in precinct	–	Initial number—441. Number remains unchanged throughout the simulation period
2	Occupants in precinct	Indian Human development Survey (IHDS), World Bank	Calculated w.r.t average household size of a city. The number changes with processes of migration, birth and death of occupants in model
3	Weather	ASHRAE Weather Data Center	Mean monthly weather data of a city is fed into the model, which changes at the start of every month
Households			
1	Household size	–	Occupants in households are distributed randomly at initialization. Number changes every month as per the population change processes
2	Household income	–	Collective income of all economically active residents in a household
3	Appliances owned	IHDS	Distributed based on the percentile ownership of household appliances, given in IHDS survey data. The number changes w.r.t household income and household size
Occupants			
1	Age of occupants	Census of India	Age of occupants is programmed based on the age-sex pyramid of India. Ageing process in the model increases an occupant's age every year
2	Occupant income	IHDS	Percentile distribution of per capita income, given in IHDS survey is adopted for programming occupant's income for a city. Income growth process is based on the GDP growth rate of India

In this model, the study area is a hypothetical urban precinct in an Indian city, and hence the number of occupants is decided based on the mean household size of the city in question. The number of households in the precinct is set at 441, which can be adjusted as per the researcher's convenience. Using the actual number of households in a city is not recommended in the model as this may result in an ocean of numbers to be dealt with, as mentioned earlier. The urban precinct in the model is intentionally kept hypothetical so that it can represent the diverse characteristics of the entire city, and not of an actual neighbourhood, which will result in skewed results.

The appliances that are represented in the model are the most electricity-consuming appliances in the households of Indian cities—air-conditioners, air-coolers, geysers and ceiling fans, as mentioned earlier. Another criterion that was considered while selecting appliances is that the usage of these appliances can be directly associated with outdoor weather conditions. The ownership of these appliances in a household is programmed based on a conditional probability on the household's total income. That is, if the household's total income is above a certain income level, then the household is considered to have enough monetary savings to buy that particular appliance. The income level for each appliance is calculated w.r.t the percentile ownership of different appliances in households of Indian cities, as identified in IHDS 2011 survey data (Desai et al. 2018b).

In NetLogo models, time proceeds in discrete steps, called 'ticks', where each tick represents a certain unit of time in the real world. In this study, each time-step in the model represents a month, and the model is set to run for 120 time-steps, making a simulation period of ten years. The reason for configuring one time-step in the model, as one month in real time is that household electricity bill in Indian cities are generally charged on a monthly basis. Further, changing weather conditions in the model is more meaningful, when it is done on a monthly basis. The simulation period is set for ten years—a time period that the authors believe to be sufficient enough to examine the influence of changing variables and intervention policies, on HEC patterns of the precinct.

3.1.3 ODD: Process Overview and Scheduling

The central process taking place in the model is households consuming electricity, when the individual occupants operate electrical appliances, to achieve indoor thermal comfort w.r.t outdoor weather conditions. Apart from this central process, numerous auxiliary processes are also taking place in the model, which changes the system dynamics in a recursive manner. These include processes such as birth, death and migration of residents, purchasing appliances and adopting energy efficiency measures, growth in income and household savings, all of which are programmed into the model in a scientifically sound manner using appropriate theories and authentic data sources.

3.1.4 ODD: Design Concepts

Interconnected and Interdependent Elements and Dimensions: HEC as an urban phenomenon constitutes of multiple elements that are characterized by multiple variables or dimensions that are interconnected and interdependent to each other and the environment. The four principal elements are electrical appliances, individuals or operators (occupants in households), households and the surrounding environment. These elements at different levels, their characteristics and control variables are interlinked to each other, which will prompt changes amongst each other, and to the overall system.

Feedback Processes: 'Feedback' in a complex system can be described as a message or influence that conveys information about the outcome of a process or activity back to its source (Capra 1996). Such feedback will escalate changes and destabilize the system, or suppress changes in the system, bringing it back to its initial state. In case of HEC, these feedbacks are primarily about the consumption patterns of households, costs incurred in the installation of different appliances, and potential monetary savings that can be achieved through the adoption of energy efficiency measures.

Emergence: As mentioned earlier, in this study the process of HEC is perceived as an emergent property—a phenomenon that emerges from various micro-level interactions: (i) the use of appliances based on the personal decisions of individuals as a response to adapt to the changes in surrounding environment; (ii) a household's decision to own an appliance based on the household's income and affordability, and the number of people in a household; and (iii) other processes and variables that affect these factors.

Adaptive Agents and Nonlinearity: Adaptive agents in a system respond to forces in their environment via feedback processes. This adaptive behaviour may be goal directed, which may take diverse forms such as desired local states, end goals, rewards or motivations. The concept manifests that human systems do not work in simple linear fashion; that the feedback processes between interconnected elements and dimensions leads to dynamic, nonlinear and unpredictable changes (Stacey 1996). In the model, this heterogeneous behaviour of individuals are programmed based on Ajzen's theory of planned behaviour (Ajzen 1991), where each individual makes decisions on performing a behaviour, based on their personal goals and expectations, and collective decision as a family.

3.1.5 ODD: Submodels

Calculating HEC

For calculating HEC of the precinct, electricity consumption of individual appliances in a household, say h_l for each month will be calculated as:

$$h_{1m} = m_u * m_n * m_r \tag{1}$$

where h_{1m} is the electricity consumed by appliance m in household h_1; m_u, the mean number of usage hours in a day for each month; m_n, the total number of appliances that are simultaneously used in the household; and m_r, the number of units of electricity (in KWh) that the appliance will consume in a month, when it is used for one hour per day. Therefore, the total amount of electricity consumed by household h_1 can then be calculated as:

$$\mathrm{HEC}_{h_1} = h_{1m} + h_{1n} + h_{1o} + h_{1p} \tag{2}$$

where $h_{(1m,\,1n,\,1o,\,1p)}$ represents the electricity consumed by the four different appliances in a household, as mentioned earlier.

The monthly mean HEC of the urban precinct will therefore be:

$$\mathrm{HEC}_T = \frac{\sum_1^n \mathrm{HEC}_h}{n} \tag{3}$$

where HEC_T is the mean HEC of precinct; HEC_h is the electricity consumed by a single household; and n is the total number of households.

Installation of Energy Efficiency Measures

Installation of energy efficiency measures in a household is classified into two as 'enforced behaviour' and 'voluntary behaviour'. Enforced behaviour is a situation where an energy efficiency measure is installed by households as mandated by the governing body.

Voluntary behaviour is formulated based on a hypothesis that a cost-related decision on performing an energy efficiency behaviour is based on a trade-off between the immediate cost incurred in implementation, and the future decrease in expenses expected from increased efficiency (Caille et al. 2008). Each household gets information on the cost-related savings of energy efficiency behaviour from other individuals in their social-circle through feedback process. It is assumed that a household will perform such a measure if their expectations on achievable cost savings are met, in the feedbacks they receive.

Based on this hypothesis, a household's decision to perform an energy efficiency behaviour is programmed into the model using the capital investments returns formula (Beattie 2022) given below:

$$A = \left[(1 + \mathrm{ROI})^{\frac{1}{n}} - 1\right] * 100 \tag{4}$$

where A is monthly cost savings from the energy efficiency measure, multiplied by twelve; ROI is the required percentage return of investment, which in this case is 100 so that the entire capital investment is returned; n is the number of years required for getting the capital investment returned, based on the savings of current month. From this, n can be derived as:

$$n = \frac{1}{\log_{\left(\frac{A}{100}+1\right)} 2} \quad (5)$$

3.2 Model Validation

The model validation in this study is carried out using 'predictive-validation' approach, making use of *BehaviourSpace* tool in NetLogo. In predictive validation, the model is initialized with state-variable values of a specific year, which the model simulates and projects to another year. The data used at model initialization is from IHDS-2005 (Desai et al. 2018a), which the model simulates for a time period of six years (72 time-steps), and the results are cross-checked with IHDS-2011 (Desai et al. 2018b); both of which are considered to represent the real-world scenario in the respective years.

The validation of the model is carried out for the case of Bhopal city, the results of which are demonstrated in Table 2. The results of the model simulation run yielded convincing evidence that the model can predict real-world scenario with considerable accuracy. The growth rate observed for different state variables in the simulation data was consistent with the growth that was witnessed in the real world. The population growth seen in the model is 13%, which is consistent with the 14% growth witnessed in Indian cities during the same period, as stated by the World Bank (2022).

Table 2 Simulation results and IHDS survey data in model validation run

Sl	Element	IHDS data			Simulation data		
Bhopal city		2005	2011	Growth rate	2005	2011	Growth rate
1	Average household income (in rupees)	56,597	148,146	0.62	53,812	125,928	0.57
2	Number of occupants	–	–	–	1707	1961	0.13
3	Air-conditioner ownership	0.0	0.0	0.0	0.0	0.0	0
4	Air cooler ownership	0.64	0.65	0.01	0.66	0.68	0.02
5	Ceiling fan ownership	0.96	0.97	0.01	0.94	0.98	0.04

3.3 Flexible and User-Friendly Design

As mentioned earlier, the model is expected to serve as a comprehensive tool for different cities and hence, it is essential to make the model interface user-friendly and flexible to use. To achieve this, different tools that are available in NetLogo like, *Sliders*, and *Choosers* were added to the model interface. Each state variable can be assigned values by sliding or choosing from these tools, by which the variables shall have the respective values of a specific city or study area in question. Further, the model design is scalable so that it may be used for conducting studies of different scales. A snapshot of the model interface is given in Fig. 2.

4 Results and Discussion

The four case cities selected for this study are Delhi, Mumbai, Bangalore and Bhopal. The characteristics of these cities, which are relevant for the model, are extracted from IHDS-2011 survey data (Desai et al. 2018b) and ASHRAE Weather Data Center. These characteristics are listed down in detail in Table 3.

It can be observed from Table 3 that the four case cities exhibit sufficient variations amongst them to examine whether the model is giving generalized outputs for different cities. The model simulations for the case cities are carried out in a systematic order—an examination of the consumption patterns at first, followed by various experiments that are formulated based on the simulation results.

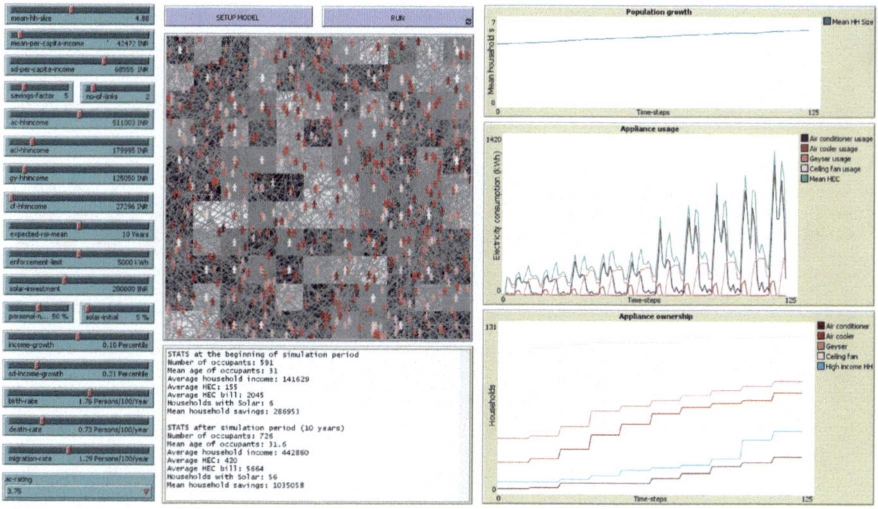

Fig. 2 A snapshot of HEC 4.0 model interface, its plots, outputs and user-friendly design. *Source* Author

Table 3 Characteristics of Delhi, Mumbai, Bangalore and Bhopal city, for the model

City	Climate type	Mean household size	Mean per capita income (INR)	SD per capita income (INR)	Migration rate (persons/100/year)
Delhi	Cwa, BSh	4.89	55,994	67,064	1.84
Mumbai	Aw	4.71	49,286	38,938	0.6
Bangalore	Aw	4.59	57,382	57,382	2.14
Bhopal	Csa	4.73	13,526	13,526	1.4

4.1 Consumption Patterns

The electricity consumption patterns of different appliances in households, for the case cities are demonstrated in Fig. 3. From the figure it can be observed that the growth in HEC patterns is most striking for the cases of Delhi and Bangalore, with a growth rate of 83.6% and 77.7% respectively, for the ten-year simulation period. The growth rate in HEC observed for the case of Mumbai and Bhopal was moderate. The mean HEC of the precinct, at the end of simulation period, was highest in Delhi (584.1 KWh), followed by Bangalore (456.7 KWh), Bhopal (362 KWh) and Mumbai (289.2 KWh).

It can also be observed from the figure that the appliances that are consuming the highest amount of electricity appear to be air-conditioners for the case of Delhi, geysers for the case of Bangalore and Bhopal, and a mix of appliances for the case of Mumbai. The degree of usage of different appliances in the case cities is dependent on

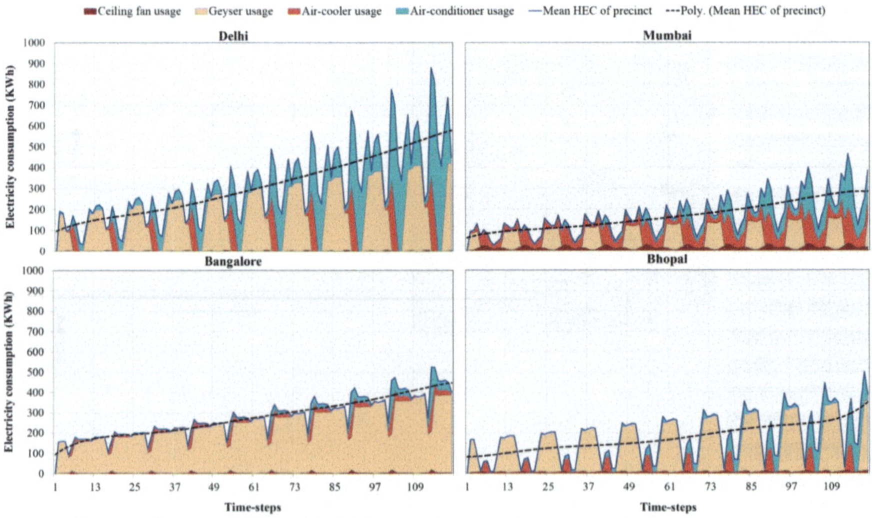

Fig. 3 Electricity consumption patterns of different appliances for Delhi, Mumbai, Bangalore and Bhopal city. *Source* Author

the unique socio-economic and environmental characteristics of each city. The case of Delhi with a comparably high-income level, it is clear that the growth in ownership of air-conditioners, is highest. This along with the extreme summers experienced in Delhi leads to very high consumption of electricity from air-conditioners. The households in Bangalore city on the other hand are also having high ownership of air-conditioners but its usage is for far fewer hours than the other cities due to the pleasant weather conditions. These initial observations though are not sufficient to deduce the driving forces and influential factors of HEC in these cities.

4.2 Sensitivity Analysis

The objective of this experiment is to identify the most 'sensitive variables' that impact HEC of each precinct. Sensitivity analysis (Railsback and Grimm 2019) is adopted for conducting the same, which explores how sensitive a model's outputs are, to changes in variable values; i.e. if a small change is made in one particular variable, do the model outputs change a lot, a little, or not at all?

In sensitivity analysis, when a variable's reference value (P) is varied by an amount (dP), the sensitivity of the model is calculated through a currency (C). The model therefore carries out simulation runs for the reference value P, and for $P + \mathrm{d}P$ and $P - \mathrm{d}P$; and the corresponding currency values of C, C^+ and C^- are determined. Sensitivity (S) will then be calculated using the formula:

$$S^+ = (C^+ - C) * \left(\frac{\mathrm{d}P}{P}\right) \tag{6}$$

$$S^- = (C - C^-) * \left(\frac{\mathrm{d}P}{P}\right) \tag{7}$$

The currency (C) in this model will be the mean HEC of the precinct, calculated through the simulation period of ten years. It may be noted that the significance of the values of S^+ and S^- for different variables are relative to each other, and not based on a definite value of 1 or 100.

The results of sensitivity analysis are demonstrated in Fig. 4. It can be observed from the figure that the two most sensitive variables determining HEC patterns for all four cities are household size and air-conditioner ownership. The degree of sensitivity of these variables though varies across the cities. Both the variables are most sensitive for the case of Delhi. Air-conditioner ownership is least sensitive for the case of Bangalore, as expected, because of the fewer hours its usage. It can also be noted that household size is least sensitive for the case of Bhopal. This can be attributed to the low per capita income that is characteristic to Bhopal, which means that the growth in population will not immediately lead to increased appliance ownership and elevated electricity consumption like the other cities.

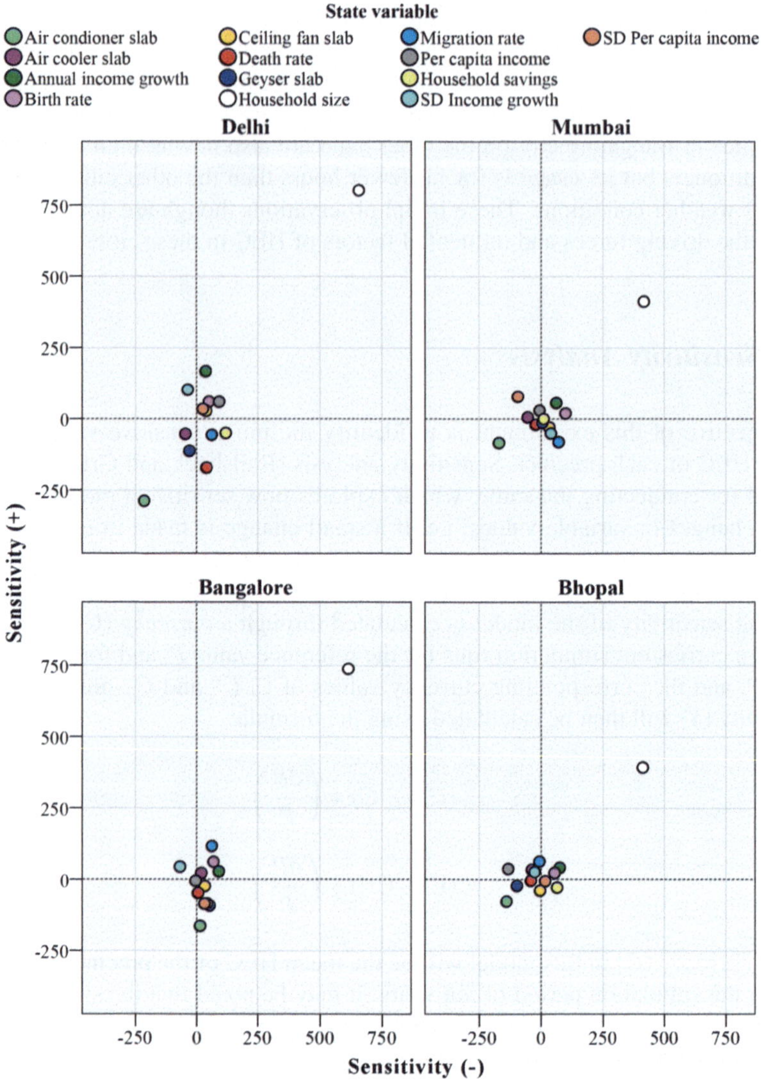

Fig. 4 Sensitivity analysis of HEC patterns in Delhi, Mumbai, Bangalore and Bhopal

Figure 4 also depicts that the relative distribution of the state variables' sensitivity values is more spread out for the case of Delhi and Bangalore. This points to the relative uncertainty of HEC as an emergent property in these two cities, where a small change in many of the state variables can have a significant impact on the model currency. This also means that there are more number of variables to focus on while addressing HEC in these cities.

Table 4 Specifications of different energy-efficient air-conditioners

	12,000 BTU (1.0 Ton) capacity	EER	Wattage	KWh per month (for one-hour usage per day)	Change (%)
a.	1 Star	3.2	3516	26.37	−17.2
b.	2 Star	3.4	3516	24.82	−10.31
c.	3 Star	3.75	3516	22.5	Mean
d.	4 Star	4.25	3516	19.85	11.77
e.	5 Star	4.75	3516	17.76	21.06

4.3 Scenario Generation

'What-if' scenarios of the case cities are then generated to determine the beneficial impacts of addressing HEC in each precinct, through different intervention policies. It was identified from sensitivity analysis that household size and air-conditioner ownership are the two most sensitive variables that determine HEC of each precinct. Based on this, an experimental intervention policy on the use of energy-efficient air-conditioners is formulated to examine its potential benefits in containing the future demand of HEC in each precinct.

4.3.1 Energy-Efficient Air-Conditioners

Energy-efficient air-conditioners in India are marked by a star rating system devised by the Bureau of Energy Efficiency (BEE), with a range from 1 to 5 stars. The specifications of these five different star-rated 'split' air-conditioners are given in Table 4.

In the table, the electricity consumption of a 3-Star air-conditioner is taken as a mean value to calculate the relative difference in the consumption of other air-conditioners.

4.3.2 Rooftop Solar Panels

As part of the global energy pathway set towards reducing GHG emissions and limiting the average global temperature rise to less than 2 °C, India has formulated numerous state-level and national-level policies. This includes, non-fossil fuel-based electricity generation, shift towards emerging fuels, increasing renewable power capacity, amongst others. At household level, the most widely practised one in various cities and states is 'Rooftop Solar panel' (RTS) installation (International Energy Agency 2021). For the same reason, the second intervention policy that'll be simulated using the model is RTS installation.

Installation of RTS in households is programmed based on the enforced and voluntary behaviour that was explained earlier. The enforced behaviour occurs when a household's annual HEC exceeds a 5000 KWh limit, while the voluntary behaviour happens as a result of feedback processes amongst the residents on the monetary benefits of installing RTS.

4.3.3 Results

Based on the aforementioned policies, three scenarios are generated for each city: (i) Scenario 1 showing 'business as usual' without any intervention policies; (ii) Scenario 2 showing the consumption patterns with RTS installation in households; and (iii) Scenario 3 with RTS installation and the use of energy-efficient air-conditioners. The same has been demonstrated in Fig. 5.

From the experiment it was identified that for the simulation period of ten years, Scenario 3 is most effective for the case of Delhi with a potential reduction of 14% from 'business as usual'; while the case of Bangalore is exhibiting lowest potential with 7.99% reduction. That being said, it is also visible from the figure that for all the four case cities, the three scenarios are not exhibiting a significant difference between them till a certain point of the simulation period. At a certain 'tipping-point' in the simulation period the scenarios can be seen diverging into different paths (refer Fig. 5). This occurs when there are sufficient changes in the values of various state variables, which prompts the model currency to take a different path.

Identifying these tipping points is essential because these are the instants when an issue grows out of control and becomes difficult to manage. This is also a point when intervention policies for addressing an issue are to be strictly in place for containing the situation.

Identifying such critical points for intervention policies is essential for Indian cities because energy efficiency measures are generally costly, and the capital investment required for households to perform these practices will always be a barrier to cross. This is further exacerbated by the scarcity of financial means and provisioning in Indian cities (Gupta et al. 2017). For which reason, it is essential to identify the apt time for the implementation of various policy measures, and for the provisioning of financial incentives in performing those measures.

To test this hypothesis, another scenario is generated for the case of Delhi, where the intervention policies are introduced in the model design at the 50th time-step, instead of at model initialization. The same has been shown in Fig. 6.

It is visible from the figure that the introduction of intervention policies at the 50th time-step is roughly having the same impact as introducing them at model initialization. In percentages, the New Scenario 3 can reduce HEC of the Delhi precinct by 11.34% for the ten year simulation period, and 15.2% in the final year; as against 14% and 16.12% when introduced at model initialization.

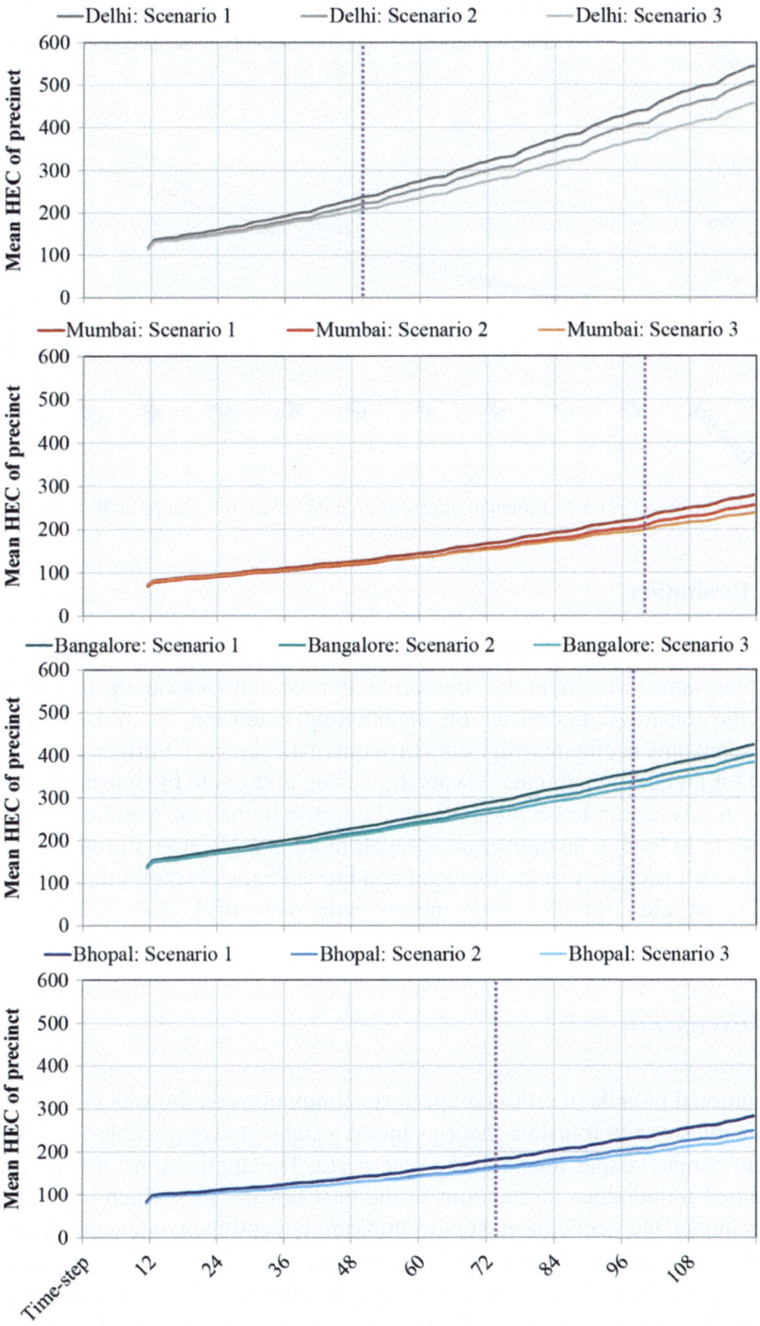

Fig. 5 What-if scenarios generated for the case cities

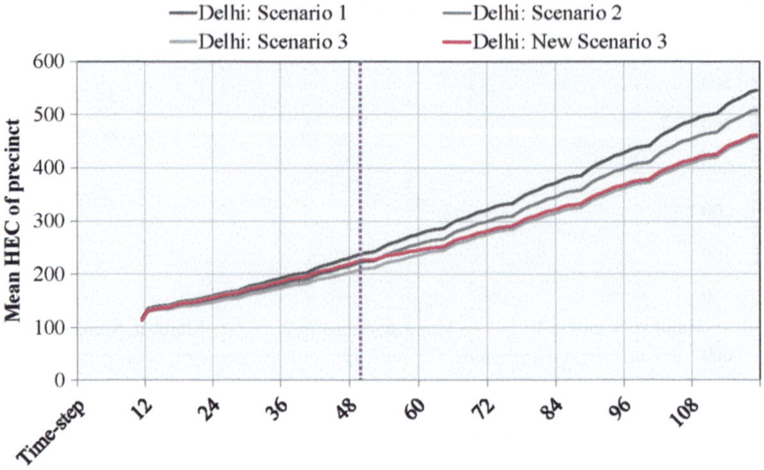

Fig. 6 What-if scenario for Delhi when intervention policies are introduced at 50th time-step

4.3.4 Limitations

In the model world, patches represent households and not buildings, whose numeral remains the same throughout the simulation. For the same reason, the model doesn't address the construction, destruction, abandoning, extension, etc. of buildings, over the years; but only occupants migrating in or out of the precinct. Further, the technique adopted for modelling appliance ownership, using household income and household size may not be accurate for all appliances because in the real world, the behaviour of a household buying an appliance is much more complicated. Further, the model design doesn't address various financial and bureaucratic obstacles that are faced in the implementation of policy interventions in the real world.

4.4 Discussion

Computational models of cities can perform simulations on the way cities function, which enables one to translate theories into a testable and applicable form, without having to conduct experiments in the real world. The applications of this approach have gained tremendous momentum in the past few decades. Such tools can help urban planners and decision-makers in thinking beyond their intellectual reach, by extending their mental abilities to better understand urban systems (Calder et al. 2018). Further, these models are handy in generating 'what-if?' scenarios, or situations that are not yet realized for the future, through controlled manipulation of the constituent variables of the phenomenon in question (Batty and Marshall 2012).

The computational model demonstrated in this study was developed using the theoretical concepts of complexity science and human behaviour, which were then blended with multiple agent-based modelling techniques. The model was validated at first, and the simulation experiments conducted on the case cities gave evidence that the model outputs are not generic. That it can efficiently simulate HEC patterns of different cities in India, which belongs to distinct geographic regions and climatic conditions, and are comprised of diverse socio-economic strata.

Detailed examination of HEC patterns of the four case cities, through sensitivity analysis and scenario generation experiments manifested on the proficiency of the model in identifying the unique characteristics, sensitive variables, tipping points, specific issues and potentials of HEC as an urban emergent phenomenon for each city. Such detailed experiments and analysis can come in handy for identifying robust, befitting and lucrative solutions for addressing HEC in different cities.

At this point, it can be stated with confidence that an integrated planning process using conceptual inputs from complexity science, and the various opportunities offered by ABM techniques can offer an effective framework to better explain the naturally arisen complexity in cities. Further, the approach was efficient in providing opportunities to generate functional complexities, by demonstrating the means and potentials of addressing the issues faced in cities using various planning interventions.

5 Conclusions

A detailed examination of consumption patterns and sensitivity analysis conducted on variables using the computational model has highlighted the significance of population growth and air-conditioner usage in determining HEC patterns in the four case cities selected for this study. Furthermore, 'what-if' scenario generation experiments conducted on the achievable potentials of widely practised domestic RTS installation policies suggested that it is a rather short-term approach towards addressing HEC for the cases of Delhi, Mumbai and Bangalore. The potential benefits due to electricity generation from RTS are limited to two years, and the potential reduction in HEC is not justifiable given the capital cost of INR 200,000 incurred in the installation of a standard domestic RTS unit. In other words, RTS installation policies in these cities do not address the root cause of the situation but are rather a mere attempt to achieve immediate results. On the other hand, the experimental policy on the implementation of energy-efficient air-conditioners proved to be much more effective in containing the future demand for HEC than RTS installation schemes for the case of Delhi.

It is clear from the model results that addressing HEC in cities is not as straightforward as installing solar panels and energy-efficient smart appliances at every nook and corner of each city. Every city needs a unique approach, which can only be formalized based on a detailed examination of its constituent elements, underlying processes and behaviours. The potentials of every planning intervention have significant variations amongst cities; thus, it is essential to identify and understand the suitability and effectiveness of every planning policy for each city.

Addressing energy efficiency in cities in the present day is extremely challenging to say the least; contemplating the nexus between the demand for affordable and clean energy and attaining social, economic and environmental development in society. Addressing such multifaceted problems necessitates the need for futuristic and avant-garde approaches. The model presented in this study is one such attempt. Consequently, the model design is deliberately kept user-friendly and open-ended, so that it can be extended or modified as per the needs of a user—whether for assessing the energy consumption patterns of different cities or to augment the design on the analytical front.

The model can be further extended to integrate it with existing building energy simulation tools to perform detailed building-level simulations. This will enable more accurate energy use predictions in the future. Furthermore, the author intends to overcome the limitations of the current model by making the model world closer to the real world—adding movement of occupants, periodic changes to buildings and including all household appliances in the model design.

References

Ajzen I (1991) The theory of planned behavior. Organ Behav Hum Decis Process 50(2):179–211

Barnes DF, Krutilla K, Hyde WF (2010) The urban household energy transition: social and environmental impacts in the developing world. Routledge

Batty M, Marshall S (2012) The origins of complexity theory in cities and planning. In: Complexity theories of cities have come of age. Springer, pp 21–45

Beattie A (2022) How to calculate Return on Investment (ROI). Retrieved from https://www.investopedia.com/articles/basics/10/guide-to-calculating-roi.asp

Bose RK (2010) Energy efficient cities: assessment tools and benchmarking practices. The World Bank, Washington

Caille A et al (2008) Energy efficiency policies around the world: review and evaluation. World Energ Counc 2008:9–36

Calder M et al (2018) Computational modelling for decision-making: where, why, what, who and how. R Soc Open Sci 5(6):172096

Capra F (1996) The web of life. Flamingo. Harper Collins, London

Chunekar A, Sreenivas A (2018) Towards an understanding of residential electricity consumption in India. Build Res Inf 47(1):75–90. https://doi.org/10.1080/09613218.2018.1489476

Desai S, Vanneman R, National Council of Applied Economic Research, New Delhi (2018a) India Human Development Survey (IHDS), 2005. Retrieved from https://doi.org/10.3886/ICPSR22626.v12

Desai S, Vanneman R, National Council of Applied Economic Research, New Delhi (2018b) India Human Development Survey (IHDS), 2011. Retrieved from https://doi.org/10.3886/ICPSR36151.v6

Dooley JJ, Runci PJ (2000) Developing nations, energy R&D, and the provision of a planetary public good: a long-term strategy for addressing climate change. J Environ Dev 9(3):215–239

Gupta P, Anand S, Gupta H (2017) Developing a roadmap to overcome barriers to energy efficiency in buildings using best worst method. Sustain Cities Soc 31:244–259

IEA (2015) Key world energy statistics. International Energy Agency

International Energy Agency (2021) India Energy Outlook 2021

Jacobs J (1961) The death and life of great American cities. Vintage Books, New York

Khanna M, Rao ND (2009) Supply and demand of electricity in the developing world. Annu Rev Resour Econ 1(1):567–596

Khosla R, Chunekar P (2017) Plugging in: electricity consumption in Indian homes. Centre for Policy Research (CPR) and the Prayas (Energy Group)
Khraief N, Shahbaz M, Mallick H, Loganathan N (2018) Estimation of electricity demand function for Algeria: revisit of time series analysis. Renew Sustain Energy Rev 82:4221–4234
Mahmood I, Quair-tul-ain, Nasir HA, Javed F, Aguado JA (2020) A hierarchical multi-resolution agent-based modeling and simulation framework for household electricity demand profile. Simulation 96(8):655–678
Marshall S (2012) Planning, design and the complexity of cities. In: Complexity theories of cities have come of age. Springer, pp 191–205
Olaniyan K, Mclellan BC (2018) Estimating residential electricity consumption in Nigeria to support energy transitions. Sustainability. https://doi.org/10.3390/su10051440
Portugali J (2012) Complexity theories of cities: achievements, criticism and potentials. In: Complexity theories of cities have come of age. Springer, pp 47–62
Railsback SF, Grimm V (2019) Agent-based and individual-based modeling: a practical introduction. Princeton University Press
Ramalingam B, Jones H, Reba T, Young J (2008) Exploring the science of complexity: ideas and implications for development and humanitarian efforts (No. 285). London
Singh J, Mantha SS, Phalle VM (2018) Characterizing domestic electricity consumption in the Indian urban household sector. Energy Build 170:74–82. https://doi.org/10.1016/j.enbuild.2018.04.002
Stacey RD (1996) Complexity and creativity in organizations. Berrett-Koehler Publishers
Steven FR, Grimm G (2019). Agent-based and individual-based modeling: a practical introduction. Princeton University Press
Tewathia N (2014) Determinants of the household electricity consumption: a case study of Delhi. Int J Energy Econ Policy 4(3):337–348
The World Bank (2022) The World Bank | Data. Retrieved from https://www.worldbank.org/en/home
UNSDG (2015). Sustainable Development Goals. United Nations
Verma A, Jaiswal YK, Wani A (2012) Energy consumption behaviour of an urban residential sector in the northern province of Madhya Pradesh (India). Indoor Built Environ 21:703–709. https://doi.org/10.1177/1420326X11425964
Vijayan M, Patil A, Kapse V, Bahadure S (2021) Assessing the energy efficiency practices in cities across the world. Curr Sci 121(3):372–383. https://doi.org/10.18520/cs/v121/i3/372-383
Weaver W (1991) Science and complexity. In: Facets of systems science. Springer, pp 449–456

Branding Sustainable Megacities with Innovative Governance and Management Strategies

Water Security Through Improved Planning, Policies, and Governance: Opportunities of U20 for Delhi and Bengaluru Under India's G20 Presidency

Maheshwari Gupta, Rabidyuti Biswas, and Ashok Kumar

1 Introduction

India is facing major challenges and issues related to water security and sustainable development of its urban settlements. The U20 recognizes water security challenges as major urban challenges. The improved and integrated planning with inclusion of water policies gets importance in the discussions and formulation of strategies in G20 and U20 held in India during the G20 presidency between December 2022 and November 2023. It serves as the premier leaders-led platform for short-term as well as long-term solutions to human growth and the challenges of the global economy. G20 provides the platform for both developed and developing countries to manifest the 'economic, political and global leadership' with rotating presidency (Kathuria and Kukreja 2019). The G20 focused on agendas including sustainable development, health, energy, climate change, trade, agriculture, environment, and anti-corruption (Government of India 2023a). U20 is a platform that facilitates discussions and engagement for G20 countries on various urban issues including water security and sustainable development.

Urban 20 (U20) is one of the engagement groups volunteered by the Ministry of Housing and Urban Affairs, Government of India in December 2022. Important urban issues discussed in U20 include climate change, finances for urban infrastructure,

M. Gupta (✉) · R. Biswas · A. Kumar
Department of Physical Planning, School of Planning and Architecture, New Delhi 110002, India
e-mail: maheshwari.gupta1990@gmail.com

R. Biswas
e-mail: r.biswas@spa.ac.in

A. Kumar
e-mail: a.kumar@spa.ac.in

M. Firoz C et al. (eds.), *Urban Planning and Design for Megacities in the Global South*, Advances in 21st Century Human Settlements,
https://doi.org/10.1007/978-981-97-8370-0_19

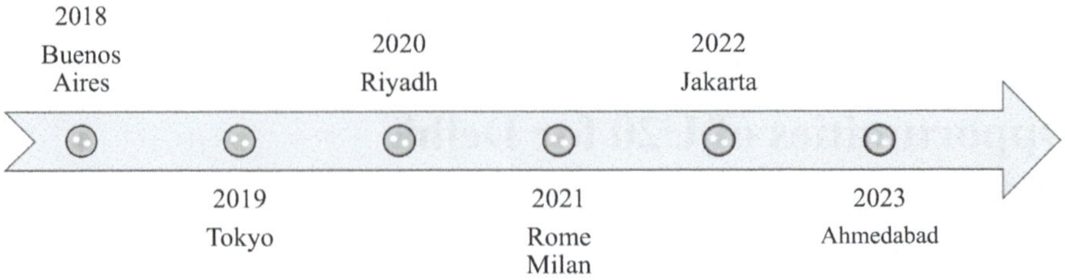

Fig. 1 Previous hosts for U2O. *Source* Authors

affordable housing, mobility, and social inclusion, etc. The events under U20 provide a platform to exchange constructive dialogue with the national and local governments promoting the need to investigate urban issues and push them forward in the G20 agenda (Government of India 2023a) India has got the opportunity to become the sixth host country for U20 (Fig. 1).

U20 India Team includes Ahmedabad as the Chair city led by the Honorable Mayor as chairperson, National Institute of Urban Affairs (NIUA) as the Technical Secretariat, Ministry of Housing and Urban Affairs as the Nodal Ministry from the Government of India and knowledge partners—International Council for Local Environmental Initiatives (ICLEI), South Asia (Government of India 2022; UCLG 2023).

The first event called the 'Sherpa Meeting' was held on 9 and 10 February 2023, and detailed thematic discussions were held on the issues faced in urban development (Fig. 2). U20 Mayors Summit was held in July 2023 in Ahmedabad to address the alarming urban issues, and to develop a roadmap for bridging the gap between policy and practice with the idea of 'intention to action' (Government of India 2023b). There are six priority areas drawn under the U20 city diplomacy as below.

1. Encouraging environmentally responsive behavior

Fig. 2 U20 inception meet, February 2023. *Source* G20 India, 2023

2. Ensuring water security
3. Accelerating Climate Finance
4. Leveraging 'local' potential and identity
5. Reinventing urban governance and planning frameworks
6. Catalyzing digital urban futures.

Some critical urban issues coming under these six priority areas were deliberated in the next Sherpa Meeting. The priority areas emphasize the strategies for ensuring water security and reinventing urban governance and planning frameworks (Government of India 2023c). The water governance, planning frameworks, and water security of urban settlement can be assessed together to provide comprehensive and solutions to the challenges discussed in the meeting held in Ahmedabad (Fig. 3).

It also emphasizes the role of sustainable urbanization and scientifically driven planned growth to complement economic growth. The delegates were encouraged to become ambassadors of 'deglobalizing, decarbonizing and digitizing' cities of the world. It was urged that the U20 team of engagement group provide precise and actionable suggestions that are doable and can be considered by the G20 leaders. Emphasis is also given to the reforms required in the urban planning frameworks of cities with a requirement of infrastructure finance for sustainable water and waste management, public transport, and climate change.

Fig. 3 U20 Priorities meet Ahmedabad, 2023. *Source* G20 India, 2023

Table 1 Focus areas of thematic sessions

Thematic sessions	1. Rethinking Planning and Governance Frameworks	6. Ensuring Water Security
Focus area	Adoption of innovative tools for planning and governance of cities to solve urban issues causing through urban sprawl, rapid growth in population and degrading built environments	Strategies required to deal with water scarcity, pollution of water systems, flooding. Ways to address these challenges: integrated urban water management, nature-based solutions, and use of technologies for resilient cities

Source U20 India (2023a), summarized by authors

A Mayoral Summit was also held on 7 and 8 July 2023 with the thematic sessions on (1) Rethinking Planning and Governance, (2) Encouraging Environmentally Responsible Behavior, (3) Catalyzing Digital Urban Futures, (4) Accelerating Climate Finance, (5) Championing 'Local' Culture and Economy, and (6) Ensuring Water Security (U20 India 2023a). The focus areas from session one and six are taken further for this research are summarized in Table 1. The communique is a collaboratively prepared document describing the role of the cities in achieving the G20 agenda including recommendations from various cities.

2 Literature Review

The growing world's population is continuously increasing the demand for water requirement. In addition to the development needs there are inevitable factors leading to the water stress situation such as climate change, floods, droughts, pandemics. Recently, World Bank 2022 reports about 2 billion people still lack access to safe drinking water globally. Experts and policymakers are continuously working together to find a sustainable solution to tackle these uncertainties and meet the growing demand of water. Worldwide water scarcity is due to the continuous reduction in freshwater availability because of climate change, groundwater exploitation, and poor water management. Referring to the examples of The Millennium Drought in Australia and Day Zero in South Africa, this section ponders two main goals adapted from Los Angeles's 'One Water Los Angeles 2040' Plan (NIUA 2023a, b):

1. Development of implementation strategy for sustainable and cost-effective water management.
2. Identified ways for departments and agencies at the city as well as regional level for integrated water management strategies.

Access to safe drinking water is still a far dream for many dwellers living in urban areas. To support the goal of equitable water supply service, one of the ways indicated is to allow the management of services by the government-owned drinking water-providing companies which is regulated by the Act like the case of the Netherlands. This can ensure durable supply, demand management, and water provision during times of emergency. One of the significant challenges in countries like India is the lack of adequate supply systems and outdated infrastructure leading to water loss through leakages and supply interruptions. Urban water is mostly polluted because of contamination of untreated wastewater. It has been estimated by the UN World Water Development Report 2020 that about 70% of the wastewater is treated in high-income countries and upper-middle countries treat only 38%, lower-middle countries treat 28%, and low-income countries treat only 8% of their wastewater. This means that there is a huge potential for developing countries to utilize their treated wastewater and reduce the stress on freshwater (UN 2020).

The key actions toward 'ensuring water security' listed in U20 secretariat report are:

- Bringing integration of sustainable urban water management practices in front.
- Investment toward technologies to bring change and solve water problems.
- Adaption of early warning systems and effective management strategies to reduce the impact of water hazards.
- Improving urban planning by promoting green–blue infrastructure to fight situations of floods, droughts, etc.
- Government initiatives to improve water service delivery.
- Institutional and regulatory frameworks to protect water ecosystems and aquatic habitats.
- Accountability in service delivery to ensure equitable and safe access to water for all the consumers.

Cities are continuously evolving and growing organically under several economic, social, and cultural factors. Dynamic planning can play a vital role in capturing the temporal changes through evolution and so can the strategies. Strategic planning is an important tool to shift the focus from outputs rather than outcomes during plan formulation. The institutionalization of a regional framework is the need of an hour to manage large urban systems with regional strategies. A few good objectives can be referred to 'the Strategic London Plan 2021' (NIUA 2023b).

Quantitative as well as quality data sets are important ingredients to plan a city. Most cities from developing countries lack comprehensive and common data sets. Lack of reliable data affects the analysis of the issues and hence can lead to inefficient strategies. In addition to data, measuring tools and well-worked indicators are also important to analyze the outcomes of plan implementation. So, the key actions toward 'frameworks for Urban Governance and Planning' listed in U20 secretariat report are:

- Development of planning frameworks to keep pace with evolving needs of production, usage, and maintenance within limited space.
- Collaboration with multiple stakeholders for technical and operational inputs including residents to create public ownership of plans. Breaking the stereo of the top-down approach of the planning process.
- Promoting regeneration and urban design tools for enabling contextual solutions in the city piece by piece.

- Flexible development control regulations to effectively implement strategic city plans.
- Invite the private sector to support and facilitate urban regeneration projects.
- Shifting from a 'one-size-fits-all' approach to context-based strategies.
- Promoting regional-level strategic planning equipping regional-level economic clusters, transportation, etc.

Three goals underlined for Ensuring Water Security are (1) mainstream sustainable water management practices, (2) protect and revitalize the water ecosystem, and (3) strengthen effective water governance. The objectives of setting these goals are to make water-secure cities. Enactment of regulations for maximum use of treated wastewater and water conservation is suggested. The adoption of green city planning which supports the recharge of groundwater through increased retention areas is underlined. Mitigation for water-related hazards is prioritized through innovative means of water plazas and water-sensitive urban design. For the protection of the water ecosystem, the creation of public databases of existing water resources is suggested through the regulatory framework for the protection of water bodies and other water resources in biodiversity. Application of contemporary concepts such as nature-based solutions under environmental and climate funds is promoted. To strengthen water governance, water sharing, monitoring of wastage, and safe and equitable access to water for all are suggested through the institutional mechanism. To support cross-sectoral awareness about water values and collective water saving, participation of cities for knowledge exchange and collaboration in international forums are encouraged.

Three goals drawn for reinventing frameworks for urban governance and planning are (1) promote frameworks for planning and governance of emerging urban patterns, (2) shift to a 'strategic' planning framework, and (3) facilitate equitable and sustainable urban regeneration of existing areas (Government of India 2023d). Formulation of planning legislation that supports multi-level planning and collaborative governance is given as the key importance. To deal with the emerging urban patterns, reshaping the frameworks and multi-level institutional cooperation is suggested. Identification of strategic spatial, economic, cultural clusters, functional areas, and urban agglomerations should be done at the national scale along with peer learning and cooperation among cities. Reformation of planning to accommodate contemporary strategies such as green–blue infrastructure, net-zero built environment, affordable housing for all, economic growth. Participatory planning, rational planning, and design-thinking approaches are promoted. Adoption of mixed-use, citizen or private sector-led eco-friendly design, are encouraged through these goals. Participatory land pooling is promoted to support old build-up and vulnerable areas to improve climate resilience through equitable urban regeneration.

The white paper released by U20 in 'Ensuring Water Security' draws eight-point recommendations to strengthen water management in cities (U20 India 2023b). The pointers are (1) re-thinking the roles and responsibilities of water managers, (2) making use of urban planning tools and techniques in promoting integrated urban water management, (3) common data storage and sharing for evidence-based decision-making, (4) promoting nature-based solutions in water management, (5) building of multifunctional infrastructure propagated in green infrastructure, (6) investment in social and citizen investment, (7) exploring of non-traditional sources for infrastructure finance such as Hybrid Annuity Model, and (8) encouraging strong national and local partnerships. The pointers are important reminders toward the allocation of resources in the right direction and inculcating better policies, planning, and governance to ensure water security (*Indian Express* 2023). Similarly, the four focus areas to achieve water-sensitive urban development are categorized as (1) Water Sensitive Communities, (2) Water Sensitive Urban Planning and Design, (3) Water Sensitive Technology, and (4) Water Sensitive Governance (T20 India 2023). A beginning toward a planning reform could be by integrating these pointers and focus areas on the planning process of cities through plans as a statutory medium.

3 Methodology

For this research, the case study method was selected. The cases chosen were Delhi and Bengaluru, India. Both the primary and the secondary data was collected. Secondary data included the reviews of the U20 Technical Secretariat Report, U20 Communique based on water security and urban planning policies (Section 1.2), detailed review of two specific documents, i.e., the Urban Development Plans and the State water policies for Delhi and Karnataka (Section 1.4). Firstly, the important water provisions through the master plans and lower order plans of Delhi and Bengaluru were critically reviewed. Secondly, the important objectives in the water policies applicable to the planning strategies of Delhi and Bengaluru were discussed. For the primary data collection experts from two specific groups, i.e., the officers involved in plan preparation and the officers responsible for the formulation of water policies for both the cities were interviewed. Interviews were conducted for Delhi in person between August and November 2022 and through telephonic conversation for Bengaluru during the same period. Total interview included for this research is nine, and five for Delhi and four for Bengaluru. The interviews lasted for 45–60 min with the top officials. The designation of the interviewees is kept anonymous. The unstructured interview included questions about the constraints faced by the urban planners and the service providers in institutional integration during the formulation of the water strategies, and the internal and external forces which drives the decision-making specific to the water sector. The interviews were recorded with consent and later transcribed for coding and analyses. The output of the interviews relevant for this study mostly fell under two categories of challenges, i.e., the procedural and the technical challenges during policy integration between the institutions. The procedural and the technical challenges of institutional integration between the planning agencies and the services providers were discussed with examples for both the cities. Based on the findings from two megacities, a few recommendations are made in the conclusion section to achieve water security through city initiatives under U20.

4 Findings from Development Plans and Water Policies of Indian Megacities

4.1 Case of National Capital Territory of Delhi

The National Capital Territory of Delhi in the North of India is surrounded by Haryana from the north, west, and south sides and Uttar Pradesh from the east. The Yamuna River enters Delhi from the west and exits to the eastern side acting as one of the major natural water resources. The other source of water in Delhi is the river Ganga and water from Haryana. The Delhi Jal Board (DJB) responsible for potable water supply in Delhi is a state government agency constituted under the Delhi Jal Board Act of 1988 (Government of Delhi 1998). Out of the total 903 million gallons per day (MGD) sourced by DJB through the processing of the raw water sources from the River Yamuna, River Ganga, and the underground tube wells with a total quantity of 863 MGD, and rest 40 MGD is obtained from the recycling plants. The total percentage of houses supplied by piped water is 83.42 and un-piped water is reported at 16.58% (NIUA-DDA 2020). The percentage of piped supply is inclusive of the frequent supply through the tankers to the housing societies. The quality of water is not satisfactory in many parts of the city resulting in high dependency on bottled water. Water supply situations in Delhi can be seen through (Fig. 4), Photographs clicked in the year 2023.

Master Plan for Delhi 2021 (MPD 2021) was published in February 2007. Major problems affecting the quality of life of the population and their access to basic infrastructures are discussed under the Physical Infrastructure section such as water supply, sanitation, power, and solid waste management. To overcome the problem of water management, the integration of water-sewage-stormwater for recycling, harvesting, and more efficient use was proposed.

The Delhi Development Authority (DDA) mentions the quantity of water supply in the city, however, fails to discuss equity in supply and quality of supply. The per capita demand (225 L) is not distributed uniformly. The inadequacy of non-uniform water distribution has resulted in a twenty-four hours' water supply in some areas, while one to two-hour water supply a day in others (Government of India 2007a). Moreover, the plan does not make any reference to the national or any other water policy applicable to Delhi. A few broad water strategies that are included in revised MPD 2021 are accessibility to safe drinking water, water conservation through decentralized management and rainwater harvesting, natural filtration and storage of stormwater, protecting river zone from encroachments through land-use regulations and recycle-reuse of wastewater.

The lower-order plans of Delhi are prepared for zones, sectors, and layouts. The zonal plans or any other lower-order plan refrains from discussing the water policies applicable for the area except mentioning the estimated quantity proposed as per the MPD. Moreover, none of the lower-order plans attaches to the proposed water service plans for the area. This makes the process of planning water supply and infrastructure a blur phenomenon. Thus, phase-wise implementation stages don't form part of the

Fig. 4 **a** Water supply tanker from Delhi Jal Board, **b** High dependency on bottled water due to poor quality of supply, **c** Use of water containers for storage in absence of piped supply. *Source* Authors

plan, and it is unclear of the method used for water distribution based on population size and density of the settlement.

The draft water policy for Delhi was posted on the DJB website in 2016 (Government of Delhi 2016). The prime objective of the policy draws major attention on long-term water security of Delhi with five major principles: demand management, optimization of available resources, recycling, augmentation of internal resources, and building resilience to climate change and equity. Based on these five principles policy draws few objectives are:

- Access to water for all as per Article 21 of the Constitution and UN Resolution.
- Reduction of per capita consumption norm in every 5 years.
- Increasing the use of recycled wastewater to 70% by 2024 and minimum of 80% by 2026.
- Implementation of water device efficiency rating system.
- Curtailing of distribution losses at all levels of distribution up to 10% by 2025 and 5% by 2030.

- Regulation of groundwater extraction.
- Closing of gaps in the dispatch of data.
- Establishment of Water Resources Commission.
- Restoring the river water quality.
- Cooperation of the public upon duties of state and of citizens.
- Implementation of National Sustainable Habitat Parameters on Urban Stormwater Management

4.2 Case of Bengaluru

Bengaluru is the capital city of Karnataka state. Bengaluru is known for its beautiful water bodies and green spaces. This rapid urbanization is the result of the fastest-growing industries based on Information and Communication Technology. The growing construction activity has resulted in filling up water bodies and low-lying areas. The number of small water bodies within the city has reduced from 400 to 64 interrupting natural valleys and groundwater recharge. This has led to reduced capacity of drains during monsoons causing frequent floods. Water supply is one of the key concerns branching out from the lack of integrated planning and poor resource management.

The currently operational Revised Master Plan (RMP) 2015 was notified in 2007 by the Bangalore Development Authority (BDA) (Government of India 2007b). The upcoming RMP 2031 projects a population of about 20 million for the year 2030 making it the third most populous city of India (Government of Karnataka 2017). As of now, Bangalore Water Supply and Sewerage Board draws about 19 thousand million cubic feet of water equivalent to 1470 Million liters per day (MLD) from the Cauvery River flowing through the states of Karnataka and Tamil Nadu. The total water available for Bengaluru from the Cauvery River is 29 TMC. Despite the availability of such quantities, the city faces a shortage of supply. One of the major reasons reported is the high level of unaccounted flow of water equivalent to more than 45% and due to the limited water network in the city. The water supply in many parts of the city is done through the water tankers forcing many residents, especially the poor community, to struggle for the regular piped supply. The quality of water is also found to be poor in many areas. Local street vendors carry their plastic pots filled with tanker water or buy it from local sellers to perform the basics for their shops and other activities. They also wait in long queue to fetch a few pots from community stand post to perform their daily rituals. The fetching of water is mostly done by young girls and women.

As per RMP 2015 quantity of water supply is 73 Litters Per Capita per Day (LPCD) which is lower than the minimum standards with only 53% having tap connections (Government of India 2007b). A significant water loss of around 40% has been recorded during the water supply due to wastage and pipeline leakages. The water distribution is uneven 'being better in the core areas and poor in the peripheral areas' with the frequency of 5–8 h on alternative days (Government of India 2007b:

7, 19). Adding to the stress, illegal private pumping from industries is depleting the groundwater table. RMP promises to provide direction to the public water supply ensuring accessibility, optimum quantity, and quality of supply to all the citizens conforming to the Central Public Health & Environmental Engineering Organisation (CPHEEO) (Government of Karnataka 2017: VII), however, seems unable to do so.

RMP 2015 underlines that Bangalore's growth in the context to water supply is dependent on the achievements of Bangalore Water Supply and Sewerage Board to achieve the following three strategies:

- Continuous water access to all.
- Quantity and quality of supply conforming to CPHEEO standards.
- Cost recovery of water service by 2015 to be utilized for asset strengthening and management.

The water sector strategic outlook includes a few steps to improve the water supply service in the city such as leak detection in the distribution system scaling from zones to city-level, holistic planning for water supply services by an integrated approach to water demand management, linking of information on GIS city maps, augmentation of investments to improve existing water treatment facilities, promotion of water conservation practices such as rainwater harvesting through awareness campaigns, knowledge co-production to empower public, and institutional reform measures to improve cost recovery and extension of technology.

The idea of Planning District (PD) was first introduced in CDP 1984 to represent the master plan's proposals in greater detail. The PDs in RMP 2015 are categorized based on administrative boundaries i.e., wards of BMP and villages except the core area. The details about wards, zones, local body boundaries, revenue numbers, physical entities, and utilities are visualized through PD. It is used for enforcement purposes only and does not serve statutory functions (Government of Karnataka 2017). These can be understood as the extension/elaborations of master plans. RMP 2015 has not discussed the section of water for specific districts or at the zone level. Although the land-use maps are developed with the inclusion of water bodies, there are no specific policies drawn through the spatial maps at the micro level.

Karnataka State Water Policy, 2002 with the highest priority given to drinking water at the rate of 55 LPCD for rural, 70 LPCD for towns, 100 LPCD for city municipal council areas, and 135 LPCD for city corporation areas (Government of Karnataka 2022). The updated state water policy has been submitted to the Government of Karnataka by a special Task Group in 2019 which awaits notification. A few important objectives covered in Urban Water Supply and Sanitation Policy are:

- Universal coverage for water supply and sanitation needs and willingness to pay for the services.
- Provide services that preserve the sustainability of the water resources of the state while also enhancing economic sustainability.
- Ensure optimum levels of service for all citizens. Water supply situations in Bengaluru can be seen through photographs clicked in the year 2023 (Fig. 5).

Fig. 5 **a** Plastic pot water containers used by the small vendors. *Source* Authors, 2023. **b** Supply through Government water tankers. *Source* Authors, 2023. **c** Fetching of water in pots from stand post. *Source* Manjunath, 2015

5 Discussions

The literature reviews from latest U20 secretariat report, events proceedings from U20 Summit 2023 and official Communique U20, 2023 clearly indicates the need of moving from one-size-fits-all to context-based strategies. Emphasis has been put toward ensuring water security through effective water governance and using urban planning tools and techniques. Therefore, the most effective tool can be thought of is the master plan of a city. To ensure water security through planning strategies, the roles and responsibilities of the urban planners, water managers, local partners and most importantly the community needs clear direction. The master plans of Delhi and Bengaluru are revised with intension of water security but fails to indicate the clear actions required to convert policies into practice. Similarly, the lower-order plans for Delhi and Bengaluru lack the basis for calculating per capita consumption in relation to the carrying capacity of a particular zone or planning district. Brief remarks are covered in Table 2.

The collaboration of the water service providers and the planning agencies is reflected in the ground realities of the cities like Delhi and Bengaluru. The task of implementing the policies on the ground and monitoring needs much more attention than given at present. However, government agencies, municipal bodies, service providers, and other officials are bringing change in the system and cooperating for good. The planning agencies are inviting several stakeholders during the plan preparation process of the upcoming master plan including cross-sectoral representatives, experts from the field, NGOs, project consultants, representatives from the Resident Welfare Associations, academia, youth, and community representatives from special groups. During the formal conversations with the officials from the planning authority and the water service providers from Delhi and Bengaluru, there were significant inputs provided. In Table 3, the challenges that occur during the spatial integration of water policies through the plan are summarized. We observe that the challenges of spatial integration are rooted in the challenges of institutional integration. The challenges are categorized in procedural and technical types.

The interviews with the officials partially explained the position of urban planners in not being able to include several goals in the plan documents. At the same time, they also provided a vision and space for institutional collaboration among stakeholders. The arguments raised by the officials are important value addition to the planning process and the need for inclusive planning. Although there are inevitable reasons behind the faced challenges in planning such as political pressure, financial dependency, change of officials midway, however, still there are significant collaborative opportunities highlighted by the interviewees such as the involvement of the beneficiaries as a stakeholder, collaborative frame of work between planning agencies and water service providers in the cases such as Delhi, Bengaluru, and other metropolitan cities with similar goals in a shared space, and promotion of transparency in planning through public involvement from the beginning.

6 Conclusions

India contributes significantly in global population and economy, the G20 presidency 2023 provided the unique platform to showcase the development achieved in the past years and opened the gates for infinite potentials for learnings and collaborations around the globe. U20 being one of the most important theme supports in expanding the tools and techniques in the existing planning practices. The efforts under U20 opens opportunities for stakeholders from the fields of water management and urban development to work toward making water-secure cities. The need

Table 2 Water policies in the urban development plans of Delhi and Bengaluru

Indicative policies to water supply covered in the development plan	Master Plan for Delhi 2021 and Zonal Plan	Revised Master Plan for Bengaluru 2015 and Planning District	Remarks
Water supply standard per capita per day in LPCD	✓	✓	The population density varies across the city, hence not justified
Quality of supply	No	✓	Quality of supply has not given due importance in the plans, and residents fall sick of water borne diseases in many parts of Delhi
Uniformity in supply	No	No	Some areas get 24 h supply some get 1–2 h
Water conservation policies	✓	✓	Water conservation policies are well defined and promoted in both the plans
Recycle-reuse policies	✓	✓	Recycle and reuse policies are general and decentralized methods for best practices are not included
Spatial allocation of water resources natural and man-made (water bodies, river branches, storm water drains, and STPs, etc.)	No	No	No such spatial maps included at any order of plans
Zone-wise population density and water demand based on carrying capacity of particular area	No	No	Water supply service zones and planning zones are different and hence are not integrated spatially or institutionally at any scale. This forbids the purpose of equal access to supply for all
Spatial service linkage to water supply-sewage-stormwater in local area plans	No	No	The integrated water resource management is held as a key but fails to clearly indicate the real purpose and need to link water supply-sewage-storm spatially
Referral to existing national and state water policies	No	No	There is no mention of any existing and applicable policies in the plans irrespective of having one

(continued)

Table 2 (continued)

Indicative policies to water supply covered in the development plan	Master Plan for Delhi 2021 and Zonal Plan	Revised Master Plan for Bengaluru 2015 and Planning District	Remarks
Clear set of actions for lower-order plans	No	No	No pointers included for the functioning of lower-order plans nor any set of rules for monitoring and implementation of strategies on ground
Protection of water bodies through regulations	✓	✓	Clear regulation of protection of rivers and other water bodies are done
Regulation for groundwater extraction	No	No	Groundwater protection policies are covered in general by not strict regulations or penalty for exploitation is covered

Source Authors

for city development plans to include social, economic, and environmental imperatives of the cities more than just land-use assignment is evident through the issues highlighted in the case studies. Hence, better planning, policies, and governance for inclusive and integrated decision-making serves as the core idea in designing the solutions to the water-related challenges of cities. Contemporary approaches such as Water Sensitive Urban Planning and Nature-Based Solutions should be adopted as a significant tool to move forward the agenda of sustainable development of cities. Therefore, action plans and frameworks that invites stakeholders and attracts finance can support institutional, spatial, and physical integration among planning agencies and water service providers. The uniqueness of the paper lies around its do ability and application in the field of urban planning. The discussion through findings indicates that there is a huge potential of achieving water-secure cities by reinventing the planning, policies, and governance in water resource management. The reinvention of policies and governance is emphasized through urban development plans as a statutory tool for collaboration between the institutes. The research was limited to discuss the policies specific to water supply and not sewage. Future research can be of significant importance if taken ahead in the direction of preparing model action plans for collaborating the objectives of 'ensuring water security' and 'reinventing framework for planning, policies and governance' for upcoming cities. Any small steps toward bringing these two themes together can be fruitful in the field of planning. This research can be used as a beginning point for planners, engineers, scholars, professionals, and consultants for the learnings and takeaways from U20, G20 2023.

Table 3 Challenges of institutional integration occur during the consideration of water policies in plan formulation

Challenge type	Case	Interviewee's association	Challenges of institutional integration
Procedural	Delhi, Bengaluru	Water service providers	• Policy formulation and budgetary investment in large infrastructure is given top priority. Multiple actors have multiple visions and hence the priority for allocation of resource varies accordingly. For example: the framework for the river-basin perspective, there are interstate institutions, political- non-rational principles, and the multiple conception of justice • There are multiple layers of politics and institutional stress over planning and implementation of policies
Technical	Delhi, Bengaluru	Water service providers	• Water infrastructure is mostly designed based on the population size and not on the population density or local water utility ratio. Even the service stations many times have no idea about the total no of connections and volume of water supplied. If we downscale the census data, water utility and requirements vary significantly • Lack of understanding and lack of beneficiary involvement from the beginning of strategies formulation
Procedural	Delhi	Water service providers	• Agreement between both the water service providers and the planning agency is not always smooth • Due to the independent targets of different agencies, the system lacks integrated planning and every other resource suffers in absence of institutional and political support *Water service providers are under state government and planning authority comes under the central government hence have independent targets
Technical	Bengaluru	Water service providers	• Planning policies are mostly based on market dynamics, how the private developers are investing, what could attract investors, and the governance structure *The current land market, real estate market response, private funding, etc. drives the important planning decisions

(continued)

Table 3 (continued)

Challenge type	Case	Interviewee's association	Challenges of institutional integration
Procedural and technical	Delhi, Bengaluru	Water service providers and planning agency	• About the issue of inequitable water supply: in some areas, there are actual technical challenges such as where to put the pipe, old infrastructure, lack of physical network. However, redevelopment and regeneration strategies can accommodate this challenge
Procedural	Delhi	Planning agency	• Financial benefits of development projects and vote bank politics play an important role *A common perception is that the river Yamuna is a very huge asset and should be utilized to make riverfront development with some decorative gardens, entertainment activities, etc. which will enhance the rate of the site and add economic value to the land
Procedural/	Delhi, Bengaluru	Planning agency	• The planning is done at three levels: Policy level, strategic level and operational level. Then comes certain factors such as, who will prepare the action plan, who will implement where will finance come from, there should not be any legal contradiction in the data, etc. Action plan is initiated on a limited scale so that it is implementable but the coordination among the stakeholders and financial management is a big issue
Procedural	Bengaluru	Planning agency	• A development plan only provides and enables the directions of growth. Not everything can be covered in the master plan. However, lower-order plans can accommodate action plans and detailed policies to be implemented • Lack of detailed action plan in the planning documents and lack of sector-wise water policies limits the scope of work for the implementing body *The Planning Districts (non-statutory) only contains the land-use demarcation of the sub-zones, and there are no details zooming out to the policies applicable to these districts

Source Authors, primary data

Acknowledgements I would like to thank all the anonymized officials from BDA, DDA, and experts who provided valuable inputs from their field experiences.

References

Government of Delhi (1998, April 2) Delhi Jal Board Act 1998. Retrieved from Delhi Jal Board: https://delhijalboard.delhi.gov.in/sites/default/files/Jalboard/generic_multiple_files/delhi_jal_board_act_1998.pdf

Government of Delhi (2016) Water Policy for Delhi, Delhi Jal Board: https://delhijalboard.delhi.gov.in/sites/default/files/Jalboard/universaltab/water_policy_21112016_0.pdf

Government of India (2007a) GOI. Master Plan for Delhi 2021, Delhi Development Authority: https://dda.gov.in/sites/default/files/planning%20files/Master_Plan_for_Delhi_2021_%28Incorporating_modifications_upto_31.08.2022%29.pdf

Government of India (2007b) Revised Master Plan for Bengaluru 2031 (Draft). Bangalore Development Authority, Bengaluru. Retrieved from https://bdabangalore.org/uploads/files/TPM_Documents/RMP_2031/Volume_1_Vision_Document.pdf

Government of India (2022, December 20) Urban-20 (U20) event being organised under G20 presidency of India. Retrieved from Press Information Bureau, Ministry of Housing & Urban Affairs: https://pib.gov.in/PressReleaseIframePage.aspx?PRID=1885167

Government of India (2023a, June 30) G20-background brief. Retrieved from G20 India: https://www.g20.org/content/dam/gtwenty/gtwenty_new/about_g20/G20_Background_Brief.pdf

Government of India (2023b, February 10) Key takeaways of the U20 City Sherpa Meeting, Ahmedabad. Retrieved from G20, Press Information Bureau: https://pib.gov.in/PressReleasePage.aspx?PRID=1898056#:~:text=The%20priority%20areas%20are%20(i,vi)%20Catalysing%20digital%20urban%20futures.

Government of India (2023c, July 16) Priority areas of the 2023 U20. Retrieved from G20 U20: https://www.u20india.org/Priorities

Government of India (2023d) U20 Communiqué 2023. Ahmedabad: G20 Secretariat, Ministry of External Affairs, Government of India. Retrieved from https://niua.in/U20/sites/default/files/Reinventing-frameworks-for-Urban-Governance-and-Planning.pdf

Government of India (2023e, June 26) U20 Engagement Group of G20 to host 2 DaysMayoral Summit in Ahmedabad. Retrieved from G20, Press Information Burue: https://pib.gov.in/PressReleaseIframePage.aspx?PRID=1935357

Government of Karnataka (2017) Revised Master Plan for Begaluru 2031 (Draft), vol 4. Bangalore Development Authority, Bengaluru. Retrieved from https://data-opencity.sgp1.cdn.digitaloceanspaces.com/Documents/Recent/Bengaluru-BDA-RMP-2031-Volume_4_PlanningDistrictReport.pdf

Government of Karnataka (2022) State water policy. Water Resources Department, Government of Karnataka. Retrieved from https://waterresources.karnataka.gov.in/storage/pdf-files/Policies/karnataka_state_water_policy.pdf

Indian Express (2023, February 10) Puri bats for better planning, policy to ensure water security in cities. The New Indian Express. Retrieved from: https://www.newindianexpress.com/nation/2023/feb/10/puri-bats-for-better-planning-policy-to-ensure-water-security-in-cities-2546113.html

Kathuria R, Kukreja P (2019) 20 years of G20: from global cooperation to building consensus. Springer, New Delhi. https://doi.org/10.1007/978-981-13-8106-5

NIUA (2023a) Ensuring Water Security at a glance. Ahmedabad: U20 G20. Retrieved from https://niua.in/U20/sites/default/files/EWS_Updated.pdf

NIUA (2023b) Reinventing frameworks for urban governance and planning, at a glance. U20 G20 Retrieved from https://niua.in/U20/sites/default/files/Reinventing-frameworks-for-Urban-Governance-and-Planning.pdf

NIUA-DDA (2020) Baseline report: Physical infrastructure. Government of India, Delhi. Available at: https://online.dda.org.in/mpd2041dda/_layouts/MPD2041FINALSUGGESTION/Baseline_Physical%20Infrastructure_%20160721.pdf

T20 India (2023) Fostering Watersensitive Urban Planning in G20 Cities. G20India, Ahmedabad. Retrieved from https://t20ind.org/content-type/policy-briefs/

U20 India (2023a) Urban 20 India Mayoral Summit. Government of India, Ahmedabad. Retrieved from https://www.u20india.org/Content/U20%20FINAL%20AGENDA%20VF2.PDF

U20 India (2023b) Urban 20 White Paper. G20, Ahmedabad. Retrieved from https://www.u20india.org/Content/WhitePaper/EWS_White%20Paper.pdf

UCLG (2023, September) Bringing the local perspective to the G20 agenda. Retrieved from Urban20: https://www.urban20.org/u20summit/2023-ahmedabad/

UN (2020) United Nations. World Water Development Report, UN: https://www.unwater.org/publications/un-world-water-developmentreport-2020

Repositioning Cities as Development Model Based on the Concept of Creative City: A Case of Kozhikode as a City of Gastronomy

Shabnam Khan, Varun Prasad, Mohammed Firoz C, and Susan Cyriac

1 Introduction

Traditionally, planning of cities and regions are generally concerned with physical, economic, and resource allocation planning. The focal point of such planning has always been to optimize the resources within the region to ensure that the population is well served and taken care of in terms of the infrastructure and services being provided (Johnson 2015). However, the concept of such traditional planning has changed over the years and recently cities are envisaged as engines of growth which accelerate the fuel for economic development through improved quality of life. Unfortunately, all the cities are not well equipped to have a core economic base. Instead may have a high legacy in terms of cultural, social, and heritage values which can be branded and marketed for the development of the city. One such emerging concept is the Creative City concept, which United Nations Educational, Scientific and Cultural Organization (UNESCO) had popularized using the Creative Cities Network (UNESCO 2023). Under the banner of the Creative Cities Network UNESCO had listed 7 themes of which Gastronomy is one theme highlighted in the present research. The notion of the "Creative City" emerged during the mid-1990s as an approach to urban renewal and revitalization, focusing primarily on culture and creativity (Černevičiute 2011; Gathen et al. 2021; Kong 2014; Mitchell 2008). It serves as a widely adopted strategy for luring fresh investments, fostering tourism, and expanding emerging markets (Camagni 2002; Gathen et al. 2021). It is assumed that the creative cities as a destination shall bring a lot of benefits like job creation and economic development often enhancing the brand building and improved image

S. Khan (✉) · V. Prasad · M. Firoz C · S. Cyriac
Department of Architecture and Planning, National Institute of Technology Calicut, Kozhikode, India
e-mail: shabnamkhan2608@gmail.com

M. Firoz C et al. (eds.), *Urban Planning and Design for Megacities in the Global South*, Advances in 21st Century Human Settlements,
https://doi.org/10.1007/978-981-97-8370-0_20

of the city (Lange et al. 2008). Such creative cities have the potential in promoting not only local or regional economic growth but also help in attracting new businesses and new labor often leading to a cumulative causation effect (Hall 2000; Power 2003; Pratt 2010; Rae 2004; Raffo et al. 2000; Skoglund and Jonsoon 2012).

Gastronomy's strong connection with tourism is widely acknowledged (Chaney and Ryan 2012; Chang and Mak 2018; Zhu and Yasami 2021). In the modern era, culinary experiences have evolved into a valuable tourism asset for destinations, serving as a means to enhance and establish their identities (Boyne et al. 2003; Lin et al. 2011; Zhu and Yasami 2021). Furthermore, these gastronomic encounters significantly influence the decision-making process of global travelers (Basil and Basil 2009; Zhu and Yasami 2021). "Given the highly competitive nature of the tourism and hospitality industry, it is of great importance for destinations and cities that have long-standing gastronomic identities to develop, expand, and optimize their existing resources to stand out from the competition" (Zhu and Yasami 2021). Due to the profound influence of culinary experiences, destinations at national, regional, and local levels have progressively recognized the imperative of integrating the advancement of gastronomic resources into their enduring, comprehensive strategies for the sustainable development of culinary tourism (Karsavuran and Dirlik 2019). To align with this evolving trend, Cities of Gastronomy, operating within the framework of the "UNESCO Creative Cities Network" (UCCN), received recognition with the aim of establishing a collaborative platform for member cities. This platform serves as a means for these cities to enhance their gastronomic resources by drawing upon local traditions, and by fostering international cooperation and networking opportunities (Rosi 2014).

Kozhikode's economy is primarily reliant on the service sector, including tourism, IT, and education, which is vulnerable to risks such as fluctuating tourist arrivals and global economic downturns. Consequently, the city urgently needs to revamp its economic model to enhance resilience and ensure economic stability for its residents. While there is existing literature on the concept of creative cities and their potential for urban development, there is a notable gap in the literature when it comes to a comprehensive examination of the practical implementation and challenges associated with repositioning cities as Creative Cities of Gastronomy, particularly in the context of emerging economies like Kozhikode. This chapter aims to bridge the gap by offering a prospective approach, a comprehensive analysis, and a strategic pathway for establishing the brand identity of a culinary city, particularly within the framework of emerging economies.

2 Literature Review

2.1 UNECSO Creative City

The Creative Cities Networks (UCCN) was created by the UNESCO in the year 2004. The purpose of this network is to promote cooperation among the member cities to enhance a sustainable development. Currently, there are more than 300 cities under the UCCN working toward this common objective: "placing creativity and cultural industries at the heart of their development plans at the local level and cooperating actively at the international level" (UNESCO 2023). Any cites to join this network should enhance the cultural activities and make life accessible to all thereby creating a plethora of opportunities linking creativity, innovation, and sustainable development (Rosi 2014; UNESCO 2023). The UCCN has a mandate to cover seven different domains namely "City of Crafts and Folk Arts, City of Media Arts, City of Film, City of Design, City of Literature, City of Music, and City of Gastronomy" (UNESCO 2023). The cities joined under this network can use the brand of UN and the UNESCO thereby enhancing their tourism potentials. They also can share their experiences and expertizes among each other thereby creating an enhanced image of the city (Cai 2002; Gathen et al. 2021; Hankinson 2007, 2001; Namyslak 2013; Rehan 2014; Riza et al. 2012; Rosi 2014; Sevin 2014).

2.2 City of Gastronomy

All the cities awarded with the tag of "City of Gastronomy" should meet a set of specific criteria namely, "a well-developed gastronomy that is characteristic of the urban center and/or region; vibrant gastronomy community with numerous traditional restaurants and/or chefs; endogenous ingredients used in traditional cooking; local know-how, traditional culinary practices and methods of cooking that have survived the industrial/technological advancement; traditional food markets and traditional food industry; tradition of hosting gastronomic festivals, awards, contests and other broadly targeted means of recognition; respect for the environment and promotion of sustainable local products; nurturing of public appreciation, promotion of nutrition in educational institutions and inclusion of biodiversity conservation programs in cooking schools' curriculum and so on." (UNESCO 2023). In fact the primary agenda of the UCCN is to convert the historical, traditional, and cultural resources of the city into an economic power house adhered within the framework of creativity (UNESCO 2023). All the cities joining the network shall submit a report on intermittent intervals demonstrating this commitment of creative economy-based development at different levels. "Research has suggested that becoming a UNESCO Creative City of Gastronomy is conducive to exchanging knowledge on promoting destinations through gastronomy, enhancing the destination images, and eventually creating economic, cultural, and social values for the local community" (Pearson and Pearson 2017).

2.3 Gastronomic Tourism

"Gastronomy tourism refers to the touristic experience of appreciating and enjoying food-related products in a particular destination". One of the broadest definitions of the notion of food tourism comes from Hall and Sharples (2003). "Food tourism is an experiential trip to a gastronomic region, for recreational or entertainment purposes, which includes visits to primary and secondary producers of food, gastronomic festivals, food fairs, events, farmers' markets, cooking shows and demonstrations, tastings of quality food products or any tourism activity related to food" (Hall and Sharples 2003). The development of gastronomy tourism necessitates a holistic approach that goes beyond the conventional notion of merely consuming local food and appreciating food culture (Karsavuran and Dirlik 2019). It should be centered on regional gastronomic culture and traditions while providing visitors with a diverse array of gastronomic experiences. Gastronomic identity is formed by combining culture and environment, which includes history, traditions, geography, climate, local ingredients, flavors, and cooking techniques (Harrington 2005; Zhu and Yasami 2021). "Gastronomy has been viewed as a creative and aesthetic cultural product, which is deeply embedded in the urban ambiance and daily lives" (Xiaomin n.d.; Zhu and Yasami 2021). Thus, destinations must adopt the concept of gastronomy in their planning and urban development process for their long-term benefits and growths (Seyitoglu and Ivanov 2020). Therefore, cities aim to enhance their image, so as to attract better growth of economy, tourism, preservation of culture and traditions, etc., through branding efforts (Cai 2002; Hankinson 2007, 2001; Rehan 2014; Riza et al. 2012; Sevin 2014).

2.4 Developing Gastronomic Resources

A holistic approach is needed for developing the Gastronomic tourism in cities. There is a need to understand the local tradition, culture, food habits, etc., so as to provide the tourists with a wide range of food experiences (Karsavuran and Dirlik 2019; Zhu and Yasami 2021). Such cities must incorporate the gastronomic concept into their strategic planning practices so as to optimize the resources for long-term basis (Seyitoglu and Ivanov 2020).

Case studies were undertaken to elucidate how cities across the globe employ gastronomy as a catalyst for economic growth, cultural exchange, and sustainable development. For instance, Bergen, Norway, hosts an annual food festival to showcase its culinary heritage and advocate sustainable food systems. Chengdu, China, capitalizes on its vibrant street food culture, attracting tourists while supporting local vendors and heritage. Such a local cultural food festivals do create an impact on the culture and the economy (Dwyer et al. 2005). "It is suggested that festivals and food-related activities create an enormous opportunity for businesses and generate income and revenue for the state" (Dwyer et al. 2005; Huang et al. 2010; Sandybayev 2018). As a result, it stimulates creative interventions and planning activities that can affect local development and regeneration processes (Di Stefano et al. 2000). Parma, Italy, leverages its renowned Parmigiano Reggiano cheese to promote regional culinary excellence and local producers. Creating awareness about the food destinations can help to protect the traditional food habits, its ingredients, culture, and associated traditions, thereby preserving its uniqueness (Boutsioukou 2018; Soderstrom 2022). These cases demonstrate gastronomy's power to propel economic development, cultural exchange, and sustainable food systems, fostering resilient and inclusive communities. Collaboration within the UNESCO Creative Cities Network further enhances knowledge sharing and innovation in addressing global food system challenges.

3 Materials and Methods

3.1 Study Area Delineation

In the south Indian state of Kerala, Kozhikode is a coastal city known for its spice trade and the landing of Portuguese traveler Vasco Da Gama in 1498. A history of foreign influences predominantly shaped the culinary heritage of Kozhikode, mainly because of successive waves of seafarers who made landfall on its shores over the centuries. The local gastronomic culture has been enriched by European, Chinese, and Arab culinary traditions. Kozhikode has evolved into a melting pot of diverse communities, attracting individuals from both the northern regions of India and other southern states. Kozhikode is an exceptional Indian locale, distinguished by its multicultural culinary offerings. The combination of cultures and their food traditions has made Kozhikode a place with exceptional culinary diversity. Figure 1 shows the location and area chosen for the study. It includes 19 Panchayat and 1 Municipal corporation.

3.2 Methodology

The purpose of this research is to synthesize a methodological framework for developing cities based on the concept of a "city of gastronomy". To investigate the role of cuisine in urban areas, a qualitative research approach was adopted, incorporating prior research on the topic and examining case examples from various cities. Figure 2 shows the flow of methodology used in the study. The culinary infrastructure within

Fig. 1 Study area

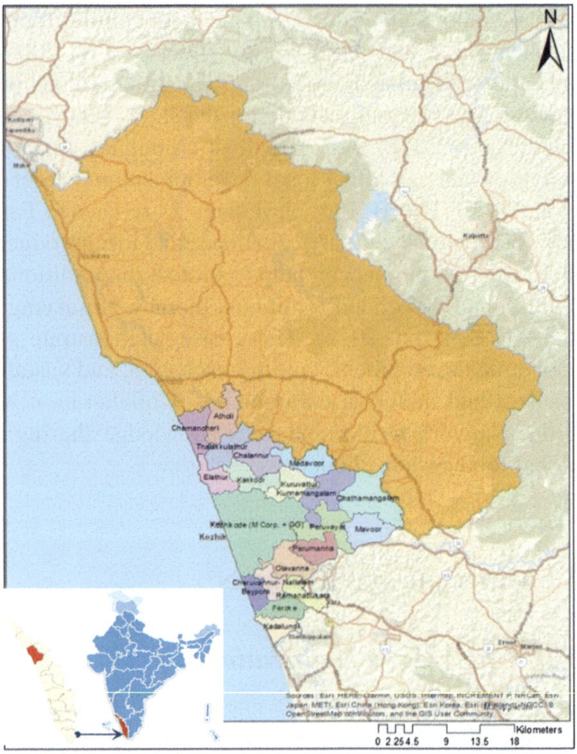

the study area was mapped using satellite imagery and verified through ground-based assessments.

The goal is to explore the gastronomic characteristics, culinary community, events, and communication methods for showcasing the traditional cuisine in the study region. This research employs a mixed methods approach, which includes conducting interviews with eminent personalities such as chefs and food outlet owners. Additionally, focus group discussions with institutions and surveys of various restaurants are utilized to gather comprehensive data. The interview focuses on the significance of culinary tourism, distinctive attributes of culinary tourism products, achievements in promotion, successful culinary tourism projects, and essential considerations for implementation. The analysis involves identifying gaps in the existing culinary infrastructure and assessing the serviceability of food outlets. In order to enhance the city's gastronomic resources alongside existing tourist offerings, prominent food clusters are identified and evaluated based on activity generators, operational hours, and proximity to existing resources, thereby providing development strategies. A comprehensive understanding of the region is attained through a SWOT analysis. Based on these analyses, recommendations are formulated for the city's repositioning as a Creative City of Gastronomy.

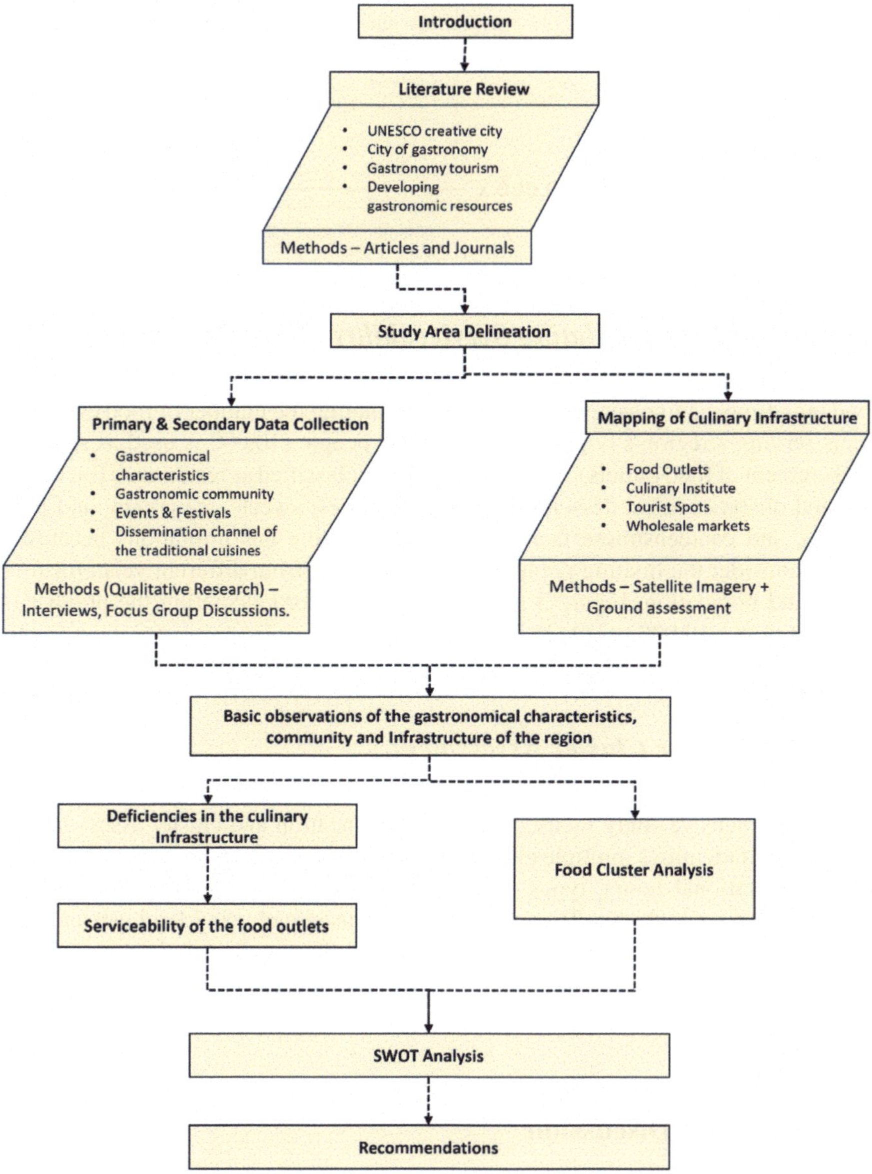

Fig. 2 Methodology for the study

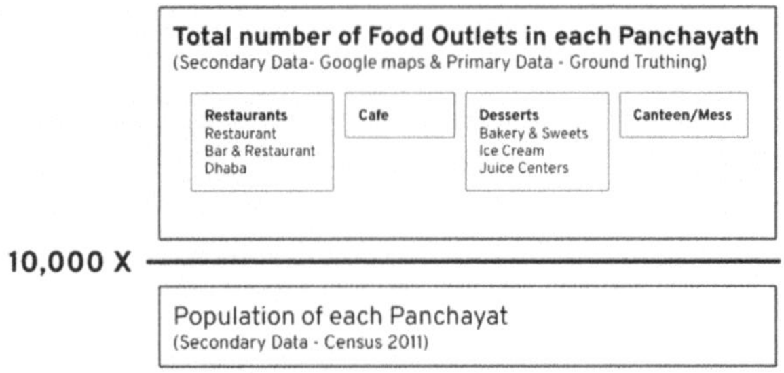

Fig. 3 Formula: food outlets per 10,000 people

3.3 Method of Calculating Serviceability of Food Outlets

Food outlets per capita were determined by calculating the number of food establishments serving a region's population per 10,000 people (10,000 is used as scale for measurement of food outlets). The food outlets are classified as restaurants (including bars and dhabas), cafes, desserts (including bakeries, sweets, ice cream, and juice centers), and canteens/mess facilities. This calculation has limitations because it doesn't consider the floating population and uses data from different years for food outlets and population. Figure 3 shows the formula for calculating the number of food outlets per 10,000 people in each panchayat.[1]

3.4 Methods for Cluster Assessment

This comprehensive study focuses on cluster delineation and cluster assessment in Kozhikode. It identifies multiple clusters and their respective characteristics, encompassing operational hours, types of food establishments, and safety and security aspects within each cluster. The urban core was discerned using land use analysis, while the clustering of food outlets was achieved by mapping their concentration along streets and roads. Figure 4 represents the methodology followed for cluster assessment.

4 Results and Discussion

4.1 Gastronomic Characteristics

For thousands of years, traders from different places like Greece, Rome, the Arab world, and Europe went to Kozhikode to find valuable spices, which were more valuable than gold and cotton in ancient times. Following foreign invasions, Kozhikode's

[1] The official organization that governs local areas in the country.

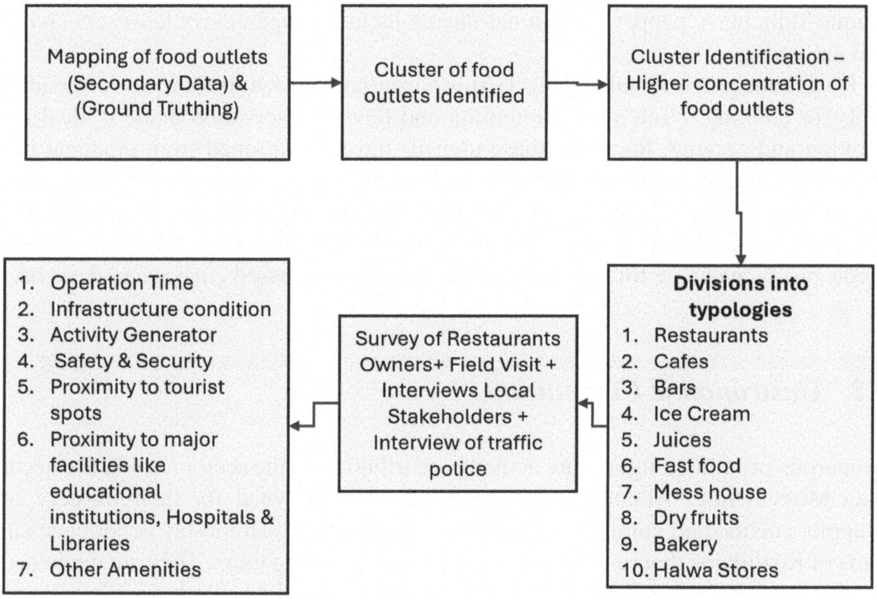

Fig. 4 Food cluster assessment methodology

culinary landscape underwent transformations, blending local and foreign flavors. Kozhikode emerged as a unique culinary hub, offering a diverse range of multicultural dishes. Mappila[2] delights like Kozhikode halwa, Pathiri (pancake made of rice flour.), and Biryani is famous worldwide.

Kerala's distinct meat-oriented cuisine sets it apart from Indian cuisine. Locals favor non-vegetarian dishes while neighboring states embrace vegetarian. For the Mappila community, mornings start with meat-based dishes (egg curry/kanji/Kadala Curry/fish curry accompanies puttu (cooked rice flour)/Ney Pathiri/delicate rice bread). Lunch features rice with meat (chicken/beef/seafood). Light dinners comprise rice or bread-based dish (kanji/pathiris). Desserts like egg pudding (mutta mala), rice cakes (kalathappam), coconut custard (watalappam), and banana rolls (Unnakaya) conclude the day.

Before modern changes, Kozhikode's food customs centered on the healing properties of local ingredients. Usage of local herbs (Takara leaves, oolanthakara, and vattathakara: commonly known as Chakramarda or Senna tora) was prevalent in curries. Mappila[3] food is spicier than West Asian cuisine, yet no spice dominates, except pepper, which initially attracted Arabs. Usage of ghee and coconut oil is favored over refined oil. Staple spices of the area include cardamom and cloves, with

[2] The Muslims of Malabar (generally known as Mappilas) are either decadents of Arab traders or of "Hindu" convert to Islam.

[3] Mappila Food—The Cuisine in the Malabar regions of Kerala.

unique Tellicherry pepper. Additional flavors include ginger, curry leaves, coconut, and aniseeds.

Kozhikode's traditional utensils (Uruli/Varpu/Kinnam/Ottukalam) were versatile tools for cooking. Uruli's[4] heat retention and flavor preservation made it ideal for cooking and serving. Recently, these utensils have transitioned from practical use to artifact. Bronze plate (Kinnam), once household staples, lost relevance due to insulation qualities. Malabar's cuisine stands distinct, shaped by Arab influence and abundant spices. Cooking techniques vary from boiling to frying. Mappila cooking pivots on spice choice for coconut milk-based, yoghurt-based curries and desserts.

4.2 Gastronomic Community

Numerous prominent figures are actively contributing to the sector (Abida Rasheed/Fiza Mossa/Ummi Abdullah). These people are renowned for their mastery of Mappila cuisine and culture. They are curating enriching Homestay experiences in parts of Kozhikode that revolve around immersive cooking tours. They are renowned for their literary works, "Classic Malabar Recipes", "Malabar Pachakam", "Classic Malabar Recipes", "Malabar Pachakam", and "Malabar Muslim Cookery". They also cater from home, conduct cooking workshops, and work as a consultant for hotels to train cooks in Mappila cuisine. Accomplished chefs (Babu Abdullah/Blair Mathieson/Vijayan Pillai) with diverse backgrounds and culinary expertise collaborated with a renowned restaurant, showcasing their inventive fusion-themed menu that attracted attention. In addition to these culinary experts, food bloggers play a crucial role in promoting the local gastronomy. They encompass diverse responsibilities, including restaurant reviews, recipe creation, culinary critique, and the exploration of various gastronomic experiences. Their engagement extends to crafting online content, social media interaction, blog promotion, and collaboration with advertisers, all of which contribute to the city's gastronomic branding and promotion. However, a recent study has underscored a decline in the transmission of regional gastronomic traditions, imperiling their distinctiveness. The potential of local ingredients remains largely untapped, constraining cuisine promotion and impeding tourism. Nevertheless, the gastronomic community, involving prominent food personalities, chefs, food bloggers, and more, can help develop the city's gastronomic resources and elevate its branding and promotion to an international level.

[4] Uruli—"The Uruli was a traditional cookware extensively used in South India and was commonly made of clay, copper and bronze".

4.3 Events and Festivals

Kozhikode's cultural diversity is exemplified through its various events and festivals, featuring regional and delectable cuisines. The city hosts three major food festivals (Mathrubhumi food festival/The Malabar Mahotsavam/Parisons government food festival) each contributing to the exploration of India's diverse food cultures. Beyond gourmet festivals, Kozhikode hosts various cultural, literary, artistic, and musical events throughout the year, incorporating the gastronomy sector. Additionally, traditional festivals like Onam[5] and Ramadan[6] in Kozhikode play a pivotal role in showcasing local delicacies, offering insights into the diverse culinary traditions of the studied communities.

These events serve as an open invitation to the public, providing an opportunity to delve deeper into the city's diverse culture. While some notable culinary figures are actively engaged in developing and promoting food-related events and festivals, the region sees a paucity of gourmet festivals, despite an abundance of food specialties and products to showcase. Furthermore, even if such events occur, public awareness is lacking, and adequate media coverage is absent. Survey results indicate that respondents are generally uninformed about the city's gourmet and cooking festivals and the presence of renowned food bloggers and chefs. Additionally, there is a notable absence of participation from marginalized communities in these events.

4.4 Dissemination Channel of Traditional Cooking Techniques

Effective dissemination channels for traditional cooking techniques play a pivotal role in preserving and perpetuating culinary heritage. These channels encompass various mediums such as culinary schools, workshops, online platforms, and cultural events. Notably, the region boasts a rich cultural tapestry, marked by events like traditional Mappila Weddings,[7] as well as popular festivals like Onam and Ramadan, all of which serve as invaluable platforms for showcasing traditional cooking methods. Gastronomic literature is pivotal in marketing culinary tourism. It showcases cultural ties, tastes, and narratives, inviting travelers to discover local cuisine and experiences. By blending storytelling, visuals, and education, it shapes a destination's culinary identity, sparking curiosity, collaboration, and digital sharing, ultimately driving gastronomic adventures. It's worth mentioning that a substantial body of gastronomic

[5] Onam: Celebrated as a Hindu festival traditionally associated with the harvest of crops. Currently it is celebrated as a Kerala state festival.

[6] Ramadan: Traditionally celebrated as a Muslim festival all over the world as per Islamic. Generally associated with religious rituals, prayers, fasting, and community gatherings.

[7] Mappila Wedding—Mappilas, the traditional Muslim population of Malabar coast, India has a unique wedding with the bride having ornate wedding with gold ornaments. A group of associates dance around the bride for the spectators surrounding her.

literature has been authored by prominent culinary figures in the Kozhikode region, although most of these publications are available in the local language. Despite the city's flourishing festival scene and gastronomic literature, it faces a deficiency in culinary schools and the organization of culinary power workshops.

4.5 Culinary Infrastructure

4.5.1 Food Outlets

From both secondary data collection and primary surveys, approximately 2330 food joints were identified in the study area, which were categorized as restaurants, cafes, canteens, dessert shops, street vendors, food courts, takeaways, and malls. Kozhikode houses 602 restaurants and 167 cafes spread across the corporation as shown in Fig. 5. A concentration of food outlets near the beach, central corporation area, and major roads was observed. The western side exhibits a high cluster of restaurants, while major clusters of cafes are located near the beach area, prominent educational institutes (Medical College, JDT Islam Institute), and the Corporation bus stand. The area has 14 bars, particularly near transit hubs, and popular bars and restaurants are near the beach area. There are 50 canteen/mess facilities near transit hubs, 221 bakery and sweets, 59 ice cream centers, and 32 juice centers. Moreover, 9 malls, primarily eight in the city center, and the regions biggest Mall (HiLITE) located at the outskirts of the city, enrich the culinary landscape. Chain catering businesses such as McDonald's, Domino's, and Subway are also prevalent. Informal vendors, often transitioning into permanent stalls, operate during the evening from 4 to 10 pm. A preference survey was conducted to understand the popular dining spots in the area. The survey identified that within the Corporation area, 19% of consumers preferred beach street food, 25% favored high-end restaurants, and another 19% expressed a preference for seaside restaurants.

4.6 Food Markets

The primary commercial hubs in the study area are the Markets (Valiyangadi/Big Bazar and Palayam), constituting the Central Business District (CBD). The major roads (Wayanad road/Mavoor road/Mini bypass/Pantheerankavu road) also host substantial trading activities. Palayam Market, the city's major wholesale and retail vegetable and fruit market was established in the 1960s and is managed by the Municipal Corporation. It accommodates 120 vegetable shops, 50 fruit shops, licensed street vendors, and daily visits from around 100 trucks. Operating 24 hours, it peaks from 4 am to 6 am. However, the market is lacking efficient warehousing and parking systems. Big Bazaar (Valiyangadi), the Malabar region's significant wholesale and retail market, faces infrastructural challenges and lacks warehouses.

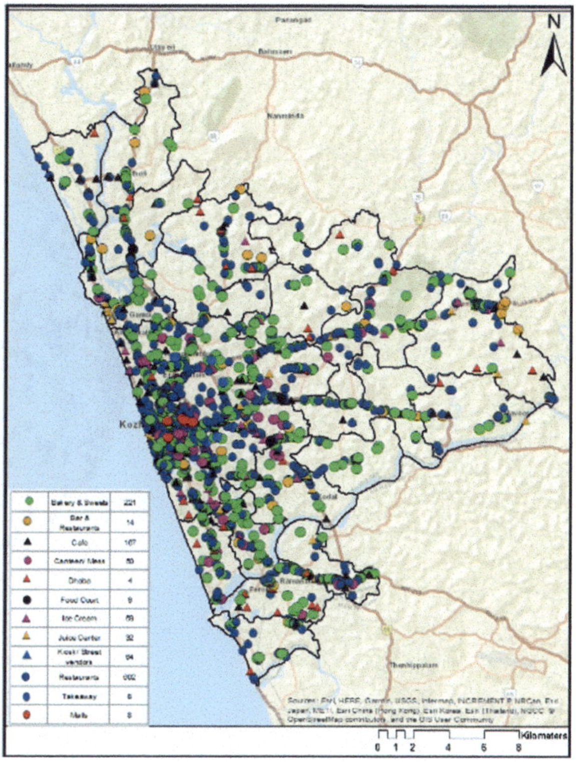

Fig. 5 Culinary infrastructure

Valiyangadi houses over 500 shops, over 10,000 laborers and drivers, and around 200 trucks. Operating 24 hours, trucks operate from 8 am to 6 pm, wholesaling from 10 am to 2 pm, where unloading starts early morning, and retail hours are 10:30 am to 10:30 pm. Fish markets in the study area struggle with insufficient space for loading/unloading, truck and vehicle parking, impacting traders and consumers.

4.7 Gap in the Culinary Infrastructure

The study emphasizes opportunities and gaps in the culinary scene of Kozhikode. Notably, there is an observed gap in food outlets toward the north side of the beach area and certain tourist spots like Sarovaram Bio-Park and Kalipoika Lake as seen in Figs. 6, 7, 8, 9. The selected area has a large population and many food consumers, making it an ideal place to attract culinary enthusiasts. However, there is a need to improve Canteen/Mess facilities within the city, as the current serviceability is inadequate with 0.9–1 canteen or mess for every 10,000 residents. The study recommends developing beachfront areas and improving food facilities while introducing combined tourist activities. Introducing culinary-focused events and festivals can

Fig. 6
Kozhikode-restaurants

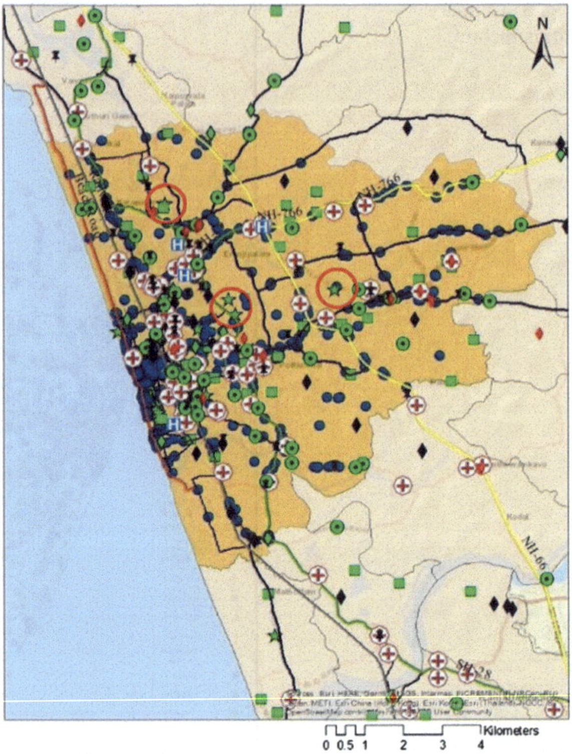

improve the city's food experiences during its frequent major activities like cultural celebrations, sports events, and tourism initiatives, thereby enhancing the overall vibrancy and appeal of the city.

The analysis reveals notable variations in the distribution of restaurants, cafes, and dessert shops across different LSGs in Kozhikode district. In Kakkodi, there are 0–3 restaurants for every 10,000 people which is spatially shown in Figs. 10, 11, 12, and 13. In Feroke and Peruvayal, there are 9–12 restaurants for every 10,000 people. However, Chathamangalam has the highest number with 15 restaurants for every 10,000 people. In terms of cafes, 11 LSGs have 0–1 café/10,000 people, 3 LSGs have 2–3 cafes/10,000 people, and 1 LSG has more than 5 cafes/10,000 people. Dessert shops show the highest ratio among all food outlets, with 8 Panchayats having 4–6 desserts shops/10,000 people, 2 LSGs with 6–8 desserts shops/10,000 people, and one LSG with more than 10 dessert shops/10,000 people.

It is evident from the analysis that 10 LSGs lack mess facilities, comprising approximately 50% of the selected LSGs. This shows the potential for canteen or community kitchens to improve food quality at a more affordable price than other food outlets. The presence of multiple activity drivers and transit hubs in the municipal corporation area leads to the highest per capita of canteen or mess.

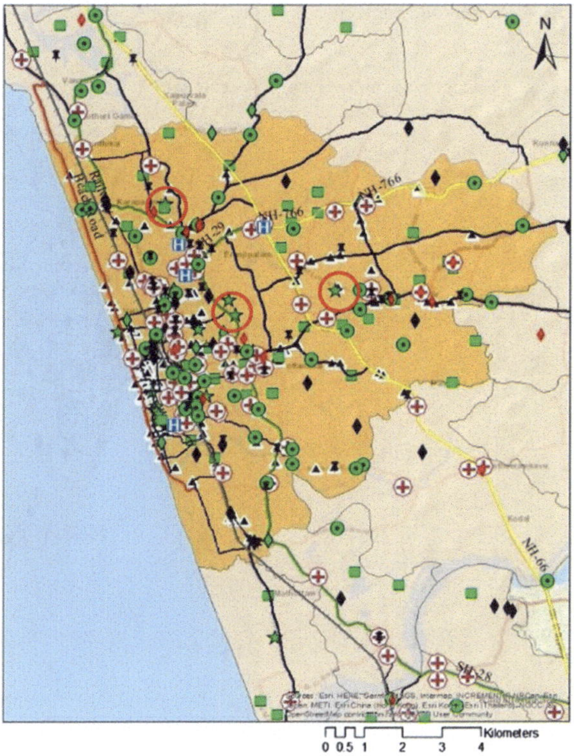

Fig. 7 Kozhikode-cafes

4.8 *Serviceability of the Food Outlets*

When analyzing the accessibility for food outlets, we considered a kilometer radius, considering 500 m as walkable and 1000 m as cyclable distance. Most restaurants fall within the accessibility limit, with only the fringe areas lacking restaurants within walkable and cycling distances as shown in Fig. 14. We observed a significant cluster of restaurants within the municipal corporation area and along major roads such as National Highway, State Highway, and other major routes. Small clusters of cafes are noticeable in each Panchayat[8] Fig. 15, while a major cluster is present in the corporation area. Madavoor Panchayat lacks cafes. Furthermore, there are several accessibility gaps in the Panchayats located at the fringe of the study area. In Fig. 16, there is a substantial gap in the accessibility of canteen or mess facilities toward the north and east sides of the study area. However, canteen or mess establishments in the corporation area form a major cluster, all of which fall within the accessible limit.

We analyzed food outlet accessibility near key sectors such as educational institutions, libraries, and hospitals. Variations in the presence of food outlets within 500 m and 1000 m radii were observed, with some institutes lacking options (Fig. 17).

[8] A village council in India.

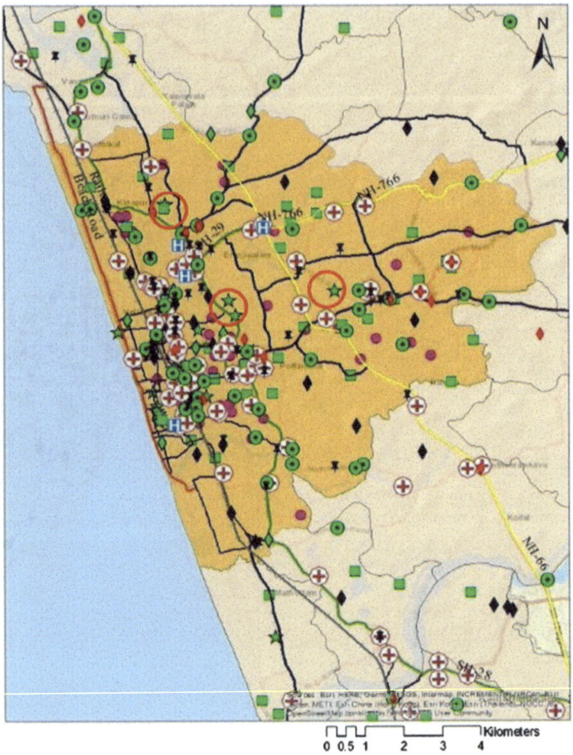

Fig. 8 Kozhikode-canteen/mess

Surveyed libraries showed fewer food establishments within the buffer (Fig. 18), suggesting potential for culinary and literary development. Hospitals generally had a higher concentration of food outlets within a 1000 m radius (Fig. 19), hinting at the prospect of improved canteen facilities. These findings provide valuable insights for enhancing amenities in these areas.

Tourism and gastronomy are intertwined forces that mutually shape experiences. Gastronomy, the art of food, tourism, and the exploration of places, converge to create a distinctive synergy. When travelers encounter new regions, they immerse themselves in local flavors and culinary traditions, deepening their understanding of cultures. Unique dishes, culinary events, and food markets define a destination's character, captivating tourists seeking authenticity. Economically, the demand for diverse food experiences fuels restaurants, markets, and businesses, fostering local growth. Additionally, tourism aids in safeguarding traditional culinary practices, while cross-cultural interactions lead to innovative fusions of taste. Through cooking classes and food tours, travelers engage actively with local cuisines, gaining insight into preparation techniques. In essence, tourism's role in gastronomy extends beyond taste, fostering cultural exchange, economic vitality, and culinary education. This symbiotic relationship ensures that the flavors of a destination continue to enchant and educate, enriching both tourists and local communities.

Fig. 9 Kozhikode-desert shops

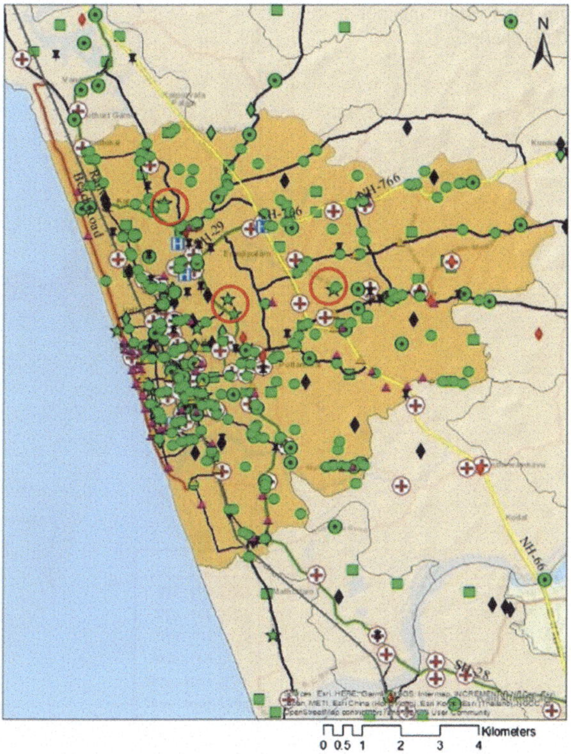

The study revealed a gap in food outlet infrastructure near popular tourist spots in the designated area. Figure 20 shows the major tourist spots in Kozhikode (i.e. [1] Kappad Beach, [2] Elathur, [3] Tasar Creative Center, [4] Beypore Beach, [5] Kadalundi Bird Sanctuary, and [6] Tourist spots in Kozhikode Municipal Corporation). At Kappad Beach, there's a notable scarcity of food outlets within the 1000-m radius, with a higher concentration of kiosks and street vendors (Fig. 21). There's a need for improvement in culinary infrastructure, either through upgrades or the addition of new facilities. Tasara creative center boasts a decent number of food outlets within the 500–1000-m range, including canteens, dessert shops, and mess facilities (Fig. 22). At Beypore Beach, informal shops dot the shoreline, yet there's a lack of food outlets within the 500-m radius, suggesting an opportunity for beachfront development (Fig. 23). Meanwhile, Kadalundi Bird Sanctuary also features informal shops nearby, with a better distribution of food outlets within the 1000-m radius, signaling a need for culinary infrastructure improvement (Fig. 24). In contrast, Mananchira, SM Street, and CSI Catholic Church have relatively few food outlets nearby, presenting an opportunity to integrate food with cultural and heritage walks (Fig. 25). Finally, both Sarovaram Biopark and Kalpoika Lake lack significant food outlets, indicating a potential for development in these areas (Fig. 25).

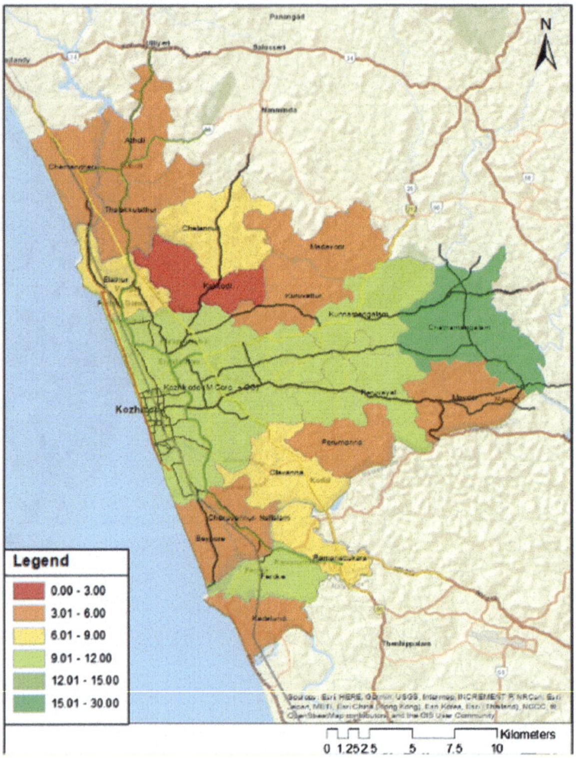

Fig. 10 Restaurants per 10,000 people

4.9 *Food Cluster Assessment*

A total of 193 food outlets were surveyed and 11 clusters were identified. The identification of cluster was done on the basis of concertation of food outlets, variety of food offered, operating time, and prominence of activity generator in proximity. Figure 26 shows the identified clusters with the main activity generator feature of the clusters. The assessment indicates that Mavoor Road stands out as a bustling culinary hub, with restaurants, cafes, bars, fast food outlets, mess houses, and dry fruit shops catering to the varied tastes of residents and visitors. Restaurants operate for an average of 17 hours, while fast food outlets attract a younger crowd during peak evening hours. The PT Usha Road is characterized by cafes, ice cream outlets, juice shops, and restaurants, with many of them staying open into the late hours. This cluster offers a mix of culinary options for day and night-time diners. Red Cross Road hosts a selection of restaurants and cafes, with a few cafes staying open at night to serve late-night customers. The beach area is a bustling gastronomic destination with numerous restaurants, cafes, ice cream shops, bars, and fast food outlets. The beach is most active during the day, with restaurants serving from 12 pm to 12 am, while cafes have a similar schedule and are busiest in the evening. SM Street is a shopper's paradise, offering a variety of goods, while food outlets mostly cater

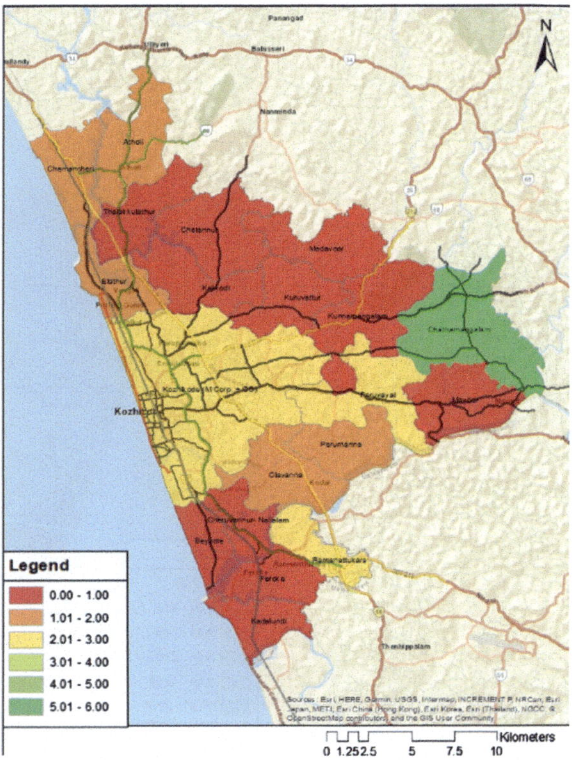

Fig. 11 Cafes per 10,000 people

to daytime crowds, with some activity continuing until 11 pm. The Market Area (Palayam Market) is noteworthy for its 24 hours operation, driven by the Palayam market. Restaurants and juice shops are open an average of 20 hours, and some even operate around the clock. Kuttichira showcases Arab-influenced cuisine, particularly known for its snacks and biryani. Restaurants in this area cater to early morning and evening diners. Kozhikode provided a diverse gastronomic experience across its various clusters. Each cluster catered to different preferences and offered a wide range of cuisines, with varying operating hours to suit both day and night-time diners, making it a dynamic culinary destination.

5 SWOT Analysis

Kozhikode, with its rich culinary heritage influenced by various cultures, including Mappila traditions, possesses several strengths in the context of gastronomic tourism. The city's culinary diversity is a significant asset, offering a wide range of unique and diverse dishes that attract food enthusiasts and tourists seeking authentic culinary experiences. Kozhikode's position as a prominent trading hub draws diverse culinary

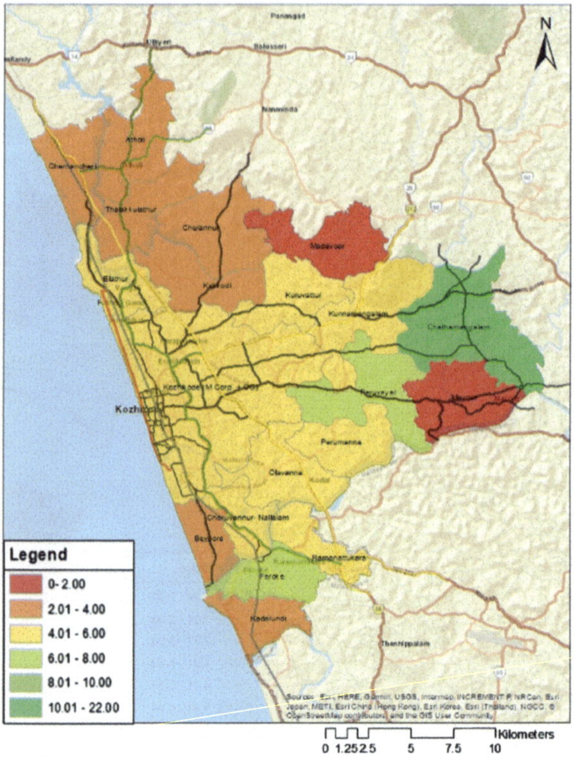

Fig. 12 Desert shops per 10,000

influences from various regions, enriching its local food scene. Additionally, the connection between food and festivals highlights the city's cultural significance, making it a vibrant gastronomic destination. Moreover, the presence of renowned culinary experts like Abida Rasheed, Fiza Mossa, and Ummi Abdullah contributes to the preservation and promotion of traditional cooking techniques and Mappila cuisine, further enhancing the city's culinary appeal. The informal food sector adds vibrancy to Kozhikode's gastronomic landscape. Restaurants near Kozhikode beach cater to both locals and tourists, enriching the city's overall culinary experience.

While Kozhikode has these strengths, it also faces certain weaknesses. One notable deficiency is the lack of culinary schools and workshops, hindering the formal training and development of local chefs and culinary professionals. Additionally, despite the presence of gourmet festivals and renowned culinary figures, there is a lack of adequate media coverage, which results in limited public awareness and participation in food-related events. Furthermore, certain areas, particularly those near popular tourist spots, lack sufficient food establishments, potentially affecting the overall dining experience for visitors.

However, there are several opportunities for the city to enhance its culinary tourism sector. Kozhikode can explore the development of culinary education by establishing culinary schools and workshops, fostering a new generation of culinary experts.

Fig. 13 Canteen/mess per 10,000 people

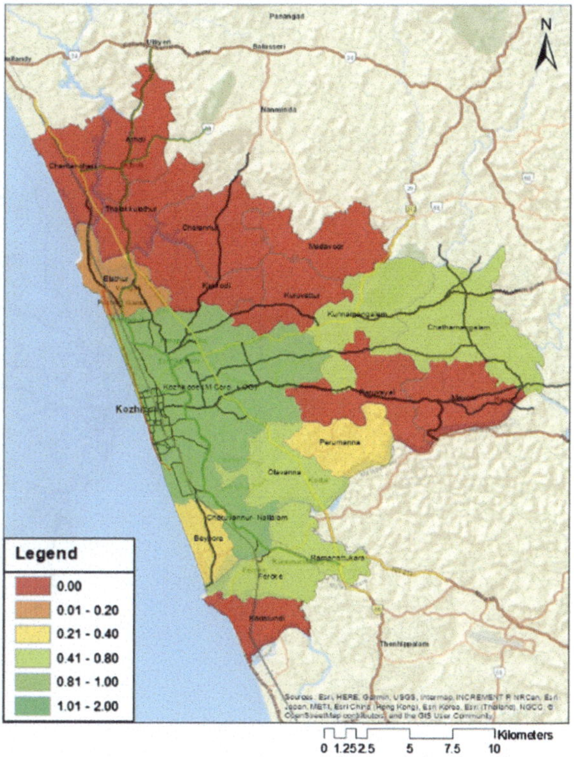

Emphasizing local ingredients in the cuisine presents an opportunity to create a unique selling point, attracting those interested in authentic and locally sourced food. Increasing the number of gourmet festivals and improving media coverage can also promote the city's culinary culture and attract a wider audience. Moreover, enhancing culinary infrastructure in areas lacking food outlets, especially those near tourist spots, can create new culinary hubs and improve the overall tourism experience.

In terms of threats, studies have indicated a decline in the transmission of regional gastronomic traditions, which poses a threat to the preservation of the city's culinary heritage. Additionally, the presence of global chain restaurants like McDonald's, Domino's, and Subway may pose competition to local, traditional eateries. The absence of participation from marginalized communities in food-related events may hinder inclusivity and cultural diversity in the culinary sector. Finally, challenges related to food markets, such as inefficient warehousing and parking systems, may impact the quality and availability of ingredients, affecting the overall food quality and serviceability in the city. Overall, Kozhikode has the potential to further develop its culinary tourism sector by capitalizing on its strengths and opportunities while addressing its weaknesses and mitigating threats.

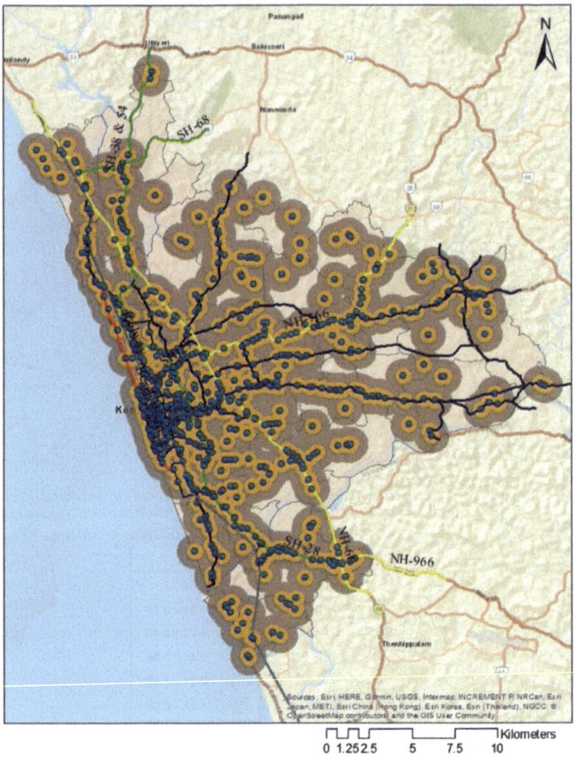

Fig. 14 Restaurants gap assessment

6 Recommendations

In light of the research findings and analysis presented in the preceding sections, this section describes a series of recommendations aimed at addressing the identified challenges and harnessing the opportunities within the study area. Achieving sustainable and inclusive development in gastronomy tourism necessitates the advocacy of a governance model characterized by transparency, participation, and effective leadership. Engaging the local community and involving stakeholders in decision-making processes will foster a sense of ownership and collective responsibility (de Heredia and Gaztelumendi 2019).

To attract gastronomy tourists and enhance the destination's identity as a gastronomy hotspot, it is imperative to foster the creation of innovative and diverse gastronomy tourism products and experiences. These offerings should be distinctive, varied, and tailored to the preferences of gastronomy enthusiasts. Implementing projects like a kitchen incubator, where culinary entrepreneurs can develop unique dishes and concepts, and establishing a monthly calendar for the organization of

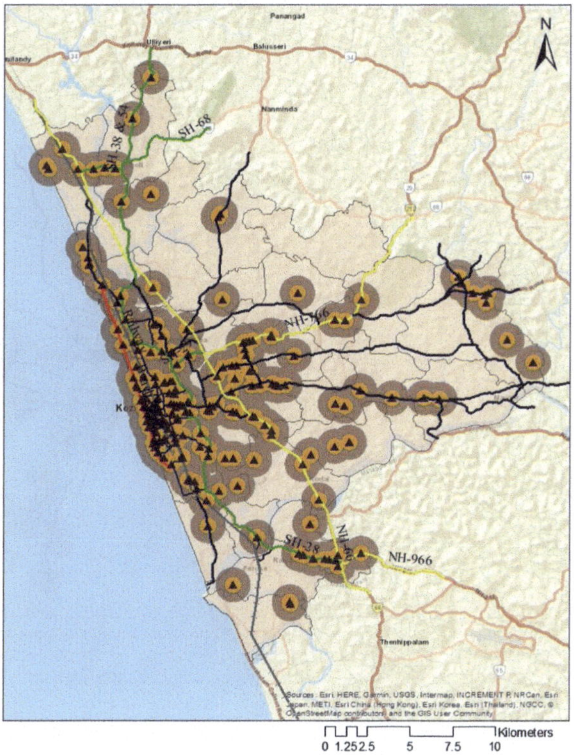

Fig. 15 Cafe gap assessment

various food festivals can play a significant role in achieving this goal. These initiatives not only showcase the local culinary talent but also create memorable experiences for visitors, contributing to the overall appeal of the destination as a gastronomic hub.

SWOT analysis helps us to look in the gastronomic characteristics and gastronomic community and understand the culinary tradition and culture of the region. To preserve culinary traditions and attract tourists interested in cultural experiences, it is recommended to promote the local gastronomy as an integral part of the destination's cultural heritage. This can be achieved by reinforcing culinary education, highlighting local products, and showcasing unique gastronomic traditions. Projects, such as the Gastronomy Lab, initiatives in the realm of food startups, and the establishment of training institutes, can play a significant role in this preservation.

As highlighted in the SWOT, enhancing the overall environment, venues, and services for hosting gastronomy tourists is of paramount importance. Infrastructure improvements and enhanced accessibility are essential to create a welcoming and enjoyable experience for visitors, encouraging their returns. Projects such as a Food Hub, Lunch & Dinner Cruise options, Beachfront Restaurants, and Food Museums showcasing traditional utensils, along with the development of food streets, are recommended to achieve this.

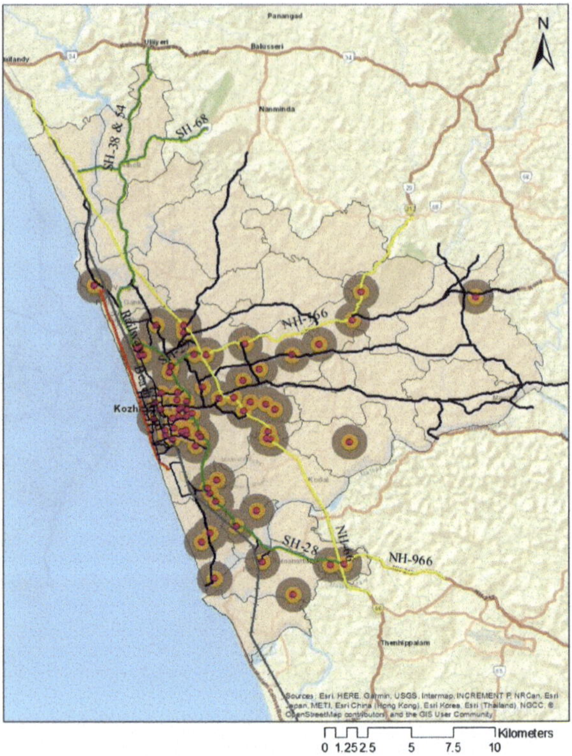

Fig. 16 Canteen/mess gap assessment

Developing an authentic and credible storyline that highlights the destination's gastronomy as a unique brand argument is key. This gastronomic branding will serve as a powerful selling point, attracting tourists seeking distinctive culinary experiences and contributing to the overall appeal of the destination (de Heredia and Gaztelumendi 2019). The findings from SWOT, focused on the food cluster assessment, provide insights into strategic locations and storylines for the city's branding. The development of projects like Food Trails along existing tourist stops, coupled with storytelling and cultural guidance, ensures that tourists are provided with a holistic experience.

To align with global sustainability efforts and demonstrate a commitment to responsible tourism, it is recommended to emphasize the role of gastronomy tourism in advancing the United Nations Sustainable Development Goals (SDGs). By showcasing the positive impact of tourism on the SDGs, the destination can position itself as a responsible and forward-thinking choice for tourists.

Fig. 17 Accessibility of food outlets for educational institutes

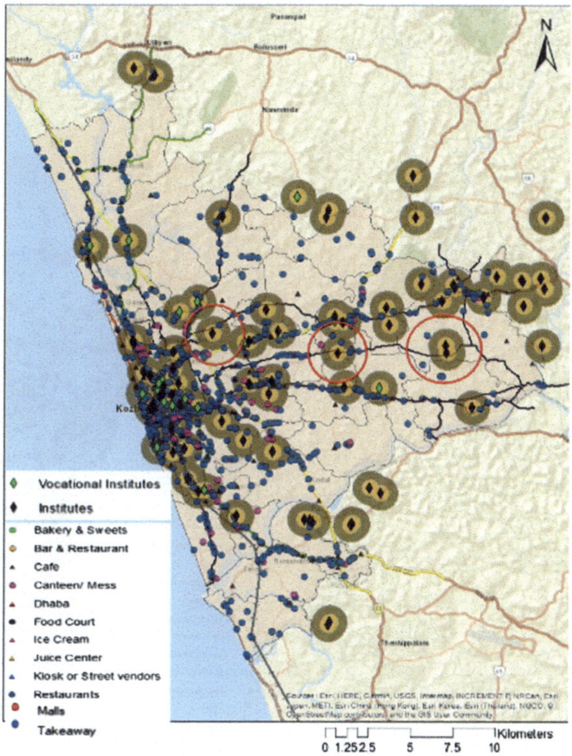

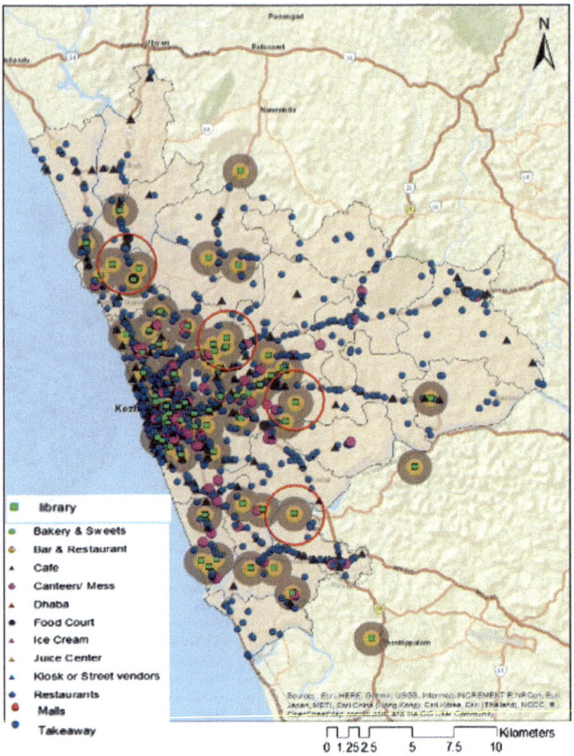

Fig. 18 Accessibility of food outlets for libraries

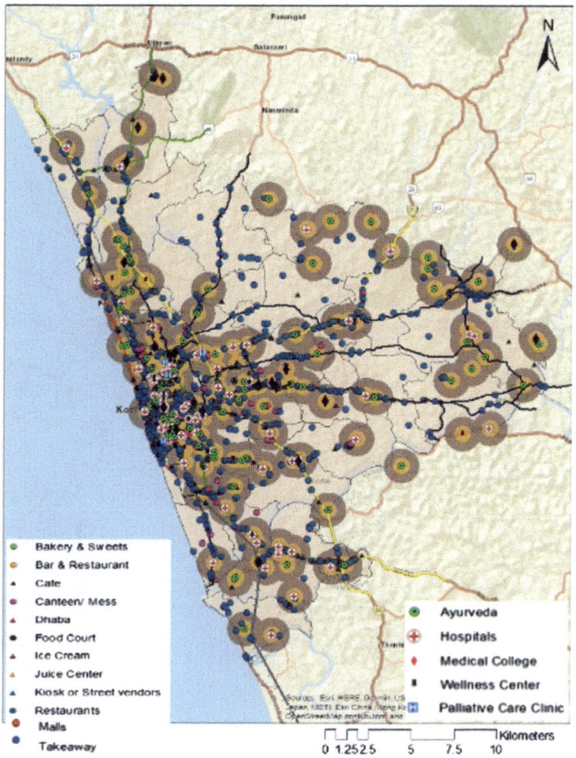

Fig. 19 Accessibility of food outlets for health care

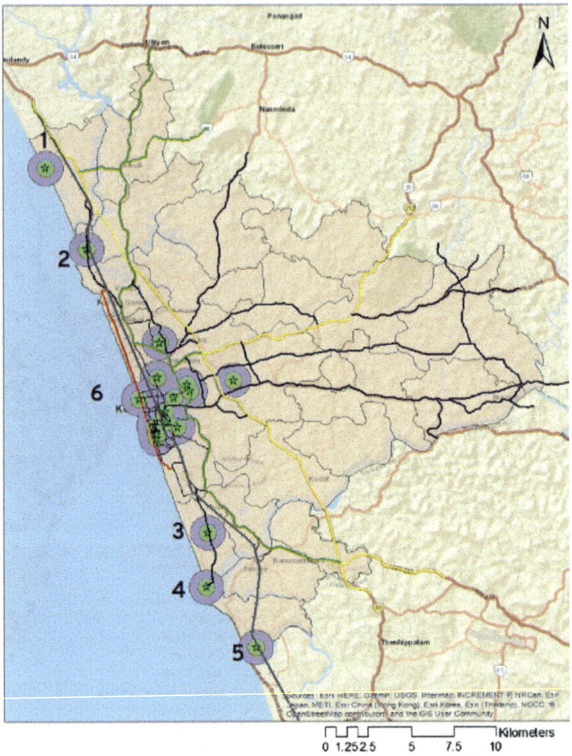

Fig. 20 Location of major tourist spots

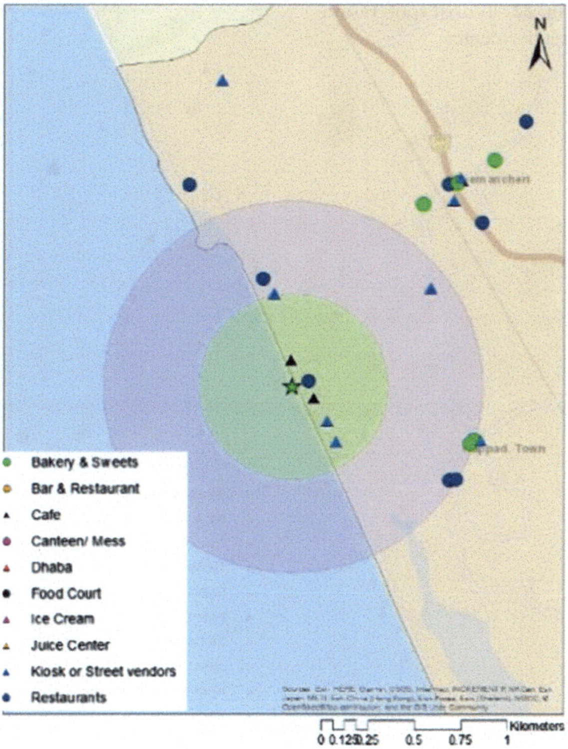

Fig. 21 Tourist spot-Kappad beach

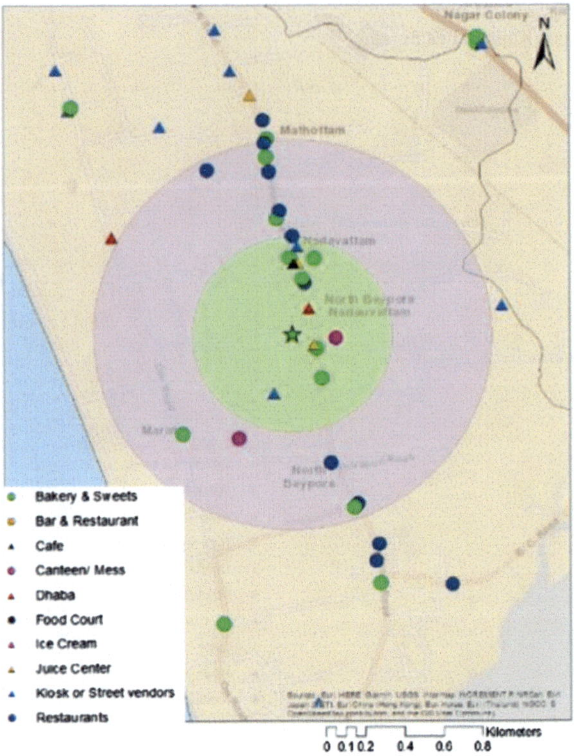

Fig. 22 Tourist spot-Tasara creative center

Fig. 23 Tourist spot-Beypore beach

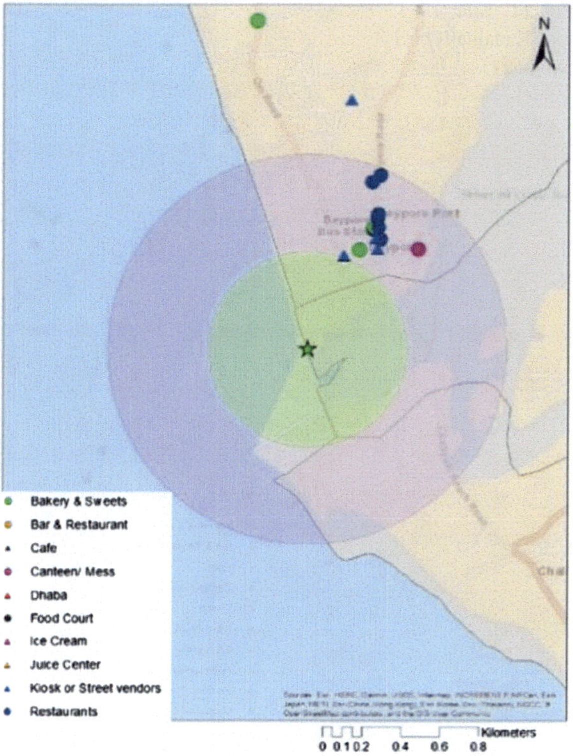

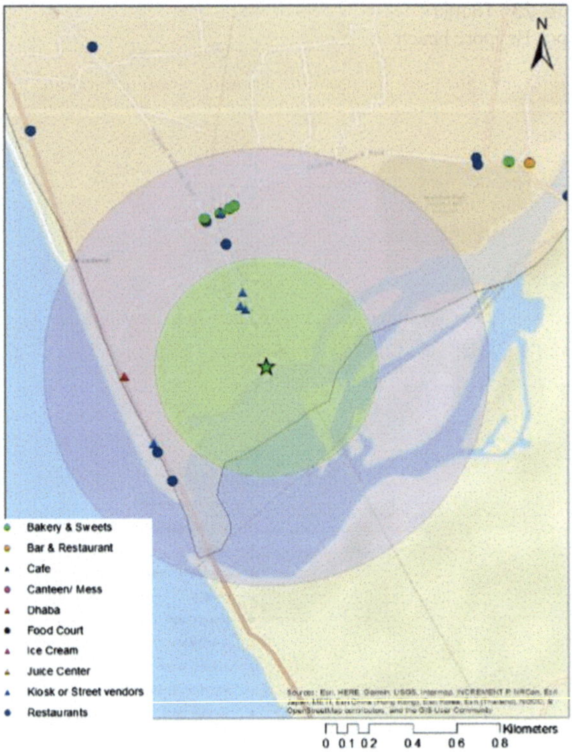

Fig. 24 Tourist spot-Kadalundi bird sanctuary

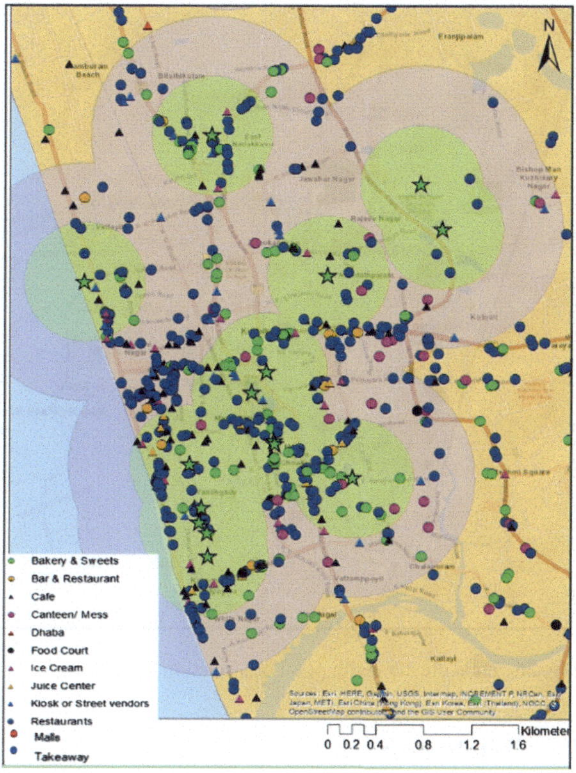

Fig. 25 Tourist spot-Multiple spots inside Kozhikode Corporation area

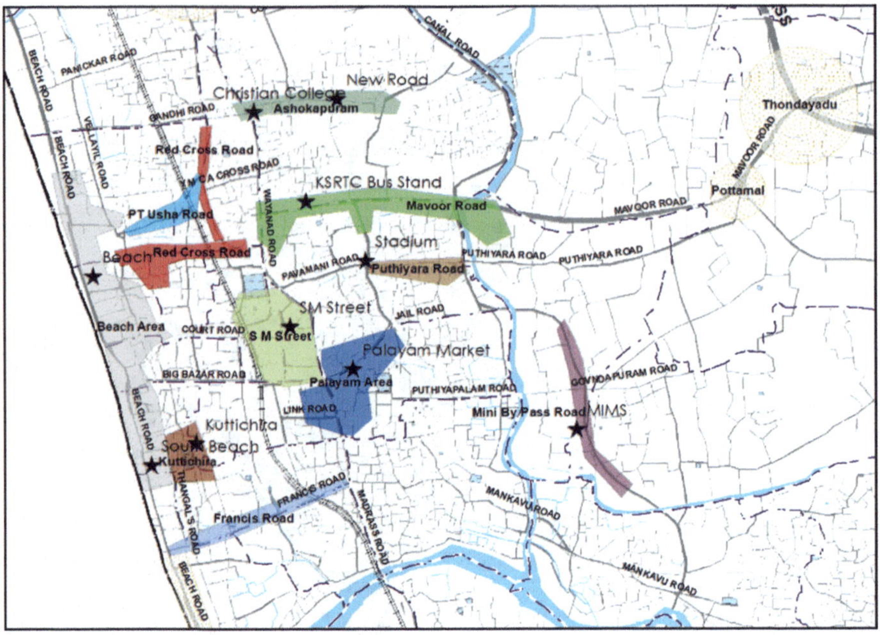

Fig. 26 Food clusters with point of attraction

7 Conclusion

The UNESCO Creative Cities Network has provided a framework for cities like Kozhikode to leverage their culinary traditions as a means of fostering economic resilience and cultural enrichment. By highlighting the city's unique blend of multicultural culinary offerings and rich history, Kozhikode can attract a new wave of tourists seeking authentic and diverse gastronomic experiences.

This chapter has explored the dynamic intersection of gastronomy and tourism within the context of Kozhikode, an emerging economy with a rich and diverse culinary heritage. It has shed light on the significant potential of gastronomy as a catalyst for economic growth, cultural exchange, and sustainable development in the city. The research has also revealed the various challenges and opportunities that Kozhikode faces in its journey toward establishing itself as a Creative City of Gastronomy.

Furthermore, the chapter has emphasized the importance of community involvement, education, and infrastructure development to unlock the potential of gastronomy tourism fully. By investing in culinary education, promoting local ingredients, and enhancing venues and services for gastronomy tourists, Kozhikode can help to create an ambiance of repeated visits through a word of mouth and other recommendations.

In the face of globalization, it is crucial for Kozhikode to preserve its regional gastronomic traditions. This can be achieved through the promotion of local products, the celebration of unique gastronomic traditions, and the inclusion of marginalized communities in food-related events.

To successfully establish Kozhikode as a culinary destination, a comprehensive approach that combines branding, infrastructure development, cultural preservation, and sustainability is required. By taking these recommendations into account and actively working toward their implementation, Kozhikode can position itself as a Creative City of Gastronomy and reap the economic, cultural, and social benefits that come with this prestigious designation. This endeavor will not only enhance the city's global reputation but also contribute to the well-being and economic stability of its residents, making Kozhikode a shining example of a city that embraces and celebrates its culinary heritage.

Acknowledgements The authors extend their gratitude to NIT Calicut for providing the opportunity to undertake this study. Additionally, the authors wish to acknowledge the contributions of Mr. Anand Joshi (DAP, NIT Calicut) and Ms. Jyoti Srivastava (DAP, NIT Calicut) toward the study. Special thanks are to the external evaluators Dr. Anil Kumar Roy, Dr. Deepak S. Kumar, and Dr. Ajith Kaliyath along with reviewers Dr. Shyni Anil Kumar for their valuable comments and suggestions, which significantly enhanced the content and structure of the chapter.

References

Basil M, Basil DZ (2009) Reflections on ultra-fine dining experiences. In: Memorable customer experiences. Routledge, p 13

Boutsioukou K (2018) Key trends in culinary tourism–insight into the key trends in culinary tourism, the profile of culinary travelers and what the future holds for the industry. https://www.globaldata.com/store/report/key-trends-in-culinary-tourism-insight-into-the-key-trends-in-culinary-tourism-the-profile-of-culinary-travelers-and-what-the-future-holds-for-the-industry/ (accessed 10.28.23)

Boyne S, Hall D, Williams F (2003) Policy, support and promotion for food-related tourism initiatives. J Travel Tour Mark 14:131–154

Cai LA (2002) Cooperative branding for rural destinations. Ann Tour Res 29:720–742. https://doi.org/10.1016/S0160-7383(01)00080-9

Camagni R (2002) On the concept of territorial competitiveness: sound or misleading? Urban Stud 39:2395–2411. https://doi.org/10.1080/0042098022000027022

Černevičiute J (2011) Mapping Vilnius as creative city. Creat Stud 4:89–100. https://doi.org/10.3846/20290187.2011.577141

Chaney S, Ryan C (2012) Analyzing the evolution of Singapore's World Gourmet Summit: an example of gastronomic tourism. Int J Hosp Manag 31:309–318. https://doi.org/10.1016/j.ijhm.2011.04.002

Chang RCY, Mak AHN (2018) Understanding gastronomic image from tourists' perspective: a repertory grid approach. Tour Manag 68:89–100. https://doi.org/10.1016/J.TOURMAN.2018.03.004

de Heredia AL, Gaztelumendi I (2019) Guidelines for the development of gastronomy tourism

Dwyer L, Forsyth P, Spurr R (2005) Estimating the impacts of special events on an economy. J. Travel Res 43. https://doi.org/10.1177/0047287505274648

Gathen C, Skoglund W, Laven D (2021) The UNESCO creative cities network: a case study of city branding. In: Smart innovation, systems and technologies. Springer, Cham, pp 727–737. https://doi.org/10.1007/978-3-030-48279-4_68
Hall CM, Sharples L (2003) The consumption of experiences or the experience of consumption? An introduction to the tourism of taste. Food Tour World: Dev Manag Mark: 1–24. https://doi.org/10.1016/B978-0-7506-5503-3.50004-X
Hall SP (2000) Creative cities and economic development. Urban Stud 37:639–649. https://doi.org/10.1080/00420980050003946
Hankinson G (2007) The management of destination brands: five guiding principles based on recent developments in corporate branding theory. J Brand Manag 14:240–254. https://doi.org/10.1057/PALGRAVE.BM.2550065
Hankinson G (2001) Location branding: a study of the branding practices of 12 English cities. J Brand Manag 9:127–142. https://doi.org/10.1057/PALGRAVE.BM.2540060
Harrington RJ (2005) Defining gastronomic identity. J Culin Sci Technol 4:129–152. https://doi.org/10.1300/J385V04N02_10
Huang JZ, Li M, Cai LA (2010) A model of community-based festival image. Int J Hosp Manag 29:254–260. https://doi.org/10.1016/J.IJHM.2009.10.010
Johnson DA (2015) History of regional planning. In: International encyclopedia of the social & behavioral sciences, 2nd edn. Elsevier. https://doi.org/10.1016/B978-0-08-097086-8.74069-4
Karsavuran Z, Dirlik O (2019) Gastronomy as a tourism resource: the case of Gaziantep. In: 2nd National and 4th International Tourism Congress Proceedings Book. Anadolu University Press
Kong L (2014) Transnational mobilities and the making of creative cities. Theory Cult Soc 31:273–289. https://doi.org/10.1177/0263276414549329
Lange B, Kalandides A, Stober B, Mieg H (2008) Berlin's creative industries: governing creativity? Ind Innov 15:531–548. https://doi.org/10.1080/13662710802373981
Lin Y-C, Pearson TE, Cai LA (2011) Food as a form of destination identity: a tourism destination brand perspective. Tourism Hosp Res 11:30–48
Mitchell T (2008) Culture and economy. In: The SAGE handbook of cultural analysis. Sage
Namyslak B (2013) Cooperation and forming networks of creative cities: Polish experiences. Eur Plan Stud 22:2411–2427. https://doi.org/10.1080/09654313.2013.843652
Pearson D, Pearson T (2017) Branding food culture: UNESCO Creative Cities of Gastronomy. J Food Prod Mark 23:342–355. https://doi.org/10.1080/10454446.2014.1000441
Power D (2003) The Nordic 'cultural industries': a cross-national assessment of the place of the cultural industries in Denmark, Finland, Norway and Sweden. Geogr Ann Ser B Hum Geogr 85:167–180. https://doi.org/10.1111/J.0435-3684.2003.00139.X
Pratt AC (2010) Creative cities: tensions within and between social, cultural and economic development: a critical reading of the UK experience. City Cult Soc 1:13–20. https://doi.org/10.1016/J.CCS.2010.04.001
Rae D (2004) Entrepreneurial learning: a practical model from the creative industries. Education+Training 46:492–500. https://doi.org/10.1108/00400910410569614
Raffo C, O'Connor J, Lovatt A, Banks M (2000) Attitudes to formal business training and learning amongst entrepreneurs in the cultural industries: situated business learning through "doing with others." J Educ Work 13:215–230. https://doi.org/10.1080/713676987
Rehan RM (2014) Urban branding as an effective sustainability tool in urban development. HBRC J 10:222–230. https://doi.org/10.1016/J.HBRCJ.2013.11.007
Riza M, Doratli N, Fasli M (2012) City branding and identity. Procedia - Soc Behav Sci 35:293–300. https://doi.org/10.1016/J.SBSPRO.2012.02.091
Rosi M (2014) Branding or sharing?: The dialectics of labeling and cooperation in the UNESCO Creative Cities Network. City Cult Soc 5:107–110. https://doi.org/10.1016/j.ccs.2014.05.002
Sandybayev A (2018) The impact of street and food festivals in gastronomic tourism through visitor's emotions and satisfaction. A case of Abu Dhabi food festival. Int J Res Tour Hosp 4:27–32. https://doi.org/10.20431/2455-0043.0401004

Sevin HE (2014) Understanding cities through city brands: city branding as a social and semantic network. Cities 38:47–56. https://doi.org/10.1016/J.CITIES.2014.01.003
Seyitoglu F, Ivanov SH (2020) A conceptual study of the strategic role of gastronomy in tourism destinations. Int J Gastron Food Sci 21. https://doi.org/10.1016/J.IJGFS.2020.100230
Skoglund W, Jonsoon G (2012) The potential of cultural and creative industries in remote areas. Hogsk I Boras Nord Kult Tidskr 15:181–191
Soderstrom A (2022) Attracting tourists with the use of local food: using a multiple case study to investigate the role of local food in the marketing of tourist destinations in the UK.
Di Stefano E, Klaic D, Bollo A, Bacchella U, Hansen K (2000) Festivals: challenges of growth, distinction, support base and internationalization. https://www.fitzcarraldo.it/ricerca/pdf/ricerca_festival_tartu.pdf
UNESCO (2023) Creative cities network. https://en.unesco.org/creative-cities/ (accessed 10.27.23).
Xiaomin C (n.d.) "City of gastronomy" of UNESCO Creative Cities Network: from international criteria to local practice. Soc Syst Stud: 55–67
Zhu H, Yasami M (2021) Developing gastronomic resources: practices of UNESCO Creative Cities of Gastronomy. GeoJ Tour Geosites 39:1406–1414. https://doi.org/10.30892/gtg.394spl11-784

Exploring City Branding through Place-Making in Megacities of Indian Subcontinent: Case of Delhi, Mumbai, Karachi, and Dhaka

Deepanjan Saha, Vidhu Bansal, Shivangi Singh Parmar, and Somi Sareen

1 Introduction

The Global South is expected to have 34 out of the 41 megacities by 2030, a majority of which will be added to the continents of Asia and Africa (Alakshendra 2022). Megacities in the Global South are facing unprecedented challenges in terms of exploding population, uncontrolled urbanization, unplanned growth, widening income inequality, informality, rise in natural calamities owing to climate change, and an enormous demand for resources required to sustain the population growth. Consequently, these megacities are competing with each other, more than ever before, to attract global investors. To achieve the same, the city governments are increasingly adopting city or place-branding strategies to strengthen their unique identities and transform themselves into global cities or 'world-class' cities, thus, intensifying inter-city competition in a globalized world (Hall and Hubbard 1996; Harvey 1989; as cited in Joo and Seo 2018).

Historically, iconic structures have worked as an effective branding strategy for global cities, e.g., Eiffel Tower being synonymous with Paris and the Statue of Liberty with New York City. Rejuvenation of the port town of Bilbao in the post-industrial era with the building of the Guggenheim Museum exemplifies how iconic architecture

D. Saha (✉)
Department of Planning and Architecture, National Institute of Technology Rourkela, Rourkela, India
e-mail: sahad@nitrkl.ac.in

V. Bansal · S. S. Parmar
School of Architecture, Anant National University, Ahmedabad, India
e-mail: shivangisingh.parmar@anu.edu.in

S. Sareen
Department of Architecture and Planning, National Institute of Technology Patna, Patna, India

M. Firoz C et al. (eds.), *Urban Planning and Design for Megacities in the Global South*, Advances in 21st Century Human Settlements,
https://doi.org/10.1007/978-981-97-8370-0_21

can even catapult cities onto the global map. The case of Dubai with Burj-al-Arab and Burj Khalifa is a similar example. Moreover, redevelopment of urban quarters, historic precincts, and open spaces has also helped create new 'place-brands', e.g., 'Gaudi Barcelona'. However, most of these examples are from the Global North, particularly, the Atlantic Rim.

In pursuit of becoming 'world-class', megacities of the Global South are heavily looking into examples from their counterparts in the Global North, trying to conjure 'worldly landscapes' and 'global imaginaries' which are evocative of the influential cities from the developed 'West' (Joo and Seo 2018). For instance, the waterfront development along Singapore river was envisioned to '*rival the panoramic views of San Fransisco Bay and Sydney Harbour*' (*The Straits Times*, 19 January 1998; as cited in Chang and Huang 2008). Such trends have resulted into repetitiveness or 'sameness' in the urban built environment, contributing to 'placelessness', degeneration of distinctiveness in cities, loss of civic identity, traditional skills, and livelihoods.

Since literature on city or place branding is mostly limited to cities in the Global North, particularly, Europe (Joo and Seo 2018) and countries in the Global South have been largely emulating those of the North, it is highly imperative to identify strategies which are most suitable for the region. Hence, this chapter attempts to collate city branding experiments adopted by megacities in the Global South, particularly focusing on urban spatial interventions. Delhi, Mumbai, Karachi, and Dhaka, being important socio-cultural and political centres and economic powerhouses in respective nation-states in the Indian subcontinent, have been selected for a closer examination.

A brief review of literature on city branding through placemaking has been conducted to identify the key objectives. It is followed by a classification of such strategies into four types: iconic architecture, urban regeneration and placemaking, creative and cultural quarters, events, and festivals. Accordingly, strategies adopted by the selected cities have been classified and analysed.

2 City Branding through Placemaking

2.1 City Branding vs. Placemaking

Conceptually, place branding or place-marketing essentially refers to the process of symbolizing a place, i.e., a site, city, region, or nation, to create a desired impression on its target audience, and thereby gain recognition, as well as help create civic pride and social cohesion (Chang and Huang 2008). 'Place promotion' (Ward 1998; Paddison 1993), 'city marketing' (Hall and Hubbard 1998), and 'branding strategies' (Ashworth and Voogd 1990; Evans 2003) are some of the related terms (as cited in Bonakdar and Audirac 2020). Commonly adopted branding strategies include catchy slogans, logos, advertising campaigns, building landmarks, and hosting international mega-events (Kavaratzis and Ashworth 2005; as cited in Joo and Seo 2018).

Place branding is more than a persuasive promotional activity and incorporates a complex set of processes which is targeted at creating a narrative specific to the place (Ashworth et al. 2015, p. 5). In such processes, the city as a place gets commodified and urban landscapes start mimicking global city imageries (Chang and Huang 2008; Zukin et al. 2009; as cited in Joo and Seo 2018).

In contrast to physical place-branding efforts (i.e., iconic architecture, development of 'smart' infrastructure, etc.) as promotional strategies adopted by city authorities, urban placemaking refers to the processes through which public, typically citizens, get to understand, perceive, and utilize spaces on a daily basis in unique ways (Chang and Huang 2008). Identities created through place-branding initiatives are mostly imposed by institutions and marketers, considering 'place' as a product; whereas those involved in placemaking interventions are largely shaped out of public perceptions (ibid.), and how individuals experience the urban environment instead of 'using' it as a consumer (Evans 2015). Outcomes of place-branding initiatives are typically 'worldly landscapes', which can be compared across cities and are aimed at impressing a global audience, whereas those of placemaking integrate people with local history and environment (ibid.).

2.2 *Need for City Branding*

Global cities have increasingly been identified as powerful political and economic units (Sassen 2001; as cited in Joo and Seo 2018). Remaining competitive in a rapidly globalizing world economy, i.e., being able to attract a large share of both financial and human capital, has been one of the key challenges faced by city governments (Bonakdar and Audirac 2020). City branding strategies have been used by cities across the world to attract local and international investors, upper-middle-class residents, tourists, the 'creative class', i.e., individuals and/or firms engaged in creative industries, and to strengthen political influence (Capone and Lazzeretti 2016, p. 169; Dinnie 2011, p. 3; Middleton 2011, p. 16; as cited in Lim 2020, p. 28). Besides being a strategy to cope with the intense inter-city competition, place branding is also considered as a tool for imagining an aspirational future for the place, thus, shaping what people—both residents and visitors—may expect from the place, which would further shape their experiences (Ashworth et al. 2015, p. 4).

2.3 *Limitations of City Branding*

Since 1980s, the concept of city branding has largely evolved from the model of product marketing and has no specific theoretical underpinnings apart from the classical product-based branding which fails to consider the multifunctionality of places to a multitude of stakeholders (Hankinson 2004). Hence, city branding relies a lot on strong visuals, emotion-based communicative power, and construction of symbolic images (Joo and Seo 2018), without much exploring the concept of brand as a relationship with multiple stakeholders. Such an insistence on imposing a 'brand image' for a city essentially disregards the intrinsic heterogeneity of the city as a place (Sethi 2021). Typically, it has been understood to be an expression of neo-liberalism, driven by market-oriented discourses which thrive upon commodification of place (Joo and Seo 2018), which is targeted at attracting the 'other', i.e., global investors or tourists instead of focusing on its citizens (Sethi 2021). Popular city branding strategies are also criticized for selectively focusing on positive aspects of the city while leaving out the less attractive realities (Greenberg 2009; Paddison 1993; as cited in Joo and Seo 2018), and even squandering the speculated long-term benefits (Kassens-Noor 2012). While planning and implementing city branding projects, city authorities often tend to ignore the underlying challenges of urban poverty, racial segregation, and disinvestment (Gotham 2007; Boland 2013), gentrification resulting out of redevelopment policies (Gibson 2005; Evans 2015), and commodification of physical and cultural space (Zukin 1996).

Such challenges are embedded in the planning and decision-making processes which are largely top-down, exclusionary, and non-participatory. City branding strategies often fail to address the issue of social justice while executing cultural flagship projects proposed to attract global capital (socio-economic, cultural, and human) (Bonakdar and Audirac 2020). Moreover, as cautioned by Chang and Huang (2008), since global cities compete in providing with amenities and attractions, the desire to offer the best of what is offered in other cities, i.e., creating a 'geography of everywhere', often leads to a 'geography of nowhere'. Hence, city branding strategies need to seek a balance between commodification of a place and preservation of its local character. Such a balance is, typically, attempted through reinforcing local identity, nurturing a stronger belongingness of citizens to the city (i.e., place attachment), and by focusing on community development (Kavaratzis 2004, p. 70; as cited in Joo and Seo 2018).

3 City Branding in the Global South

Although city branding, as an exercise, started in the developed economies of the Global North, the Global South has recently picked it up as a medium to enter the foray of global marketplace. City branding strategies in the Global South have mostly been attempts to create positive urban imaginaries as one of the potential ways for economic upliftment (Bonakdar and Audirac 2020), and to 'catch-up' with the West (Joo and Seo 2018). However, such cases are not bereft of the limitations of social exclusion and commodification of culture, as met by city branding initiatives throughout the world. For instance, the image makeover of Dubai was accompanied by a high cost of social disintegration (Elsheshtawy 2009; as cited in Bonakdar and Audirac 2020); branded as a UNESCO Creative City of Design, Bangkok draws a heavy traffic of international tourists, whereas the issues of traffic, inequality, poverty, and flooding largely go unaddressed (Chuangchai 2021).

Although several non-spatial interventions abound in branding literature, the significance of physical environments, particularly in the form of iconic architecture and urban design interventions, has been highlighted by scholars (Evans 2015). The following sub-sections briefly discuss a variety of placemaking initiatives which aid to city branding, focusing on spatial interventions adopted by cities in the Global South.

3.1 Iconic Architecture

Iconic architecture, particularly those designed by 'starchitects', has been the most noticeable manifestation of city branding (Evans 2015), e.g., Guggenheim Museum designed by Frank Gehry in Bilbao had a multiplier effect in drawing global attention to the post-industrial port town in Spain, widely known as the 'Bilbao Effect' high (refer Fig. 1). 'Gaudi Barcelona' and 'Mackintosh Glasgow' highlight the strength of architecture in building an identity for the city. Though such signature buildings help in promoting the visually iconic 'image' of the city, these often stand in contrast to their immediate physical and cultural environs (Sethi 2021). Several signature projects—hoping to ride on the wave of 'Bilbao Effect'—have failed to accrue comparable benefits to the host city w.r.t. the investment made (Griffiths 2022).

Cities outside the Global North have also toyed with the strategy recently, e.g., Marina Bay Sands—a hotel and museum complex in Singapore—designed by Israeli

Fig. 1 Guggenheim in Bilbao. Image by: Jorge Fernández Salas, unsplash.com

Fig. 2 Marina Bay Sands, Singapore. Image by: Guo Xin Goh, unsplash.com

architect Moshe Safdie (refer Fig. 2). Oil rich nations in the Arab world rose to prominence with the new ambitious architecture, particularly, Burj Al-Arab and Burj Khalifa. The new cultural district in Abu Dhabi, UAE houses a number of projects designed by 'starchitects' notably, the franchise of Louvre designed by Jean Nouvel, Guggenheim Museum by Frank Gehry, and the Zayed National Museum by Norman Foster (Evans 2015).

3.2 Urban Regeneration and Placemaking

Besides iconic architecture, urban spatial interventions, in the form of regenerated parts of an old city or new extensions, add to the brand identity of a city. Such interventions range from publicly owned spaces in redeveloped town squares (e.g., Times Square in New York City, see Fig. 3) to historic districts or heritage precincts (e.g., Barri Gothic in Barcelona), cultural quarters (e.g., Chinatowns and Jewish quarters in several global cities), and even creative industry quarters (e.g., 'loft-living' quarters in SoHo of NYC) (Evans 2015). These places typically draw success for being highly distinctive in terms of the experiences they offer.

Placemaking, as a way of citizen-led approach augmenting the quality of public spaces to promote liveability of cities and well-being of citizens, has gained relevance in urban design. Pedestrianization of public squares is one of the examples of placemaking which unlocks public spaces for large gatherings and events, e.g., Trafalgar Square in London. Similarly, urban industrial districts from the past are getting reborn with new usage, thus, adding to the much-needed public spaces in the

Fig. 3 Times Square in New York City. *Source* Author

city, and thereby, adding identity to it, e.g., High Line in Meatpacking District in New York City or Distillery District in Toronto (Evans 2015).

In the Global South, although truly citizen-led placemaking initiatives are few, owing to the prevalence of largely formal and top-down urban planning systems (Jamadar 2021), the significance of public spaces in cities is slowly gaining traction. For instance, while the development of Sabarmati Riverfront in Ahmedabad, India, was aimed at environmental improvement and economic upliftment, it was also intended to reconnect citizens with the river (SRFDCL 2020).

3.3 Creative and Cultural Quarters

With the growth of cultural and creative industries, cities across the globe are trying to rebrand themselves as 'creative cities' (Wynn and Yetis-Bayraktar 2016). It has been one of the mechanisms to put cities on the world map (Lim 2020, p. 27) as it helps attract the '*creative class*' (Florida 2003). Since its inception in 2004, about 300 cities across the world have joined the UNESCO's Creative City Network (UCCN) under seven creative fields, namely Crafts and Folk Art, Media Arts, Film, Design, Gastronomy, Literature, and Music (UNESCO and World Bank 2021).

Ethnic quarters are also getting rebranded as cultural districts, thus drawing attention of tourists and residents. For instance, Chinatowns and Jewish quarters in historic cities, with substantially large migrant communities living in the past, add to the city's diverse character which it might lack today. Even, new districts are 'born' with conscious rebranding, e.g., 'Banglatown' in East London got developed as a new commercial concept centred on Brick Lane 'Curry Quarter', presently gentrified by migrants from Bangladesh and dotted with restaurants (Evans 2015). Historic districts being accorded with 'World Heritage' status by the UNESCO is another way to brand cities, even if only a part of it might have been awarded the status (ibid.).

Despite being late, cities in the Global South have picked up the concept of creative city from the Global North with several cities having joined the UNESCO Creative Cities Network over the last decade (Lim 2020, p. 27). However, scholars have questioned the very process of replication and translation of the concept as what is considered to be 'creative' in the Global South differs from that of the Global North (Gu et al. 2020, pp. 4–5).

In India, there has been an unprecedented focus on developing 'heritage cities' in the last two decades. Titled as 'Heritage City Development and Augmentation *Yojana* (HRIDAY)', this central government scheme adopted in selected 'heritage cities' had a central focus on urban infrastructure development based on cultural heritage and conservation. Currently, several temple precincts in India are getting revamped to turn them into 'world-class' pilgrimage destinations, such as the Kashi Vishwanath Corridor in Varanasi and Mahakal Corridor in Ujjain. However, such large-scale remodelling has resulted in eviction of residents and a controversial change of 'heritage' character of the area.

3.4 Events and Festivals

Events have the potential of creating images of places where they are held and act like catalyst for urban regeneration. Recently, cities are not only becoming venues for events, but also new urban spaces are getting 'produced' by the virtue of 'eventalisation', thus helping the city to brand itself. For instance, '*Bosch 500*', commemorating the 500th death anniversary of the painter Hieronymus Bosch, who had lived and died in the provincial Dutch city of Den Bosch, helped the city get global attention, despite having no original works of the painter surviving in the city (Richards 2017). Similarly, music festivals have been adopted as an affordable alternative to creating costly cultural infrastructure while successfully promoting the city image (Wynn and Yetis-Bayraktar 2016).

Sporting mega-events like Olympic Games provide the host city with an opportunity for catalytic urban transformation, such as enhanced mobility, a 'world-class image', and a slew of urban regeneration projects. For instance, the hosting of Commonwealth Games 2010 in Delhi led to a host of infrastructural developments in Indian capital city (Baviskar 2011). However, besides the projected benefits, a study comparing the experiences of India, Brazil, and South Africa (Maharaj 2015) has noted several negative impacts of hosting sporting mega-events, i.e., widening of socio-economic inequalities, violations of human rights, and communities, and diversion of public spending priorities without public consultation (Kassens-Noor 2012). Global South has also attempted to brand themselves as global destinations by hosting international art events. One such example is Kochi-Muziris Biennale, an international exhibition of contemporary art to 're-invent' the local for a 'global' audience (Varghese 2017).

4 Study Design

Having reviewed the literature on city branding through placemaking, rest of the article examines the initiatives taken in selected megacities to appraise the state of city branding in Global South better. The selection of cases from Global South, particularly South Asia, has been governed by the city population and the ranking of megacities in recent global indices on city branding.

Delhi, Karachi, Mumbai, and Dhaka (in descending order) have been identified as the most populated urban areas in South Asia (Sawe, 2018). With Delhi and Dhaka being respective national capitals of India and Bangladesh, and Mumbai and Karachi being financial hubs of India and Pakistan, respectively, these cities are essentially major political and cultural centres of respective countries, being part of the larger Indian subcontinent, and share a common geo-political history (refer Fig. 4). Separately, as per the Brand Finance City Index 2023 on place branding, New Delhi, Bengaluru, Dhaka, Mumbai, and Karachi are at the end of the list of world's top 100 cities, ranked at 95, 96, 97, 98, and 100, respectively (Brand Finance, 2023). Hence, the quartet of Delhi, Karachi, Mumbai, and Dhaka have been selected for a closer assessment of city branding strategies adopted in each city, particularly, in the past two decades.

As elucidated in the literature reviewed, city branding initiatives and outcomes in the selected cases have been looked at through the lens of urban spatial interventions, i.e., the physical manifestation of branding which include, but are not limited to, the following (refer Fig. 5).

- Iconic architecture
- Urban regeneration and placemaking
- Creative and cultural quarters
- Events and festivals

The four categories taken above have shown most catalytic activity in urban transformation and hence taken as the part of urban enquiry.

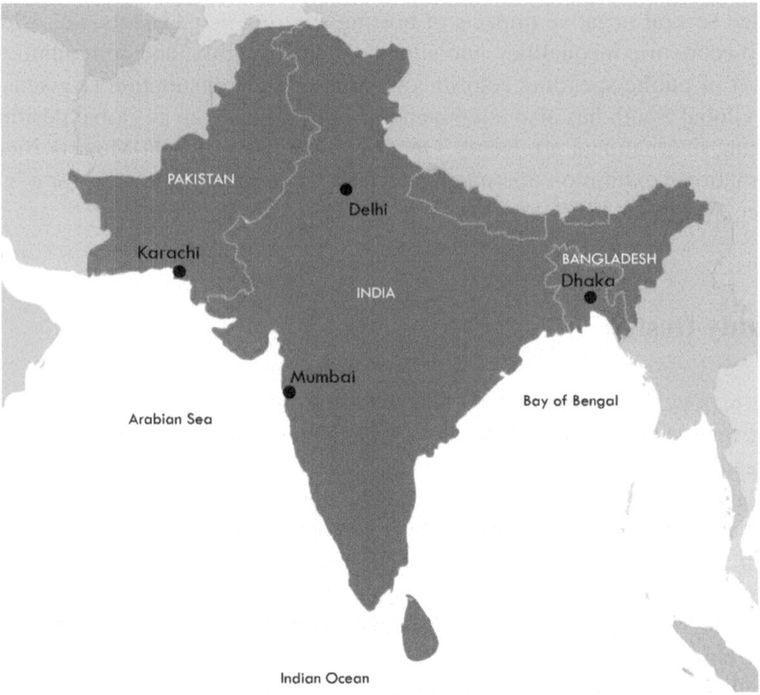

Fig. 4 Location map of selected megacities in Indian subcontinent. *Source* Author

4.1 Delhi

The Master Plan for Delhi, with the perspective year of 2021, had read that the '*Vision-2021 is to make Delhi a global metropolis and a world-class city*' (DDA 2007). As per the Master Plan, '*pollution, degraded built environment, congestion, lack of safety and disparate living conditions*' have been identified to be a threat to the city's global attractiveness. To counter these issues and to put itself on the global map, Delhi has taken a number of steps in the last two decades. Delhi's bid to host the Commonwealth Games in 2010 is widely noted as one. Opened in 2010, Delhi's Airport Terminal-3 (T3), having an annual capacity of 34 million passengers, has substantially improved Delhi's reputation as a centre for international aviation. Expanding since 2002, Delhi Metro has now become the world's 12th longest mass rapid transit system, serving around 2.76 million people each day (Bajaj and Singh 2019). It has completely transformed the way people move and has greatly lowered traffic congestion (NIUA 2022).

'*From Walled City to World City*'—a marketing campaign launched by the Times of India, one of the country's largest English dailies—is a testimony to the conscious effort to rebrand Delhi which has struggled to carve itself an identity beyond that of an administrative capital. Despite being tagged as '*Dilwalon ki Dilli*', which literally

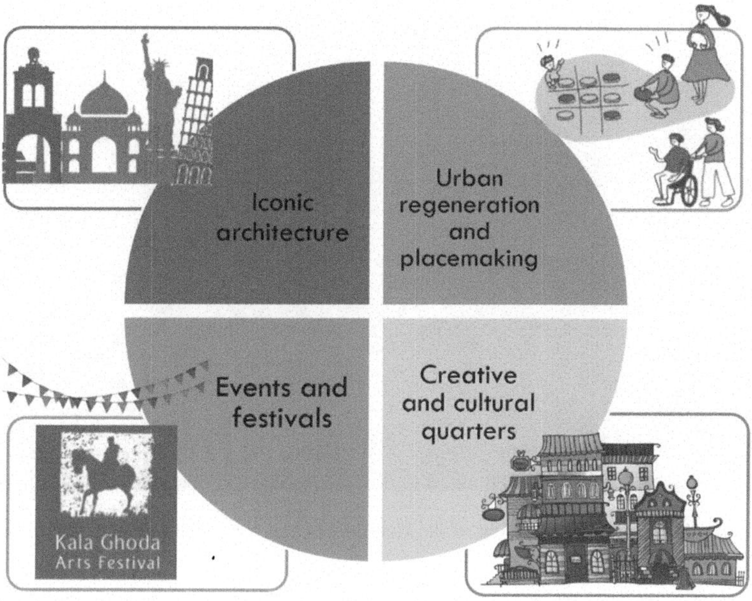

Fig. 5 Urban spatial intervention lens. *Source* Author

means 'Delhi belongs to the large hearted', the city frequently faces challenges in altering the unfavourable attitude about public safety, particularly for women, and poor air quality.

Architecture of Delhi is a living witness to the city's more than thousand-year-old history and cultural legacy, with many renowned buildings playing an important role in establishing the city's image, with three of them being inscribed as UNESCO World Heritage, namely Qutab Minar, Humayun's Tomb, and the Red Fort (Mehta 2022). India Gate, a massive archway, is revered as one of the city's most imposing symbols, comparable to the Eiffel Tower in Paris (Mabin and Harrison 2023). While Delhi is proud of its architectural heritage, the city has largely trodden the path of creating 'modernist spectacular architecture' since economic liberalization in the 1990s. Consequently, the city has witnessed a concentration of high-rises, gated housing complexes (some even being named as 'Beverly Park' and 'Sentosa City' to signify their 'modern' character), and shopping malls made of glass and steel in its satellite cities (Moser 2015), devoid of any distinctive character and ending up in creating new exclusionary spaces of 'urban apartheid'.

The city branding of Delhi includes urban interventions intended at reinventing the city's character. The ongoing Central Vista Project (see Fig. 6) is a major urban redevelopment project at the administrative heart of Delhi, with building of a new parliament house and several administrative offices. It is a daring attempt to establish a new, ambitious image for the city, one that distances itself from its colonial past

Fig. 6 Ongoing urban redevelopment of the Central Vista in New Delhi. *Source* centralvista.gov.in

(Meskell 2021), to symbolize the making of an '*Atma Nirbhar Bharat*' or a self-reliant India.

The redevelopment of Chandni Chowk, an iconic commercial street in the historic city-core of Shahjahanabad, illustrates Delhi's dedication to preserving its built heritage. This project not only revitalizes the area by decongesting the street but also encourages heritage tourism by enhancing its imageability (Sengupta and Sen 2022). Besides such mega redevelopment projects, grassroots initiatives such as *Raahgiri,* in which locals reclaim the streets from vehicles to engage in recreational activities, demonstrate Delhi's potential for community participation and placemaking (Delhi Urban Arts Commission 2018).

Delhi's rich cultural past has bestowed itself with many historic quarters which draw both tourists and citizens, particularly, the historic precincts of Mehrauli, Nizamuddin, Shahjahanabad, and Lutyen's Delhi. Delhi had also tried getting the coveted tag of UNESCO 'world heritage city' by nominating an area for protection under the banner of 'Imperial Cities of Delhi' which comprised of the Mughal City of Shahjahanabad and New Delhi—the seat of British Empire in India (Jain 2016). However, the nomination was later withdrawn in 2015, thus, taking it off the race to become India's first world heritage city (Meskell 2021), thereby, losing its potential to brand itself.

Besides such protected conservation areas, some historic parts of Delhi are getting gentrified with new commerce. For instance, Hauz Khas, a quaint 'urban village' from the fifteenth century, witnessed a remarkable transformation into a commercial district bustling with cafés, restaurants, discotheques, and antique stores. On the one hand, it has successfully created a new place-brand by blending its historical appeal with contemporary 'buzz', thus attracting affluent youngsters and tourists alike; and on the other, it has sparked contradictions between the local and the global (Bernroider 2015).

Hosting of the Commonwealth Games in 2010 has demonstrated Delhi's ability to organize large-scale events. However, it has been criticized for allowing large-scale urban restructuring based on visual aesthetics of order and cleanliness, without considering the prevalent informality in the cityscape and ignoring the needs of socially marginalized sections of the society. For instance, about 300,000 people living in squatters on Yamuna riverfront were evicted between 2004 and 2006 to pave way for the Games Village (Dupont 2011).

Having hosted several International Trade Fairs in the past, redevelopment of the exhibition-cum-convention centre at Pragati Maidan in 2023 is expected to make Delhi a destination for Meetings, Incentives, Conventions, and Exhibition (MICE). Delhi is also trying to create a niche for itself in the cultural scene by hosting modern and contemporary art exhibitions (e.g., India Art Fair—one of the largest art fairs held in South Asia), and literary fests (e.g., Jashn-e-Rekhta—world's largest Urdu language literary festival held since 2015).

4.2 Mumbai

Mumbai, formerly known as Bombay, is the capital of the Indian state of Maharashtra and is located on the western coast of India. Its metropolitan area stretches along the sea and includes the satellite towns of Navi Mumbai and Thane (Bharadwaj et al. 2017). Sitting on a peninsula, Mumbai is almost an 'island city' which restricts its physical spread. With a staggering population of more than 20 million and an exceptionally high density, Mumbai's limited island area is teeming with life and aspirations of a global megacity. Being home to the Bombay Stock Exchange, a hub for large multinational companies, and an international gateway for exports, the city has an indisputable impact on the world economy. Mumbai is truly the 'Maximum City' (Mehta 2006).

Mumbai is also home to an outstanding collection of historic buildings, each serving as a reminder of various events that have shaped the city into the busy metropolis it is today. For instance, the Victorian Gothic Revival architecture of Chhatrapati Shivaji Maharaj Terminus, formerly Victoria Terminus, is an imposing reminder of colonial Mumbai. The Gateway of India, a colossal arch overlooking the Arabian Sea, is an enduring representation of Mumbai as a crucible of cultures. Art deco buildings along the Marine Drive represent the modernity which the city witnessed in early twentieth century.

Built in 2009, Bandra-Worli Sea Link—a 5.6 km long 8-lane wide bridge above sea, a mega infrastructure project, has added to the visual identity of the city and an icon representing Mumbai's tenacity in overcoming geographical challenges posed by the island city. The project has become as one of the architectural landmarks of the city. Yet the problem of decongesting the connection lines of the city is still the same. The sea link has become exclusively for the car owners whereas a majority of

Mumbai population consists of pedestrians, two wheelers, and three wheelers who are not really the target population of this project (Patel 2020).

Several urban design projects, like Bandra Bandstand, the Sassoon Dock Art Project, and Kala Ghoda Mural Art Project (see Fig. 7) are examples of placemaking-led city branding. With a beachside boardwalk, planting, seating spaces, an amphitheatre, and illumination, Bandra Bandstand, a 1.2 km long promenade in the neighbourhood of Bandra, has transformed into a popular hangout for leisure and events, thus, promoting social interaction.

Inscription of Victorian Gothic and Art Deco ensembles as UNESCO World Heritage in 2018 has given an impetus to the city's brand identity and celebrates its distinctive architectural character and a unique legacy.

Besides being a commercial centre, Mumbai boasts of a collection of cultural quarters, each with its distinct personality and history, thus, adding to the cosmopolitan character of the city. Over time, such neighbourhoods have turned into cultural destinations for tourists and visitors alike. For instance, Kala Ghoda neighbourhood in Fort area has grown into an art district and a venue for cultural activities, particularly the annually held Kala Ghoda Arts Festival. The neighbourhood is dotted with multiple institutions and art galleries from the first half of twentieth century. A host of restaurants, cafes, and designer boutiques got added to the precinct over time, thereby, transforming the area into an epicentre for creativity.

Fig. 7 Kala Ghoda Mural Art Project. *Source* Author

Sassoon Docks sets an example of creative makeover as the bustling fish market turned into an open-air art gallery (Ansari 2021). Vibrant murals depicting the lives of the local fishing community not only made the streets more attractive, but also delivered strong messages about social issues and environment that includes unhygienic conditions caused by the fishing activity in the wet dock area (*Indian Express* 2023). Coupled with the Mumbai Urban Art Festival, the dockyard has turned into a cultural destination. Moreover, ethnically diverse neighbourhoods like Khotachiwadi in Girgaon, a heritage village with eighteenth-century Portuguese-styled houses (Bautès 2017), and Dadar's Parsi Colony exhibit the rich cultural tapestry of community life in Mumbai.

Kala Ghoda Arts Festival, held since 1988, and the recently launched Mumbai Urban Art Festival held every two years have helped the city build a distinctive cultural scene. These festivals exemplify how public art can transform places, thus, generating the benefits of placemaking.

4.3 Karachi

Karachi, Pakistan's largest and most populous city as well as its financial, economic, and cultural centre, holds significant potential to establish itself as a dynamic and influential global metropolis. With a population of over 20 million, its rich history contributes to a vibrant cultural melting pot, attracting people from various regions and backgrounds seeking economic opportunities and a cosmopolitan lifestyle (Aamir et al. 2019; Mangi et al. 2020).

Karachi employs its iconic architectural landmarks, including Frere Hall and Mazar-e-Quaid, as central elements in its city branding strategy, shaping a unique identity. These landmarks convey a sense of national pride and portray a modern image while representing the city's historical significance and its role in the foundation of Pakistan. By promoting artistic events at Frere Hall and highlighting architectural diversity as a reflection of its adaptability, Karachi portrays itself as a thriving cultural hub (Akbar et al. 2023). These sites also act as landmarks and gathering places for locals, encouraging a sense of belonging and community. Spaces like Frere Hall are kind of elastic spaces which provide the much-needed transience to the exponential layers of complexity (Rehman 2014). It is a symbol of inclusivity which brings people from many walks of life to come together and help in alleviating the exclusionary nature of large cities like Karachi (Haroon 2023).

Karachi strives to enhance its urban spaces and infrastructure by engaging in various spatial interventions to uplift different districts and neighbourhoods like Clifton Beach and Port Grand. With the construction of a modern promenade and recreational facilities, the Clifton Beach area has transformed into a popular destination for leisure and entertainment, attracting both locals and tourists. Port Grand, a unique food and entertainment complex developed around the historic Native Jetty Bridge, has become a symbol of Karachi's revival (Anwar and Viqar 2014). It is an ambitious project based on the Karachi Strategic Plan 2020 which focuses on adopting the characteristics of 'world cities' that 'foster competitive activities' and aspires to match Quincy Market in Boston, 3rd Street in Los Angeles or Clarke Quay in Singapore.

Among the diverse neighbourhoods which lend Karachi its unique charm, Saddar stands out as the heart of the city's shopping district, representing Karachi's commercial character, cultural heritage, and identity (Akbar et al. 2023). Streets of Saddar are filled with bustling energy, lined with historical buildings that showcase an intriguing blend of colonial and modern architecture (Soomro et al. 2020). Karachi Parsi Colony stands as a testament to the Parsi community's profound cultural contributions to the city's identity. This locality preserves its architectural elegance with well-maintained heritage buildings and serene gardens, symbolizing the community's rich history and traditions (Sadiq et al. 2010). In such cultural quarters, Karachi's identity thrives in its ability to celebrate its heritage while embracing a dynamic and contemporary spirit of global city ambition.

Karachi hosts a variety of cultural events celebrating the city's artistic and creative spirit, e.g., Karachi Biennale and Karachi Literature Festival. A contemporary art exhibition that showcases the works of local and international artists, Karachi Biennale (KB) is the flagship initiative of the Karachi Biennale Trust, constituted in 2016, to utilize art as an approach of exploring, debating, and responding to Karachi as Pakistan's centre of art and culture (Kettle 2019). Founded by the Oxford University Press in Pakistan and the British Council in 2010, Karachi Literature Festival is an annual international literary event that brings together writers, poets, and intellectuals from across the country and beyond (Bilal 2022). It fosters intellectual discourse and promotes literature and arts in the city. Such event-led branding initiatives focus on showcasing the city's cultural diversity and its artistic prowess, thus drawing from its intrinsic character instead of aping the 'West'.

4.4 Dhaka

Being the country's political, economic, and cultural centre, Dhaka plays a crucial role in shaping the global image of Bangladesh. With a population of over 20 million and a population density of 23,234 per sq.km (according to World Population Review 2024), the city is one of the largest and densest across the world. Despite the city being prone to flooding, congestion, and unplanned urbanization, Dhaka is a growing economic powerhouse and a focal point for regional diplomacy.

Dhaka is investing in modern infrastructure to support its development. The construction of new commercial buildings, retail malls, and transit networks demonstrates Dhaka's desire to deliver world-class facilities and improve connectivity; the ongoing construction of Padma Bridge is an example. Besides being an established cloth manufacturing hub (Mia and Akhter 2019), the city is also emerging as a hub for information technology (IT) with the setting up of new tech parks and enactment of enabling policies, e.g., 'Digital Bangladesh' (Mazumdar and Alharahsheh 2020).

With these advancements, the city has taken a string of efforts to brand itself, as discussed in the following paragraphs.

Dhaka's satellite towns, like Gulshan and Banani, have witnessed an unprecedented growth in recent decades (Islam 2010). These areas are dotted with skyscrapers, business parks, residential townships, and shopping malls in glass and steel, representing a globalized modernity. Dhaka's aspiration to become a global metropolis necessitates a careful balance between protecting its historic landmarks and constructing new ones that contribute to the city's modern image. Hence, while the city has stressed upon preserving its cultural past by restoring the iconic structures of Lalbagh Fort, Ahsan Manzil, and the Star Mosque (Mowla 2011), Dhaka is also devoted to developing a modern character by constructing new icons, such as the *Jatiya Sangsad Bhaban* or the National Parliament House.

The aspiration of Dhaka to become a global city is visible in its urban redevelopment efforts. Initiatives such as the Hatirjheel Development Project, which converted an abandoned water body into a dynamic urban lake and recreational area, demonstrate the city's dedication to improve public spaces and people's quality of life (Saha et al. 2017). Other examples include Jatrabari Flyover, Gulshan-Banani-Baridhara Lake Development Project, the Dhaka Elevated Expressway, and the Suhrawardy Udyan revamp. The city's parks and squares are being rebuilt to offer accessible and friendly spaces for inhabitants to socialize and participate in cultural events. However, such practices are not completely devoid of any harmful impacts; for instance, the earth-filling process in the Elevated Expressway project near Hatrijheel drew a lot of furores from various sections of the society (TBS Report 2023).

Dhaka seeks to promote itself as a distinctive and appealing destination for heritage tourism by protecting historical sites, lively marketplaces, and traditional architecture (Siddika et al. 2018). For instance, the fifteenth century artisan neighbourhood of Shankhari Bazar in Old Dhaka has been conserved while ensuring modern conveniences for visitors (Islam 2013).

Dhaka has embraced the notion of city branding by staging of mega-events. With Dhaka Art Summit, a biannual event held since 2012, Dhaka is emerging as a hotspot for contemporary art in South Asia. Dhaka International Film Festival (DIFF), Dhaka International Trade Fair (DITF), and Dhaka Fashion Week are similar events which try to put Dhaka on the global map.

5 Conclusion

Case studies have indicated that cities in the Global South are obsessed with becoming 'world class' and in matching with those from the Global North. In that process, they tend to adopt branding strategies aiming to create 'worldly environments' or 'landscapes' which are evocative of the cities in the developed West (Chang and Huang 2008) and thereby contribute to the issue of 'sameness' in the design of urban built environments across the globe. A review of the city branding strategies, involving spatial interventions and placemaking, adopted by the selected set of megacities from the Indian subcontinent, namely Delhi, Mumbai, Karachi, and Dhaka, endorse the same.

All the four cities reviewed are megacities teeming with 20+ million population and are one of the densest urban footprints in the world. All are grappling with similar challenges of informality, housing shortage, widening income inequality, and climate change—be it hazardous air quality to urban flooding (Swapan et al. 2017; Teklemariam 2022). In fact, in terms of liveability, Dhaka and Karachi are placed at the bottom ten in the list of 173 cities worldwide (EIU 2023). However, despite such challenges, all the four cities have strongly suggested independent aspirations of becoming 'world-class', particularly these being important urban centres in respective countries which are playing a larger role in the global economy, more than ever before.

As observed, interventions ranged from adding to iconic architecture (e.g., one of the world's largest international exhibition-cum-convention centre came up in Delhi in 2023) to large-scale urban regeneration projects, such as the Central Vista project in Delhi, Hatirjheel Development in Dhaka, or the ambitious Bandra-Worli Sea link in Mumbai. As expected, such 'world-class' infrastructure has not only become new icons of the city, but also added 'sameness' to the built environment.

As city branding has largely been accepted by city authorities as a response to the demand of inter-city competition in this globalizing world, care shall be taken to ensure that such attempts to brand a city shall not reduce the diversity which the city offers. The city authorities need to look beyond the creation of a singular 'brand image', instead focus on creating a 'vision' which takes all the stakeholders into account. To minimize the ill effects of the present-day city branding, megacities in the Global South shall focus on approaching place as a 'character' instead of a 'product'. Hence, instead of pressing on creating a 'spectacle' through 'iconic' architecture and implementation of so-called modern and smart infrastructure, a city shall plan for development of its cultural and creative quarters, thus building on the rich and distinctive cultural heritage it is bestowed with. Rejuvenation and conservation of Chandni Chowk in Old Delhi, Shankhari Bazar in Old Dhaka, Kala Ghoda precinct in South Mumbai, and the Port Grand of Karachi as cultural districts are examples in that direction, thus, trying to find a balance between the city's traditional past and its modernity. It helps in creating cheaper alternatives to expensive cultural infrastructure and standardized urban forms, besides celebrating the distinctive place identity. Moreover, cultural festivals and grassroots artistic initiatives represent the diversity in megacities more inclusively, thus fostering a stronger sense of civic identity (Baffoe and Roy 2023; Criqui 2015).

Since planning systems in the cities of the Global South are mostly top-down and borrow heavily from the developed West, most of these branding exercises are governed by bureaucratic decisions, which are heavily criticized for being exclusionary and elitist. However, gradually though, citizen-led bottom-up placemaking initiatives are gaining strength. Hence, it may be argued that in their quest to brand themselves for a global audience, megacities in the Global South need to shed the notion of aping their counterparts in the Global North; and instead focus on their distinctiveness by exploiting the rich cultural heritage, so that Mumbai need not turn into Shanghai or Kolkata into London, and rather become better versions of themselves—more liveable, sustainable, and inclusive.

References

Aamir M, Masroor S, Ali ZA, Ting BT (2019) Sustainable framework for smart transportation system: a case study of Karachi. Wireless Pers Commun 106(1):27–40. https://doi.org/10.1007/S11277-019-06259-4/FIGURES/6

Akbar SH, Iqbal N, Van Cleempoel K (2023) Saddar Bazar Quarter in Karachi: a case of British-Era protected heritage based on the literature review and fieldwork. Heritage 6(3):3183–3210. https://doi.org/10.3390/HERITAGE6030169

Alakshendra A (2022) The Routledge handbook of planning megacities in the Global South. J Urban Technol 29(4):127–130. https://doi.org/10.1080/10630732.2022.2130612

Ansari M (2021) Creative cities: a case of Mumbai. Int J Appl Arts Stud (IJAPAS) 6(1):7–16. http://ijapas.ir/index.php/ijapas/article/view/329

Anwar NH, Viqar S (2014) Producing cosmopolitan Karachi: freedom, security and urban redevelopment in the post-colonial metropolis. South Asian Hist Cult 5(3):328–348. https://doi.org/10.1080/19472498.2014.905322

Ashworth, GJ, Voogd, H (1990) Selling the City: Marketing Approaches in Public Sector Urban Planning. London, UK: Belhaven Press.

Ashworth GJ, Kavaratzis M, Warnaby G (2015) The need to rethink place branding. In: Ashworth GJ, Kavaratzis M, Warnaby G (eds) Rethinking place branding: comprehensive brand development for cities and regions. Springer, Cham, pp 1–11

Baffoe G, Roy S (2023) Colonial legacies and contemporary urban planning practices in Dhaka, Bangladesh. Plan Perspect 38(1):173–196. https://doi.org/10.1080/02665433.2022.2041468

Bajaj G, Singh P (2019) Understanding preferences of Delhi metro users using choice-based conjoint analysis. IEEE Trans Intell Transp Syst 22(1):384–393

Bautès N (2017) Caught between metropolitan renovation and resistance: the dilemma of Khotachi-wadi, Mumbai (India). In: Neighbourhood & city-between digital and analogue perspectives, p 185. https://www.researchgate.net/publication/330366879

Baviskar A (2011) Spectacular events, city spaces and citizenship: the commonwealth games in Delhi. In: Urban navigations: Politics, space and the city in South Asia. Routledge, pp 138–161

Bernroider L (2015) Dynamics of social change in South Delhi's Hauz Khas village. SOAS South Asia Institute Working Papers, 1, pp 1–16

Bharadwaj S, Ballare S, Chandel MK (2017) Impact of congestion on greenhouse gas emissions for road transport in Mumbai metropolitan region. Transp Res Procedia 25:3538–3551. https://doi.org/10.1016/j.trpro.2017.05.282

Bilal M (2022) Pakistani literature festivals and a scopaesthesiac consciousness. J World Lit 7(3):348–368. https://doi.org/10.1163/24056480-00703004

Boland J (2013) Curriculum development for sustainable civic engagement. In: Emerging issues in higher education III: from capacity building to sustainability. Educational Developers of Ireland Network

Bonakdar A, Audirac I (2020) City branding and the link to urban planning: theories, practices, and challenges. J Plan Lit 35(2):147–160

Capone F, Lazzeretti L (2016) Fashion and city branding: an analysis of the perception of Florence as a fashion city. J Glob Fash Market 7(3):166–180. https://doi.org/10.1080/20932685.2016.1166715

Chang TC, Huang S (2008) Geographies of everywhere and nowhere: place-(un)making in a world city. Int Dev Plan Rev 30(3):227–247. https://link.gale.com/apps/doc/A243789018/AONE?u=anon~2ba1fb2f&sid=googleScholar&xid=cd04e167

Chuangchai P (2021) Rhetoric of the creative city in the Global South: the case of Bangkok city of design. In: Proceedings of 16th International Conference on Humanities and Social Sciences

Criqui L (2015) Infrastructure urbanism: roadmaps for servicing unplanned urbanization in emerging cities. Habitat Int 47:93–102. https://doi.org/10.1016/j.habitatint.2015.01.015

DDA (2007) Master Plan for Delhi-2021. Delhi Development Authority. https://dda.gov.in/master-plan-2021

Delhi Urban Arts Commission (2018) The enabling of safe public spaces in Delhi. Delhi Urban Arts Commission. https://duac.org.in/Upload/City%20Level%20Studies/Vision%20of%20Delhi/653209398198444.pdf

Dinnie K (2011) Introduction to the practice of city branding. In: City branding: theory and cases. Palgrave Macmillan, London, pp 93–98

Dupont VD (2011) The dream of Delhi as a global city. Int J Urban Reg Res 35(3):533–554

Elsheshtawy Y (2009) Dubai: behind an urban spectacle, 1st edn. Routledge. https://doi.org/10.4324/9780203869703

EIU (2023) The Global Liveability Index 2023. Economist Intelligence Unit (EIU). https://dev-services.eiu.com/n/campaigns/global-liveability-index-2023/

Evans, G (2003) Hard-branding the cultural city–from Prado to Prada. Int J Urban Reg Res 27(2):417–440.

Evans G (2015) Rethinking place branding and place making through creative and cultural quarters. In: Rethinking place branding: comprehensive brand development for cities and regions. Springer, Cham, pp 135–158. https://doi.org/10.1007/978-3-319-12424-7_10

Florida R (2003) Cities and the creative class. City Community 2(1):3–19

Gibson RS (2005) Principles of nutritional assessment, 2nd edn. Oxford University Press Inc., New York

Gotham KF (2007) (Re)Branding the big easy: tourism rebuilding in post-Katrina New Orleans. Urban Aff Rev 42(6):823–850. https://doi.org/10.1177/1078087407300222

Greenberg M (2009) Branding New York: how a city in crisis was sold to the world. Routledge

Griffiths A (2022, May 18) Frank Gehry's Guggenheim Museum Bilbao is "the greatest building of our time". Dezeen. Retrieved from https://www.dezeen.com/2022/05/18/frank-gehry-guggenheim-museum-bilbao-deconstructivism/

Gu X, Lim MK, O'Connor J (eds) (2020) Re-imagining creative cities in twenty-first century Asia. Palgrave Macmillan, London

Hankinson G (2004) Relational network brands: towards a conceptual model of place brands. J Vacat Mark 10(2):109–121

Haroon A (2023) History of Frere Hall. Retrieved from htv.com: https://htv.com.pk/health/history-of-frere-hall/

Indian Express (2023, August 30) Sassoon docks' modernisation essential to remove pollution due to solid waste from fishing activities': HC seeks authorities to take measures. Retrieved from indianexpress.com: https://indianexpress.com/article/cities/mumbai/sassoon-docks-modernisation-pollution-solid-waste-fishing-activities-8916661/

Islam MS (2010) Effects of real estate development on the built environment: a study on Banani residential area in Dhaka City. Unpublished Master's Dissertation, University of Dhaka.

Islam SMR (2013) Renewal of Old Dhaka: a case study on Shankhari Bazar. Dhaka Metropolitan Development Area and It's Planning Problems, Issues and Policies', Bangladesh Institute of Planners, Dhaka

Jain AK (2016) The significance of Shahjahanabad and Lutyens' Delhi. Int J Environ Stud 73(4):651–672. https://doi.org/10.1080/00207233.2016.1179013

Jamadar A (2021) How is global south shaping urban design. Zeyka [Online]. https://medium.com/zeyka/how-is-global-south-shaping-urban-design-fc411983edad

Joo YM, Seo B (2018) Transformative city branding for policy change: the case of Seoul's participatory branding. Environ Plan C: Polit Space 36(2):239–257

Kassens-Noor E (2012) Planning Olympic legacies: transport dreams and urban realities. Routledge

Kavaratzis M (2004) From city marketing to city branding: towards a theoretical framework for developing city brands. Place Brand 1:58–73. https://doi.org/10.1057/palgrave.pb.5990005

Kavaratzis M, Ashworth GJ (2005) City branding: an effective assertion of identity or a transitory marketing trick? Tijdschr Econ Soc Geogr 96(5):506–514. https://doi.org/10.1111/j.1467-9663.2005.00482.x

Kettle A (2019) Stitch A Tree, Karachi. Manchester Metropolitan University's Research Repository. https://e-space.mmu.ac.uk/624616/. Accessed 29 July 2023

Lim MK (2020) Global city as place branding strategy: the case of Bonifacio Global City (Philippines). In: Re-imagining creative cities in twenty-first century Asia. Springer, pp 27–39

Mabin A, Harrison P (2023) Contemporary planning and emergent futures: a comparative study of five capital city-regions on four continents. Prog Plan 169(August 2021):100664. https://doi.org/10.1016/j.progress.2022.100664

Maharaj B (2015) The turn of the south? Social and economic impacts of mega-events in India, Brazil and South Africa. Local Econ 30(8):983–999. https://idrc-crdi.ca/sites/default/files/sp/Documents%20EN/pat-10010786.pdf

Mangi MY, Yue Z, Kalwar S, Lashari ZA (2020) Comparative analysis of urban development trends of Beijing and Karachi Metropolitan Areas. Sustainability 12(2):451. https://doi.org/10.3390/SU12020451

Mazumdar DA, Alharahsheh HH (2020) Digital Bangladesh—Vision 2021: what is the digital Bangladesh concept? South Asian Res J Eng Technol **02**(01):6–9. https://doi.org/10.36346/sarjet.2020.v02i01.002

Mehta S (2006) Maximum city: Bombay lost and found. Penguin Books India

Mehta R (2022) Architectural study of the Humayun's Tomb and its restoration. Int J Creat Res Thoughts (IJCRT) **10**(8):723–728. https://ijcrt.org/papers/IJCRT2208338.pdf

Meskell L (2021) A tale of two cities: the fate of Delhi as UNESCO World Heritage. Int J Cult Prop 28(1):27–42. https://doi.org/10.1017/S0940739121000102

Mia S, Akter M (2019) Ready-made garments sector of Bangladesh: its growth, contribution and challenges. Econ World 7(1):17–26. https://doi.org/10.17265/2328-7144/2019.01.004

Middleton AC (2011) City branding and inward investment. In: City branding: theory and cases. Palgrave Macmillan UK, London, pp 15–26. https://doi.org/10.1057/9780230294790_3

Moser S (2015) New cities: old wine in new bottles? Dialogues Hum Geogr 5(1):31–35. https://doi.org/10.1177/2043820614565867

Mowla QA (2011) Urban aesthetics: a study on Dhaka. Hist Herit Urban Issues Cap Dhaka 3:169–262

NIUA (2022) Transforming urban landscape in India—a compendium of 75 public spaces. National Institute of Urban Affairs, New Delhi

Paddison R (1993) City marketing, image reconstruction and urban regeneration. Urban Stud 30(2):339–349. https://doi.org/10.1080/00420989320080331

Patel A (2020, May 20) Mumbai's Bandra-Worli Sea Link is a symbol of aspiration—and reckless development. Retrieved from Scroll.in: https://scroll.in/article/962964/mumbais-bandra-worli-sea-link-is-a-symbol-of-aspiration-and-reckless-development

Rehman A (2014, January 10) Urban problems: 'with elasticity of space, let cities find their own solutions'. Retrieved from The Express Tribune: https://tribune.com.pk/story/657537/urban-problems-with-elasticity-of-space-let-cities-find-their-own-solutions

Richards G (2017) From place branding to placemaking: the role of events. Int J Event Festiv Manag 8(1):8–23. https://www.emerald.com/insight/content/doi/10.1108/IJEFM-09-2016-0063/full/pdf

Sadiq A, Shakir MM, Ahmed S (2010) Gated living in the context of Karachi. J Res Arch Plan 9(1):50–64

Saha D, Ahmed S, Shahriar ATM, Mithun SNH (2017) A journey through the history: introduction to heritage tourism and tourist trails for the renewal of Old Dhaka. Am J Civ Eng Arch 5(3):98–107. https://doi.org/10.12691/ajcea-5-3-4

Sassen S (2001) The global city: New York. Princeton University Press, London, Tokyo. https://doi.org/10.2307/j.ctt2jc93q

Sengupta S, Sen A (2022) Re-creating Chandni Chowk: place-making and place-marketing of 'Old' Delhi's Historic City Centre. Indian J Spat Sci 13(2):47–53

Sethi R (2021) From walled to world city: creating brand Delhi. Dialog 38(Autumn):262–276. Retrieved December 31, 2023, from https://dialog.puchd.ac.in/wp-content/uploads/2022/03/15.-From-Walled-to-World-City-Creating-Brand-Delhi.pdf

Siddika A, Zaman M, Das JB (2018) Regeneration of the oldest and prime Central Business District (CBD) at Dhaka, Bangladesh. Int J Sci Eng Res 9(4):1422–1429. http://www.ijser.org

Soomro TA, Shah AA, Pasha YN (2020) The current state of built heritage in Karachi. J Art Arch Built Environ 3(1):58–83. https://doi.org/10.32350/JAABE.31.04

SRFDCL (2020) Sabarmati Riverfront: a catalyst for Ahmedabad's economic growth [Online]. Sabarmati Riverfront Development Corporation Ltd. Retrieved July 29, 2023, from https://sabarmatiriverfront.com/wp-content/uploads/2020/07/SRFDCL-presentation.pdf

Swapan M, Zaman A, Ahsan T, Ahmed F (2017) Transforming urban dichotomies and challenges of South Asian megacities: rethinking sustainable growth of Dhaka, Bangladesh. Urban Sci 1(4):31. https://doi.org/10.3390/urbansci1040031

TBS Report (2023, May 14) Urban planners call for stopping earth-filling Hatirjheel for expressway. Retrieved from tbsnews.net: https://www.tbsnews.net/bangladesh/urban-planners-call-stopping-earth-filling-hatirjheel-expressway-631698

Teklemariam N (2022) Sustainable development goals and equity in urban planning: a comparative analysis of Chicago, São Paulo, and Delhi. Sustainability 14(20). https://doi.org/10.3390/su142013227

UNESCO and World Bank (2021) Cities, culture, creativity: leveraging culture and creativity for sustainable urban development and inclusive growth World Bank

Varghese RA (2017) Past as a metaphor in the new utopian imaginations of heritage in Kerala. In: Urban utopias: excess and expulsion in neoliberal South Asia. Springer, pp 169–187. https://doi.org/10.1007/978-3-319-47623-0_9

Ward, S (1998) Selling Places: The Marketing and Promotion of Towns and Cities 1850–2000 (1st ed.). Routledge. https://doi.org/10.4324/9780203976746

World Population Review (2024) Dhaka, Bangladesh Population 2024. Worldpopulationreview.com. https://worldpopulationreview.com/cities/bangladesh/dhaka

Wynn JR, Yetis-Bayraktar A (2016) The sites and sounds of placemaking: branding, festivalization, and the contemporary city. J Pop Music Stud 28(2):204–223

Zukin S (1996) Space and symbols in an age of decline. In: Re-presenting the city: ethnicity, capital and culture in the twenty-first century metropolis. Palgrave, London, pp 43–59. https://doi.org/10.1007/978-1-349-24439-3_3

Zukin S, Trujillo V, Frase P, Jackson D, Recuber T, Walker A (2009) New retail capital and neighborhood change: boutiques and gentrification in New York City. City Community 8(1):47–64. https://doi.org/10.1111/j.1540-6040.2009.01269.x

Conclusions

Conclusion: Towards Sustainable and Resilient Megacities in the Global South

Lakshmi Priya Rajendran, Mohammed Firoz C, Bharat Dahiya, Lalit Kumar Dashora, and Uday Chatterjee

1 Conclusion

The concluding chapter of this edited volume, *Urban Planning and Design for Megacities in the Global South: Strategies for Smart and Sustainable Urban Development*, serves as a synthesis of the diverse perspectives, insights, and recommendations presented throughout the preceding chapters. As we reflect on the wealth of knowledge shared by contributors from various disciplines and geographical contexts, it becomes evident that the challenges and opportunities facing megacities in the Global South are complex and multifaceted. From environmental resilience to social inclusion, from energy sustainability to governance innovation, the chapters in this book offer a comprehensive exploration of the key issues shaping the future of urban development in megacity contexts.

At the heart of this discussion lies a shared commitment to promoting smart and sustainable urban development strategies that can address the pressing needs of rapidly growing urban populations while safeguarding the long-term well-being of

L. P. Rajendran (✉)
Bartlett School of Architecture, University College London, London, UK
e-mail: l.rajendran@ucl.ac.uk

M. Firoz C
Department of Architecture and Planning, NIT Calicut, Kozhikode, India

B. Dahiya · L. K. Dashora
Research Center for Sustainable Development and Innovation, School of Global Studies, Thammasat University, Bangkok 10200, Thailand

B. Dahiya
School of Public Leadership, University of Stellenbosch, Stellenbosch 7602, Western Cape, South Africa

U. Chatterjee
Department of Geography, Bhatter College Dantan, Paschim Medinipur, West Bengal, India

M. Firoz C et al. (eds.), *Urban Planning and Design for Megacities in the Global South*, Advances in 21st Century Human Settlements,
https://doi.org/10.1007/978-981-97-8370-0_22

communities and ecosystems. As we navigate the dynamic landscape of urbanisation in the Global South, it is essential to recognise the interconnectedness of urban challenges and the importance of holistic, integrated approaches to urban planning and design.

In this concluding chapter, we aim to distil key insights and lessons learned from the evidence-based case studies and discussions presented in the preceding chapters. We will briefly explore the overarching themes and trends that emerge from the diverse array of contributions, identifying common challenges, innovative solutions, and areas for further research and action. By synthesising the collective wisdom shared by our contributors, we hope to offer a roadmap for policymakers, decision-makers, planners, practitioners, experts, scholars, and researchers engaged in the pursuit of smart, sustainable, and resilient megacities in the Global South.

As we embark on reflection and synthesis, it is important to acknowledge the urgency of the urban challenges facing megacities in the Global South. Rapid urbanisation, population growth, climate change, environmental degradation, social inequality, and economic disparities present formidable obstacles to sustainable development and inclusive prosperity, as emphasised by the *New Urban Agenda* (United Nations 2017), the *2030 Agenda for Sustainable Development* (United Nations 2015), UNFCCC (United Nations Framework Convention on Climate Change) (2016) and the Sustainable Development Goals or SDGs (United Nations 2024). However, amidst these challenges lie opportunities for innovation, collaboration, and positive transformation.

By harnessing the collective intelligence and creativity of diverse stakeholders, we can chart a course towards more liveable, equitable, and resilient megacities. This requires a paradigm shift in urban governance, planning, and design—one that prioritises the needs of people and the planet, embraces diversity and inclusion, fosters collaboration and partnership, and empowers communities to shape their own destinies.

The compilation of case studies and discussions presented in this edited volume offers valuable insights into the multifaceted challenges and opportunities associated with urban planning and design in megacities of the Global South. Across the book's thematic parts addressing environmental resilience, spatial planning, urban expansion and regeneration, social resilience, energy planning, and innovative governance, a common thread emerges—the urgent need for smart and sustainable strategies to address the complex dynamics of urban growth and development.

One of the overarching themes that resonates throughout the book is the imperative of resilience in the face of climate-induced hazards and disasters. Case studies from cities such as Bandung, Thiruvananthapuram, and Kochi underscore the importance of proactive planning and mitigation measures to enhance urban resilience and safeguard communities against the impacts of climate change and natural disasters. These case studies highlight the critical role of environmental planning strategies and technological innovations in building resilient urban infrastructure and ecosystems.

Furthermore, the discussions on spatial planning and urban expansion emphasise the importance of inclusive and sustainable urban development practices. By reclaiming streets for pedestrians, revitalising inner urban cores, and promoting compact urban forms, cities like Surat, Ahmedabad, and Kolkata demonstrate how thoughtful urban design can foster social inclusion, economic vitality, and environmental sustainability. Moreover, the comparative analysis of commercial centres in traditional and planned cities sheds light on the importance of adaptive planning approaches that respect local contexts and cultural heritage.

The chapters addressing social resilience delve into the housing preferences of slum dwellers, the vitality of ethnic migrant communities, and conceptual frameworks for designing socio-cultural ecosystems within cities. These discussions underscore the significance of community participation, cultural diversity, and social cohesion in building resilient urban communities that can withstand socio-economic shocks and disruptions.

In the realm of energy planning and services, innovative approaches such as institutional frameworks for service provision, energy retrofitting, and agent-based modelling offer promising avenues for enhancing energy efficiency, reducing carbon emissions, and promoting renewable energy integration in megacity contexts. These approaches align with the broader goal of transitioning towards low-carbon, sustainable energy systems to meet the growing demands of urban populations.

Finally, the chapters on governance and management strategies highlight the importance of effective policy frameworks, participatory decision-making processes, and city branding initiatives in shaping the future trajectory of megacities. By leveraging international partnerships, harnessing creative industries, and embracing place-making strategies, cities in the Global South can position themselves as vibrant hubs of innovation, culture, and sustainable development.

In conclusion, this edited volume serves as a comprehensive resource for researchers, policymakers, practitioners, and stakeholders engaged in the planning and development of megacities in the Global South. By sharing best practices, lessons learned, and innovative solutions, the contributors to this book contribute to the collective effort towards building smart, resilient, and sustainable megacities that can thrive in the face of global challenges and uncertainties. As we look towards the future, let us continue to collaborate, innovate, and advocate for a more equitable and sustainable urban future for all.

References

UNFCCC (United Nations Framework Convention on Climate Change) (2016) The Paris Agreement. UNFCCC, Bonn

United Nations (2015) Transforming our world: the 2030 agenda for sustainable development. A/RES/70/1. United Nations, New York. https://sustainabledevelopment.un.org/post2015/transformingourworld/publication

United Nations (2017) The New Urban Agenda. A/RES/71/256. Habitat III and United Nations, New York

United Nations (2024) Sustainable Development Goals (SDGs). United Nations. https://sdgs.un.org/goals

The manufacturer's authorised representative in the EU is Springer Nature Customer Service Centre GmbH, Europaplatz 3, 69115 Heidelberg, Germany. If you have any concerns regarding our products, please contact ProductSafety@springernature.com

Printed and bound by CPI Group (UK) Ltd, Croydon, CR0 4YY
01/10/2026
02225930-0019